Jukka Lehto and Xiaolin Hou

Chemistry and Analysis of Radionuclides

Related Titles

Lambert, J. D. B. (ed.)

Nuclear Materials

2011

ISBN: 978-3-527-32352-4

Atwood, D. (ed.)

Radionuclides in the Environment

2010

ISBN: 978-0-470-71434-8

Prussin, S. G.

Nuclear Physics for Applications

2007

ISBN: 978-3-527-40700-2

Lieser, K. H.

Nuclear and Radiochemistry

Fundamentals and Applications

2001

ISBN: 978-3-527-30317-5

Jukka Lehto and Xiaolin Hou

Chemistry and Analysis of Radionuclides

Laboratory Techniques and Methodology

WILEY-VCH Verlag GmbH & Co. KGaA

The Authors

Prof. Jukka Lehto
University of Helsinki
Laboratory of Radiochemistry
A.I.Virtasen aukio 1
00014 Helsinki
Finland

Dr. Xiaolin Hou
Technical University of Denmark
Risö National Laboratory for Sustainable Energy
Radiation Research Division
Frediksborgvej 399
4000 Roskilde
Danmark

Library of Congress Card No.: applied for

British Library Cataloguing-in-Publication Data
A catalogue record for this book is available from the British Library.

Bibliographic information published by the Deutsche Nationalbibliothek
The Deutsche Nationalbibliothek lists this publication in the Deutsche Nationalbibliografie; detailed bibliographic data are available on the Internet at http://dnb.d-nb.de.

Composition Thomson Digital, Noida
Printing and Binding Strauss GmbH, Mörlenbach
Cover Design Adam Design, Weinheim

Printed on acid-free paper

ISBN: 978-3-527-32658-7

Contents

Chemistry and Analysis of Radionuclides. Jukka Lehto and Xiaolin Hou

ISBN: 978-3-527-32658-7

Preface

I started to give a lecture course on radionuclide analysis to students of radiochemistry in 2001. Two problems quickly became apparent. The first was that I could not properly lecture on this subject if the basic chemistry underlying the behavior of radionuclides in separation procedures was not understood. There seemed to be no sense in talking about precipitation of hydrolyzable metals with ferric hydroxide, for example, if the hydrolysis of metals was not understood. I had to go back to basics and teach the chemistry of the elements. This was a good choice – not least for myself – since I had to refresh my understanding of basic chemical phenomena. The second problem was that there was no adequate textbook on the chemistry of radionuclides. I had a handout from my predecessor to give to the students, but it had been written in the 1970s: it was good but outdated and short. Most books on radiochemistry available at that time, though comprehensive enough, contained little actual chemistry of the radionuclides and concentrated on their radioactive decay processes and the detection and measurement of radiation. In 2005 I began to write a text of my own, in Finnish. Then, seeing a broader need for such a text, I decided to write in English.

After working on the book for three years, I realized that analytical methods cannot be properly described if one has not done the analysis oneself, as was true in my case. I therefore asked Dr. Xiaolin Hou, of Risö National Laboratory, Denmark, to join me in the project. I knew him as a most experienced analytical radiochemist who had personally analyzed a great number of radionuclides in environmental and nuclear waste samples and developed new separation methods. During the past two years we have collaborated in writing this book, and I have learned a host of new things, not just from reading papers but also from extensive discussions with Dr. Hou.

Our book describes the basic chemistry needed to understand the behavior and analysis of radionuclides of most groups of elements, and the analytical methods required to separate the most important alpha- and beta- decaying radionuclides from environmental and nuclear waste samples (e.g., ^{90}Sr and plutonium isotopes). Many new radionuclides have become important in radiochemistry in the past ten to fifteen years. Most of these are very long-lived, appearing in spent nuclear fuel and

Chemistry and Analysis of Radionuclides. Jukka Lehto and Xiaolin Hou

ISBN: 978-3-527-32658-7

nuclear reactor structures and are relevant to safety analysis of the final disposal of the nuclear fuel and decommissioning waste. Mass spectrometric techniques are well suited for the measurement of these radionuclides (^{135}Cs, ^{129}I, etc.) because of their low specific activities. Traditionally, radiometric methods have been used to measure radionuclides, but the development of mass spectrometric techniques has opened up new avenues for the analysis of radionuclides, in particular for their analysis in much lower concentrations. Mass spectrometric measurements also create new requirements for radionuclide analyses, because the interfering radionuclides and other elements which need to be separated before measurement are mostly not the same ones that affect radiometric measurements.

My first intention was to write a book for undergraduate and post-graduate students, but now that the book is finished I see that it could also serve as a handbook for more experienced radiochemists – at least I hope so.

November 1, 2010

Jukka Lehto
Professor in Radiochemistry
University of Helsinki
Department of Chemistry
Finland

Acknowledgments

The authors thank Dr. Kathleen Ahonen for translating part of the book from Finnish and Dr. Shannon Kuismanen, Mr. Howard McKee, and Mr. Stewart Makkonen-Craig for language revision. The comments of Professor Markku Leskelä and Mr. Martti Hakanen from the University of Helsinki, Dr. Sven P. Nielsen and Dr. Jussi Jernstöm from Risø-DTU, Denmark, and Dr. Iisa Outola from STUK, Finland, have led to many improvements in the text, and we warmly thank them for their help. We are grateful to Mr. Lalli Jokelainen for careful preparation of figures and also to Dr. Steffen Happel from Triskem International, France, for providing some of the figures. Finally, we thank Wiley-VCH and Dr. Eva-Stina Riihimäki for publishing the book.

Chemistry and Analysis of Radionuclides. Jukka Lehto and Xiaolin Hou

ISBN: 978-3-527-32658-7

1
Radionuclides and their Radiometric Measurement

1.1
Radionuclides

The first radioactive elements – radium and polonium – were discovered by Marie Curie at the end of the nineteenth century. During the first decades of the twentieth century, tens of natural radioactive elements and their various isotopes in the uranium and thorium decay chains were identified. The first artificial radionuclide, ^{30}P, was produced by Frédéric and Irène Joliot-Curie in an accelerator by bombarding aluminum with protons. Today, more than two thousand artificial radionuclides have been produced and identified, especially after the discovery and use of nuclear fission of uranium and plutonium. This book focuses on radionuclides found in the environment and in nuclear waste. This chapter presents an overview of radionuclides, radioactive decay processes, and the radiometric measurement of radionuclides, which are categorized according to their sources and ways of formation; in later chapters they are classified based on their chemical nature and are discussed in more detail. Radionuclides can be primarily categorized into natural and artificial radionuclides.

1.1.1
Natural Radionuclides

In nature there are three types of radionuclides: those belonging to the decay chains of uranium and thorium, single very long-lived radionuclides, and cosmogenic radionuclides.

The decay chains of uranium and thorium start with two isotopes of uranium and one of thorium, ^{235}U, ^{238}U, and ^{232}Th, which were formed at the birth of the Universe some 13.7 billion years ago, and, since they are so long-lived, they have survived in the earth since its birth 4.5 billion years ago. These three primordial radionuclides each initiate a decay chain leading to the stable lead isotopes ^{207}Pb, ^{206}Pb, and ^{208}Pb, respectively. In between, there are altogether 42 radionuclides of 13 elements, of which nine elements, those heavier than bismuth, have no stable isotopes at all. The three decay chains are depicted in Tables 1.1–1.3. The determination of radionuclides

Chemistry and Analysis of Radionuclides. Jukka Lehto and Xiaolin Hou

ISBN: 978-3-527-32658-7

Table 1.1 Uranium decay chain.

Nuclide	Decay mode	Half-life	Decay energy (MeV)	Decay product
^{238}U	α	4.4×10^9 y	4.270	^{234}Th
^{234}Th	β^-	24 d	0.273	^{234}Pa
^{234}Pa	β^-	6.7 h	2.197	^{234}U
^{234}U	α	245 500 y	4.859	^{230}Th
^{230}Th	α	75 380 y	4.770	^{226}Ra
^{226}Ra	α	1602 y	4.871	^{222}Rn
^{222}Rn	α	3.8 d	5.590	^{218}Po
^{218}Po	α 99.98%	3.1 min	6.874	^{214}Pb
	β^- 0.02%		2.883	^{218}At
^{218}At	α 99.90%	1.5 s	6.874	^{214}Bi
	β^- 0.10%		2.883	^{218}Rn
^{218}Rn	α	35 ms	7.263	^{214}Po
^{214}Pb	β^-	27 min	1.024	^{214}Bi
^{214}Bi	β^- 99.98%	20 min	3.272	^{214}Po
	α 0.02%		5.617	^{210}Tl
^{214}Po	α	0.16 ms	7.883	^{210}Pb
^{210}Tl	β^-	1.3 min	5.484	^{210}Pb
^{210}Pb	β^-	22.3 y	0.064	^{210}Bi
^{210}Bi	β^- 99.99987%	5.0 d	1.426	^{210}Po
	α 0.00013%		5.982	^{206}Tl
^{210}Po	α	138 d	5.407	^{206}Pb
^{206}Tl	β^-	4.2 min	1.533	^{206}Pb
^{206}Pb		stable		

Table 1.2 Actinium decay chain.

Nuclide	Decay mode	Half-life	Decay energy (MeV)	Decay product
^{235}U	α	7.1×10^8 y	4.678	^{231}Th
^{231}Th	β^-	26 h	0.391	^{231}Pa
^{231}Pa	α	32,760 y	5.150	^{227}Ac
^{227}Ac	β^- 98.62%	22 y	0.045	^{227}Th
	α 1.38%		5.042	^{223}Fr
^{227}Th	α	19 d	6.147	^{223}Ra
^{223}Fr	β^-	22 min	1.149	^{223}Ra
^{223}Ra	α	11 d	5.979	^{219}Rn
^{219}Rn	α	4.0 s	6.946	^{215}Po
^{215}Po	α 99.99977%	1.8 ms	7.527	^{211}Pb
	β^- 0.00023%		0.715	^{215}At
^{215}At	α	0.1 ms	8.178	^{211}Bi
^{211}Pb	β^-	36 min	1.367	^{211}Bi
^{211}Bi	α 99.724%	2.1 min	6.751	^{207}Tl
	β^- 0.276%		0.575	^{211}Po
^{211}Po	α	516 ms	7.595	^{207}Pb
^{207}Tl	β^-	4.8 min	1.418	^{207}Pb
^{207}Pb		stable		

Table 1.3 Thorium decay chain.

Nuclide	Decay mode	Half-life	Decay energy (MeV)	Decay product
^{232}Th	α	1.41×10^{10} y	4.081	^{228}Ra
^{228}Ra	β^-	5.8 y	0.046	^{228}Ac
^{228}Ac	β^-	6.3 h	2.124	^{228}Th
^{228}Th	α	1.9 y	5.520	^{224}Ra
^{224}Ra	α	3.6 d	5.789	^{220}Rn
^{220}Rn	α	56 s	6.404	^{216}Po
^{216}Po	α	0.15 s	6.906	^{212}Pb
^{212}Pb	β^-	10.6 h	0.570	^{212}Bi
^{212}Bi	β^- 64.06%	61 min	2.252	^{212}Po
	α 35.94%		6.208	^{208}Tl
^{212}Po	α	299 ns	8.955	^{208}Pb
^{208}Tl	β^-	3.1 min	4.999	^{208}Pb
^{208}Pb		stable		

in the decay chains has been, and still is, a major topic in analytical radiochemistry. They are alpha and beta emitters, most of which do not emit detectable gamma rays, and thus their determination requires radiochemical separations. This book examines the separations of the following radionuclides: U isotopes, ^{231}Pa, Th isotopes, ^{227}Ac, ^{226,228}Ra, ^{222}Rn, ^{210}Po, and ^{210}Pb.

In addition to ^{235}U, ^{238}U, and ^{232}Th, there are several single very long-lived primordial radionuclides (Table 1.4) which were formed in the same cosmic processes as those that formed uranium and thorium. The most important of these, with respect to the radiation dose to humans, is ^{40}K. However, as this emits readily detectable gamma rays and does not require radiochemical separations, neither this nor the others are discussed further in this book.

The third class of natural radionuclides comprises cosmogenic radionuclides, which are formed in the atmosphere in nuclear reactions due to cosmic radiation (Table 1.5). These radionuclides are isotopes of lighter elements, and their half-lives vary greatly. The primary components of cosmic radiation are high-energy alpha particles and protons, which induce nuclear reactions when they impact on the nuclei of the atmospheric atoms. Most of the cosmogenic radionuclides are attached to aerosol particles and are deposited on the ground. Some, however, are gaseous, such as ^{14}C (as carbon dioxide) and ^{39}Ar (a noble gas), and thus stay in the atmosphere. In

Table 1.4 Some single primordial radionuclides.

Nuclide	Isotopic abundance (%)	Decay mode	Half-life (y)
^{40}K	0.0117	β^-	1.26×10^{9}
^{87}Rb	27.83	β^-	4.88×10^{10}
^{123}Te	0.905	EC	1.3×10^{13}
^{144}Nd	23.80	α	2.1×10^{15}
^{174}Hf	0.162	α	2×10^{15}

Table 1.5 Some important cosmogenic radionuclides.

Nuclide	Half-life (y)	Decay mode	Nuclide	Half-life (y)	Decay mode
^{3}H	12.3	beta	^{7}Be	0.15	EC
^{10}Be	2.5×10^{6}	beta	^{14}C	5730	beta
^{22}Na	2.62	EC	^{26}Al	7.4×10^{5}	EC
^{32}Si	710	beta	^{32}P	0.038	beta
^{33}P	0.067	beta	^{35}S	0.24	beta
^{36}Cl	3.1×10^{5}	beta/EC	^{39}Ar	269	beta
^{41}Ca	3.8×10^{6}	EC	^{129}I	1.57×10^{7}	beta

primary nuclear reactions, neutrons are also produced, and these induce further nuclear reactions. Two important radionuclides are produced in these neutron-induced reactions: ^{3}H and ^{14}C (reactions 1.1 and 1.2), whose chemistry and radiochemical separations are described in Chapter 13. These radionuclides – tritium and radiocarbon – are generated not only by cosmic radiation but also in other neutron activation processes in nuclear explosions and in matter surrounding nuclear reactors.

$$^{14}N + n \rightarrow {}^{14}C + p \tag{1.1}$$

$$^{14}N + n \rightarrow {}^{12}C + {}^{3}H \tag{1.2}$$

1.1.2 Artificial Radionuclides

Artificial radionuclides form the largest group of radionuclides, comprising more than two thousand nuclides produced since the 1930s. The sources of artificial radionuclides are:

- nuclear weapons production and explosions;
- nuclear energy production;
- radionuclide production by reactors and accelerators.

A wide range of radionuclides are produced in nuclear weapons production, where plutonium is produced by the irradiation of uranium in reactors and in nuclear power reactors. Most are *fission products*, and are generated by the neutron-induced fission of ^{235}U and ^{239}Pu. In nuclear power reactors, they are practically all retained in the nuclear fuel; however, in nuclear explosions they end up in the environment – on the ground in atmospheric explosions or in the geosphere in underground explosions. The spent nuclear fuel from power reactors is stored in disposal repositories deep underground. The radionuclide composition of nuclear explosions and the spent fuel from nuclear power reactors differ somewhat for several reasons. Firstly, the fissions in a reactor are mostly caused by thermal neutrons, while in a bomb fast neutrons are mostly responsible for the fission events, and this results in differences in the radionuclide composition. Secondly, fission is instantaneous in a bomb, while in a reactor the fuel is irradiated for some years. This allows the ingrowth of some activation products, such as ^{134}Cs, that do not exist in weapons fallout. ^{90}Sr and ^{137}Cs are the most important fission products because of their relatively long half-lives and high fission yields. In addition to these, there is range of long-lived fission products, such as ^{79}Se, ^{99}Tc, ^{126}Sn, ^{129}I, ^{135}Cs, and ^{151}Sm, the radiochemistry of which is discussed in this book.

Along with fission products, activation products are also formed in side reactions accompanying the neutron irradiation. The intensive neutron flux generated in the fission induces activation reactions both in the fuel or weapons material and in the surrounding material. These can be divided into two categories, the first comprising the transuranium elements – a very important class of radionuclides in radiochemistry. These are created by successive neutron activation and beta decay processes starting from ^{238}U or ^{239}Pu (Figure 1.1). Of these, the most important and the most radiotoxic nuclides are ^{237}Np, $^{238,239,240,241}Pu$, $^{241,243}Am$, and $^{243,244,245}Cm$, which are discussed further in this book. In addition to transuranium elements, a new uranium isotope ^{236}U is also formed in neutron activation reactions.

Another activation product group comprises radioisotopes of various lighter elements. In addition to tritium and radiocarbon, a wide range of these activation products are formed in nuclear explosions and especially in nuclear reactors. Elements of the reactor's construction materials, especially the cladding and other metal parts surrounding the nuclear fuel, the steel of the pressure vessel and the shielding concrete structures are activated in the neutron flux from the fuel. Part of these activation products, such as elements released from the steel by corrosion, end up in the nuclear waste disposed of during the use of the reactor. A larger part,

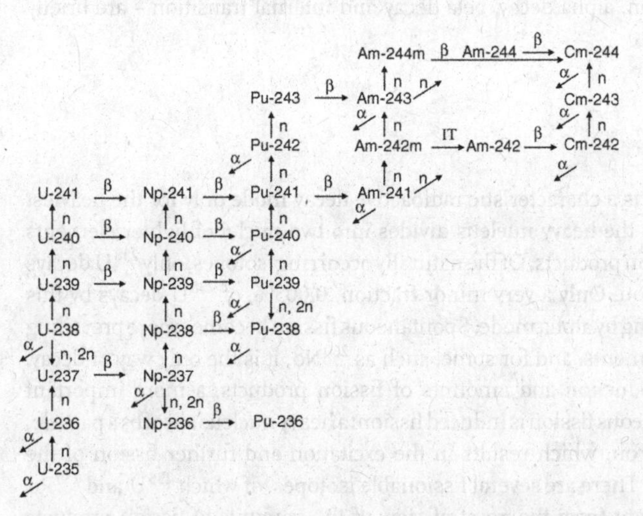

Figure 1.1 Formation of transuranium elements in nuclear fuel and nuclear weapons material (Holm, E., Rioseco, J., and Petterson, H. (1992) Fallout of transuranium elements following the Chernobyl accident. *J. Radioanal. Nucl. Chem. Articles*, **156**, 183).

however, remains in the steel and concrete and ends up in the waste when the reactor is decommissioned. This category has many important radionuclides, such as ^{14}C, ^{36}Cl, ^{41}Ca, ^{55}Fe, ^{59}Ni, and ^{63}Ni, which are discussed later in the book. These are all purely beta-decaying radionuclides that require radiochemical separations. In addition to these, there is a range of activation products, such as ^{60}Co, ^{54}Mn, ^{65}Zn, which emit gamma rays and are thus readily detectable and measurable. In addition to the reactor steel and shielding concrete, the spent fuel, its metal cladding, and other metal parts surrounding the fuel and ending up in the final disposal, contain large amounts of the long-lived beta decaying activation products ^{93}Zr, ^{94}Nb, and ^{93}Mo (together with ^{14}C, ^{36}Cl, ^{59}Ni, ^{63}Ni), which are also discussed in this book.

There are also a number of *radionuclides produced by neutron and proton irradiations in reactors and in cyclotrons*. Their properties are later described only if they are used as tracers in radionuclide analysis. An example is a fairly short-lived gamma-emitting strontium isotope, ^{85}Sr, which is used as a tracer in model experiments for studying the behavior of the beta-emitting fission product ^{90}Sr or as a yield-determinant in ^{90}Sr determinations.

1.2
Modes of Radioactive Decay

This book describes the chemistry and analysis of radionuclides – nuclei which are unstable, that is, radioactive. The instability comes from the fact that the mass of the nucleus is either too high or its neutron to proton ratio is inappropriate for stability. By radioactive decay, the nucleus disposes of the mass excess or adjusts the neutron to proton ratio more closely to what is required for stability. The four main radioactive decay modes – fission, alpha decay, beta decay and internal transition – are briefly described below.

1.2.1
Fission

Spontaneous fission is a characteristic radioactive decay mode only for the heaviest elements. In fission, the heavy nucleus divides into two nuclei of lighter elements which are called fission products. Of the naturally occurring isotopes, only ^{238}U decays by spontaneous fission. Only a very minor fraction, 0.005%, of ^{238}U decays by this mode, the rest decaying by alpha mode. Spontaneous fission becomes more prevailing with the heaviest elements, and for some, such as ^{260}No, it is the only way of decay. Considering the production and amounts of fission products, a more important process than spontaneous fission is induced fission: a heavy nucleus absorbs a particle, most usually a neutron, which results in the excitation and further fission of the nucleus (Figure 1.2). There are several fissionable isotopes, of which ^{235}U and ^{239}Pu are the most important from the point of view of the amounts of fission products generated. These two nuclides are not only fissionable but also fissile, that is, they undergo fission in the presence of thermal neutrons, which enables their use as

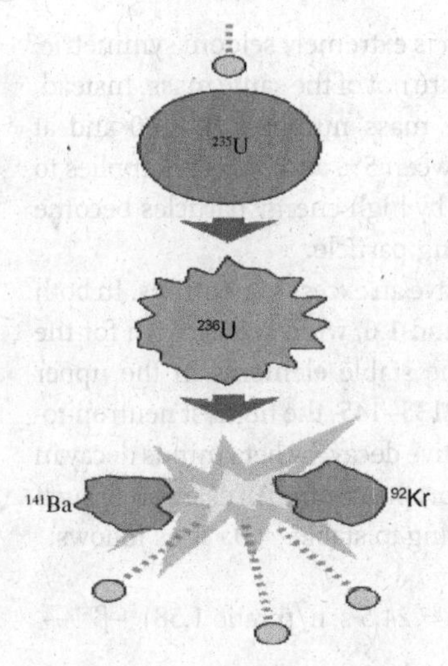

Figure 1.2 An example of a neutron-induced fission of ^{235}U. The reaction is $^{235}U + n \rightarrow {}^{236}U \rightarrow {}^{141}Ba + {}^{92}Kr + 3n$ (http://en.wikipedia.org/wiki/Nuclear_fission).

nuclear fuel in nuclear reactors. ^{235}U is obtained by isotopic enrichment from natural uranium and ^{239}Pu by the irradiation of ^{238}U in a nuclear reactor and subsequent chemical separation of plutonium from the irradiated uranium.

A large number of fission products are generated in fission processes. Figure 1.3 gives, as an example, the distribution of fission products for ^{235}U from thermal

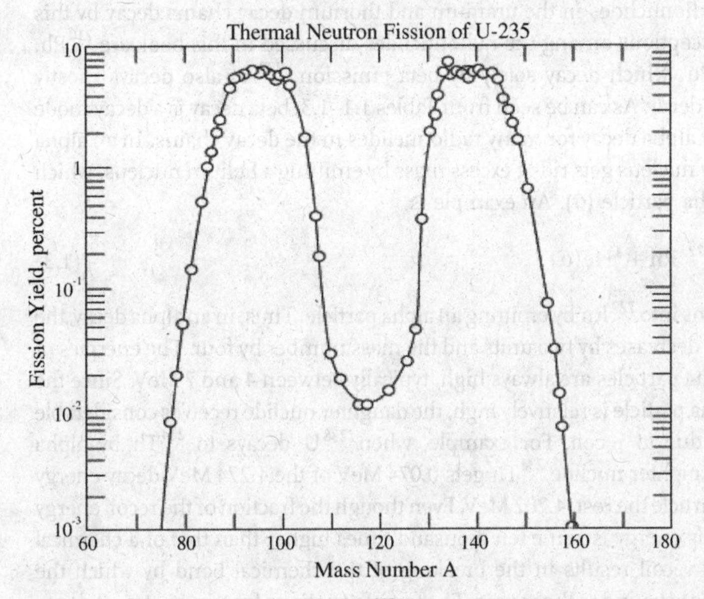

Figure 1.3 Fission yield distribution of ^{235}U as a function of the mass number of the fission product.

neutron-induced fission. As can be seen, the fission is extremely seldom symmetric, that is, the two fission products of one fission event are not of the same mass. Instead, the maxima of fission products are found at the mass numbers 90–100 and at 135–145. At these ranges, the fission yields are between 5% and 7%. This applies to thermal neutron-induced fission; fissions induced by high energy particles become more symmetric with the energy of the bombarding particle.

Most fission products are radioactive since they have an excess of neutrons. In both ^{235}U and ^{239}Pu, the neutron to proton ratio is around 1.6, which is too high for the lighter elements to be stable. For example, for the stable elements in the upper maximum of the fission yield at the mass numbers 135–145, the highest neutron-to-proton ratio is around 1.4, and, through the radioactive decay, by beta minus decay in this case, the nucleus transforms the ratio into an appropriate one. An example of such a decay chain of neutron-rich fission products leading to stable ^{137}Ba is as follows:

$$^{137}\mathrm{Te}(t_{1/2} = 3.5\ \mathrm{s};\ \mathrm{n/p\ ratio}\ 1.63) \rightarrow {}^{137}\mathrm{I}(t_{1/2} = 24.5\ \mathrm{s};\ \mathrm{n/p\ ratio}\ 1.58) + \beta^- \rightarrow$$

$$^{137}\mathrm{Xe}(t_{1/2} = 3.82\ \mathrm{min};\ \mathrm{n/p\ ratio}\ 1.54) + \beta^- \rightarrow$$

$$^{137}\mathrm{Cs}(t_{1/2} = 30\ \mathrm{y};\ \mathrm{n/p\ ratio}\ 1.49) + \beta^- \rightarrow {}^{137}\mathrm{Ba}(\mathrm{stable};\ \mathrm{n/p\ ratio}\ 1.45) + \beta^-$$

1.2.2
Alpha Decay

Alpha decay is also a typical decay mode for the heavier radionuclides. Most actinide isotopes and radionuclides in the uranium and thorium decay chains decay by this mode. A few exceptions among the radionuclides discussed in this book are ^{210}Pb, ^{228}Ra, and ^{241}Pu, which decay solely by beta emission. ^{227}Ac also decays mostly (98.8%) by beta decay. As can be seen from Tables 1.1–1.3, beta decay is a decay mode competing with alpha decay for many radionuclides in the decay chains. In an alpha decay, the heavy nucleus gets rid of excess mass by emitting a helium nucleus, which is called an alpha particle (α). An example is

$$^{226}\mathrm{Ra} \rightarrow {}^{222}\mathrm{Rn} + {}^{4}\mathrm{He}(\alpha) \tag{1.3}$$

where ^{226}Ra turns into ^{222}Rn by emitting an alpha particle. Thus, in an alpha decay, the atomic number decreases by two units and the mass number by four. The energies of the emitted alpha particles are always high, typically between 4 and 7 MeV. Since the mass of the alpha particle is relatively high, the daughter nuclide receives considerable kinetic energy due to recoil. For example, when ^{238}U decays to ^{234}Th by alpha emission, the daughter nuclide ^{234}Th gets 0.074 MeV of the 4.274 MeV decay energy and the alpha particle the rest, 4.202 MeV. Even though the fraction of the recoil energy is only 1.7%, this energy is some ten thousand times higher than that of a chemical bond, and thus recoil results in the breaking of the chemical bond by which the daughter nuclide is bound to the matrix. The transformations from parent nuclides to daughter nuclides take place between defined energy levels corresponding to defined

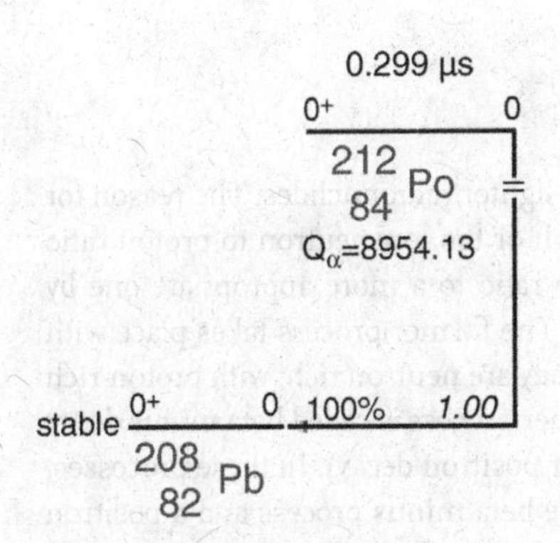

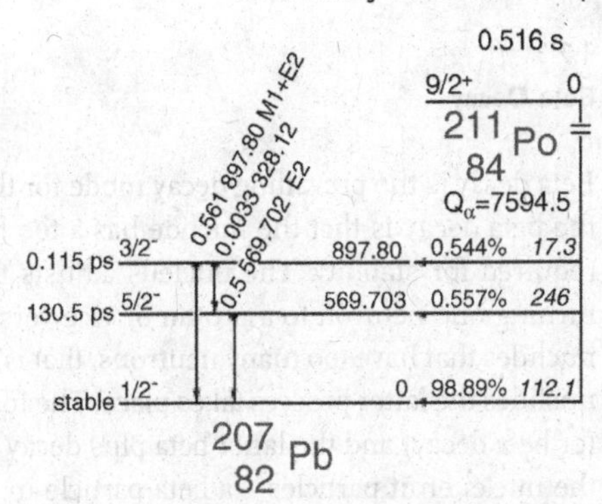

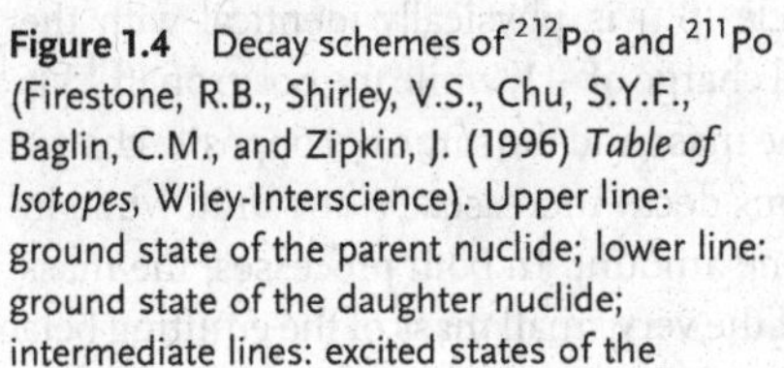

Figure 1.4 Decay schemes of ^{212}Po and ^{211}Po (Firestone, R.B., Shirley, V.S., Chu, S.Y.F., Baglin, C.M., and Zipkin, J. (1996) *Table of Isotopes*, Wiley-Interscience). Upper line: ground state of the parent nuclide; lower line: ground state of the daughter nuclide; intermediate lines: excited states of the daughter nuclide; Q = decay energy (keV); half-life of the parent nuclide at the top; half-lives of the excited states on the left, energies of the excited states in the middle (keV); arrows represent internal transitions and their energies are at their top, percentages are proportions of transitions.

quantum states. Thus, the emitted alpha particles have always the same energy, and peak spectra are obtained when measuring the alpha particles. Often the decay processes lead not only to the ground level of the daughter nuclide but also to its excited energy levels. The alpha particles leading to excited levels have thus lower energies. Since the excited levels, also representing defined quantum states, have defined energy levels, these lower energy alpha particles have defined energies. Because the same energy should be released in each decay process, the rest of the energy, that is, the energy difference between the ground and the excited levels, is emitted as electromagnetic gamma radiation or as conversion electrons. Figure 1.4 gives two schemes of alpha decay processes. The first is for ^{212}Po leading to the ground level of the daughter nuclide only; the other is for ^{211}Po leading also to the excited levels. Later in the text, decay schemes are given for all important radionuclides discussed in this book. The most important information relevant to decay schemes is set out and explained in Figure 1.4.

The fractions of the decay processes going each way, called intensities, have certain probabilities. For example, in the case of ^{211}Po, 98.89% of the decay events go directly to the daughter ground state, while 0.544% go to the upper excitation level and 0.557% to the lower. As the relaxation of the excited levels by gamma emission takes place almost instantaneously after alpha decay in most cases, the excited levels can be considered to decay at the same rate as the parent nuclide. If the intensities of decay processes leading to the gamma ray emissions and the gamma energies are high enough, it is advantageous to measure the activities of the parent nuclides by these gamma emissions since they are often easier to measure than the alpha particles. There are, however, only a few such nuclides. This book discusses in detail those alpha-decaying radionuclides that do not have detectable gamma emissions, since they require radiochemical separations.

1.2.3
Beta Decay

Beta decay is the prevailing decay mode for the lighter radionuclides. The reason for the beta decay is that the nuclide has a too high or too low neutron to proton ratio required for stability. The nucleus adjusts the ratio to a more appropriate one by turning one neutron to a proton or vice versa. The former process takes place with nuclides that have too many neutrons, that is, they are neutron rich; with proton-rich nuclides the latter process takes place. The former process is called beta minus decay (or beta decay) and the latter beta plus decay (or positron decay). In these processes, the nuclei emit particles – a beta particle in the beta minus process and a positron particle in beta plus process. A beta particle (β^-) is physically identical with the electron having the same mass and electrical charge of -1, while the positron (β^+) is an antiparticle of an electron, having the same mass and the same but opposite charge of $+1$. The atomic number (Z) in beta minus decay increases by one unit, while in the beta plus process it decreases by the same amount. In both processes, the mass number (A) remains unchanged. Because of the very small mass of the emitting beta and positron particles, the fraction of the decay energy which the daughter nuclide receives in recoil is very small. The beta particle and the positron do not, however, receive the rest of the decay energy since there is also another particle emitted along with them: a neutrino (ν) with the positron and an antineutrino ($\bar{\nu}$) with beta particle. The complete decay equations are thus:

$$\text{Beta plus decay}: \quad {}^{A}_{Z}\text{M} \rightarrow {}_{Z+1}^{\ \ A}\text{M} + \beta^- + \bar{\nu}$$

$$\text{Beta plus decay}: \quad {}^{A}_{Z}\text{M} \rightarrow {}_{Z-1}^{\ \ A}M + \beta^+ + \nu$$

The decay energy is randomly distributed between the particles in the pairs $\beta^-/\bar{\nu}$ and β^+/ν. Since the neutrino and the antineutrino only very weakly interact with the matter, they do not interact with normal detection systems; therefore, when measuring beta-emitting radionuclides, only the beta and positron particles are detected. Since only a fraction of the decay energy goes to these particles, continuous spectra are obtained instead of single peaks corresponding to the decay energy (Figure 1.5). In beta minus decay, the median beta particle energy is about 30% of the maximum energy, while in beta plus decay, the median positron energy is about 40% of the maximum energy. In both spectra the end points of the spectra represent maximum energy, that is, the decay energy.

The positron particles emitted in beta plus decay are not stable. After they have lost their kinetic energy they annihilate by combining themselves with an electron. These two antiparticles are turned into two gamma quanta of 0.511 MeV, an energy corresponding to the mass of an electron. The two gamma rays are emitted in opposite directions.

A competing process for the positron emission in beta plus decay is electron capture (EC). In this process, the proton-rich nuclide, instead of emitting a positron, captures an electron from its atomic electron shell, usually from the innermost K shell and less often from the L shell. This leads to the transformation of a proton into a

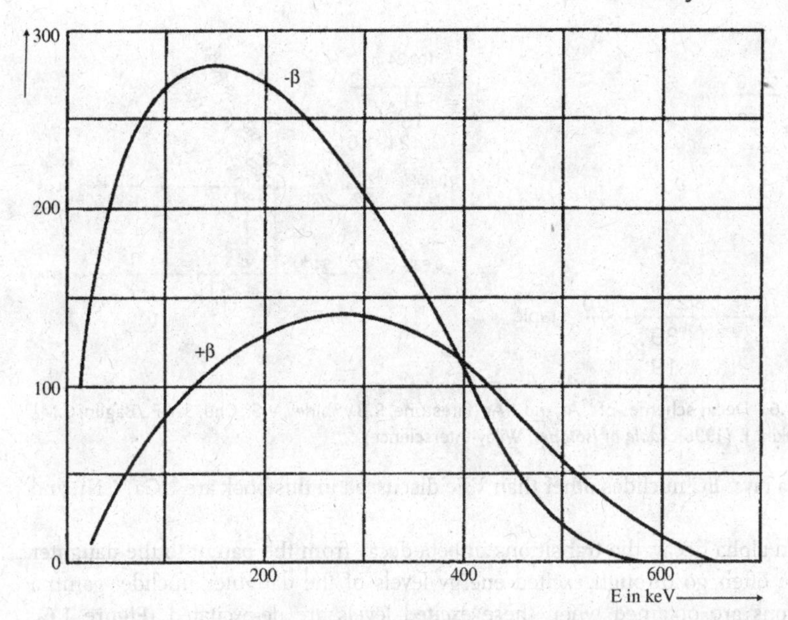

Figure 1.5 Energy spectra of beta and positron particles of ^{64}Cu (Keller, C. (1988) *Radiochemistry*, Ellis Horwood Limited, Chichester).

neutron within the nucleus. In the primary process only neutrinos are emitted from the nucleus. Since these are not detectable, it would seem that it would not be possible to measure the radioactivity of EC-decaying radionuclides. However, EC nuclides can be measured by detecting the X-rays which are formed when the hole in the electron shell is filled by electrons from the upper shells. Though these X-rays are characteristic for the daughter nuclide, they can be used to measure the parent nuclide's activity, since their formation takes place practically simultaneously with the electron capture, and thus their rates are the same. In addition to characteristic X-rays, there are also Auger electrons emitted after electron capture. After the formation of the X-rays, some of these do not leave the atom but transfer their energy to an electron on an upper shell. These Auger electrons, departing from the atom, are monoenergetic and have rather low energies. They can be detected by liquid scintillation counting and thus can be used, in addition to X-rays, to measure activities of the EC nuclides. EC is most typical for the heaviest elements with $Z > 80$, while positron emission is the prevailing decay mode with the lighter elements with $Z < 30$. In between, $30 < Z < 80$, both processes take place in parallel. An example of an EC nuclide dealt with in this book is ^{55}Fe. Its daughter nuclide, ^{55}Mn, emits X-rays with energies of 5.9 keV and 6.5 keV, which are measured for the activity determination of ^{55}Fe. The primary EC process can also lead to an excited state of the daughter nuclide. For example, in the EC decay of ^{125}I, all the primary transitions go to the 35.5-keV excited state of the ^{125}Te daughter nuclide. This excited state will relax by emission of gamma rays, which can be used to measure this and other similar EC nuclides. Thus, there are several options to measure EC nuclides, X-rays, Auger electrons and in some cases

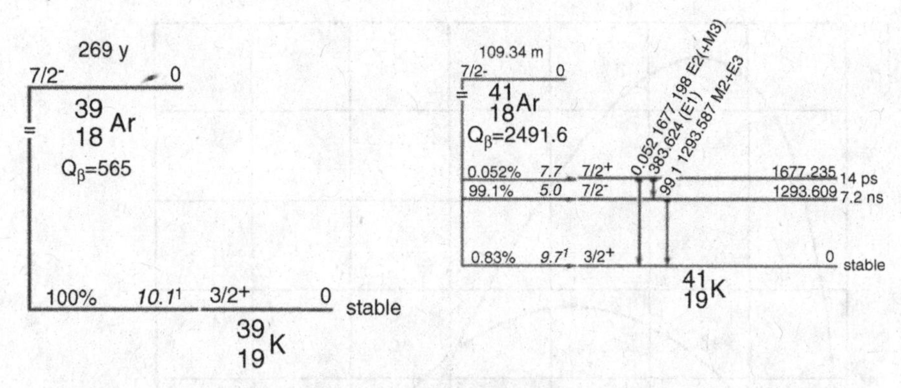

Figure 1.6 Decay schemes of ^{39}Ar and ^{41}Ar (Firestone, R.B., Shirley, V.S., Chu, S.Y.F., Baglin, C.M., and Zipkin, J. (1996) *Table of Isotopes*, Wiley-Interscience).

gamma rays. EC nuclides other than ^{55}Fe discussed in this book are ^{41}Ca, ^{59}Ni, and ^{93}Mo.

As in alpha decay, the transitions in beta decay from the parent to the daughter nuclide often go through excited energy levels of the daughter nuclide; gamma emissions are obtained when these excited levels are de-excitated (Figure 1.6). Whenever these gamma emissions have high enough intensities and energies so that they can be readily detected, they are used to measure the activities of beta-emitting radionuclides. This book mostly deals with the pure beta emitters, since they require radiochemical separations. These include ^{3}H, ^{14}C, ^{63}Ni, ^{79}Se, ^{90}Sr, ^{93}Zr, ^{94}Nb, ^{99}Tc, ^{126}Sn, ^{129}I, ^{135}Cs, ^{228}Ra, ^{227}Ac, and ^{241}Pu.

1.2.4 Internal Transition

As mentioned above, alpha and beta decays often go through the excited states of the daughter nuclide. De-excitation of these states takes place in two ways: by emission of gamma rays or by internal conversion. The collective term for these is internal transition. As already described, the excited states represent the defined quantum levels of the daughter nuclide, and therefore the transitions between them have defined energies. Thus, the emitted gamma rays are monoenergetic and the spectrum obtained is a peak spectrum. For example, when ^{41}Ar decays by beta decay (Figure 1.6), only 0.8% of the transitions lead directly to the ground state of the daughter nuclide ^{41}K, the beta decay energy being 2.492 MeV. The remaining 0.05% and 99.1% of the transitions, in turn, lead to the excited states with energy levels of 1.677 MeV and 1.294 MeV, respectively. When these states are de-excited with gamma emissions, the energies of the gamma rays are the same as the difference of the energy of the quantum levels between which the de-excitation takes place. The energies of the beta transitions in Figure 1.6 are the differences between the total beta decay energy (2.492 MeV) and the energies of the excited states (1.677 MeV and 1.294 MeV), that is, 0.815 MeV and 1.198 MeV. Consequently, the gamma rays originating from the de-excitation of these states are 1.677 MeV and 1.294 MeV.

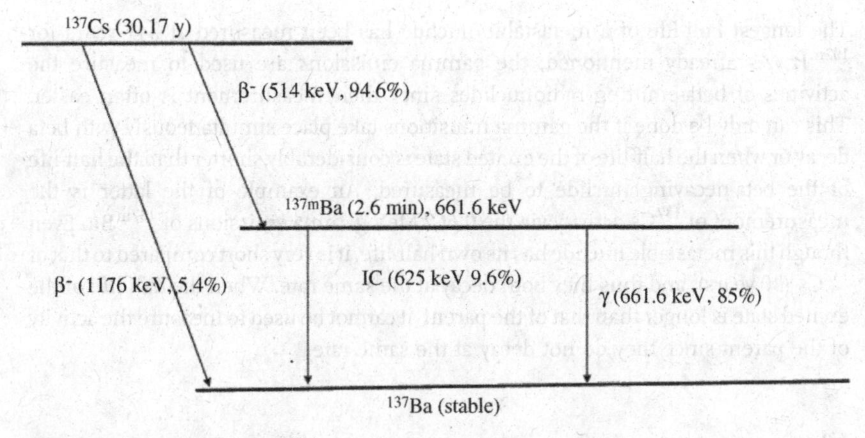

Figure 1.7 Decay scheme of ^{137}Cs.

Many beta-decaying radionuclides emit gamma rays, which are detected to determine the activities of these nuclides.

A competing process for the gamma emission is internal conversion, in which the excited state is not relaxed by the emission of gamma rays; rather, the excitation energy is transferred to an electron at the electron shell of the daughter atom. These electrons are monoenergetic, the energy being the difference between the de-excitation energy and the binding energy of the electron. An example of internal transition is the decay of ^{137}Cs, which partly takes place through the excited state of the daughter nuclide ^{137}Ba (Figure 1.7). Only 5.4% of the transitions go directly to the ground level of ^{137}Ba, the remaining 94.6% going to the 0.662 MeV excited level. This excited state does not solely decay by gamma emissions; 10.2% of the excitations are relaxed by internal conversion and the formation of conversion electrons. The radiations obtained in this decay are:

- 5.4% beta particles with a maximum energy of 1.176 MeV;
- 94.6% beta particles with a maximum energy of 0.514 MeV (1.176–0.662 MeV);
- 85.0% (94.6% × 0.898) gamma rays with an energy of 0.662 MeV;
- 9.6% (94.6% × 0.102) conversion electrons with an energy of 0.625 MeV (0.662–0.037 MeV, where 0.037 MeV is the binding energy of an electron at the K shell of ^{137}Ba). Conversion electrons, the total intensity of which is the above mentioned 9.6%, also have energies of more than 0.625 MeV, since some part of the conversions take place from upper electron shells with lower binding energies.

In addition, there are also X-rays and Auger electrons formed when the electron hole is filled with an electron from an upper shell.

In most cases, the lifetimes of the excited states of the daughter nuclides are very short; they decay simultaneously with the decay of the parent nuclide. In some cases, the lifetimes of the excited states are so long that they can be reasonably easily measured. In these cases, the excited states are considered as independent nuclides and marked with the letter m, meaning metastable. For example, the 0.662 keV excited state of ^{137}Ba has a half-life of 2.55 min and the nuclide is marked as ^{137m}Ba.

The longest half-life of a metastable nuclide has been measured at 241 years for ^{192m}Ir. As already mentioned, the gamma emissions are used to measure the activities of beta-emitting radionuclides since their measurement is often easier. This can only be done if the gamma transitions take place simultaneously with beta decay or when the half-life of the excited state is considerably shorter than the half-life of the beta-decaying nuclide to be measured. An example of the latter is the measurement of ^{137}Cs activity via the 0.662 MeV gamma emissions of ^{137m}Ba. Even though this metastable nuclide has its own half-life, it is very short compared to that of ^{137}Cs (30 years), and thus they both decay at the same rate. When the half-life of the excited state is longer than that of the parent, it cannot be used to measure the activity of the parent since they do not decay at the same rate.

1.3 Detection and Measurement of Radiation

Measurement of radionuclides can be accomplished either by measuring the radiation they emit, that is, radiometrically, or by measuring the number of radioactive atoms by mass spectrometry. The latter methods are described in Chapter 17. In this section, radiometric methods are only summarized. More detailed descriptions of them can be found in a number of textbooks. Radiation measurements are taken in two modes: pulse counting and spectrometry. In pulse counting, the number of pulses is recorded – not their heights. In spectrometry, both the number of pulses and their heights are measured; the pulses are then sorted to the channels of a multichannel analyzer according to their sizes. Pulse counting can only be used for samples containing one single radionuclide or when gross pulse rates are measured. Whenever information on the energy of the pulses is needed, for example in the identification of radionuclides from a mixture, spectrometry is used.

1.3.1 Gas Ionization Detectors

Gas ionization detectors are among the oldest methods still in use for radiation detection and measurement. A gas ionization detector is a chamber filled with an ionizable gas, such as Ar, with an electric field applied across the chamber. The detector is typically an argon gas-filled metal tube, where a metal wire in the middle of the tube acts as the anode, while the tube wall is the cathode (Figure 1.8). When an alpha particle, beta particle, or gamma ray hits the filling gas atoms it loses its energy by ionizing the argon atoms to Ar^+ cations. Because of the electric field applied across the chamber, these cations are collected onto the walls while the electrons go to the anode wire. The electrons are directed out of the tube to an external current circuit and are led back to the tube walls, thus neutralizing the argon cations to form argon atoms. Thus, for each ionization event in the tube, an electric pulse is formed in an external electric circuit. If the pulse height is proportional to initial energy of the particle or gamma/X ray, the system can be used in radiation spectrometry.

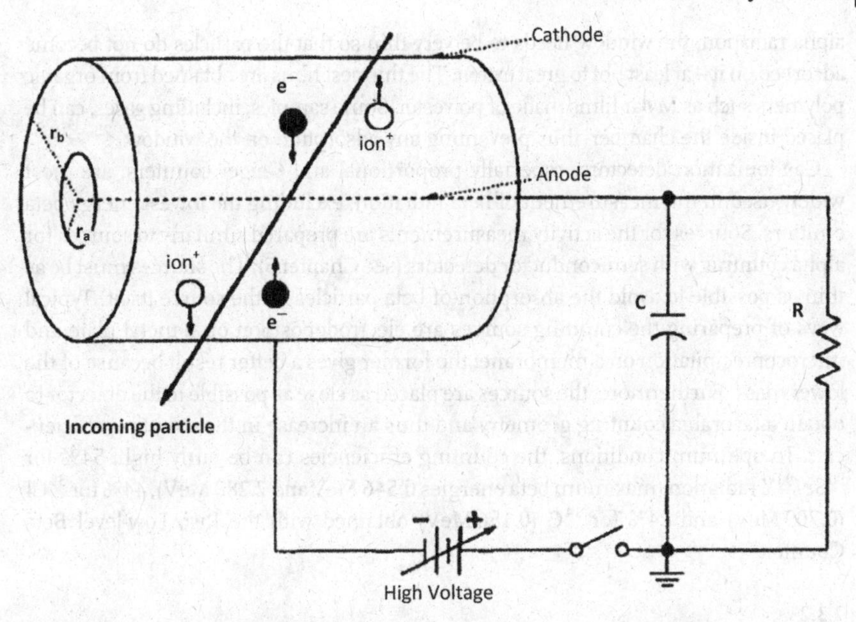

Figure 1.8 Gas ionization detector.

There are basically three types of gas ionization detectors: ionization chamber, proportional counter, and Geiger counter. In the ionization chamber, the electric field applied through the tube is the lowest at about 100–200 volts; the Geiger counter has the highest at some 1000 volts, while in proportional counters the voltage is in an intermediate range. In both ionization chambers and proportional counters, the electric pulse obtained is proportional to the energy lost by a particle or a ray in the filling gas. They can therefore be used to measure energy spectra. The pulse height response in the proportional counter is several orders of magnitude higher than that in the ionization chamber. In the ionization chamber, only ions and electrons in the primary ionization events are recorded, while in the proportional counter the number of ions and electrons is multiplied by a factor of 100–10 000 depending on the voltage used. Because of the very high voltage applied in Geiger counters, they produce pulses that always have the same size, since the filling gas is ionized to its maximum ionization state no matter what the energy of the particle or ray is. For this reason, they can only be used in pulse counting mode. Compared to the proportional counter, even higher voltage pulses are obtained from Geiger counters due to the high voltage, which results in the ionization multiplication by a factor of 10^7–10^{10}. In addition to radiometric measurements, Geiger counters have a wide application area as radiation dose meters used in radiation protection.

In order to be detected, the radiation must get into the chamber. This is usually no problem for gamma radiation since it penetrates the chamber wall. For beta and alpha counting, the other end of the chamber has a thin window through which the particles go into the gas chamber. For low-energy beta radiation, and especially for alpha radiation, the window needs to be very thin so that the particles do not become adsorbed on it – at least not to great extent. The thinnest films are obtained from organic polymers such as Mylar films made of polyester. Some samples, including gases, can be placed inside the chamber, thus preventing any adsorption on the window.

Gas ionization detectors, especially proportional and Geiger counters, are most widely used in the measurement of beta radiation, excluding the lowest energy beta emitters. Sources for the activity measurements are prepared similarly to sources for alpha counting with semiconductor detectors (see Chapter 5). The sources must be as thin as possible to avoid the absorption of beta particles in the source itself. Typical ways of preparing the counting sources are electrodeposition on a metal plate and microcoprecipitation on a membrane; the former gives a better result because of the lower mass. Furthermore, the sources are placed as close as possible to the detector to obtain a favorable counting geometry and thus an increase in the counting efficiencies. In optimum conditions, the counting efficiencies can be fairly high: 54% for $^{90}Sr/^{90}Y$ radiation (maximum beta energies 0.546 MeV and 2.280 MeV), 44% for ^{36}Cl (0.709 MeV) and 24% for ^{14}C (0.156 MeV) obtained with the Risö Low-level Beta Counter.

1.3.2 Liquid Scintillation Counting

Liquid scintillation counting (LSC) is a widely utilized radiometric method especially in the measurement of beta-emitting radionuclides. LSC is based on the radiation-induced light formation and transformation of the light into electric pulses. Radionuclide-containing liquid (or solid in a few special cases) is mixed with a liquid scintillation cocktail consisting of an organic solvent, such as toluene, and a scintillating agent. Typically, the radionuclide-cocktail mixture is in 20-mL polyethylene or glass vials, but smaller vials are also available. Scintillating agents are organic aromatic molecules that emit light photons by fluorescence when their excited states are relaxed. Excitations of the scintillator molecules are created by the transfer of the kinetic energy of beta or alpha particles or electrons via solvent molecules to the scintillator molecules. An example of such a scintillating molecule is 1-phenyl-4-phenyloxazole (PPO), which emits light photons at a wavelength of 357 nm. The light pulses penetrate the vial wall and are detected with a photomultiplier tube (PMT) (Figure 1.9).

In a PMT, first a photocathode, consisting of a photosensitive compound such as Cs_3Sb, releases electrons when hit by light pulses. The electrons are multiplied in the PMT with ten or more dynodes, each multiplying the number of electron by a certain factor. This is accomplished by an electric field of about one thousand volts applied through the tube. At the end of the tube, the electrons have been multiplied by a factor of about one million, and the pulse is now high enough to be analyzed further in the system. The height of the electric pulse obtained is proportional to the initial energy of the emitted particle, and for each keV of the particle/ray energy 5–7 light photons are created. For example, when tritium (E_{max} 18 keV) is measured, an average of 100 light photons per beta particle are created in the cocktail; when radiocarbon (E_{max}

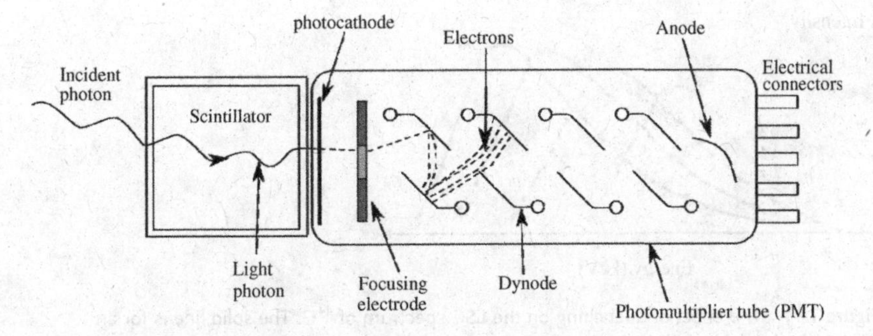

Figure 1.9 Photomultiplier tube coupled to a scintillation detector.

156 keV) is measured, some 300 light photons on average are created. These light pulses release electrons from the photocathode in the same proportion, and since the PMT multiplies these electrons by the same factor, the final electric pulse is proportional to the initial energies of the beta particles.

The liquid scintillation counter has two PMTs facing the sample vial from opposite directions (Figure 1.10) in order to suppress background. The two PMTs are connected in a coincidence mode so that pulses are accepted only if they come from the two multipliers at the same time. Pulses coming from single events on one PMT only, caused by cosmic radiation for example, can thus be avoided. After the coincidence unit, the pulses are summed, further amplified, and turned into digital form in an ADC unit. Finally, the digital pulse goes into a multichannel analyzer which sorts the pulses into different channels based on their pulse height. In this way, an energy spectrum is obtained.

The most challenging tasks in LSC are sample preparation and efficiency calibration. The former is discussed in Chapter 5. Efficiency calibration is needed to take into account the quenching that affects the counting efficiency. Quenching is a decrease in the efficiency of energy transfer from the beta particles, for example, to the scintillator, and a decrease in light photon intensity before they reach the PMTs. The effect of quenching on ^{14}C spectrum can be seen in Figure 1.11 which shows how pulses are lost and the spectrum is shifted to lower channels by quenching.

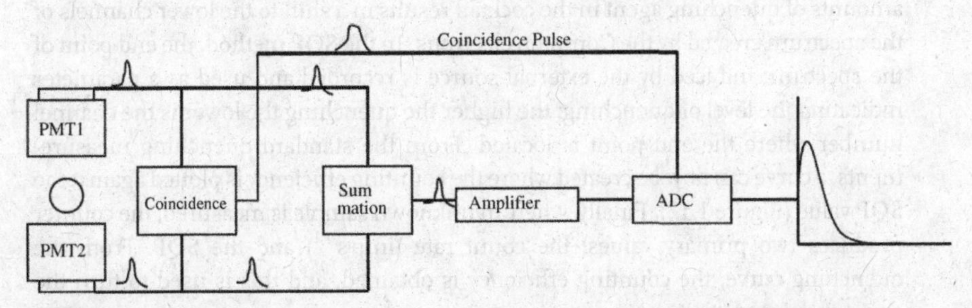

Figure 1.10 Scheme of a liquid scintillation counter (Kessler, M.J. (ed.) (1989). *Liquid Scintillation Analysis, Science and Technology*, Publ. No.169–3052, Perkin-Elmer Life and Analytical Sciences, Boston).

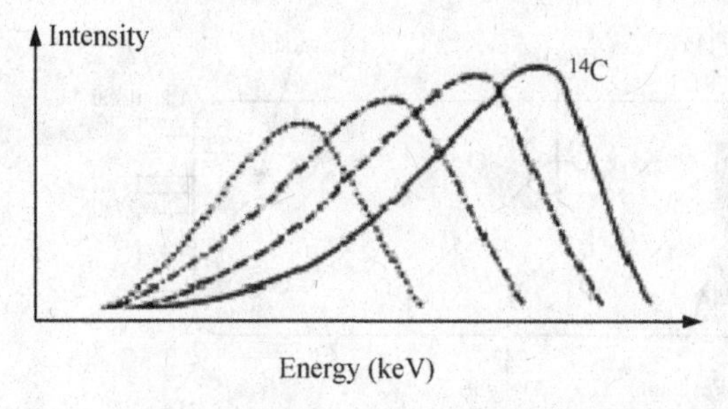

Figure 1.11 The effect of quenching on the LSC spectrum of ^{14}C. The solid line is for an unquenched sample and the broken lines for quenched samples.

There are a number of agents that cause quenching such as acids, alkalis, heavy metals, and alcohols. Since the composition of samples varies, the quenching of different samples also varies. Thus, the counting efficiency must be determined for all samples individually in order to correctly calculate the activity from the measured count rate. This can be done in various ways, of which only a few are described here. The most accurate way to determine the counting efficiency is to use an internal standard. The sample is measured twice: firstly as it is, and secondly after the addition of a known activity amount of the same nuclide. From the difference in count rates, the counting efficiency can be calculated. However, this method is rather laborious because of the double measurement and is thus only seldom used. Other methods include: the sample channels ratio method (SCR), the external standard channels ratio method (ESCR), and the external standard spectral quench parameter (SQP) method. In these methods, a quenching standard curve is created by measuring the counting efficiency for a set of samples containing the same known activity of the studied radionuclide and an increasing amount of quenching agent, such as CCl_4. In the ESCR and SQP methods, the samples are automatically measured twice in the counter: once as it is and then for a short period with an external gamma-emitting source (e.g., ^{226}Ra) positioned below the sample. The gamma rays from the external source cause Compton electrons in the scintillation cocktail, these cause further light emissions in the scintillator and finally electrical pulses in the counter. Increasing the amounts of quenching agent in the cocktail results in a shift to the lower channels of the spectrum created by the Compton electrons. In the SQP method, the end point of the spectrum induced by the external source is recorded and used as a parameter indicating the level of quenching: the higher the quenching the lower is the channel number where the end point is located. From the standard quenching measurements, a curve can now be created where the counting efficiency is plotted against the SQP value (Figure 1.12). Finally, when an unknown sample is measured, the counter produces two primary values: the count rate (imp s^{-1}) and the SQP. From the quenching curve, the counting efficiency is obtained, and this is used to turn the count rate into activity.

Besides beta measurements, LSC can also be effectively used for alpha measurement. The disadvantage of LSC in alpha measurements is its poor resolution. Even at

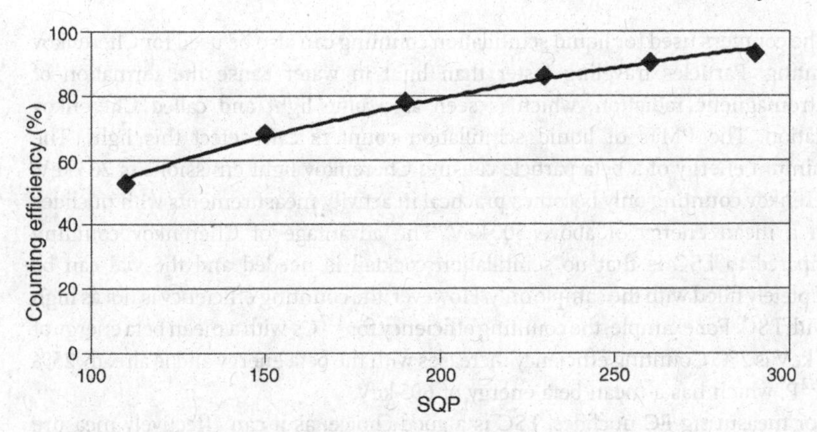

Figure 1.12 Quenching curve for standardization of LSC count rates.

best the resolution is about 200 keV, being ten times higher than in semiconductor measurements. Isotopic information is lost in many cases due to poor resolution. Thus LSC can be mainly used for gross alpha measurements and in cases where high resolution is not needed. Many modern LS counters are able also measure beta and alpha spectra simultaneously from the same sample. This is based on the discrimination of electric pulses caused by beta and alpha radiation due to their different life times: beta pulses are short and alpha pulses last several tens of nanoseconds longer. An example of such a spectrum, where both beta and alpha components are shown, can be seen in Figure 1.13. The alpha/beta discrimination feature facilitates measurement of radionuclide mixtures from natural decay chains in which there are always both beta and alpha emitters present.

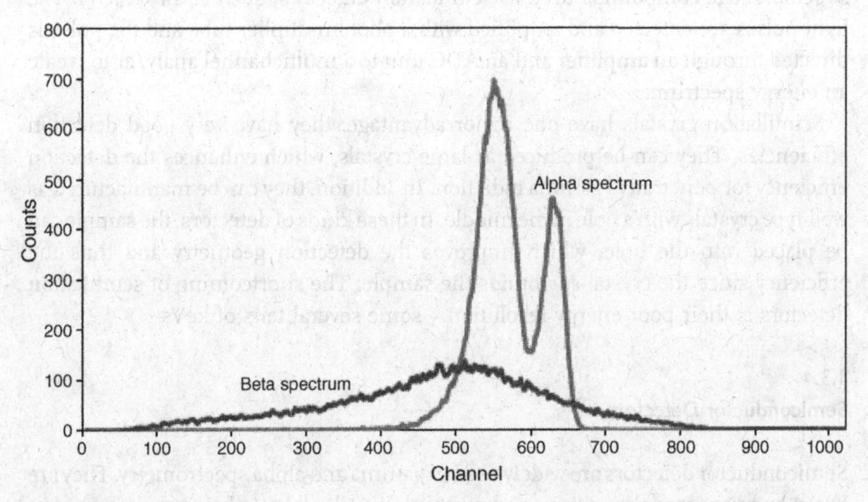

Figure 1.13 LSC spectrum of ground water divided into beta and alpha spectra using alpha/beta pulse discrimination.

The counters used for liquid scintillation counting can also be used for Cherenkov counting. Particles traveling faster than light in water cause the formation of electromagnetic radiation, which is seen as a blue light and called Cherenkov radiation. The PMTs of liquid scintillation counters can detect this light. The minimum energy of a beta particle causing Cherenkov light emissions is 263 keV. Cherenkov counting only becomes practical in activity measurements with nuclides with a mean energy of above 500 keV. The advantage of Cherenkov counting compared to LSC is that no scintillation cocktail is needed and the vial can be completely filled with the sample only. However, the counting efficiency is not as high as with LSC. For example, the counting efficiency for ^{137}Cs with a mean beta energy of 427 keV is 2%. Counting efficiency increases with the beta energy and is already 25% for ^{32}P, which has a mean beta energy of 695 keV.

For measuring EC nuclides, LSC is a good choice, as it can effectively measure Auger electrons created as a secondary radiation in the EC process. The activity of ^{55}Fe, a pure EC nuclide, is typically measured by LSC.

1.3.3 Solid Scintillation Detectors

Solid scintillation detectors are most typically used for gamma counting and spectrometry, and can also be used to measure other types of radiation; however, this is not discussed here. Solid scintillation detectors used in gamma detection typically consist of an inorganic crystal capable of producing light emissions when its excitation states are relaxed. Gamma rays are absorbed in the crystal and cause excitations in the crystal material – most usually NaI doped with a small fraction of thallium ions and thus denoted as NaI(Tl). The purpose of the thallium ions is to act as luminescent centers by which the light emissions take place. There are also stoichiometric compounds used as scintillation detectors, such as $Bi_3Ge_4O_{12}$. The light pulses are detected and amplified with a photomultiplier tube and the pulse is directed through an amplifier and an ADC unit to a multichannel analyzer to create an energy spectrum.

Scintillation crystals have one major advantage: they have very good detection efficiencies. They can be produced as large crystals, which enhances the detection efficiency for penetrative gamma radiation. In addition, they can be manufactured as well-type crystals with a hole in the middle. In these kinds of detectors, the sample can be placed into the hole, which improves the detection geometry and thus the efficiency since the crystal surrounds the sample. The shortcoming of scintillation detectors is their poor energy resolution – some several tens of keVs.

1.3.4 Semiconductor Detectors

Semiconductor detectors are widely used in gamma and alpha spectrometry. They are based on two materials – silicon and germanium – the former being mainly used in alpha detectors and the latter in gamma detectors. In a semiconductor detector, two semiconducting parts are attached together. One part is an n-type semiconductor with mobile electrons, which is produced by introducing phosphorus atoms with a valence of +V into a framework of germanium or silicon crystals – the valence of both being +IV. The other part is a p-type semiconductor with positive holes, achieved by having indium atoms with a valence of +III in the framework of germanium or silicon crystals. When an electric field is applied across the system in a reverse bias mode, a region depleted of holes and electrons is formed at the interface of the n and p type semiconductors. When a gamma ray or alpha particle hits this depleted region, electron-hole pairs are formed in the region and the system becomes conducting. The electric field then produces an electric pulse which can be recorded in the external circuit.

In germanium, the mobility of electrons and positive holes is tens of times higher than that in silicon. The depleted zone is thus much deeper in germanium (several centimeters) than it is in silicon, where it is less than one millimeter. Germanium is thus more suitable as a gamma detector: there is more mass in the depleted zone to absorb the energy of the very penetrating gamma radiation. In addition, germanium has a higher atomic number ($Z = 32$) than silicon ($Z = 14$), so that the formation of photoelectrons, by which the gamma ray energies are principally detected, is more than ten times as probable. Silicon, however, is more suitable for the detection of alpha particles (or low-energy gamma rays and X-rays) since these lose their energy over a very short range and interfering gamma rays penetrate the detector. For these reasons, germanium detectors are made as big as possible for gamma detection, while silicon detectors for alpha measurements are small, only a few mm thick, and have only a very thin depletion layer. In a silicon detector for alpha measurement, the p-type layer facing the alpha source is only tens of nanometers deep to allow the easily absorbable alpha particles to reach the depletion zone.

Germanium semiconductor detectors have superior energy resolution compared to scintillation detectors. The resolution can be as low as 0.5% for 122-keV gamma energy, while it is fifteen times higher for a NaI detector (Figure 1.14). Germanium detectors can thus be used to analyze samples containing even tens of gamma emitting radionuclides at the same time – process which is not possible with a NaI detector. Germanium detectors are therefore valuable tools to identify radionuclides in complex nuclear waste and environmental samples. Special attention is paid in this book to the radionuclides that require radiochemical separations, mostly alpha and beta-emitting radionuclides. For this reason, gamma spectrometry of the samples without radiochemical separations is not discussed in detail.

In conventional gamma detectors, the actual germanium detector is shielded with an aluminum cover which absorbs low-energy gamma rays and often makes their detection impossible. To detect low-energy gamma rays, new detectors have been developed with very thin windows that allow these rays to reach the germanium detector. The windows in these low-energy detectors and broad-energy detectors consist of beryllium or carbon composite. The use of these low-energy gamma detectors enables some radionuclides, such as ^{210}Pb and ^{241}Am, to be detected without radiochemical separation. The detection limit in direct gamma measurement, however, is lower than that in beta or alpha measurement following a

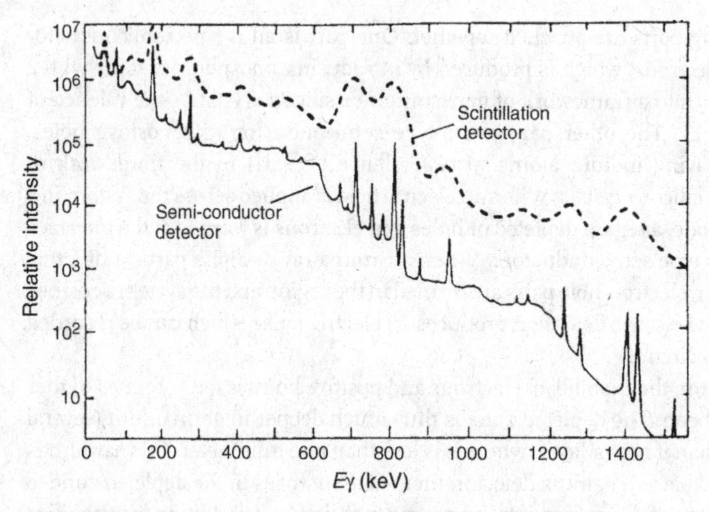

Figure 1.14 Gamma spectrum of ^{166}Ho measured by a scintillation detector and by a semiconductor detector (Keller, C. (1988) *Radiochemistry*, E. Horwood, Halsted Press, Chichester, New York).

radiochemical separation. Thin-window germanium detectors can also be used to detect X-rays, the measurement of which can be important, as in the case of electron capture radionuclides. As well as thin-window germanium detectors, silicon detectors are used to measure X-rays, but these differ from the silicon detectors used for alpha counting: the X-ray detectors are Li drifted, that is, they contain Li ions as donors.

Unlike gamma spectra, alpha spectra obtained with silicon detectors are, in general, simple to interpret, consisting only of peaks caused by the absorption of alpha particles and only a few pulses due to the background which can be subtracted after separate background measurement; in gamma spectra, however, the background is complicated by the formation of Compton electrons in the detector. Alpha radiation is easily absorbable, and therefore the counting source needs to be prepared with as low mass as possible to prevent self-absorption, which results in a loss of alpha particles and the broadening of the alpha peaks. In addition, the measurements have to be carried out in a vacuum to prevent the absorption of alpha particles in the air between the source and the detector.

1.3.5
Summary of Radiometric Methods

Of the four detector types described above, liquid scintillation counters and gas ionization detectors are the most suitable for beta measurements. The dominant method today in beta measurements is LSC because of its easy sample preparation, high counting efficiency, and high sample throughput rate. The most important advantage of proportional and Geiger counting over LSC is the considerably lower

Table 1.6 Suitable measuring methods for alpha- and beta-emitting radionuclides in the environment and in nuclear waste.

Nuclide	Decay mode	Measurement method
^{3}H	beta	liquid scintillation counting
^{14}C	beta	accelerator mass spectrometry
		liquid scintillation counting, gas ionization detectors
^{36}Cl	beta	accelerator mass spectrometry
		liquid scintillation counting
^{41}Ca	EC	accelerator mass spectrometry
		liquid scintillation counting
^{55}Fe	EC	liquid scintillation counting
^{63}Ni	beta	liquid scintillation counting, gas ionization detector
^{59}Ni	EC	X-ray spectrometry, accelerator mass spectrometry
^{79}Se	beta	liquid scintillation counting, accelerator mass spectrometry
^{85}Kr	beta	gas ionization detectors
^{90}Sr	beta	liquid scintillation counting
		gas ionization detectors
^{93}Zr	beta	liquid scintillation counting, ICP mass spectrometry
^{94}Nb	beta	gamma spectrometry, liquid scintillation counting
^{93}Mo	EC	gamma spectrometry, liquid scintillation counting
^{99}Tc	beta	liquid scintillation counting, ICP mass spectrometry
		gas ionization detectors
^{126}Sn	beta	gamma spectrometry, accelerator mass spectrometry
^{129}I	beta	accelerator mass spectrometry
		neutron activation analysis, liquid scintillation counting
^{135}Cs	beta	ICP mass spectrometry, thermal ionization mass spectrometry
Xe-isotopes	gamma	gamma spectrometry
	beta	gas ionization detectors
^{147}Pm	beta	liquid scintillation counting, ICP mass spectrometry
^{151}Sm	beta	liquid scintillation counting, ICP mass spectrometry
^{210}Pb	beta	liquid scintillation counting
	(gamma)	alpha spectrometry (via ^{210}Po), gamma spectrometry
^{210}Po	alpha	alpha spectrometry
		liquid scintillation counting
^{226}Ra	alpha	alpha spectrometry
		liquid scintillation counting
^{228}Ra	beta	liquid scintillation counting (via ^{228}Ac)
		gas ionization detectors (via ^{228}Ac)
^{227}Ac	beta	alpha spectrometry (via ^{227}Th)
Th-isotopes	alpha, beta	alpha spectrometry
		ICP mass spectrometry, liquid scintillation counting
^{231}Pa	alpha	alpha spectrometry
$^{234,235,238}U$	alpha	alpha spectrometry
		ICP mass spectrometry, liquid scintillation counting
^{237}Np	alpha	ICP mass spectrometry
$^{238,239,240}Pu$	alpha	alpha spectrometry
		ICP mass spectrometry (except ^{238}Pu)
^{241}Pu	beta	liquid scintillation counting

(Continued)

Table 1.6 (*Continued*)

Nuclide	Decay mode	Measurement method
^{241}Am	alpha	alpha spectrometry
	(gamma)	liquid scintillation counting, gamma spectrometry
^{242}Cm	alpha	alpha spectrometry

background, which enables lower activities to be measured. For example, the detection limit of the gas ionization method for ^{99}Tc using a Risø Low Level beta counter is 2 mBq, with a counting efficiency of 37% and a background count rate of 0.15 CPM, while the detection limit of LSC for ^{99}Tc using Quantulus ultra low level LSC is 55 mBq, with a counting efficiency of 85% and a background count rate of 5.5 CPM in ^{99}Tc window. Liquid scintillation counting is also a good method to measure radionuclides decaying by electron capture by measuring Auger electrons. Liquid scintillation counting can be used to measure alpha-emitting radionuclides with almost 100% detection efficiency; however, the resolution is poor compared to that of silicon semiconductor detectors, which are most often utilized in alpha counting. For gamma spectrometry, the best choices are germanium detectors, which have very good energy resolution and enable a number of radionuclides at the same sample to be the measured. Solid scintillation detectors can also be used in gamma spectrometry; however, their energy resolutions are poor compared to germanium detectors, and therefore they are mostly used in a single-channel mode to measure only one radionuclide at a time. For most long-lived radionuclides with a low specific activity, mass spectrometric methods give better detection limits than radiometric methods. Table 1.6 gives a list of the radionuclides dealt with in this book together with suitable measuring methods.

2
Special Features of the Chemistry of Radionuclides and their Separation

Radionuclides nearly always chemically behave identically to their corresponding stable isotopes, and the chemical separations are mainly based on the same traditional methods used in inorganic analysis: precipitation, ion exchange, and liquid extraction. Several special features must, however, be taken into consideration in the chemistry of radionuclides and the separations. Naturally, protecting against radiation emitted by the radionuclides (radiation protection) is an important special feature, but this is not dealt with in this context.

2.1
Small Quantities

The amounts of the radionuclides in the samples to be examined are nearly always extremely low, in other words trace amounts. Table 2.1 shows the masses of some radionuclides which correspond to 1 Bq activity.

Based on the table, it can be seen that only the amounts of the most long-lived radionuclides can be at the macro level. The dependence of activity and mass can be derived from the radioactive decay law:

$$A = \lambda \times N = (\ln 2/t_{1/2}) \times N \rightarrow A = (\ln 2/t_{1/2}) \times (m/M) \times N_{av}$$

At constant activity (A), the mass of a radionuclide (m) is directly proportional to the half-life ($t_{1/2}$) of the nuclide. N = number of atoms, N_{av} = Avogadro number, and M = molar mass.

In nature, the only radionuclides found in reasonably weighable macro quantities are ^{40}K and the isotopes of uranium and thorium. Uranium occurs in the lithosphere at an average abundance of about 3 mg kg^{-1}. The most important uranium minerals are uraninite and pitchblende. In the ores based on these, the content of the uranium can be as high as 10%. The abundance of Th in the lithosphere is almost three times higher than that of uranium, that is 8 mg kg^{-1}. The most important thorium mineral is monazite, and in the monazite-bearing ores the thorium content can be several percent. Potassium is a very generally occurring element; for example, in oceanic

Chemistry and Analysis of Radionuclides. Jukka Lehto and Xiaolin Hou

ISBN: 978-3-527-32658-7

Table 2.1 Masses and numbers of atoms of some radionuclides corresponding to 1 Bq activity.

Radionuclide	Half-life	Number of atoms	Mass (g)	Concentration in one liter (mol/L)
^{238}U	4.5×10^{9} y	2.0×10^{17}	8.0×10^{-5}	3.4×10^{-7}
^{237}Np	2.1×10^{6} y	9.3×10^{13}	3.7×10^{-8}	1.6×10^{-10}
^{226}Ra	1600 y	7.3×10^{10}	2.7×10^{-11}	1.2×10^{-13}
^{90}Sr	28.8 y	1.3×10^{9}	2.0×10^{-13}	2.2×10^{-15}
^{60}Co	5.3 y	2.4×10^{8}	2.4×10^{-13}	4.0×10^{-16}
^{210}Po	138 d	1.7×10^{7}	6.0×10^{-15}	2.9×10^{-17}
^{32}P	14.3 d	1.8×10^{6}	9.5×10^{-17}	3.0×10^{-18}
^{28}Al	2.24 min	1.9×10^{2}	9.0×10^{-21}	3.2×10^{-22}
^{20}F	11 s	16	5.3×10^{-22}	2.6×10^{-23}

water the potassium concentration is 0.07%. Since 0.0117% of all potassium is ^{40}K, the concentration of ^{40}K in oceanic water is 0.08 mg kg^{-1}.

The amounts of radionuclides which have ended up in the environment as pollution are extremely low. The deposition of ^{137}Cs from the nuclear weapons tests in the 1950s and 1960s is about 2 kBq m^{-2} in Finland, corresponding to 20 pg m^{-2} expressed as mass. Supposing that it is distributed in the 10-cm uppermost soil layer, the density of which is 2 g cm^{-3}, the concentration of ^{137}Cs in the soil is about 0.1 pg kg^{-1} (8×10^{-16} mol kg^{-1}). $^{239,240}Pu$ deposition from the nuclear weapons tests in turn is about 50 Bq m^{-2}, which corresponds to a concentration of 20 ng m^{-2} or 0.1 ng kg^{-1} (8×10^{-11} mol kg^{-1}). The contents of the radionuclides are also generally at trace levels in nuclear waste. The exceptions are spent nuclear fuel and the highly active waste accumulated in the reprocessing of the spent fuel. In addition to uranium (and plutonium in mixed oxide fuel (MOX)), the spent fuel contains 2–3% fission products and transuranium elements; in other words, there are 20–30 g of these in each kilogramme of fuel. In the highly active waste solutions generated in fuel reprocessing, the concentrations of some radionuclides are in the mmol/L range, which are extremely high solution concentrations for radionuclides.

2.2 Adsorption

When one handles very small amounts of matter, the danger exists that they will be adsorbed onto surfaces of the vessels and other tools, particles, and precipitates. On glass surfaces, ion exchange adsorption takes place on the surface –Si–OH groups. This silanol group is a weak acid and its hydrogen dissociates in neutral and alkaline solutions and can bind a metal ion. The glass surfaces typically have 10^{-10} mol cm^{-2} ion exchange capacity. Thus, for example, in a 100 milliliter beaker there is 10^{-8} mol of ion exchange capacity. If a glass contains 100 mL of a solution, in which the concentration of the radionuclide is 10^{-7} mol L^{-1}, the whole radionuclide content can be adsorbed onto the surface of the beaker. As seen from the Table 2.1, only a few radionuclides will have such high contents even at very high activities. Thus the adsorption is usually a problem and it must be considered. Alkali metal cations with low electric charge are least adsorbable to the glass surfaces, and the adsorption will increase with the charge of the metal ion. This is caused by the increase in the coulombic interaction with increasing metal charge density. With the nonhydrolyzable metal cations, that is, alkali metals and most alkaline earth metals, ion exchange absorption to the glass will increase with pH until it levels off at a constant value. Instead, the ion exchange absorption with the hydrolyzable metals will first increase with pH to a fixed maximum after which it starts to decline with further pH increases. The decrease in the absorption is caused by the competing effect of the hydrolysis and carbonate complexation with the ion exchange adsorption; in other words metals are desorbed from the glass surface into the solution because of an increasing tendency to form hydroxide and carbonate complexes at higher pH values. This kind of behavior is typical, among others, for the adsorption of Th(IV) to glass, seen in Figure 2.1. Adsorption also takes place to plastic surfaces, but not with ion exchange as to a glass. On plastic surfaces, atoms appear with free electron pairs, such as nitrogen and oxygen, which bind metals with co-ordination bonds. Thus, transition elements are adsorbed most sensitively to plastic surfaces. Plastic surfaces also adsorb organic molecules more sensitively than glass surfaces. Figure 2.1 shows that a polyethylene surface adsorbs Th more effectively than a glass surface.

Harmful absorption can also take place to the surfaces of particles suspended in a solution. This is dealt with in Chapter 3 in connection with colloids. Adsorption

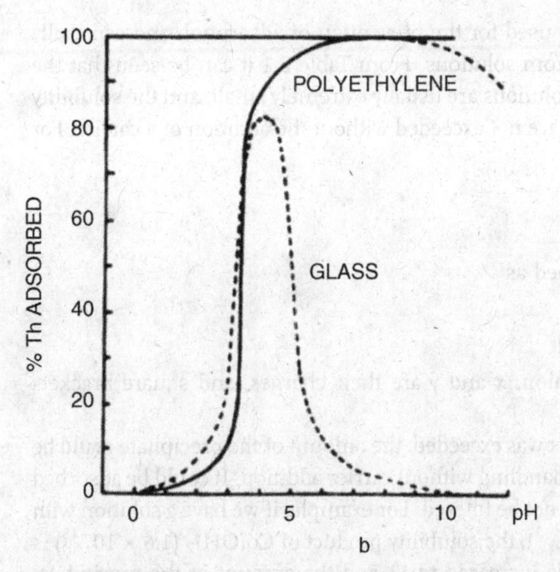

Figure 2.1 Adsorption of Th on glass and plastic vials as a function of pH. 2×10^{-8} M $Th(ClO_4)_4$ solution (Choppin, G., Linjenzin, J.-O., and Rydberg, J. (2002) *Radiochemistry and Nuclear Chemistry*, 3rd edn, Butterworth-Heineman).

can likewise take place when some materials are precipitated. This may in fact be desirable, and the adsorption can be used as a means of separation. This will be dealt with in depth in the chapter on separation methods. In connection with desired precipitation, undesired adsorption can also occur. If ^{140}La is, for example, separated from its parent nuclide ^{140}Ba by precipitating lanthanum as LaF_3, nearly all the ^{140}Ba is adsorbed on the LaF_3 precipitate despite the fact that barium does not coprecipitate with LaF_3. The adsorption will transpire to the surface of crystals as their size increases, and the absorbed ^{140}Ba partly ends up in the inner parts of the LaF_3 crystals. The adsorption is prevented with the help of hold-back carriers, discussed in the next section.

Adsorption can be avoided by the following methods:

- Radionuclides are stored in acid solutions in which the glass –OH groups do not dissociate but stay in the hydrogen form and thus cannot bind metal ions.
- Radionuclides are stored and handled in hydrophobic plastic vials, thus avoiding ion exchange adsorption.
- A stable isotope of the same element is added to the radionuclide solution to such a large extent that the ion exchange sites of surfaces are saturated.

This last method, using a stable isotope, exemplifies the use of carriers, which is important in radiochemistry and is not only restricted to the prevention of adsorption.

2.3 Use of Carriers

In radiochemistry a carrier is used for the prevention of adsorption and especially to precipitate radionuclides from solutions. From Table 2.1 it can be seen that the contents of radionuclides in solutions are usually extremely small, and the solubility products of their compounds are not exceeded without the addition of a carrier. For the precipitation reaction

$$yM^{x+} + x\,A^{y-} \rightarrow M_yA_x$$

the solubility product is defined as

$$k_s = [M]^y \times [A]^x,$$

where M is a metal, A an anion, x and y are their charges, and square brackets represent concentrations.

Even if the solubility product was exceeded, the amount of the precipitate could be impracticably low for further handling without carrier addition. It could be adsorbed on various surfaces and could not be filtered. For example, if we have a solution with 4 MBq per liter of ^{60}Co (1 μg L^{-1}), the solubility product of $Co(OH)_2$ (1.6×10^{-15}) is exceeded only when the pH is increased to 11 and the amount of the precipitate, 1.6 μg, is so small that the solid is invisible. Adding 60 mg of stable cobalt to the solution allows the cobalt to precipitate as a hydroxide at pH 8.1, and the amount of precipitate created is 94 mg. Such amounts can be reliably weighed and filtered, which is not the case without carrier addition.

As a carrier for the aforementioned ^{60}Co, a stable isotope (^{59}Co) of the same element as the radionuclide was added. The carrier in this case is called an isotopic carrier (which should be used whenever possible). In case of an isotopic carrier the radionuclide and the carrier behave chemically in an identical manner since they are same element. The added isotopic carrier, however, must be in the same chemical form as the examined radionuclide. For example, if the ^{60}Co in the system is in the +II oxidation state, as a Co^{2+} ion, the cobalt carrier must also be in this form and not, for example, as Co^{3+} ion. When the radioisotopes of iodine are correspondingly determined, the carrier must not be in the iodate form (IO_3^-) if the radioisotope to be examined exists as iodide (I^-). If the oxidation state of the radionuclide to be examined is unclear, the oxidation states of both the carrier and radionuclide, are adjusted by the addition of a suitable oxidant or reductant. In studies where oxidation states are to be determined, this kind of carrier usage and adjusting of oxidation states cannot be done.

The carrier is added at as early a stage as possible in the radiochemical separation processes. If the sample is aqueous, the carrier will be dissolved before other stages. Hydrolyzable metal carriers cannot be added to neutral or alkaline solutions, however, because they would cause precipitation of hydrolysis products before the carrier has had time to be evenly mixed into solution. For example, in the aforementioned precipitation of ^{60}Co with the aid of a stable cobalt carrier, the pH must be lowered to such a low value that the cobalt hydroxide does not precipitate. After the carrier has been added, the pH is raised to precipitate cobalt hydroxide. If cobalt is suspected to appear as some complex, EDTA for example, in the solution, the pH must be decreased prior to carrier addition to a value low enough to destroy the complex. This ensures that both ^{60}Co and the carrier cobalt are in the same chemical form and behave identically in the separation process.

A carrier can be added to a solid sample, for example a plant sample, as a solution which is absorbed into the solid material and dries during heating. Ashing (in other words burning of the sample) does not guarantee the conversion of the carrier to the same chemical form as the analyte, but additional adjustment of the chemical form of the analyte and carrier may be needed after ashing and acid leaching.

If several analytes occur in the same sample to be separated, their individual carriers might need to be added. In addition, the carrier is not merely used for the target radionuclide but also for the interfering radionuclides and other trace elements. When ^{90}Sr, for example, is separated using traditional nitrate precipitation in fuming nitric acid, a stable barium carrier will be added along with the strontium carrier at the beginning of the separation, ensuring that barium can effectively be removed with the chromate precipitation at a later stage.

As already mentioned, ^{140}Ba adsorbs on LaF_3 precipitate when ^{140}La is removed from its parent ^{140}Ba by fluoride precipitation. To precipitate the lanthanum fluoride, the addition of a stable lanthanum carrier is necessary. The ^{140}Ba does not form a fluoride, but is almost completely adsorbed onto the formed LaF_3 precipitate. This can be prevented by adding a stable barium carrier to the solution which almost totally prevents adsorption. In this case, the carrier is called a 'hold-back' carrier: a stable element is not added to precipitate the interfering radionuclide, but instead to keep it in solution while the target nuclide is precipitated. A hold-back carrier does not usually have a hold-back function at every stage of a separation procedure, and it may follow the target nuclide in some of these stages. The primary function of the hold-back carrier, however, is to prevent the interfering radionuclide from following the target radionuclide into the final sample used for activity measurements. Thus all carriers, other than that of the target radionuclide, are hold-back carriers.

A stable isotopic carrier cannot be used in all cases. Technetium, promethium, and all elements heavier than bismuth are radioactive, and no stable isotopes usable as carriers exist. In these cases a chemically analogous stable element can be used as a carrier. For example, Ba is generally used as a carrier for ^{226}Ra. Both are alkaline earth metals and barium is the next heaviest. Both form a divalent ion in solutions and behave chemically, more or less, in the same way. This is an example of the use of a nonisotopic carrier. In the same way trivalent lanthanides can be used as the carrier of trivalent actinides, for example Eu^{3+} can be used as a carrier for ^{241}Am since they are chemically very much alike: americium also forms Am^{3+} ion in solution. Table 2.2 shows the most important nonisotopic carriers.

A nonisotopic carrier is also used for radionuclides having a suitable stable isotope if that carrier needs to be removed at a later stage. For example, when a carrier-free ^{88}Y tracer is produced, a stable yttrium carrier cannot be used. Instead, Fe^{3+} is used, since it can be separated. In the process, the deuterium-irradiated strontium target is dissolved into acid and the Fe^{3+} carrier is added. ^{88}Y is coprecipitated with $Fe(OH)_3$ by raising the pH to 9. The ferric hydroxide precipitate is then dissolved in 9M HCl and the iron is extracted into di-isopropyl ether while the carrier-free ^{88}Y remains in the aqueous phase.

Furthermore, the carrier is also needed to 'force' the radionuclides in extremely diluted systems to obey thermodynamic and kinetic laws. If, for instance, only a few thousand radioactive atoms are present in the system, their mutual distances are so large that they do not 'feel' the presence of next identical atom and thus do not obey the laws of the thermodynamics in the same manner as atoms in macro quantities.

Table 2.2 The most important nonisotopic carriers.

Radionuclide to be separated	Nonisotopic carrier	Typical chemical forms
^{99}Tc	rhenium	TcO_4^-/ReO_4^-
$^{226,228}Ra$	barium	Ra^{2+}/Ba^{2+}
^{210}Po	tellurium	Po^{4+}/Te^{4+}
^{223}Fr	cesium	Fr^+/Cs^+
Pr	lanthanum, cerium, neodymium	Pr^{3+}/La^{3+}, Ce^{3+}, Nd^{3+}
At	iodine	
Ac, Pa, Pu, Np, Am, Cm	lanthanum, cerium, neodymium	$An^{3+,4+}/La^{3+}$, Ce^{3+}, Nd^{3+}

2.4
Utilization of Radiation in the Determination of Radionuclides

An important feature of radiochemical analyses is that radiation, as a rule, is utilized for the measurement of radionuclides. Mass spectrometry is also used for radionuclide measurements although its use is mainly restricted to the determination of long-lived radionuclides. The radiation can be used to make both qualitative and quantitative analysis:

- The energy of the radiation emitted by the radionuclide is characteristic of the nuclide and the energy spectrometry can be used to identify radionuclides. Gamma spectrometry, especially, is used for radionuclide identification, even for samples with tens of different radionuclides present. Also, the decay rate (half-life) is characteristic for each nuclide, and determination of half-lives can also be utilized in identifying radionuclides.
- Quantitative analysis can be carried out by determining the pulse rate obtained with some detection system. The activity can be calculated from the pulse rate when the efficiency of the detection system is known, that is, the percentage of decay is transformed into electrically recordable pulses. Another method of quantitative analysis is comparison of the pulse rate of a standard sample with a known activity with that of an unknown sample in the same circumstances.

2.5
Consideration of Elapsed Time

In radiochemical separations, the elapsed time needs to be considered for two reasons:

- If a radionuclide is short-lived, the separation procedure must be quick enough to ensure enough measurable activity throughout. The new heaviest elements that have been produced with large particle accelerators are extremely short-lived and require extremely quick separations.
- Even if the separation processes does not need to be quick to prevent their complete decay during separation, the decrease in their activity during the separation process needs to be calculated for radionuclides with such short half-lives that an essential part of them decays during the separations.

2.6
Changes in the System Caused by Radiation and Decay

The radiation emitted by the radionuclide may change the system if it causes radiolysis, that is, breaking of the chemical bonds in the medium. More detail on this and other chemical changes caused by radioactive decay can be found in Chapter 19.

2.7
The Need for Radiochemical Separations

Until the 1960s, most radionuclides could be measured only after they had been separated from other radionuclides. With great progress in gamma spectrometry and the application of semiconductor detectors (Ge(Li), HPGe), separation is now seldom required for gamma-emitting radionuclides. Tens of these can be simultaneously measured in the same sample. For pure beta- and alpha-emitting radionuclides, the need for separation still persists. Table 2.3 shows the most important radionuclides that still require separation. Natural and anthropogenic nuclides present in the environment and radionuclides found in nuclear waste are included.

The separation of radionuclides is required in the following situations:

- Since the spectrum of energies produced by beta particles is continuous, a beta-emitting nuclide can only be measured after it has been separated from other beta nuclides, as well as from alpha nuclides. Most beta nuclides are measured with a liquid scintillation counters (LSC), which allow only two beta emitters to be measured at the same time, but only if their beta energies are sufficiently different.
- Radionuclides decaying by electron capture (EC) do not emit any observable radiation in the actual decay process. In the rearrangement of a daughter nuclide's electron shell, X-ray photons and Auger electrons are emitted and used to measure the activity of EC nuclides. The X-ray photons and the Auger electrons have fairly low energies (a few keV to a few tens of keV), and when measuring them with liquid scintillation counters all coexisting beta- and gamma-emitting radionuclides prevent their measurement. Thus EC nuclides need to be separated in a very pure form.
- An alpha nuclide must be separated from other alpha nuclides with similar alpha energies, which would produce overlapping peaks in spectra. Semiconductor detectors, the most widely used detectors in alpha determinations, have an energy resolution of 20–30 keV. The resolution of liquid scintillation counters, which are also used in alpha counting, is ten times poorer. For measurements with semiconductor detectors, alpha nuclides must not only be separated from other alpha nuclides but also from most of the matrix elements present in the sample. These stable elements may be deposited/precipitated with the target alpha radionuclide on the alpha measurement source, which will cause self-absorption of alpha radiation and result in reduction of the counting efficiency and broadening of the alpha peaks. Modern liquid scintillation counters allow discrimination of a beta spectrum from an alpha spectrum (Figure 1.13). This technique, relying on pulse-shape analysis, takes advantage of the approximately hundred times longer pulse for an alpha particle than that for a beta particle. While facilitating the measurement of alpha nuclides, particularly in environmental samples, the poor resolution of the alpha spectrum in LSC makes it mainly suitable for the measurement of total alpha activity.

Table 2.3 Important radionuclides in the environment and in nuclear waste requiring radiochemical separation.

Nuclide	Decay mode	Half-life (y)	Source
^{3}H	beta	12.3	nature, nuclear explosions, nuclear energy
^{14}C	beta	5730	nature, nuclear explosions, nuclear energy
^{36}Cl	beta	3.01×10^{5}	nature, nuclear energy
^{41}Ca	EC	1.03×10^{5}	nature, nuclear energy
^{55}Fe	EC	2.7	nuclear explosions, nuclear energy
^{63}Ni	beta	96	nuclear energy
^{59}Ni	EC	76 000	nuclear energy
^{79}Se	beta	1.13×10^{6}	nuclear energy
^{90}Sr	beta	29	nuclear explosions, nuclear energy
^{93}Zr	beta	1.53×10^{6}	nuclear energy
^{94}Nb	beta	20 300	nuclear energy
^{93}Mo	EC	4000	nuclear energy
^{99}Tc	beta	2.13×10^{5}	nuclear energy
^{126}Sn	beta	1×10^{5}	nuclear energy
^{129}I	beta	1.57×10^{7}	nature, nuclear explosions, nuclear energy
^{135}Cs	beta	2.3×10^{6}	nuclear energy
^{147}Pm	beta	2.6	nuclear energy
^{151}Sm	beta	90	nuclear energy
^{210}Pb	beta (gamma)	22	nature
^{210}Po	alpha	0.38	nature
^{226}Ra	alpha	1600	nature
^{228}Ra	beta	5.8	nature
^{227}Ac	beta	22	nature
Th-isotopes	alpha, beta	up to 1.4×10^{10}	nature
^{231}Pa	alpha	32 750	nature
$^{234,235,236,238}U$	alpha	2.5×10^{5}–4.5×10^{9}	nature, nuclear explosions, nuclear energy
^{237}Np	alpha	2×10^{6}	nuclear explosions, nuclear energy
$^{238,239,240,241}Pu$	alpha, ^{241}Pu beta	14–24 000	nuclear explosions, nuclear energy
^{241}Am	alpha (gamma)	433	nuclear explosions, nuclear energy
^{242}Cm	alpha	0.44	nuclear explosions, nuclear energy

- A few gamma-emitting radionuclides, such as ^{94}Nb in spent nuclear fuel, though having high energy gamma rays with high intensities, must be separated from the matrix since there is a great excess of other gamma-emitting radionuclides, such as ^{137}Cs, preventing the detection of minor gamma emitters.
- If a radionuclide is measured by mass spectrometry, which is often the case with long-lived radionuclides, elements causing interference in the same mass as the target nuclide need to be removed. Principally there are two types of interference. First, a much more abundant neighbor isotope, such as ^{238}U, when measuring ^{239}Pu, causes tailings in the mass of the target nuclide. Another type of interference is caused by molecular ions, such $^{238}UH^{+}$, when measuring ^{239}Pu, causing extra pulses in the peak of the target nuclide.

3
Factors Affecting Chemical Forms of Radionuclides in Aqueous Solutions

A high proportion of samples analyzed for radionuclides consist of aqueous solutions such as natural waters and waste effluents from nuclear facilities. Solid samples need to be brought into solution before radiochemical analysis. The solutions can have a wide range of physical and chemical properties, which have an influence on the physical and chemical form of the radionuclides. Many factors affect the form in which radionuclides appear in solution. The most important of these are solution pH, redox potential, dissolved gases, ligands forming complexes with metals, humic substances, colloids, and the source and mode of generation of the radionuclides. These are discussed below. Speciation analysis, that is, analytical methods to study the forms of radionuclides, is discussed in Chapter 16.

3.1
Solution pH

The pH of the solution influences the form of radionuclides in several ways. pH directly affects the hydrolysis and redox reactions of metals, and indirectly affects the binding of metals to oxide and silicate surfaces and the formation of complexes. In this section we examine the effect of pH on the hydrolysis of metals and only briefly the other effects.

Most of the radionuclides discussed in this book are metals; notable nonmetals are carbon, selenium, iodine, and chlorine. A metal cation M^{z+} hydrolyzes in water as follows:

$$M^{z+} + xH_2O \leftrightarrow M(OH)_x^{z-x} + xH^+$$

The higher the charge on the metal and the smaller its size, the higher is the charge density of the metal and the more easily it hydrolyzes. In practice, strong hydrolysis means increasing tendency to form sparingly soluble hydroxide precipitates, and soluble hydroxide complexes and these hydrolysis products begin to form at low pH.

In aqueous solution, metal ions are surrounded by bipolar water molecules, so that the negative part of the water molecule (the oxygen, with a free electron pair) is

Chemistry and Analysis of Radionuclides. Jukka Lehto and Xiaolin Hou

ISBN: 978-3-527-32658-7

oriented toward the positive metal ion, while the positive hydrogen atoms are directed away from it. In this way aqua complexes $M(OH_2)_x^{z+}$ are formed, typically with four or six water molecules in the first hydration shell. The higher the charge density of the metal, the more strongly it attracts the electron pair of the oxygen, and the bond becomes less ionic and increasingly covalent. For example, trivalent iron (Fe^{3+}) hydrolyzes much more easily than the divalent iron (Fe^{2+}) and more easily than lanthanides in the same oxidation state because the radius of iron(III) is 0.645 Å, whereas that of lanthanum, for example, is much larger, 1.172 Å. Another term used instead of charge density is ionic potential, which is the ratio of the charge of the metal ion to its radius, z/r, where the radius is expressed in angstroms (Å, 10^{-10} m). The higher the ionic potential of a metal, or any element, the more strongly it attracts the electrons of the oxygen in a water molecule, and the more covalent is the chemical bond. Table 3.1 presents the ionic potentials of elements having important radionuclides.

Monovalent alkali metals, such as Cs^+, do not, in principle, hydrolyze at all, nor do the heavier divalent alkaline earth metals, such as Sr^{2+}, Ba^{2+}, and Ra^{2+}. In contrast to this, divalent transition metal ions (Co^{2+}, Ni^{2+}, Fe^{2+}, Pb^{2+}), trivalent metal ions (Fe^{3+}, Ce^{3+}, Am^{3+}, Cm^{3+}), and metals of higher oxidation state (Po, Th, U, Np, Pu) readily hydrolyze. Raising the pH of a solution promotes hydrolysis since the aqua complexes of metals behave like weak acids, and their hydrogen ions dissociate more easily at higher pH. When a neutral hydroxide species is achieved, the metal

Table 3.1 Ionic radii and potentials of selected hexacoordinated ions of radionuclide elements. Other coordination numbers than six are shown in parenthesis (Shannon, R.D. (1976) Revised effective ionic radii and systematic studies of interatomic distances in halides and chalcogenides. *Acta Cryst.*, **A32**, 751).

Ion	Ionic radius (Å)	Ionic potential	Ion	Ionic radius (Å)	Ionic potential
Cs^+	1.67	0.60	Sn^{2+}	1.12	1.79
Sr^{2+}	1.18	1.69	Sn^{4+}	0.69	5.80
Ca^{2+}	1.01	2.00	Se^{4+}	0.50	8.00
Ra^{2+}	1.48(8)	1.35	Se^{6+}	0.42	14.3
Ni^{2+}	0.69	2.90	Ac^{3+}	0.94	3.19
Fe^{2+}	0.79	2.56	Th^{4+}	0.94	4.26
Fe^{3+}	0.65	4.65	Pa^{5+}	0.78	6.41
Zr^{4+}	0.72	5.55	U^{4+}	0.89	4.49
Nb^{5+}	0.64	7.81	U^{6+}	0.73	8.22
Mo^{6+}	0.59	10.2	Np^{4+}	0.87	4.60
Tc^{7+}	0.56	12.5	Np^{5+}	0.75	6.67
Sm^{3+}	0.96	3.13	Pu^{3+}	1.00	3.00
Pm^{3+}	0.97	3.10	Pu^{4+}	0.86	4.65
I^{+5}	0.95	5.26	Pu^{5+}	0.74	6.76
C^{+4}	0.17	25.0	Pu^{6+}	0.71	8.45
Pb^{2+}	1.20	1.68	Am^{3+}	0.85	3.53
Po^{4+}	0.94	4.26	Cm^{3+}	0.85	3.53

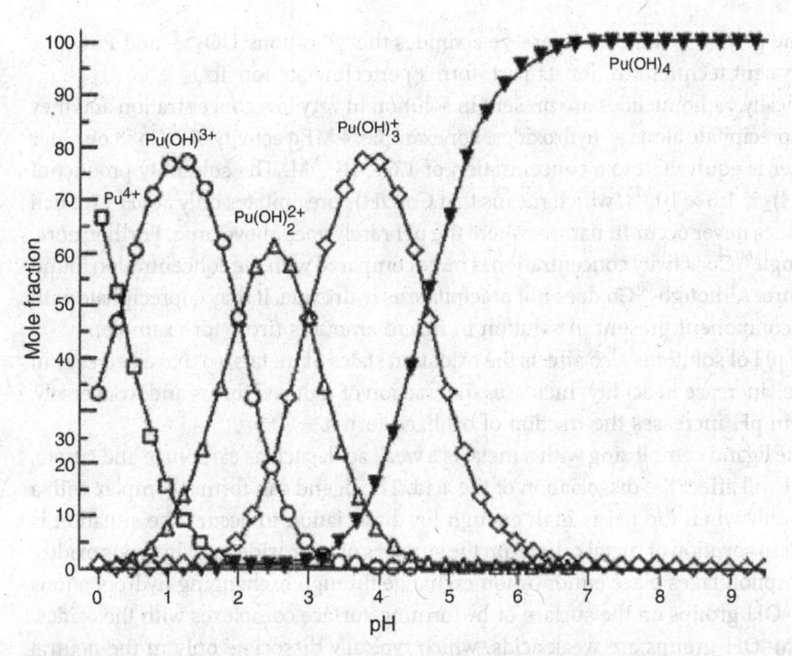

Figure 3.1 Distribution of hydrolysis products of plutonium as a function of pH. (Choppin, G.R. (2003) Actinide speciation in the environment. *Radiochimica Acta*, **91**, 645).

hydroxide will precipitate if the amount of metal is high enough to exceed the solubility product. Figure 3.1 shows the distribution of the hydrolysis products of plutonium as a function of pH. As can be seen, the Pu^{4+} ion is present only in strongly acidic solution, and hydrolysis products already appear alongside it in 1 M acid, first as monohydroxide complex $Pu(OH)^{3+}$, and then, as pH increases, as di- and trihydroxide complexes, $Pu(OH)_2{}^{2+}$ and $Pu(OH)_3{}^+$. Above pH 6, only soluble $Pu(OH)_4$ complexes exist in the solution. And when the concentration of plutonium is high enough to exceed the solubility product of $Pu(OH)_4$, the hydroxide precipitates.

When the oxidation state of the metal increases to +V, not only are simple hydroxide complexes $M(OH)_n{}^{5-n}$ formed but so also are negatively charged oxoanions, described by the 'ate' ending, and oxocations. The reason for these oxo forms is that the ionic potential of pentavalent metal ions is very high, and the positive charge of the metal attracts oxygen atoms of water molecules so strongly that all hydrogen atoms are liberated and covalent bonds are formed between the metal and the oxygen. This process takes place with nonmetals also. Elements of smaller size appear as oxoanions – among the elements having important radionuclides iodine as iodate ($IO_3{}^-$) and niobium as niobate ($NbO_3{}^-$). Ion sizes of the actinides, in turn, are so large that, in pentavalent form, they bind the oxygen atoms of only two water molecules and form 'yl' ions: uranyl $UO_2{}^+$, neptunyl $NpO_2{}^+$, and plutonyl $PuO_2{}^+$.

Elements at oxidation state six also form 'ate' and 'yl' species: smaller-size molybdenum and selenium form the negative species molybdate ($MoO_4{}^{2-}$) and selenate (SeO_4^{2-}) and the larger-size actinides the 'yl' cations UO_2^{2+} and PuO_2^{2+}. Heptavalent technetium, for its part, forms pertechnetate ion TcO_4^-.

Typically, radionuclides are present in solution in very low concentration and they rarely precipitate alone as hydroxides. For example, 4 MBq activity of ^{60}Co in one liter of water is equivalent to a concentration of 1.6×10^{-9} M. The solubility product of $Co(OH)_2$ is 1.6×10^{-15}, which means that $Co(OH)_2$ precipitates only at pH 11. Such pH values never occur in nature, where the pH rarely goes above nine. Furthermore, such high ^{60}Co activity concentration is huge compared with the concentration found in nature. Although ^{60}Co does not precipitate as hydroxide, it may coprecipitate with some component present in solution in macro amounts (iron, for example).

The pH of solutions also affects the oxidation states of metals, so that a decrease in pH (i.e., increase in acidity) increases the fraction of reduced forms and, conversely, a rise in pH increases the fraction of oxidized forms.

If the ligand complexing with a metal is a weak acid, such as carbonate and citrate, the pH will affect the dissociation of the acid. The ligand can form a complex with a metal only when the pH is high enough for dissociation to occur. The situation is similar in sorption of metal cations on the surfaces of hydroxides and hydrous oxides. The sorption takes place either by ion exchange through exchanging hydrogen ions of –M–OH groups on the surface or by forming surface complexes with the oxides. The –M–OH groups are weak acids, which typically dissociate only in the neutral range or in weakly acidic solution. Thus, attaching to oxides and hydrous oxides by ion exchange is possible only if the –M–OH groups have dissociated. pK_a values, that is, the pH values where dissociation occurs, increase for the oxides and hydroxides typically found in natural waters in the order $SiO_2 < MnO_2 < Fe(OH)_3 < Al(OH)_3$.

3.2
Redox Potential

A second highly important factor for the form of an aqueous metal is the redox potential of the system. High redox potential favors oxidized forms of metals and low redox potential reduced forms. Many radionuclides appear in more than one oxidation state, and because their chemical behavior depends on the oxidation state, the redox potential is of great significance. First of all, the oxidation state affects metal solubility in aqueous solutions (Table 3.2). In general, the oxidation state +IV is the least soluble and the solubility systematically increases as the oxidation state decreases from this value. Higher oxidation states are also more soluble than +IV but in a somewhat more complex manner. At oxidation state +VII all elements are in the highly soluble anionic MO_4^- form, while the solubility of the +V and +VI oxidation states depends on whether the elements are in oxoanionic or oxocationic form and what are the valences of these forms. For example, the actinyl forms of actinides in oxidation state +VI (AnO_2^{2+}) are less soluble than those in oxidation state +V (AnO_2^+) because of the higher charge density of the former. Secondly, the oxidation state affects complex formation, and the stability of a metal complex follows the same trend as its solubility. As can be seen in the table below, changes in hydrogen ion

Table 3.2 Effect of reductions on solubility of important redox-sensitive radionuclides. Reduced forms are to the right of the arrow.

Radionuclide	Reduction reaction	Solubility
^{55}Fe	$Fe^{3+} + e^- \rightarrow Fe^{2+}$	Increases
^{99}Tc	$TcO_4^- + 3e^- + 8H^+ \rightarrow Tc^{4+} + 4H_2O$	Decreases
I isotopes	$IO_3^- + 4e^- + 6H^+ \rightarrow I^- + 3H_2O$	Decreases
U isotopes	$UO_2^{2+} + 2e^- + 4H^+ \rightarrow U^{4+} + 2H_2O$	Decreases
^{237}Np	$NpO_2^+ + e^- + 4H^+ \rightarrow Np^{4+} + 2H_2O$	Decreases
Pu isotopes	$PuO_2^{2+} + e^- \rightarrow PuO_2^+$	Increases
	$PuO_2^+ + e^- + 4H^+ \rightarrow Pu^{4+} + 2H_2O$	Decreases
	$Pu^{4+} + e^- \rightarrow Pu^{3+}$	Increases

concentration often occur in redox reactions. In systems in which oxygen participates in redox reactions, either as oxygen gas (O_2) or as part of a species (e.g., IO_3^-), the pH decreases in reduction reactions and, conversely, increases in oxidation reactions.

Because electrons are never free in solution, the reactions shown in the table do not occur alone. In oxidation and reduction reactions, electrons are always transferred between two components, metal or ion, as in the following reaction:

$$Zn + Cu^{2+} \rightarrow Zn^{2+} + Cu$$

Here, zinc metal donates two electrons to copper ion, simultaneously being oxidized to oxidation state two, while the copper ion is reduced to copper metal. The reaction proceeds in precisely this direction and not the reverse because of the standard electrode potentials, E°, of the two pairs, Zn/Zn^{2+} and Cu/Cu^{2+}, participating in the reaction. For each component participating in the reaction (zinc and copper in this case) a half-reaction is defined so that the oxidized form is on the left side of the equation

$$Zn^{2+} + 2e^- \rightarrow Zn$$

$$Cu^{2+} + 2e^- \rightarrow Cu$$

and for both half-reactions standard electrode potential (E°) is determined. The standard electrode potential is the potential difference, or voltage, obtained when the potential of either redox pair (e.g., zinc metal electrode in zinc sulfate solution) is measured relative to the potential of the standard hydrogen electrode. The standard hydrogen electrode consists of a platinum electrode in 1 M HCl solution which is bubbled with hydrogen gas, where the pressure of the hydrogen gas in the HCl solution is 1 atm (1.013 bar). The value of the potential of the standard hydrogen electrode when it is connected to a platinum electrode in 1 M HCl solution is set to zero. The redox half-reaction of this system is thus

$$2H^+ + 2e^- \rightarrow H_2$$

When the platinum electrode in 1 M HCl is replaced with another metal- or ion-containing electrode system whose potential is to be measured, a potential value other than zero is obtained, positive or negative depending on whether the system is more reducing or oxidizing than the hydrogen ion system. If, for example, the potential of a zinc electrode in 1 M zinc sulfate solution is measured relative to the standard hydrogen electrode, the voltage obtained is −0.76 V. In the corresponding copper system the voltage is +0.34 V. These voltages are the standard electrode potentials for the redox pairs Zn^{2+}/Zn and Cu^{2+}/Cu. Based on these standard electrode potentials one can predict which redox pair reduces the other: the redox pair with a more negative standard potential reduces that with a more positive standard potential. For any two half-reactions, that with the more negative standard potential moves to the left, toward the oxidized form ($Zn^{2+} + 2e^- \leftarrow Zn$), and that with the more positive standard potential moves to the right, toward the reduced form ($Cu^{2+} + 2e^- \rightarrow Cu$).

Standard potentials are always determined in the standard state, in which the activity of each component is one. The redox potential Eh at other concentrations of the system can be calculated from these standard potentials with the help of the Nernst equation:

$$\mathrm{Eh} = E^\circ + (RT/nF) \times \ln([\mathrm{ox}]/[\mathrm{red}])$$

in which R is the gas constant (8.314 J mol^{-1} K^{-1}); T is the absolute temperature (K); F is the Faraday constant (96 485 C mol^{-1}), or one electron mole; n is the number of electrons involved in the redox reaction; [ox] is the activity/concentration of the oxidized form, and [red] is the activity/concentration of the reduced form at equilibrium. If, for example, the value of the redox potential measured at 25 °C for a solution containing 0.011 M iron is 0.712 V, then, with use of the Nernst equation, the ratio of the concentrations of Fe^{3+} and Fe^{2+} can be calculated as $([Fe^{3+}]/[Fe^{2+}]) = e^{(((0.712\,V - 0.771\,V)^* nF)/RT)} = e^{((-0.059 \times 1 \times 96485)/(298 \times 8.314))} = e^{-2.298} = 0.10$, where 0.771 V is the standard potential of the reaction $Fe^{3+} + e^- \rightarrow Fe^{2+}$. The ratio is thus 0.1; that is, 9.1% of the iron is in oxidized form and the rest in reduced form. In a redox reaction in which there are two redox pairs, for example, $5Fe^{2+} + MnO_4^- + 8H^+ \leftrightarrows 5\ Fe^{3+} + Mn^{2+} + 2H_2O$, the Nernst equation is used to calculate Eh values for the two half-reactions ($MnO_4^- + 8H^+ + 5e^- \rightarrow Mn^{2+} + 4H_2O$ and $Fe^{3+} + e^- \rightarrow Fe^{2+}$), and their difference gives the Eh value for the whole reaction.

It also holds for redox reactions that the Gibbs free energy is $\Delta G^\circ = nF \times E^\circ = RT \times \ln K$. The Gibbs free energy can, therefore, be calculated from either the standard potential or the equilibrium constant (K). The value of ΔG° indicates whether the reaction is spontaneous (negative value) or forced (positive value). As is clear from the above equation, the equilibrium constant can be calculated from the standard potential, and vice versa. Thus, if we know the standard potential, we can use it to calculate the species distribution from the equilibrium constant.

Although the standard hydrogen electrode is the normal reference electrode for measuring standard potentials, it is too inconvenient to use in measurements of redox potential. The calomel electrode (Hg/Hg_2Cl_2) or silver electrode ($Ag/AgCl$) is used instead. The redox potential of a reaction is then obtained by correcting for the potential given by the calomel or silver electrode relative to that of the standard hydrogen electrode. These correction values, at 25 °C, are 0.2444 V for the saturated calomel electrode and 0.19888 V for the saturated silver electrode. These values need to be added to obtain the values for a standard hydrogen electrode.

In natural systems, the ultimate determinants of redox potential are oxygen and organic matter. Soluble oxygen from the atmosphere raises the redox potential of a system, which is lowered through the aerobic decomposition of organic matter in water, soil, and bedrock. In deeper parts of the bedrock, where oxygen is depleted, and in sediment and other layers inaccessible to atmospheric oxygen, redox reactions are governed by secondary pairs, such as Fe^{2+}/Fe^{3+}, Mn^{2+}/Mn^{4+}, and S^{2-}/SO_4^{2-}. By acting as reducing agents for metals, microbes, too, play an important role in the redox potential. The redox potential correlates with water depth, allowing the following rough division to be made. In open surface water, the redox potential is +200 to +800 mV, in shallow groundwater down to about 100 meters and in deep lake and sea waters it ranges between +200 and −200 mV, and in deep groundwater it varies between −200 and −400 mV.

Redox potential is often presented as a function of pH in Eh–pH diagrams, or Pourbaix diagrams. In the Eh–pH diagram of plutonium (Figure 3.2), the predominant species of plutonium are displayed as a function of Eh and pH. At the typical pH and Eh values of groundwater, Pu is predominantly present in oxidation state +IV as the soluble $Pu(OH)_4$ complex. This exists in equilibrium with solid $Pu(OH)_4$ if the concentration is sufficiently high. With decrease in pH, the trivalent form Pu^{3+} becomes predominant. Under more oxidizing conditions, the pentavalent form PuO_2^+ becomes more prevailing, and this form is also predominant in sea water. When using an Eh–pH diagram it needs to be remembered that it is primarily based on the equilibrium calculations made with initial values, such as standard potential, obtained in pure reference systems. These calculations give a good starting point for the values of each species, but, particularly for complex natural systems, practical experiments will be required to identify the real forms and their distribution. Furthermore, Eh-pH diagrams present only the predominant species in an area, and in boundary areas there are always species from both sides of the line. Species that are nowhere predominant do not appear in diagrams at all, even though they might be present in significant proportions.

Among the radionuclides, the redox potential is of greatest significance for actinides uranium, neptunium, and plutonium, which have several possible oxidation states. Deep in the bedrock, hundreds of meters below the surface, and in other sites with no oxygen, the redox potential is so low that it has a reducing effect not only on the actinides but on other radionuclides as well, such as those of Fe, Se, Nb, Mo, Tc, Sn, I and Po.

At the end this chapter is an appendix where are presented typical agents used to adjust oxidation states of radionuclides in radiochemical analyses.

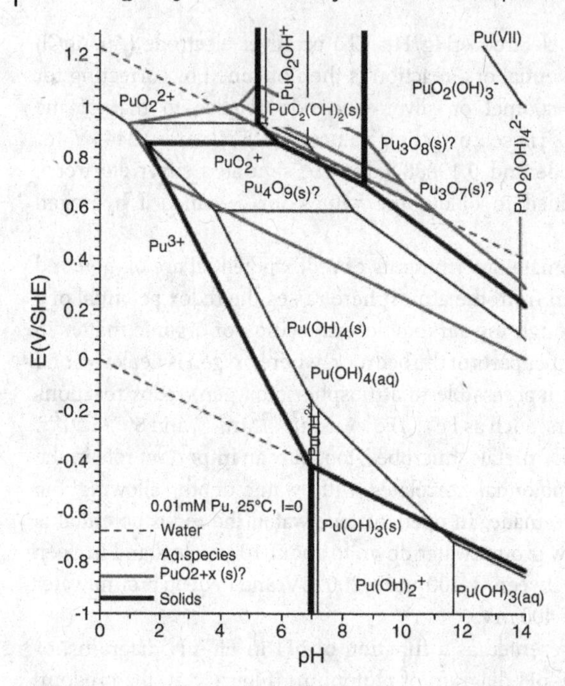

Figure 3.2 Eh-pH diagram for plutonium. Plutonium concentration 0.01 mM. The upper dashed line indicates 1 atm O_2 pressure and the lower dashed line 1 atm H_2 pressure. These lines represent stability boundaries of water. (Vitorge, P., Capdevila, H., Maillard, S., Fauré, M.-H., and Vercouter, T. (2002) Thermodynamic Stabilities of $MO_{2+x}(s)$ (M = U, Np, Pu, and Am), Pourbaix diagrams. *J. Nucl. Sc. Technol.*, (Supplement 3), 713).

3.3
Dissolved Gases

Gases dissolved from the atmosphere play a major role in the speciation of metals in natural waters. By far the most important of these gases are oxygen and carbon dioxide.

3.3.1
Oxygen

Dissolved oxygen from the atmosphere is the ultimate determinant of the redox potential of water, and it directly affects the oxidation state of radionuclides. Divalent iron, for example, oxidizes under the influence of oxygen as follows:

$$Fe^{2+} + 1/4\,O_2 + 5/2\,H_2O \rightarrow Fe(OH)_3 + 2H^+$$

Surface water is continually being replenished with oxygen from the atmosphere, and the concentration of dissolved oxygen has a maximum value of 8.25 mg L^{-1} (0.258 mmol L^{-1}) at 25 °C and 1 bar air pressure. Oxygenated rainwater carries oxygen deep into the soil and bedrock, but the amount of oxygen gradually diminishes with the depth. Oxygen is consumed first and most intensively in the oxidation of organic matter in the top layer of the soil.

$$RCHO + O_2 \rightarrow CO_2 + H_2O$$

(RCHO represents the organic matter, e.g., cellulose)

Still more oxygen is consumed lower down in the soil profile by organic matter transported there by water. Oxygen in soil and bedrock is also consumed by microorganisms, which use it in energy production. The oxidation of organic matter and microbial activity are intimately related. Other ways in which oxygen is consumed are in the oxidation of metals, for example, the oxidation of iron(II) to iron(III) and manganese(II) to manganese(IV), and the oxidation of sulfides to sulfates. In unsaturated soil (soil that is not saturated with groundwater), oxygen diffuses more deeply into the ground relatively fast, but maximally at the rate it diffuses in air (0.205 cm^2/s). The oxygen consumed in the oxidation of organic matter is then replenished by fresh oxygen from the atmosphere. The diffusion rate of oxygen in water is over ten thousand times slower than that in air. Thus, the oxygen consumed in oxidation in waterlogged soil is replenished much more slowly than that consumed in unsaturated soil, and the redox potential is lower in waterlogged soil. And if oxygenated water does not seep through fractures in the bedrock, the redox potential will decrease with increasing depth. Oxygen is also efficiently consumed by the organic matter in water: just 3 mg L^{-1} of dissolved organic matter can consume the maximal amount of oxygen that can dissolve in water. Hundreds of meters deep in the bedrock, where spent fuel from nuclear power plants will be buried, there is practically speaking no oxygen at all and the redox potential in ground water will be extremely low, as low as −400 mV.

The oxygen in rivers, lakes, and the sea is used up in the same way as that in the soil, that is, mostly in the oxidation of organic matter. If there is no vertical flow of water, as may happen in frozen lakes in winter, oxygen may become depleted and anoxic conditions develop in the bottom water. In anoxic conditions, $Fe(OH)_3$ and MnO_2 precipitated under oxidizing conditions may dissolve, with Fe^{3+} reducing to soluble Fe^{2+} and Mn^{4+} to soluble Mn^{2+}. Simultaneously, any radionuclides coprecipitated with them will be remobilized.

3.3.2 Carbon Dioxide

The other major gas important for the speciation of metals in water is carbon dioxide. Carbon dioxide reacts in water to form carbonic acid: $CO_2(aq) + H_2O \leftrightarrows H_2CO_3$. The first dissociation constant (pK_1) of carbonic acid, in the reaction $H_2CO_3 \leftrightarrows H^+ + HCO_3^-$, is 6.35, and the second (pK_2), in the reaction $HCO_3^- \leftrightarrows H^+ + CO_3^{2-}$, is 10.33. Thus, hydrogen carbonate ion is predominant when the pH exceeds 6.33 and carbonate ion when the pH exceeds 10.33 (Figure 3.3). Carbonic acid is the predominant acid in natural waters and the primary determinant of pH of natural

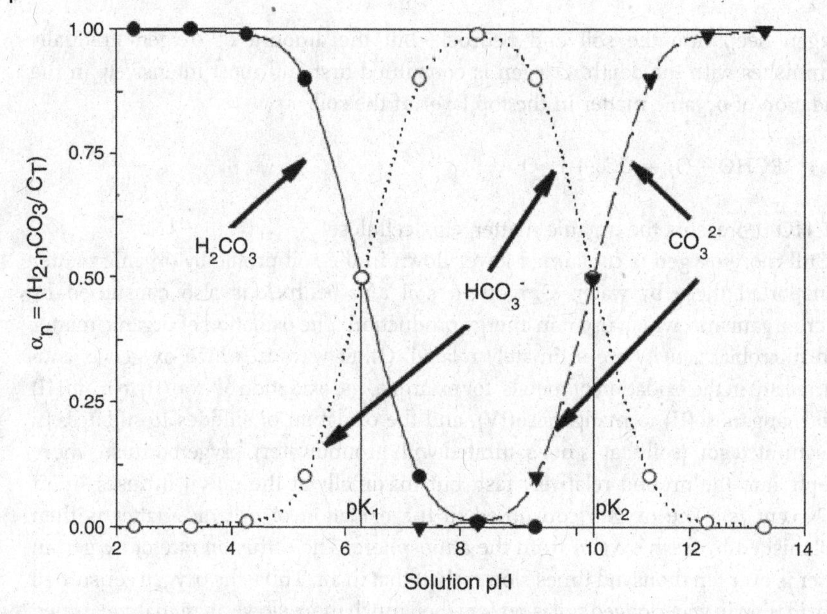

Figure 3.3 Carbonate species in water as a function of pH.

waters, while the carbonate ion, CO_3^{2-}, effectively precipitates many metals and forms complexes with them.

Carbon dioxide is dissolved into water from the atmosphere, where it makes up 0.033 volume percent. The concentration is also expressed as partial pressure, which in a gas mixture, in this case the air, is equal to the mole fraction multiplied by the total pressure. At one bar pressure, at 25 °C, the partial pressure of carbon dioxide in dry air is thus $10^{-3.5}$ bar (0.00033 bar). The partial pressure of carbon dioxide in natural waters generally ranges between 10^{-4} and 10^{-2} bar. In surface water, carbon dioxide is almost always in equilibrium with atmospheric carbon dioxide, and its partial pressure is the same as that in the atmosphere. Aquatic plants take up carbon dioxide for photosynthesis, and, at times of vigorous growth, the partial pressure of carbon dioxide will drop below the equilibrium level if the atmosphere is unable to restore the balance quickly enough. Typically, carbon dioxide builds up in the bottom water of lakes and the seas as a consequence of the decomposition of aquatic plants and the oxidation of organic matter in the bottom sediment. Carbon dioxide is also introduced into the bottom water through the influx of carbonate-rich groundwater. Increase in the amount of carbonic acid intensifies the anoxic bottom conditions.

In soil, the amount of carbon dioxide is determined by the aerobic decomposition of organic matter in the unsaturated zone ($RCHO + O_2 \rightarrow CO_2 + H_2O$), which continually adds carbon dioxide to the soil. During warm growing periods, the partial pressure of carbon dioxide in the soil may rise as high as 0.1 bar. Part of the generated carbon dioxide dissolves in water, and part escapes from the soil to the atmosphere. If the ground freezes in winter and the carbon dioxide is unable to escape to the atmosphere, the amount of carbon dioxide in the soil may increase dramatically. Carbon dioxide, and dissolved organic material in general, are carried with water to deeper levels in the soil. The amount of carbon dioxide is greater in deeper, anoxic soil and sediment layers and in the bedrock than in surface layers because of the many processes generating carbon dioxide (reduction of sulfate and nitrate, formation of methane) and the inability of carbon dioxide to escape from deep in the ground to the atmosphere. Even though the amount of carbon dioxide is high in groundwater, the pH of the water increases rather than decreases because of reactions that consume hydrogen ions (e.g., the reduction of sulfate, $2CH_2O + H^+ + SO_4^{2-} \leftrightarrows 2CO_2 + HS^- + 2H_2O$, where CH_2O represents organic material). As a result, the pH of deep groundwater is distinctly higher than that of shallow groundwater, ranging between 8 and 9.

In surface water, as well as soil water and groundwater, carbonate can precipitate as calcium carbonate, $CaCO_3$, usually in the form of calcite, if the concentrations of calcium and carbonate are high enough for the solubility product to be exceeded. The solubility product of calcite, $K_s = [Ca^{2+}] \times [CO_3^{2-}]$, is 3×10^{-9}. Along with sodium, potassium, and magnesium, calcium is one of the four cations most commonly dissolved in natural waters in macro amounts. In lakes the concentration of calcium is typically 0.25–1.25 mmol L^{-1} and in the sea 10 mmol L^{-1}. Calcite will also precipitate in soil and bedrock and in bottom water of lakes when the amount of carbon dioxide is sufficiently high.

The relevance of all this to the speciation of radionuclides is that many metal radionuclides can coprecipitate with calcite. Most of them form sparingly soluble carbonates (see Table 4.1). Alkali metals are exceptional in not forming sparingly soluble carbonates and do not coprecipitate with calcite. Strontium is closely related to calcium in chemical behavior, and ^{90}Sr, for example, strongly coprecipitates with calcite. One of the methods for speciation analysis that is introduced later (Chapter 16) is selective/sequential extraction, which is used to determine the speciation of radionuclides in soil and sediment samples. A key step in these extractions is to determine the radionuclides bound to carbonates. The carbonates are dissolved in acetic acid at a pH of about 4, which releases radionuclides associated with them. Calcite is least soluble at a pH greater than 9 but dissolves even in weak acids such as acetic acid.

Besides precipitating with calcite, carbonate affects the chemistry of radionuclides in natural waters by forming strong complexes, especially with the actinide elements, whose formation constants with carbonate are higher than those of most other ligands present in water (sulfate, phosphate, chloride). Typically fluoride forms even stronger complexes, but its concentration in natural waters is very low compared to that of carbonate. The main competitor to the formation of carbonate complexes is the formation of hydroxide complexes by hydrolysis. Particularly noteworthy in the formation of carbonate complexes is the behavior of uranium in groundwater, where uranium appears in oxidized form as uranyl ion UO_2^{2+}. In water where the pH exceeds 6, uranyl ion forms highly soluble negatively charged carbonate complexes: $UO_2(CO_3)_2^{2-}$ at pH 6–7 and $UO_2(CO_3)_3^{4-}$ at pH above 7. The other actinides also form strong carbonate complexes. The strongest complexes are formed by tetravalent actinides. The stability of carbonate complexes of the actinides typically decreases in the following order of actinide ions:

$$An^{4+} > AnO_2^{2+} > An^{3+} \approx AnO_2^{+}$$

Figure 3.4 shows the speciation of neptunium as a function of pH at a total CO_2 concentration of 2.3 mmol L^{-1}, respectively. At low pH (<7.5), Np is mainly as $NpO_2{}^{+}$, while at higher pH, NpO_2OH, $NpO_2CO_3{}^{-}$, and $NpO_2(CO_3)_2{}^{3-}$ are the major species of neptunium.

3.4 Ligands Forming Complexes with Metals

The ligands in water available for complex formation markedly affect the chemical forms of radionuclides, since most of the metal cations present in natural waters in trace concentrations appear in complexed form. Complexes are formed with inorganic and organic anions and with neutral organic molecules. The most important of the complexes formed with organic molecules in water are humic complexes, which are discussed in the following section. Of the inorganic ligands, the most important are OH^{-}, $CO_3{}^{2-}$, $HCO_3{}^{-}$, $SO_4{}^{2-}$, $HPO_4{}^{2-}$, $PO_4{}^{3-}$, Cl^{-}, and F^{-}. Of these, hydroxide complexes were covered earlier in the context of hydrolysis and carbonate complexes in the discussion of dissolved carbon dioxide. Table 3.3 shows the concentrations of inorganic ions in fresh water and in sea water. The concentrations in fresh water are clearly lower than those in sea water, and they vary over a wide range, while the concentrations in sea water are more or less constant.

As described earlier, in pure water solution, metal cations are present in the form of aqua complexes, typically surrounded by four or six water molecules in the first hydration shell closest to the metal. The above-mentioned inorganic ligands form both outer-sphere and inner-sphere complexes with metal ions. Outer-sphere complexes form between hydrated metal cations and negative ligands with no change taking place in the first hydration shell. The bond is purely electrostatic, which means that it is an ion–ion interaction. Because there are water molecules between the metal and the ligand, the metal and the charged ligand lies far apart from one another. In inner-sphere complexes, in turn, the ligand replaces one or more of the water molecules in the first hydration shell and approaches the metal, causing the covalent character of the bond to increase. The stability of the complex increases at the same time. The move from ionic to covalent bond occurs gradually, and the bonds of many complexes have both ionic and covalent character. The nature of the bond is determined by the metal and ligand charges and their size, that is, their charge densities.

In natural waters, outer-sphere complexes are mostly formed with the ions Na^{+}, Mg^{2+}, and Ca^{2+}, which occur in macro concentrations. The fourth important macro cation, potassium, does not form complexes at all. The outer-sphere complexes of sodium, magnesium, and calcium are generally called ion pairs. The weakest complexes are formed with hydrogen carbonate, chloride, and sulfate, and elements

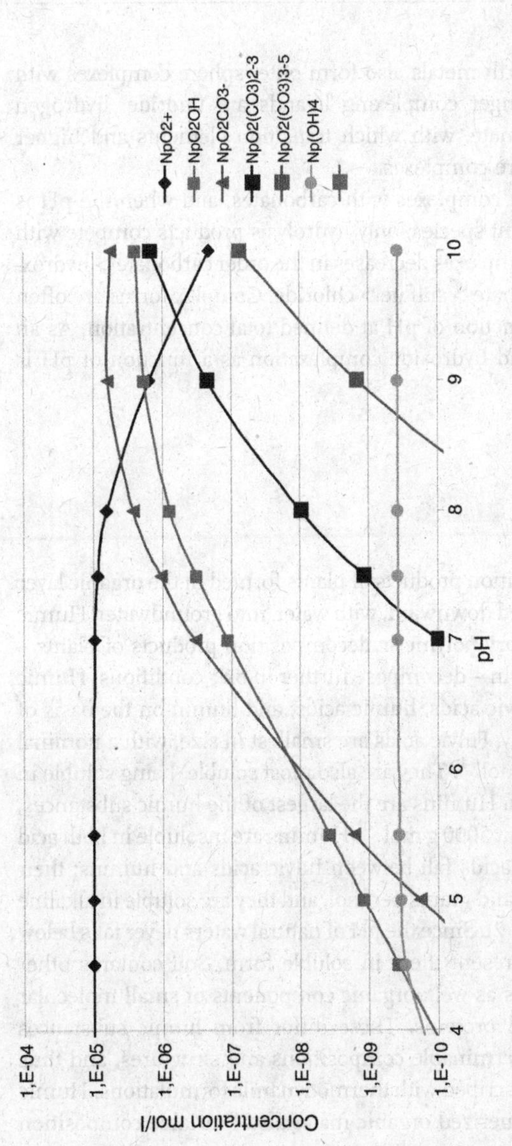

Figure 3.4 Aqueous speciation diagram of neptunium at 25 °C. Total Np concentration 0.01 mmol/L total carbonate concentration 2.3 mmol/L.

Table 3.3 Anion concentrations in fresh water and sea water (Stumm, W. and Morgan, J. (1995) *Aquatic Chemistry*, 3rd edn, John Wiley & Sons, p. 290).

Ion	Fresh water	Sea water
HCO_3^-	0.1–5 mmol/L	2.5 mmol/L
CO_3^{2-}	0.001–0.1	0.03
SO_4^-	0.01–1	28
Cl^-	0.01–1	550
F^-	0.001–0.1	0.06

other than alkali and alkaline earth metals also form outer-sphere complexes with these ligands. Among the stronger complexing ligands are fluoride, hydrogen phosphate, and especially carbonate, with which transition elements and higher charged metals form inner-sphere complexes.

Actinides form their strongest complexes with carbonates, and when the pH is above 6 these are the predominant species; only hydrolysis products compete with them. The strength of actinide complexes decreases in the order carbonate > hydroxide > fluoride ≈ hydrogen phosphate > sulfate > chloride. Complex forms are often presented in graph form as a function of pH at defined total concentration. As an example, uranium phosphate and hydroxide complexation as a function of pH is presented in Figure 3.5.

3.5 Humic Substances

Humic substances are decomposition products of plants formed in the organic layer of the soil, and they are also carried downward, with water, into groundwater. Humic substances form when the primary polymeric decomposition products of plants – cellulose, hemicellulose, and lignin – decompose further in oxic conditions. Humic substances are classified into fulvic acids, humic acids, and humin on the basis of their molecular size and solubility. Fulvic acids are smallest in size, with a nominal molecular weight of 500–1500 $g\,mol^{-1}$. They are also most soluble, being soluble in both acidic and alkaline solutions. Humins are the largest of the humic substances, with molecular weights up to about 5000 $g\,mol^{-1}$. Humins are insoluble in both acid and alkaline conditions. Humic acids fall between fulvic acids and humins: their molecular weight is several thousand grams per mol, and they are soluble in alkaline but not in acidic conditions ($pH < 2$). Since the pH of natural waters never falls below pH 2, humic acids are always present there in soluble form. Soil contains other decomposition products of plants as well: organic components of small molecular size, such as carbohydrates and proteins. These differ from humic substances chemically, in their precisely determinable compositions and structures, and thus their chemical behavior can be described with thermodynamic formulations. Humic substances, in contrast, are polymerized organic materials of variable composition

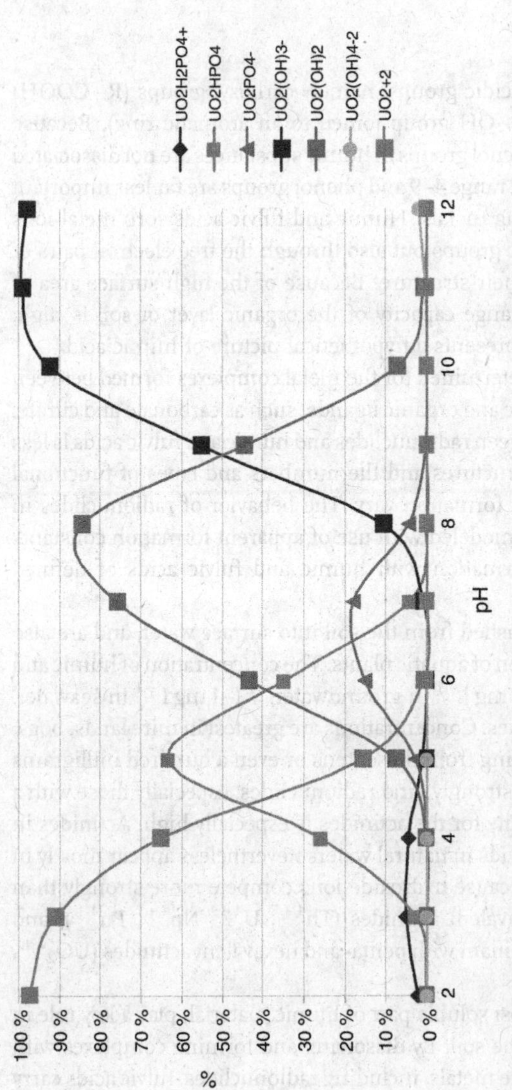

Figure 3.5 Uranium phosphate and hydroxide complexes as a function of pH. Uranium concentration 0.084 μM, phosphate concentration 2.6 μM.

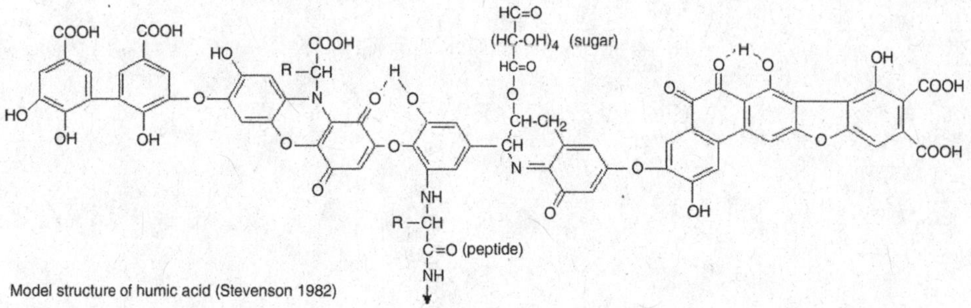

Model structure of humic acid (Stevenson 1982)

Figure 3.6 Model structure of humic acid (Stevenson, F.E. (1982) *Humus Chemistry*, Wiley-Interscience, New York).

and molecular size containing acidic groups, namely carboxyl groups (R–COOH) and less acidic phenol groups (–OH group joined to an aromatic ring). Because phenolic acid is very weak, the phenol groups in humic substances are not dissociated in natural waters with a pH in the range 4–9 and phenol groups are far less important than carboxylic groups in binding metals. Humic and fulvic acids sorb metal ions not only through their carboxylic groups but also through the free electron pairs of oxygen and nitrogen atoms in their structure. Because of the high surface area of humic substances, the ion exchange capacity of the organic layer of soil is high: 100–200 meq/100 g. Figure 3.6 presents a hypothetical picture of humic acid.

Formation constants can be determined for the metal complexes formed between most radionuclides and inorganic and organic ligands, such as carbonate and citrate. The formation of complexes between radionuclides and humic and fulvic acids is less straightforward because their structures and the numbers and types of functional groups participating in complex formation vary. The behavior of radionuclides in natural waters, however, can be modeled with use of apparent formation constants determined for the complex formation with humic and fulvic acids of defined composition.

Humic and fulvic acids are washed from the soil into surface water, and are also formed there in the decomposition of aquatic plants. The concentration of humic and fulvic acids together is about 0.1 mg L^{-1} in groundwater, 0.1–1 mg L^{-1} in seawater, and 1–10 mg L^{-1} in rivers and lakes. Concentrations are greatest in mire lands, bogs, and marshes, and in water draining from these: tens or even a hundred milligrams per liter. Humic and fulvic acids strongly bind radionuclides, especially those with a high charge, and thus their affinity for the actinides is especially high. Actinides in complexes of humic and fulvic acids in natural waters nevertheless appear mostly in trivalent form (Am^{3+}, Pu^{3+}) because hydroxide ions compete more strongly than humic and fulvic acids for tetravalent actinides (Th^{4+}, U^{4+}, Np^{4+}, Pu^{4+}), and carbonate complexes are predominant with penta- and hexavalent actinides (UO_2^{2+}, NpO_2^+, PuO_2^+, PuO_2^{2+}).

Fulvic acids, which are the most soluble part of humic material, play a key role in leaching metals downwards in the soil. By dissolving and forming complexes with iron and aluminum and with trace metals, including radionuclides, fulvic acids carry these down into the mineral layers of the soil. With time, the fulvic acids break down through oxidation, and the bound iron and aluminum (and radionuclides), are released and precipitate as oxides into the enrichment zone of the soil.

3.6 Colloidal Particles

Another important factor besides humic and fulvic acids that influences the speciation of radionuclides in natural waters is the association of radionuclides to inorganic colloidal particles. Colloidal particles are the fine particles in water varying in diameter from a nanometer to several hundred nanometers. Though 0.45 μm is often taken as the upper limit, the definition is not precise. Colloidal systems are relatively stable, or metastable, whereas larger, suspended particles will sediment out within a reasonable time. Colloidal particles are stable, or persistent in solution, because their rate of diffusion due to thermal motion is higher than their rate of sedimentation due to gravity. As the size of particles increases, the rate of sedimentation begins to exceed the rate of diffusion. The lower size limit for colloidal particles is similarly not precise. Thermodynamically, colloidal particles are distinguished from dissolved components in that chemical potential can be determined for dissolved components but not for colloidal particles. An exact composition can be assigned to truly dissolved matter, and the behavior of dissolved components can be described with conventional thermodynamic concepts. Although in many systems colloidal particles will appear to behave like dissolved matter, their true behavior differs in an essential way. If colloidal and dissolved forms of the same element are present in a system, the behavior of the two forms will be distinct. Their separation from each other is nevertheless often very difficult.

Colloidal particles abound in natural waters; there may be millions of them in one milliliter, and their total surface area is high. The composition of colloids varies widely. The humic and fulvic acids discussed in the previous section are important organic colloidal particles. Most inorganic colloidal particles are clay minerals, that is, aluminum silicates and hydrous oxides of iron, manganese, titanium, aluminum, and silicon. OH groups on the surface of the oxides are largely dissociated in the neutral pH range of natural waters, allowing metal cations to adsorb by ion exchange. Metal cations, especially those with high charge, can also attach to the surface through complexation, that is, formation of covalent bonds with surface oxygen atoms. Radionuclides not only attach to clay minerals through surface OH groups but also absorb in the interlayer space in the crystal lattice, which results in a strong bonding. Cesium sorption by clay minerals is an important example of the latter process. Though cesium is in general a very soluble element, sorption by clay mineral particles, if present, results in effective sedimentation of cesium into lake bottoms by the clay minerals and makes cesium rather immobile in clayey soils.

Colloidal particles containing radionuclides are classified into two groups. The first group is the carrier colloids, which include humic and fulvic acid colloids and colloids formed with oxides onto whose surfaces radionuclides have sorbed. Also included in this group are particles in which radionuclides are coprecipitated, isomorphically or nonisomorphically, along with the compound of some closely related element. The second group is the intrinsic colloids, or colloidal particles containing a single radionuclide, typically in hydroxide or oxide form. The formation of intrinsic colloids is rare because the concentrations of radionuclides in natural waters are generally very low, and the solubility product is seldom exceeded. Only a few radionuclides are ever present in solution in such large amounts that the solubility product is exceeded. In nature, only uranium and thorium appear in such large amounts, and even then the formation of intrinsic colloids is highly unlikely, especially in case of uranium, which is very soluble in natural waters.

3.7
Source and Generation of Radionuclides

The forms of radionuclides are also affected by their source. The major sources of radionuclides in the environment are nature's own processes, nuclear weapons tests in the atmosphere, accidents at nuclear power plants, nuclear weapons production facilities, fuel reprocessing plants, nuclear power plants and nuclear waste repositories. These have already been discussed in Chapter 1. Here we briefly focus on the effect of the formation process on the physical and chemical form of radionuclides.

The most important source of radionuclides is *nature* itself, most radionuclides being members of the decay series of uranium and thorium. Among these nuclides are U, Th, ^{226}Ra, ^{222}Rn, ^{210}Pb, and ^{210}Po. With the notable exception of radon, they are almost entirely bound in minerals present in soil and bedrock, and to some degree appear in dissolved form in ground waters and surface waters. Radon occurs in gaseous form and also diffuses into the atmosphere, from which its daughters (^{210}Pb, ^{210}Po, and short-lived lead, polonium and bismuth isotopes) deposit on the earth's surface attached to aerosol particles. Within the soil there are also a few individual long-lived natural radionuclides, the most notable being ^{40}K. A large group of lighter, cosmogenic radionuclides are generated in the atmosphere. Of these, ^{14}C and ^{3}H are mostly present in the gas phase, while many others, such as ^{7,10}Be, are attached to aerosol particles and precipitate onto the Earth's surface either as dry deposition or, in rain or snow, as wet deposition. Also, industrial plants spread natural nuclides in the environment. For example, the burning of coal and peat releases natural nuclides into the atmosphere, and smelters and fertilizer plants discharge natural nuclides in soluble form to watercourses and the sea.

A second important source of radionuclides is the *nuclear weapons tests carried out in the atmosphere in the 1950s to 1970s*. Of all the fallout radionuclides associated with these tests, the major ones still present in the environment are ^{90}Sr, ^{137}Cs, ^{238,239,240,241}Pu, and ^{241}Am. Most of the fallout was global stratospheric fallout, in which the radionuclides that deposited on the ground were bound to atmospheric aerosols. Over the years, the radionuclides have deposited onto the surface layer of the soil. At actual test sites (Nevada, Semipalatinsk, etc.) and the areas round about, however, a major part of the radionuclides, especially of plutonium and americium, is present as large, insoluble particles, some consisting of unfissioned plutonium and uranium. Most of the transuranium elements at test sites are present on the soil surface.

A third source is *accidents at nuclear power plants*, the most consequential of these being the explosion and fire at the Chernobyl plant in 1986. Most of the nuclides were the same as those released to the environment in nuclear weapons tests, but there were also notable differences. Unlike the debris from nuclear weapons tests, the fallout from Chernobyl rose in an emission plume and was carried over Europe in the troposphere at a height of 1000–2000 meters. It was not only aerosol particles with adsorbed radionuclides that were swept over Europe but also larger, hot particles – fuel fragments and condensation products of weakly vaporizing metals – where radionuclide concentrations were high. These hot particles are distinctly less soluble than the particles in global fallout from nuclear weapons tests. The proportion of hot particles in the fallout from Chernobyl was greatest closest to the accident site and decreased with distance. The nuclide composition and isotope ratios also differed from those of weapons fallout. Thus, the fallout from Chernobyl contained, in addition to the fission product ^{137}Cs and also the activation product ^{134}Cs, which was not found in deposition from weapons tests. Likewise, the ratios of plutonium isotopes in the fallout from Chernobyl and that from weapons tests were noticeably different, allowing these ratios to be used in source identification.

The emissions from *nuclear power plants* are very low compared with the previous two sources. Small amounts of ^{137}Cs, ^{60}Co, ^{54}Mn, and ^{63}Ni are discharged to rivers and the sea, and settle in sediments in the near surroundings. Tritium remains in the water phase. ^{14}C and ^{85}Kr may be released to the atmosphere, where they are diluted in air masses.

Over the decades, *nuclear weapons production facilities and civilian fuel reprocessing plants* have released high amounts of radioactive materials to the environment, and in particular to surface water. Particularly polluted are areas around nuclear weapons production complexes in Russia, the worst being the area around the Majak nuclear complex in the southern Urals. In the past two or three decades, effluents from the reprocessing facilities at La Hague in France and Sellafield in Great Britain have been reduced to a fraction of what they once were, but still a moderate amount of radioactive material is discharged to the sea.

Large amounts of radionuclides end up to the environment in the *final disposal of nuclear waste*. At present, only low and intermediate active waste generated during the operation of nuclear plants is disposed of in final repositories, which means that the waste mostly contains relatively short-lived fission and activation products, such as ^{137}Cs, ^{60}Co, and ^{63}Ni. In future, waste from decommissioned nuclear plants will be taken to repositories. This includes waste such as radioactive concrete, where the nuclide composition is much the same as that in the waste generated in plant operations. The most active nuclear waste that will be brought to repositories will, however, be the spent nuclear fuel and the waste generated in reprocessing, both containing a broad spectrum of short- and long-lived fission and activation products, such as ^{90}Sr, ^{99}Tc, ^{135}Cs, ^{137}Cs, and isotopes of plutonium and ^{241}Am. From these final repositories, the nuclides will find their way into the biosphere tens of thousands of years from now, at the earliest, and then only small amounts of those that are longest lived.

3.8 Appendix: Reagents Used to Adjust Oxidation States of Radionuclides

Radionuclide analyses often include adjusting the oxidation state of the radionuclide. A number of reagents can be used for this purpose, of which some of the most important are briefly described here.

3.8.1 Oxidants

Hydrogen peroxide H_2O_2 is an oxidizing agent in acidic medium, the oxygen in it having an oxidation state of −I and aiming at a more stable oxidation state of −II. Hydrogen peroxide, for example, oxidizes tetravalent technetium to pertechnetate in the following way:

$$2TcO_2 + 3H_2O_2 \rightarrow 2TcO_4^- + 2H_2O + 2H^+$$

H_2O_2 is also often used as an oxidizing agent to decompose organic matter to convert all radionuclide species into their inorganic forms.

Oxyanions such as NO_3^-, $Cr_2O_7^{2-}$, $S_2O_8^{2-}$, ClO^-, MnO_4^-, and ClO_4^- are oxidizing agents. However, their oxidizing powers vary greatly. The last two, permanganate and perchlorate, are the strongest oxidizers. Perchloric and nitric acids are not often used to adjust the redox states of radionuclides; rather, they are widely used to decompose organic matter by oxidation to drive the radionuclides bound in the organics into solution for analysis. The use of perchloric acid, although very efficient, is mostly avoided since it can result in an explosion if the fumes are not flushed properly. Permanganate ions are highly oxidizing and can be used to oxidize Np(V) to Np(VI), for example. Permanganate is so oxidizing that it can result even in the reduction of hydrogen peroxide to oxygen while the permanganate is reduced to manganese dioxide:

$$2MnO_4^- + 3H_2O_2 \rightarrow 2MnO_2 + 3O_2 + 2H_2O + 2OH^-$$

Nitrite NO_2^- is used as $NaNO_2$ to adjust the oxidation state of plutonium to +IV. It is very useful since it can both oxidize trivalent plutonium to the tetravalent state and, at the same time, reduce hexavalent plutonium to the same tetravalent state when Fe^{3+} is present as a catalyst. Since plutonium can exist in several oxidation states at the same time, nitrite is an important agent when only tetravalent plutonium is needed. The reduction of hexavalent plutonium to tetravalent is a slow process since it requires Pu−O bond breakage. The rate of this reaction, however, can be enhanced by addition of Fe^{3+}.

$$Pu^{3+} + NO_2^- + 2H^+ \rightarrow Pu^{4+} + NO + H_2O$$

$$PuO_2^{2+} + NO_2^- + H^+ \rightarrow Pu^{4+} + NO_3^- + OH^-$$

3.8.2 Reductants

Metal ions at their less stable lower oxidation states, such as Ti^{3+}, Cr^{2+}, Fe^{2+}, can be used to reduce radionuclides from their higher oxidation states to lower ones. An example is the reduction of tetravalent plutonium to trivalent with Ti^{3+}.

$$Ti^{3+} + Pu^{4+} \rightarrow Ti^{4+} + Pu^{3+}$$

Ascorbic acid (vitamin C) $C_6H_8O_6$ is used to reduce trivalent iron to its divalent state. Trivalent iron interferes with many radionuclide analyses, as in Am^{3+} determinations, for example.

$$2Fe^{3+} + C_6H_8O_6 \rightarrow 2Fe^{2+} + C_6H_4O_6 + 2H^+$$

Sulfamic acid H_3NSO_3 or ferrous sulfamate $Fe(NH_2SO_3)_2$ is often used to reduce plutonium to Pu^{3+}. It has been used, for example, in the separation of plutonium from uranium in the PUREX process in the reprocessing of nuclear fuel; however, while it has now been replaced by other reductants, it is still used in analytical radiochemical separations.

Sulfurous acid H_2SO_3 and sulfite Na_2SO_3 or $Na_2S_2O_5$ (becoming $NaHSO_3$ when dissolved in water) are reductants used in radiochemical separation processes. As they can rapidly reduce the high valence states of Pu (Pu^{4+}, $PO_2{}^+$, $PuO_2{}^{2+}$) to Pu^{3+} and Np ($NpO_2{}^+$, $NpO_2{}^{2+}$) to Np^{4+}, they are often used for this purpose. For example, in the separation of Pu from large volumes of water, Pu is first co-precipitated with $Fe(OH)_3$. Since the recovery of Pu in the form of $PuO_2{}^+$ and $PuO_2{}^{2+}$ is very low, Pu is reduced with sulfite to Pu^{3+}, for which the recovery is good. In chromatographic separations of Pu, all oxidation states of Pu should be adjusted to Pu(IV) because of its high affinity to chromatographic columns (strongly basic anion exchange resin or TEVA Resin). This is normally carried out in two steps: first, high oxidation states of Pu ($PuO_2{}^{2+}$, $PuO_2{}^+$, Pu^{4+}) are reduced to Pu^{3+} by sulfurous acid (or sulfite in acidic solution), and then Pu^{3+} is oxized to Pu^{4+} by nitrite (or by using $NO_2/NO_2{}^-$ in concentrated nitric acid). In addition, sulfurous acid is also often used as a reductant in the separation of ^{129}I from water, since it can rapidly reduce all high oxidation states of iodine ($IO_4{}^-$, $IO_3{}^-$, IO^-, I_2) to I^-; I^- is then oxidized to I_2 with nitrite for extraction to the organic phase. Since sulfite is a strong but unstable reductant, especially in an acidic condition, the water solution of sulfite is normally freshly prepared before utilization.

Hydroxylamine hydrochloride $NH_2OH{\cdot}HCl$ is an often used reductant to reduce elements to their lower oxidation state. In radiochemical separations of Pu, it is often eluted from an anion exchange column or an extraction chromatographic column (for example TEVA) by reducing Pu(VI) with $NH_2OH{\cdot}HCl$ to Pu^{3+} in dilute hydrochloric acid solution. Hydroxylamine is, however, not a strong reductant: it can, for example, reduce IO_3^- to I_2 but not to the lowest oxidation state of iodine (I^-).

Hydrazine N_2H_4 or a hydrazium salt (for example hydrazine sulfate, $N_2H_4{\cdot}H_2SO_4$) is also a reductant and can be used to reduce Pu to Pu^{3+}. The reducing ability of hydrazine is similar to that of hydroxylamine. The reduction rate of Pu to Pu^{3+} by hydrazine is slower than that using sulfamic acid; however, the stability of hydrazine and hydroxylamine is much better than that of sulfamic acid and sulfite, especially in acidic solution.

4
Separation Methods

Radiochemical separations make use of the traditional methods of separating elements: precipitation, ion exchange, and solvent extraction, of which precipitation has been employed from the very beginning. A new method, accepted into wide use during the last decades of the last century, is extraction chromatography, which combines solvent extraction as the separation method with column chromatography technology used earlier in ion exchange. These four methods are discussed below.

4.1
Precipitation

Precipitation is a traditional method for radiochemical separations. Marie Curie used precipitation methods in the late 1890s to separate the new radioactive elements, radium and polonium, from ore minerals. Figure 4.1 shows Curie's separation scheme for polonium, which is marked with an 'X' since the element was unknown at the time. Although the separation scheme used in the analysis was familiar, a new feature was that the route of the component X could be followed by detecting the radiation it emitted, and conclusions about its chemical properties were drawn on the basis of the elements that it accompanied. This was the very first radiochemical separation procedure.

As new radionuclides – first natural and then artificial ones – became known, separation methods, most based on precipitation, were developed for them. In the context of the nuclear weapons development projects in the 1940s, separation methods based on precipitation were developed for most of the fission products – some still remain in use. However, precipitation has since lost its favored status as the most essential step in radiochemical separation, yielding place to ion exchange and liquid extraction, and later to extraction chromatography.

Chemistry and Analysis of Radionuclides. Jukka Lehto and Xiaolin Hou

ISBN: 978-3-527-32658-7

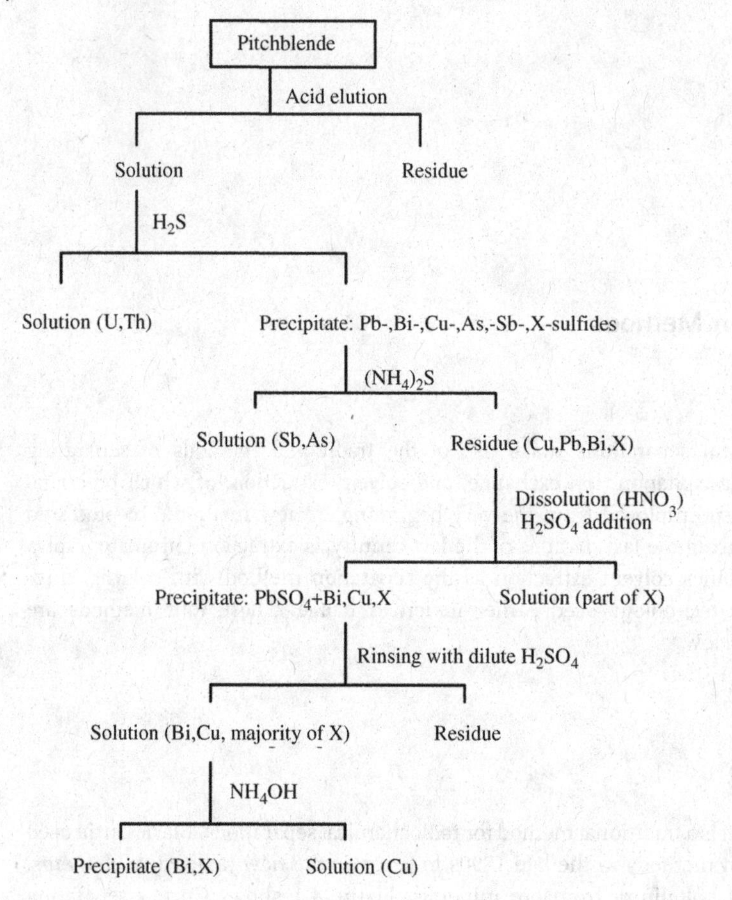

Figure 4.1 Separation scheme used in Marie Curie's separation of polonium (X) from pitchblende.

4.2 Solubility Product

For radionuclides forming sparingly soluble compounds, precipitations are used in radiochemical separation procedures. For the precipitation of a compound to occur, its solubility product (K_s) must be exceeded. The solubility product of a compound A_xM_y is expressed as:

$$K_s = [A]^x \cdot [M]^y \tag{4.1}$$

where [A] and [M] are the concentrations of the anion and cation in the solution and x and y are their charges. The smaller the solubility product, the lower is the solubility of the compound, though direct comparison of solubility products is possible only for compounds whose anion and cation charges are the same.

Because the concentrations of radionuclides in solution are generally very low, their solubility products are seldom exceeded without the addition of a carrier. Even if the solubility product were exceeded, the amount of precipitate would probably be too small to be practically handled; in this case, the addition of a carrier is also necessary. When an isotopic carrier is added to exceed the solubility product or to increase the amount of precipitate, it must be added to the solution in the same chemical form as the target radionuclide. As the chemical form of a radionuclide in the solution is often not known, the chemical form of the nuclide and the carrier must be made the same after the addition of the carrier, for example, by treating the solution with strong acids and with oxidizing or reducing agents.

4.2.1 Coprecipitation

For those radionuclides for which a suitable isotopic carrier is not available, a nonisotopic carrier must be used for the precipitation. In this case, the precipitation is called coprecipitation. Coprecipitation relies on one of the following mechanisms:

The precipitate consists of normal mixed crystals, in which the trace radionuclide is homogeneously distributed in the crystal lattice and replaces the carrier atoms without causing lattice defects; that is, it replaces the carrier atoms in the lattice isomorphically. For this to occur, the carrier and the target radionuclide must be chemically similar. For example, ^{226}Ra can be coprecipitated with sparingly soluble $BaSO_4$, because both Ra and Ba are alkaline earth metals with a charge of $+2$ and their ionic radii are not very different. If the trace component forms a less soluble compound than the carrier does with the oppositely charged precipitating ion (in the above case, sulfate), it will enrich into the precipitated compound. In other words, the ratio of the trace component to the carrier component will be larger in the precipitate than in the solution. Isomorphic replacement does not necessarily mean that the concentration of the trace component is homogeneous in the whole crystal. Precipitation conditions (temperature, rate of precipitation, concentrations, aging) affect the homogeneity. In general, a high precipitation temperature, slow precipitation, and aging promote homogeneity. A strong enrichment of a radionuclide promotes heterogeneity, so that the concentration is greater in the crystal core than on the surface region.

The precipitation processes may also give rise to anomalous mixed crystals, in which the radionuclide may be uniformly distributed in the compound formed with the carrier; however, the replacement is not isomorphic. The charges of the carrier atoms and the radionuclide to be precipitated are not necessarily identical in anomalous mixed crystals. For example, lanthanide fluorides (CeF_3, NdF_3, etc.) coprecipitate from hydrofluoric acid both tri- and tetravalent actinides (e.g., both Pu^{3+} and Pu^{4+}). Although the precipitation is efficient, Pu does not substitute Ce or Nd in the crystals isomorphically, at least not in valence state four.

The third coprecipitation mechanism is adsorption. During precipitation of the carrier compound, the radionuclide adsorbs onto the surface of the precipitating substance, for example, by ion exchange adsorption. Hydroxides of multivalent metals, for example, iron hydroxide $Fe(OH)_3$, or hydrous oxides, for example,

$MnO_2 \cdot xH_2O$, coprecipitate metals effectively, especially metals with a high charge. These hydroxides and oxides have $-OH$ groups on their surfaces, the hydrogen dissociates in neutral and basic solutions, and the generated $-O^-$ groups bind cations. Because this is a surface reaction, the adsorption is greater the larger the surface area of the adsorbing substance. The adsorbing precipitate is exploited for the separation of radionuclides in two ways. Most often, the adsorbing precipitate is formed by adding appropriate chemicals to the solution containing the radionuclides. For example, several transition elements adsorb onto $Fe(OH)_3$ precipitate, which is formed by adding trivalent iron to the acidic solution (pH about 2) and then precipitating $Fe(OH)_3$ by adding NaOH or NH_4OH to the solution. The other, less common approach is to add a preprepared adsorbing precipitate to the solution containing the radionuclide to allow sorption of radionuclides on the precipitate.

4.2.2 Objectives of Precipitation

Precipitation processes can be divided into five categories according to the objective:

- radionuclide-specific precipitations;
- group precipitations for the enrichment of a target nuclide;
- group precipitations for the removal of interfering radionuclides and stable elements.

These three categories are discussed in the next three Sections.

- microcoprecipitations for the preparation of counting sources;
- precipitations for the determination of actinide oxidation states.

These two categories are discussed in Chapters 5 and 15, respectively.

4.2.2.1 Precipitations Specific for the Investigated Radionuclide

There are only a few such precipitations: for example, nickel (^{63}Ni) is precipitated from solution with dimethylglyoxime, which is highly specific for nickel (Figure 4.2).

Sulfate precipitation can also be considered to be rather specific since sulfate only forms insoluble salts with strontium, barium, radium, and lead, and these precipitations are used in ^{210}Pb and ^{226}Ra separations. Since radium concentration is so low, barium is needed as a carrier to enable the precipitate to be formed. Iodide and chloride form insoluble salts, AgI and AgCl, with silver, and these precipitations are utilized in ^{36}Cl and ^{129}I separations. A few chromates (CrO_4^{2-}) are insoluble, such as those of barium, lead, and silver. Chromate precipitation could thus be used in ^{210}Pb separations. Chromate precipitation can also be used for removal of barium and radium as $Ba(Ra)CrO_4$ in ^{90}Sr analysis. ^{140}Ba, a fission product coexisting with ^{90}Sr in fresh nuclear fallout, for example, interferes with the beta measurement of ^{90}Sr. Since strontium does not form an insoluble chromate, it remains in the solution. This is an example of what is called 'scavenging precipitation', in which interfering nuclides are removed from the solution by precipitation while the target nuclide remains in the solution.

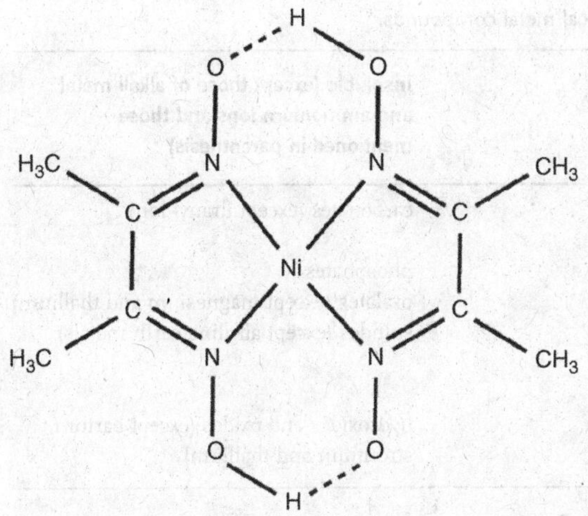

Figure 4.2 Binding of nickel to dimethylglyoxime, which is a highly specific precipitant for nickel.

In the past, when no gamma spectrometers were available for direct measurement of radioactive cesium, several cesium-specific precipitation agents were used for its separations. Today, they are used for the preconcentration of radiocesium from large volumes of natural waters. Especially ammonium phosphomolybdate (AMP) and various transition metal hexacyanoferrates such as potassium cobalt hexacyanoferrate(II) ($K_2CoFe(CN)_6$) have been utilized for this purpose.

4.2.2.2 Group Precipitations for the Preconcentration of the Target Radionuclide

Preconcentration involves separating the target radionuclide from the matrix using materials behaving in the same way in order to achieve a smaller volume. For example, iron hydroxide precipitation is used to concentrate actinides and other hydrolyzing metals present in very large sample volumes – as much as 100–200 L. The obtained precipitate is dissolved in a much smaller amount of acid for further separation and purification.

4.2.2.3 Group Precipitations for the Removal of Interfering Radionuclides and Stable Elements

Table 4.1 presents the solubility of typical metal compounds. These data can be used to design a process to remove interfering radionuclides and stable elements. More detailed information is, however, needed for actual separations.

Most of the group separations used in radionuclide separations are based on hydroxide, carbonate, phosphate, and oxalate precipitations. Precipitation as hydroxide, usually with iron hydroxide, at pH 8–9, leaves all alkali and alkaline earth metals in solution, since the alkali metals, barium, and strontium do not form insoluble hydroxides at all and the other alkaline earth metals do not precipitate as hydroxides at such a low pH value. Moreover, radionuclides in anionic form remain in solution in hydroxide precipitation. Lanthanides, actinides, and other cationic metals precipitate as hydroxides.

Table 4.1 General solubility of typical metal compounds.

Water soluble	**Insoluble (except those of alkali metal and ammonium ions and those mentioned in parenthesis)**
alkali metal and ammonium compounds (a few exceptions)	carbonates (except uranyl ion)
acetates (except silver)	phosphates
nitrates	oxalates (except magnesium and thallium)
halides (except silver, lead, copper, mercury halides and alkaline earth metal and actinide fluorides)	sulfides (except alkaline earth metals)
sulfates (except silver, lead, barium, strontium and calcium)	hydroxides and oxides (except barium, strontium and thallium)

The primary objective in many radiochemical analyses is to remove iron, which interferes with several of the stages in radiochemical separations. Iron is abundant in soil and sediment samples, and also the iron added to the sample in iron hydroxide precipitations must be separated before the following stages. In the separation of actinides, iron can be removed by coprecipitation of the actinides with calcium oxalate. Prior to oxalate precipitation, the iron in the solution is totally oxidized to trivalent Fe^{3+} ion, which does not form an oxalate precipitate. Since the precipitation is carried out at a pH less than 2, the iron does not precipitate as hydroxide either.

Carbonate and phosphate precipitations are the usual choice when alkali and alkaline earth metals are to be separated from each other: alkaline earth metals form sparingly soluble carbonates and phosphates whereas alkali metals do not. If the target nuclide is an alkaline earth metal, carbonate and hydroxide precipitations can follow each other. Hydroxide precipitation is carried out first to remove the metals that form hydroxides, while alkali and alkaline earth metals and anionic radionuclides remain in the solution. Alkali metals and anionic components are then removed by carbonate or phosphate precipitation. Table 4.2. summarizes the use of

Table 4.2 Typical precipitations used to remove interfering radionuclides and stable elements in radiochemical separations.

Precipitation	**Elements removed (remain in solution)**
hydroxide precipitation ($Fe(OH)_3$)	alkali metals alkaline earth metals anionic components
carbonate and phosphate precipitations	alkali metals anionic components
oxalate precipitation	alkali metals iron as Fe(III)

Table 4.3 The major precipitations used for radionuclide separations.

Radionuclide	Precipitation	Purpose of the precipitation
^{14}C	$CaCO_3$ precipitation of $^{14}CO_2$ formed in sample combustion and trapped in NaOH solution	precipitate $^{14}CO_3^{2-}$ as $CaCO_3$
^{36}Cl	AgCl precipitation	^{36}Cl separation/purification
^{41}Ca	$Fe(OH)_3$ coprecipitation	removal of transition metals
	carbonate precipitation	Ca separation from alkali metals
^{55}Fe	$Fe(OH)_3$ precipitations	enrichment, separation of interfering components
^{63}Ni	carbonate or hydroxide coprecipitation	separation of interfering componenents
	dimethylglyoxime precipitation	selective separation of ^{63}Ni
^{90}Sr	nitrate precipitation in 70% HNO_3	separation of strontium from calcium
	$Ba(Ra)CrO_4$ precipitation	separation of strontium from barium and radium
	carbonate precipitations	enrichment, separation of interfering components
	$Y_2(C_2O_4)_3$ precipitation	separation of ^{90}Y from ^{90}Sr
	$SO_4{}^{2-}$ precipitation	separation of strontium
^{99}Tc	$Fe(OH)_2$ coprecipitations	preconcentration of Tc from natural waters
^{129}I	AgI precipitation	^{129}I separation/purification, source preparation for AMS
^{210}Pb	$PbSO_4$ and PbS precipitations	^{210}Pb separation/purification
^{226}Ra	$BaSO_4$ coprecipitation	separation of ^{226}Ra
Th, U, Np, Pu, Am, Cm	$Fe(OH)_3$ coprecipitation	enrichment, separation of interfering components
	oxalate coprecipitation	enrichment, separation of interfering components
	CeF_3 or NdF_3 coprecipitation	source preparations for activity measurement, determination of oxidation states

typical precipitations to remove interfering radionuclides and elements, while Table 4.3. lists the most relevant precipitations in all radiochemical separations.

The most important precipitations used in radiochemistry are summarized below:

- carbonate and phosphate precipitations used for preconcentration of the target radionuclide and for the removal of alkali metals and anionic radionuclides from samples; these remain in solution;
- iron hydroxide precipitations used for preconcentration of the target radionuclide and for the removal of alkali metals, alkaline earth metals, and anionic radionuclides from samples; these remain in solution;

- oxalate precipitations used for concentrating the target radionuclide and for the removal of alkali metals, alkaline earth metals, iron, and anionic radionuclides from samples; these remain in solution;
- lanthanide fluorides used for the precipitation of tri- and tetravalent actinides;
- a specific dimethylglyoxime precipitation used to separate nickel;
- sulfate precipitation used for radium and lead separations;
- silver halide precipitations used in chloride and iodide separations;
- ammonium phosphomolybdate and transition metal hexacyanoferrates used to pre-concentrate cesium from natural waters.

4.3 Ion Exchange

4.3.1 Ion Exchange Resins

Almost all the ion exchangers employed in radiochemical analyses are common organic cation and anion exchange resins. Resins are prepared from polystyrene cross-linked with divinylbenzene (DVB) (Figure 4.3). A functional group is attached to the benzene rings of the porous polymer: either the sulfonic acid group $-SO_3H$ is added so that the resin works as a cation exchanger upon dissociation of the hydrogen ion of the sulfonic acid group, or the quaternary ammoniun ion $-N(CH_3)_3OH$ is added, which acts as an anion exchanger upon dissociation of the hydroxide group. Both are strong ion exchangers, that is, they dissociate easily and thus function as ion exchangers over the whole pH range and in strong acids.

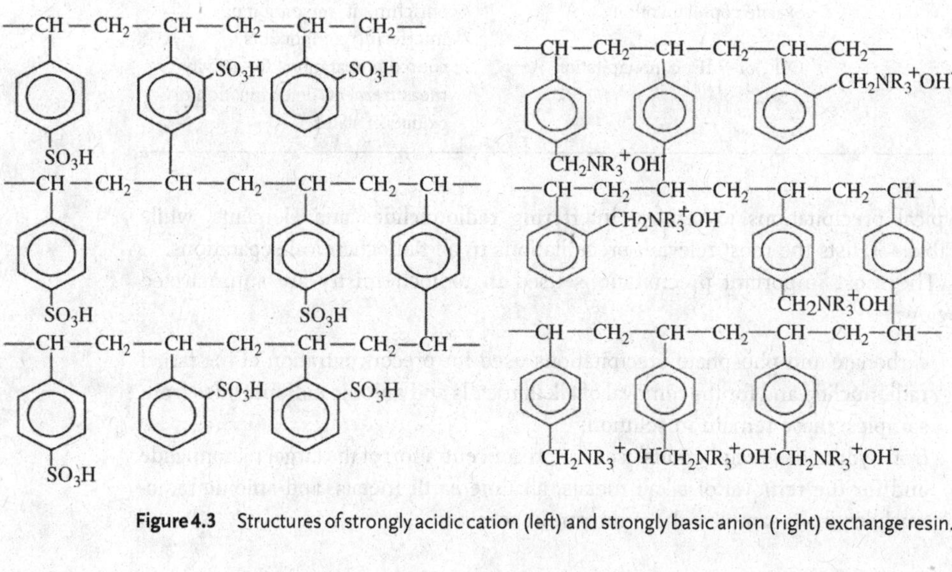

Figure 4.3 Structures of strongly acidic cation (left) and strongly basic anion (right) exchange resin.

Table 4.4 Common ion exchange resins.

Resin type	Matrix	Functional group
Strongly acidic cation resin	DVB-PS	Sulfonate $-SO_3^-$
Weakly acidic cation resin	PMMA	Carboxylate $-CO_2^-$
Strongly basic anion resin	DVB-PS	Quaternary ammoniun ion $-N(CH_3)_3^+$
Weakly basic anion resin	DVB-PS	Tertiary amino $-N(CH_3)_2$
Chelating resin (an example)	DVB-PS	Aminophosphonate $-CH_2-NH-PO_2^{2-}$

PMMA = polymethylmethacrylate.
DVB-PS = divinylbenzene-crosslinked polystyrene.

While ion exchangers are available in powdered and granular form, granules are usually used in radiochemical separations for analytical purposes in order to allow packing in columns. Typical grain sizes are 0.30–0.85 mm (20–50 mesh), 0.15–0.30 mm (50–100 mesh), and 0.07–0.15 mm (100–200 mesh). The smaller the grain size, the more favorable is the ion exchange kinetics. Another factor affecting the kinetics is the degree of cross-linking in the resin, which is a reflection of the percentage of divinylbenzene in the resin reaction mixture. The degree of cross-linking typically varies between 2% and 20%. Resins with a low degree of cross-linking offer faster ion exchange reactions but are less durable than highly cross-linked resins: low cross-linking offers more space for ions to diffuse within the bead; however, since the polystyrene chains are less attached to each other they are more vulnerable to framework breaking. Ion exchange resins also contain functional groups other than sulfonic acid or quaternary ammonium. If the functional group is a carboxylic group, the resin is a weakly acidic cation exchanger, which only functions as a cation exchanger in neutral and basic solutions. If it is a tertiary amino group, the resin is a weak anion exchanger and only functions as an anion exchanger in neutral and acidic solutions. There are also ion exchange resins in which the functional group is a chelating group, such as amino phosphonate. Types of organic ion exchange resins are listed in Table 4.4. In addition to organic resins, some highly radionuclide-selective inorganic exchangers, such as titanates and hexacyanoferrates, have been developed for the treatment of nuclear waste; however, they are not typically used in radioanalytical separations, except in preconcentration of radionuclides from large water volumes.

4.3.2 Distribution Coefficient and Selectivity

A typical way to present the selectivity of an ion exchanger is to calculate the distribution coefficient. The distribution coefficient is the concentration of an ion in the exchanger ($[M]_s$) divided by its concentration in solution $[M]_l$ at the equilibrium state.

$$D(\text{or } K_D \text{ or } K_d) = [M]_s/[M]_l(\text{mL g}^{-1} \text{ or L kg}^{-1}) \quad (4.2)$$

D is a coefficient – not a constant – and its value depends on several factors. First, it is dependent on the concentration of the ion to be separated. This does not, however, apply to trace concentrations in practice. Radionuclides are almost always present in trace concentration, and their distribution coefficients are constant with respect to their concentrations. This simplifies the use of the distribution coefficient as a separation parameter or as a measure of the selectivity of the ion exchanger for a certain ion. Second, *D* is dependent on the concentration of other ions with the same charge present in the solution – the presence of other competing ions decreases the binding of the target ion to the exchanger. A third important factor affecting ion exchange is the H^+/OH^- ratio, that is, the pH. As the pH decreases, that is, as the concentration of H^+ ions increases, the sorption of cations on a cation exchanger decreases because of hydrogen ions competing with metal ions on the ion exchange site. The reverse is true for anions and an anion exchanger: increasing pH, that is, OH^- concentration, decreases the sorption of anions on anion exchangers. For strong cation and anion exchangers the effect of H^+ and OH^- is, however, minor.

4.3.3 Cation Exchange or Anion Exchange?

Cation exchangers retain metal cations more effectively the higher their charge and the smaller their size because the Coulombic interaction with the functional groups of the exchanger is then increased. Cation exchangers generally allow the separation of metals with valence states +1, +2, and +3. Separations are based on pH adjustment, since the higher the charge on the cation, the lower is the pH at which it is retained on the exchanger (Figure 4.4). The separation of cations is not very effective, and this means that cation exchangers cannot be effectively used for the

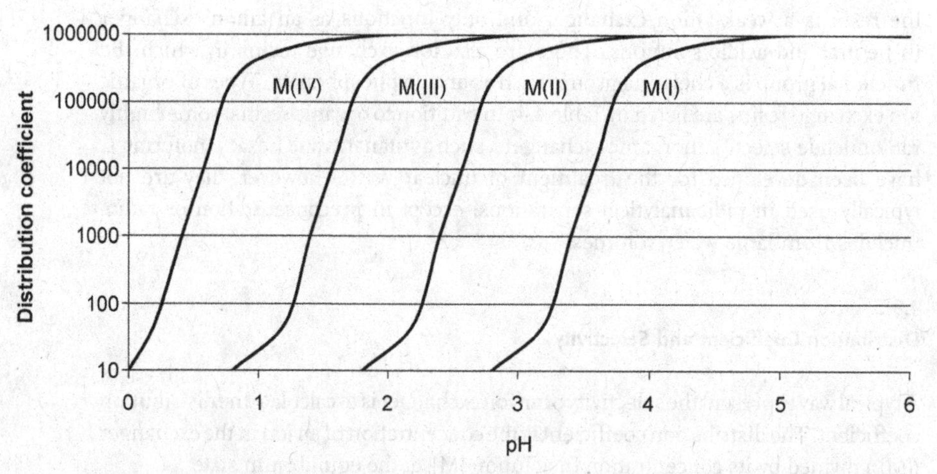

Figure 4.4 Distribution coefficients of metals in valence states +1, +2, +3, and +4 as a function of pH for a cation exchanger. The values of the distribution coefficients are suggested only and do not depict a real case.

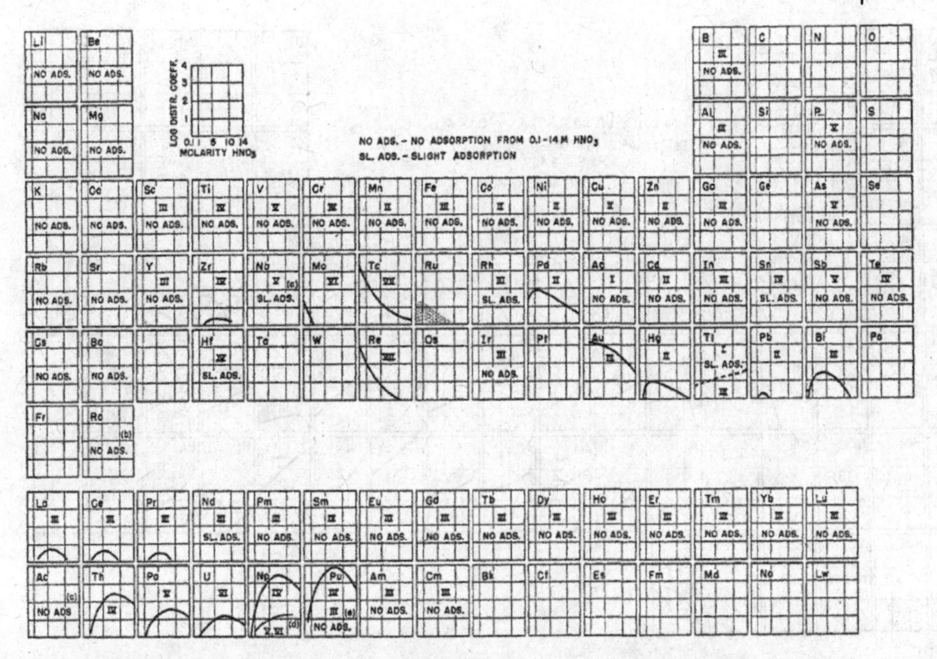

Figure 4.5 Distribution coefficients of the elements on strong anion exchange resins in HNO_3 (Faris, J.P. and Buchanan, R.F. (1964) Anion exchange characteristics of the elements in nitric acid medium. *Anal. Chem.*, **36**, 1158).

separation of metals, for example alkali metals, of the same valence state. Cation exchangers are thus not widely used in radiochemical separations.

Anion exchangers are much more commonly used than cation exchangers in radiochemical analyses. Their widest use is in the separation of actinides. They form anionic complexes in strong acids, and these complexes are retained on the anion exchanger according to their charge; that is, the higher the charge the more effectively they are bound. The most widely used acids are HCl and HNO_3, and the complexes are chloride and nitrate complexes. Figures 4.5 and 4.6 show the distribution coefficients for the elements on an anion exchanger as a function of the concentrations of HNO_3 and HCl: the x-axis shows the concentration of the acid between 0 and 14 M and the y-axis the distribution coefficient as a logarithm between 0 and 1 000 000.

4.3.4 Ion Exchange Chromatography

The normal way in which an ion exchanger is used in the separation of radionuclides is in ion exchange chromatography. A granular ion exchanger is packed into a glass – or more often a plastic – column, and is pretreated with a suitable solution. Anion exchangers are pretreated with acid at the same concentration as the sample solution. The sample solution is poured into the column in a small volume, and the column is

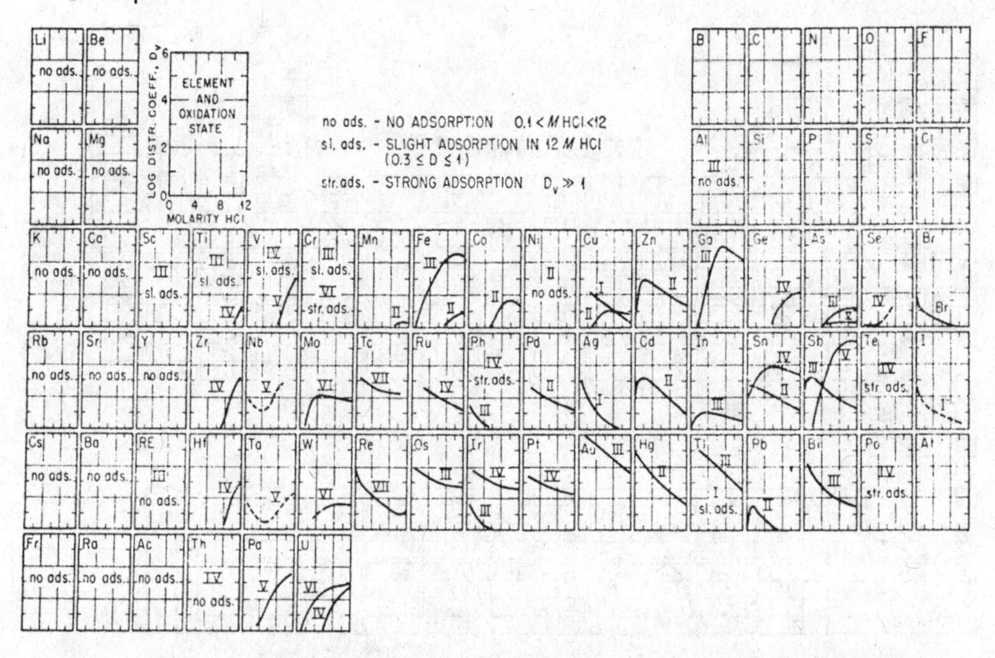

Figure 4.6 Distribution coefficients of the elements on strong anion exchange resins in HCl (Kraus, K.A. and Nelson, F. (1955) Proc. 1st Int. Conf. Peaceful Uses Atomic Energy, 7, 118).

rinsed with the same acid so that nonsorbing radionuclides and elements are flushed from the column. Any radionuclides that have sorbed to the column are then eluted out – usually with dilute acids or with complexing agents. The concentration (or composition) of the eluent can be adjusted during the elution step so that radionuclides sorbed with different strengths are separated from one another. The advantage of ion exchange chromatography over precipitation is that no carrier is required and separation performance is typically better than in precipitation.

4.3.5
Ion Exchange in Actinide Separations

A most important field of ion exchange applications in radiochemistry is actinide separations. In these separations the following features are most relevant:

- Uranium is retained efficiently in anion exchange resin in 9 M HCl.
- Thorium is retained in anion exchange resin in 8 M HNO_3 – in HCl medium no thorium uptake takes place.
- Trivalent actinides are usually not taken up by anion exchange resins.
- The separation of plutonium from other actinides is accomplished by the adjustment of its oxidation state.
- The separation of americium from lanthanides takes advantage of their different complexation behavior with thiocyanate.

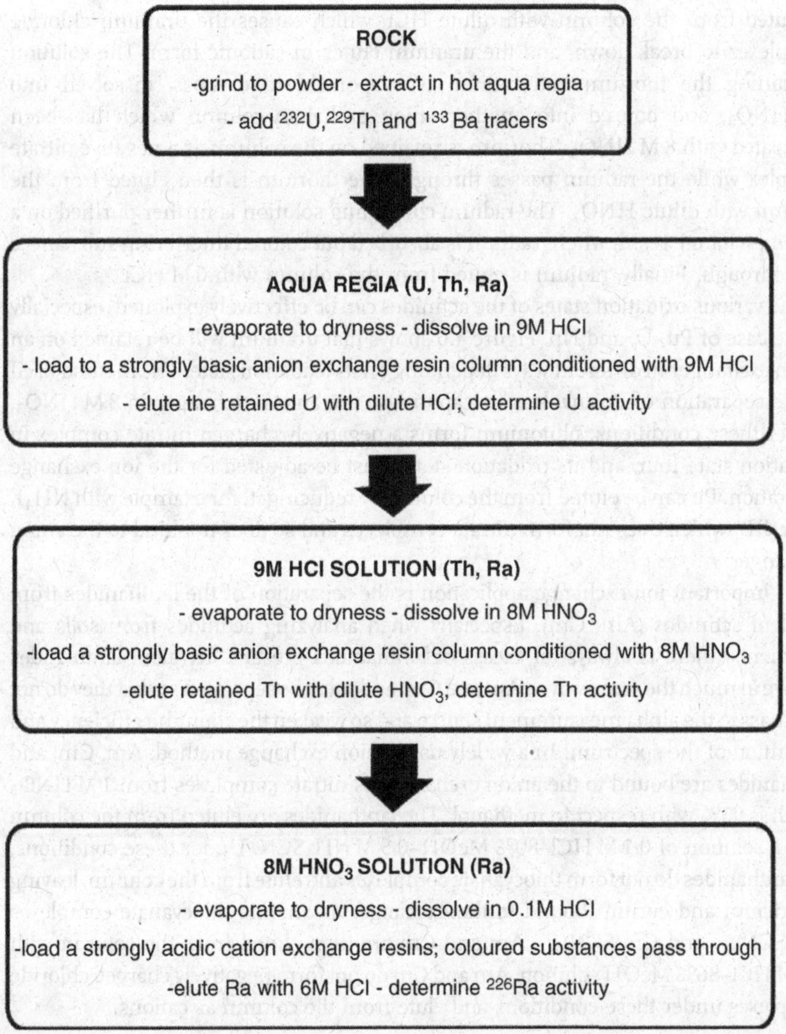

Figure 4.7 Separation of uranium, thorium, and radium by ion exchange chromatography from the acid used in extracting rock samples (Juntunen, P., Ruutu A., and Suksi, J. (2001) Determination of ^{226}Ra from Rock Samples using LSC, Proceedings of the International Conference on Advances in Liquid Scintillation Spectrometry, Karlsruhe, Germany, May 7–11, 2001, pp. 299–302).

An example of actinide separation by ion exchange is shown in Figure 4.7, in which uranium, thorium, and radium are separated from each other in the acid that was used in leaching rock samples. The ion exchange column is pretreated with 9 M HCl and the sample solution is added to the column in 9 M HCl. The uranium is retained on the exchanger as a negative uranium chloride complex, while thorium and radium, which do not form chloride complexes, pass through the column. Uranium is eluted from the column with dilute HCl, which causes the uranium chloride complexes to break down, and the uranium elutes in cationic form. The solution containing the thorium and radium is evaporated to dryness, dissolved into 8 M HNO_3, and poured into another anion exchange column which has been pretreated with 8 M HNO_3. Thorium is retained on the column as a negative nitrate complex while the radium passes through. The thorium is then eluted from the column with dilute HNO_3. The radium-containing solution is further purified on a cation exchange resin, where radium is absorbed but colored interfering substances pass through. Finally, radium is eluted from the column with 6 M HCl.

The various oxidation states of the actinides can be effectively exploited, especially in the case of Pu, U, and Np. Figure 4.6 shows that uranium will be retained on an anion exchanger in different ways depending on its oxidation state. The method used in the separation of Pu involves its capture on the anion exchanger in 8 M HNO_3. Under these conditions, plutonium forms a negatively charged nitrate complex in oxidation state four, and its oxidation state must be adjusted for the ion exchange separation. Pu can be eluted from the column by reducing it, for example with NH_4I, to Pu(III), which does not form nitrate complexes and so does not bind to the anion exchanger.

An important ion exchange application is the separation of the lanthanides from trivalent actinides (Am, Cm). Especially when analyzing actinides from soils and sediments, there is always an excess of lanthanides present. Trivalent lanthanides behave in much the same way as Am and Cm, and must be separated so that they do not add mass to the alpha measurement source and so weaken the counting efficiency and resolution of the spectrum. In a widely used anion exchange method, Am, Cm, and lanthanides are bound to the anion exchanger as nitrate complexes from 1 M HNO_3 which is 93% with respect to methanol. The lanthanides are eluted from the column with a solution of 0.1 M HCl–80% MeOH–0.5 M NH_4SCN. Under these conditions, the lanthanides do not form thiocyanate complexes and elute from the column, leaving americium and curium bound to the exchanger as anionic thiocyanate complexes $Am(SCN)_4^-$ and $Cm(SCN)_4^-$. Am and Cm are then eluted from the column with 1.5 M HCl–86% MeOH solution. Am and Cm do not form negatively charged chloride complexes under these conditions and elute from the column as cations.

4.4 Solvent Extraction

Solvent extraction involves the transfer of the investigated component, usually a metal, from one to the other of two immiscible phases. In the analysis of radionuclides, typically the metal is in an aqueous phase and is extracted into an organic phase while the interfering component remains in the aqueous phase. Not all interfering components are removed through solvent extraction; rather, solvent extraction is one step in a chain of several analytical steps. After the extraction, the radionuclide is back-extracted to the aqueous phase and the analysis continues.

4.4.1 Extractable Complexes

A metal ion in the aqueous phase cannot be extracted to the organic phase as such because its electroneutrality must be preserved in both phases. Further, the metal ions form aqua complexes $(M(H_2O)_x)^{y+}$ in the aqueous phases – these are hydrophilic and thus not soluble in organic solvents. Before it can be extracted from the aqueous phase to the organic solution, the metal ion must form a neutral molecule that is soluble in the organic phase. Such neutral molecules are:

- Simple molecules and compounds such as I_2 or RuO_4.
- Simple coordination complexes with anionic unidentate ligands (halide ions, CN^-, SCN^-, NO_3^-). For example, in hydrochloric acid, trivalent metal M^{3+} forms a coordination complex MCl_3. Conditions, in this case the acid concentration, must be carefully chosen, so that cationic or anionic coordination compounds such as MCl_2^+ or MCl_4^- are not essentially formed.
- Simple coordination complexes with neutral unidentate ligands. Typical ligands are R_3N, R_3P, and R_3S, in which R is an organic group. These ligands have a free electron pair on nitrogen, phosphorus, or sulfur, which forms a coordination bond with the metal. Because the ligand is neutral, before the complex can transfer to the organic phase anions must be added to neutralize the charge. An ion associate is formed, for example $(R_3N)M^{3+}(Cl_3)^{3-}$, in which trivalent metal M^{3+} forms a complex with R_3N, and three chloride ions balance the charge.
- Chelates – ring-structured complexes – which form with multidentate ligands. Chelates, pentagonal and hexagonal chelates in particular, are considerably stronger than simple complexes. Figure 4.8 shows the bidentate ligand 2,2′, bipyridine, which forms a five-ringed chelate with metal. Because the arms of bipyridine do not carry a charge, a neutral ion associate must be formed with the

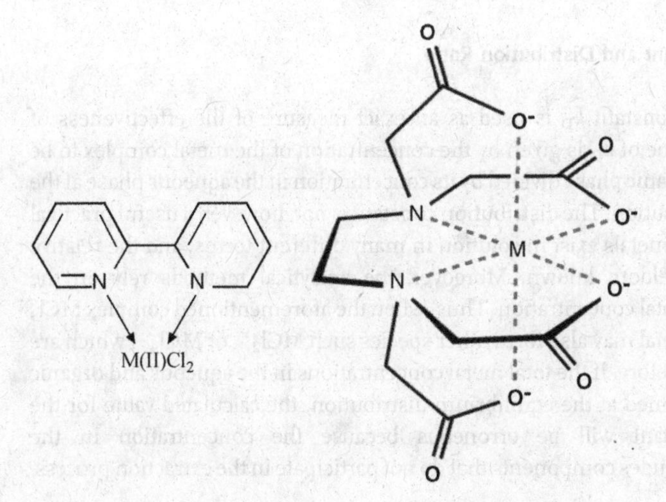

Figure 4.8 Left: 2,2′-Bipyridine chelate. Right: EDTA chelate.

anions in order for the chelate to transfer to the organic phase. Besides having free electron pairs, and through them atoms forming coordination bonds (S, N, P, O), the arms may also have weakly acidic groups which, upon dissociation, form anionic groups that bind the metal cations with the ionic bonds. An example is EDTA (Figure 4.8), which includes both coordinating nitrogen atoms and anionic carboxyl groups.

The binding of metals with both neutral and dissociated acid ligands is pH dependent. Acid ligands, such as the carboxyl group, are weak acids which dissociate in weakly acidic or neutral solutions to a degree depending on their acid strength. Thus, weak acid ligands are not able to bind metals in acidic solutions, where their acid groups are protonated. The same applies to neutral ligands; for example, in weakly acidic and neutral solutions, nitrogen atoms have a free electron pair with which they can form a coordination bond with a metal. In an acidic solution, however, nitrogen atoms protonate forming $-NH^+$ ions which, as cations themselves, are unable to bind metal cations. The binding of metals to ligands is dependent not only on pH but also on the nature of the metal. Metal ions with a higher positive charge and smaller size sorb more strongly because of the larger Coulombic interaction. Steric factors are also important: some cations fit into the ligand space more readily than others. Neutral ligands (O, S, P, N) only bind transition elements, which have partially filled d or f orbitals, allowing the formation of a coordination bond with the free electron pair of a ligand. Alkali and alkaline earth metals are typically not susceptible to solvent extraction because they do not form complexes; they do not have unfilled outer shells that allow the formation of a coordination bond. They are also large in size and have a small charge. Crown ethers are exceptional in that they can be used to extract strontium. A crown ether widely used for strontium separation is 18-crown-6, which is discussed in more detail in the next section on extraction chromatography.

4.4.2 Distribution Constant and Distribution Ratio

The distribution constant k_D is used as an exact measure of the effectiveness of extraction. The value of k_D is given by the concentration of the metal complex to be extracted in the organic phase divided by its concentration in the aqueous phase at the equilibrium distribution. The distribution constant is not, however, a useful practical measure, because metals exist in solution in many different forms, and the relative proportions are seldom known. Moreover, the analytical methods rely on the determination of total concentration. Thus, when the aforementioned complex MCl_3 is extracted, the metal may also form other species such MCl_2^+ or MCl_4^-, which are not extracted. Therefore, if the total metal concentrations in the aqueous and organic phases are determined at the equilibrium distribution, the calculated value for the distribution constant will be erroneous because the concentration in the aqueous phase includes components that do not participate in the extraction process.

In place of the distribution constant, wide use is made of the distribution ratio *d*, which measures the total concentration of the metal in the organic phase divided by its total concentration in the water phase. Yet another measure is the percentage distribution or extraction percentage. Neither measure is a constant like k_D. The distribution of metals between the two phases in solvent extraction is not dependent on the original concentration of the metal, which means that the method is suitable for the separation of both trace and macro amounts. The extraction efficiency is improved when the organic solution is divided into several small portions, the extraction is carried out several times and finally the organic fractions are combined.

4.4.3 Examples of the Use of Solvent Extraction in Radiochemical Separations

A primary purpose of most separations of radionuclides by solvent extraction is to remove alkali and alkaline earth metals as well as interfering anions which are usually not extractable. Metals can also be separated from one another through an appropriate choice of a specific complexing extraction reagent and adjusting the pH to a value such that the complexing agent binds as much as possible the target metal only.

Solvent extraction methods have been used, especially in actinide separations. These separations make use of the ability of certain extraction agents to extract actinides in various oxidation states and of adjusting their oxidations states, especially those of plutonium. A widely used extraction agent in actinide separations is bis(2-ethylhexyl) phosphoric acid (HDEHP), which efficiently extracts trivalent actinides and lanthanides, as well as Fe(III) and Y(III). Tenoyltrifluoroacetone (TTFA) makes use of pH adjustment in the separation of actinides from each other: at pH 1 it only extracts tetravalent actinides, while at pH 4–5 it also extracts tri- and hexavalent actinides. An important extraction agent in radionuclide separations is tri-*n*-butyl-phosphate (TBP), which does not extract trivalent or pentavalent actinides, but only tetra- and hexavalent. It is used not only in analytical actinide separations but also in industrial scale separations in the reprocessing of spent nuclear fuel. In the PUREX process, the fuel is dissolved in nitric acid, and uranium and plutonium are extracted as nitrate complexes with TBP, while fission products and other actinides remain in the aqueous phase. TBP is also used in the analytical separation of ^{99}Tc from nuclear waste solutions: most of the interfering substances are removed first by scavenging with $Fe(OH)_3$, and ^{110m}Ag is precipitated as AgCl; thereafter, technetium, which forms the pertechnetate ion TcO_4^- in solution, is extracted into TBP as the neutral dimeric Tc_2O_7 molecule in 3 M H_2SO_4.

One more example of solvent extraction is the separation of ^{55}Fe from nuclear waste solution and fallout samples. The iron is first concentrated through precipitation as $Fe(OH)_3$. The precipitate is dissolved in 8 M HCl, where the Fe(III) ion forms an $FeCl_3$ complex. This is then extracted fairly specifically into di-isopropyl ether, being followed only by ^{125}Sb. Iron is back-extracted into the aqueous phase, and ^{125}Sb is removed with a cation exchanger on which the Fe^{3+} ion is retained while SbO_3^- passes through the column.

4.5 Extraction Chromatography

4.5.1 Principles of Extraction Chromatography

Extraction chromatography or solid-phase extraction was developed for radiochemical separations only a few decades ago and is now widely applied. The underlying separation process is solvent extraction, which is carried out in a chromatographic column. The reagents for solvent extraction, functioning as the stationary phase, are impregnated into a porous inert support, either silica gel or an organic polymer. The space between the beads enables passage of the mobile phase, normally nitric or hydrochloric acid, which contains the radionuclides to be separated. In extraction chromatography, as in ion exchange chromatography, a small volume of the sample solution is poured into the column. Those nuclides that do not transfer to the stationary phase are flushed out of the column. The elements that were retained in the stationary phase are then eluted by adjustment of the composition of the eluent: change in the acid concentration, addition of a complexing agent or adjustment of the oxidation state within the resin. Extraction chromatography resins are particularly useful in the separation of actinides and lanthanides. In addition to these, there are commercially available resins designed for specific radionuclides: Ni Resin for ^{63}Ni, Sr Resin for ^{90}Sr, and Pb Resin for ^{210}Pb.

4.5.2 Extraction Chromatography Resins

Table 4.5 lists some of the extraction chromatography resins and resins designed for specific radionuclides that are available from Eichrom Technologies and Triskem International.

Not all resins listed in the table are extraction chromatography resins. The separation mechanism in Ni Resin is not solvent extraction occurring in the resin pores: nickel forms a very sparingly soluble solid compound with the dimethylglyoxime (see Figure 4.2) solution that is present in the resin pores. Before precipitation occurs, a nickel carrier must be added to the sample solution. Diphonix Resin is an organic ion exchanger which contains two functional groups: diphosphonate and sulfonic acid groups. MnO_2 Resin is an inorganic ion exchanger, the uptake of metals by which takes place by adsorption on its surface hydroxyl sites. Separation by TEVA Resin is not entirely an extraction process. The functionality of this resin is the same as that in strongly basic anion exchange resins: a quaternary amine. In anion exchange resins the amine is attached to the polymer framework by a covalent bond, while in the TEVA Resin it is in free molecular form and functions as a liquid exchanger.

Table 4.5 Extraction chromatography resins and resins designed for the separation of specific radionuclides supplied by Eichrom Technologies and Triskem International.

Resin	Use	Extraction Reagent	Separation Process
Nickel Resin	Ni	dimethylglyoxime (DMG)	precipitation
Pb Resin	Pb	crown ether (18-crown-6)	extraction
Sr Resin	Sr, Pb	crown ether (18-crown-6)	extraction
MnO_2 Resin	Ra	MnO_2	ion exchange
Diphonix® Resin	actinides and transition metals	diphosphonic acid and sulfonic acid	ion exchange
Ln Resin	lanthanides, Ra-228	di(2-ethylhexyl) orthophosphoric acid (HDEHP).	extraction
Actinide Resin	group actinide separations/gross alpha measurements	DIPEX	extraction
DGA Resin	actinides, lanthanides, Y, Ra	*N,N,N′,N′*-tetra-*n*-octyldiglycolamide	extraction
TEVA® Resin	Tc, Th, Np, Pu, Am, lanthanides	aliphatic quaternary amine	extraction/ion exchange
TRU Resin	Fe, Th, Pa, U, Np, Pu, Am, Cm	octylphenyl-*N,N*-diisobutyl carbamoylphosphine oxide (CMPO)	extraction
UTEVA® Resin	Th, U, Np, Pu	diamyl amylphosphonate (DAAP)	extraction

4.5.3
Pb and Sr Resins

As can be seen from Table 4.5, Pb Resin and Sr Resin are very similar: both contain crown ether (18-crown-6) as the extraction reagent (Figure 4.9). 18-Crown-6 ether is a heterocyclic compound with 12 carbon atoms and with oxygen atoms between the carbon atoms. The shape, reminiscent of a crown, gives the compound its name. The free electron pairs of the oxygen atoms at the inner rim of the crown form a strong negative field which attracts metal cations. The size of the cavity determines the

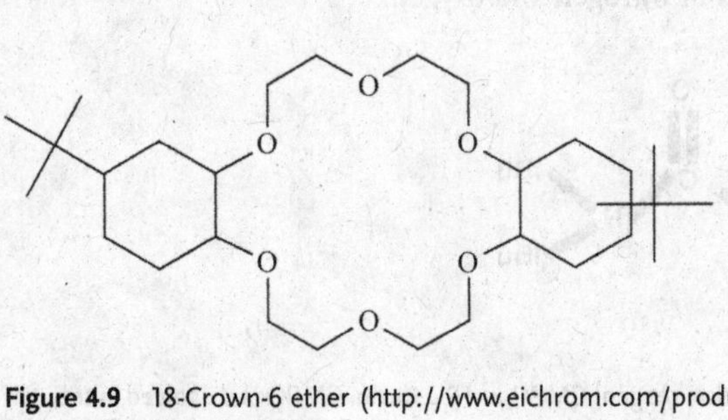

Figure 4.9 18-Crown-6 ether (http://www.eichrom.com/products/info/sr_resin.cfm) within the Eichrom/Triskem Sr Resin.

selectivity: the closer the inner space diameter of the crown to the diameter of the analyte metal ion, the greater is the selectivity. Because the crown ether has no exchangeable cations, the positively charged cation ion carries sufficient anions with it into the complex to preserve electroneutrality. The difference between Sr Resin and Pb Resin is that the latter has a lower concentration of crown ether and the solvent is a longer chain alcohol, isodecanol, instead of 1-octanol in Sr Resin. The purpose of these differences is to facilitate the elution of lead from the column. Selectivity of Sr Resin to various metals is described in Figures 7.9 and 7.10.

4.5.4
Use of Extraction Chromatography in Actinide Separations

The main application field of extraction chromatography is in actinide separations. For this purpose, the three last products in Table 4.5 are mainly used: TEVA, TRU, and UTEVA. They are mainly utilized in nitric acid solutions, where their special characteristics are:

- TEVA Resin binds only tetravalent actinides Th^{4+}, U^{4+}, Np^{4+}, Pu^{4+}.
- TRU Resin binds both tri and tetravalent actinides Th^{4+}, U^{4+}, Np^{4+}, Pu^{4+}, Pu^{3+}, Am^{3+}, and hexavalent uranium $UO_2{}^{2+}$.
- UTEVA Resin binds tetravalent actinides Th^{4+}, U^{4+}, Np^{4+}, Pu^{4+}, and hexavalent uranium $UO_2{}^{2+}$.

These essential differences make it possible to separate most actinides from each other. Americium can be effectively separated from other actinides by using TEVA and UTEVA Resins in which americium is not retained. The oxidation state of plutonium can be adjusted to +III, and it can be removed along with americium from other actinides. Plutonium and americium are, in turn, separated from each other by adjusting the oxidation state of plutonium back to +IV.

Separations are usually carried out in nitric and hydrochloric acid solutions, in which the elements to be separated form nitrate and chloride complexes. In extraction chromatography, as in normal solvent extraction, a neutral hydrophobic complex should form which can move to the organic stationary phase. In the organic phase, it forms coordination bonds with the extraction reagent; in TRU Resin, for example, in which the extraction reagent is CMPO (Figure 4.10), a chelate forms through the free electron pairs of nitrogen and oxygen.

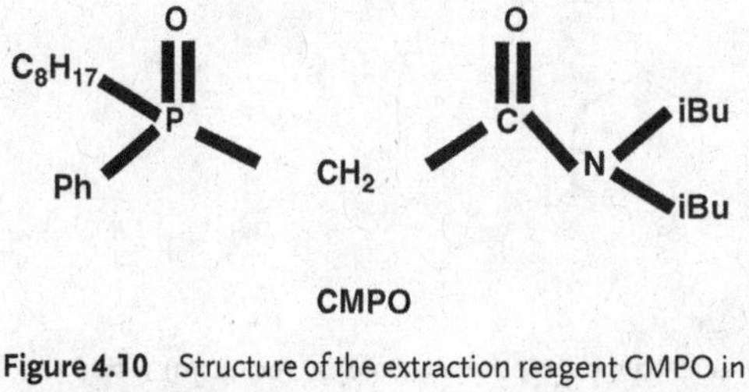

O
C_8H_{17}
P
Ph
CH_2
O
C
N
iBu
iBu

CMPO

Figure 4.10 Structure of the extraction reagent CMPO in TRU Resin. CMPO is dissolved in tributyl phosphate (TBP).

In nitric acid, the actinides form neutral nitrate complexes as follows:

$$Am^{3+} + 3NO_3^- \leftrightarrow Am(NO_3)_3 \tag{4.3}$$

$$Pu^{4+} + 4NO_3^- \leftrightarrow Pu(NO_3)_4 \tag{4.4}$$

$$UO_2^{2+} + 2NO_3^- \leftrightarrow UO_2(NO_3)_2 \tag{4.5}$$

The complexes transfer to the organic stationary phase, which, in the case of TRU, is CMPO/TBP. The formation of nitrate complexes is dependent on the acid concentration: when the concentration of the acid (nitrate) increases, the percentage of actinide forming a complex with the nitrate increases. This is reflected as better retention on the resin. Figure 4.11 shows the k' values of TRU Resin for several metals as a function of the nitric acid concentration. The k' value is the number of column volumes used to elute the target element from the column measured when the element reaches maximum concentration in the eluate. The more selective

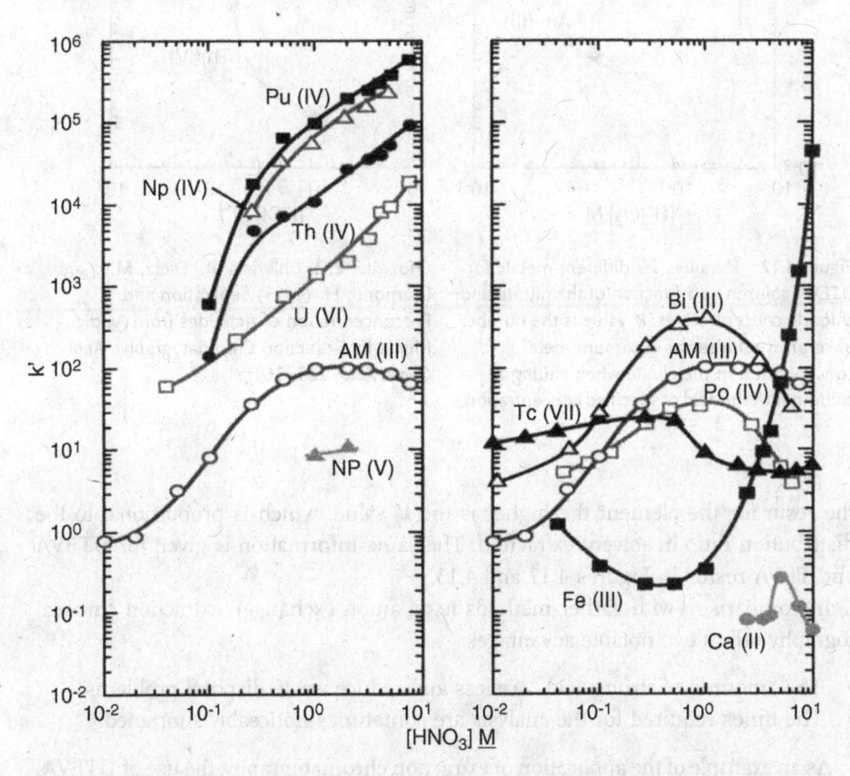

Figure 4.11 k' values for different metals for TRU column as a function of the nitrate concentration. k' value is the number of column volumes at maximum metal concentration in the eluate when eluting the column with the acid of specified concentration. (Horwitz, E.P., Chiarizia, R., Dietz, M.L., and Diamond, H. (1993) Separation and preconcentration of actinides from acidic media by extraction chromatography. *Anal. Chim. Acta*, **281**, 361).

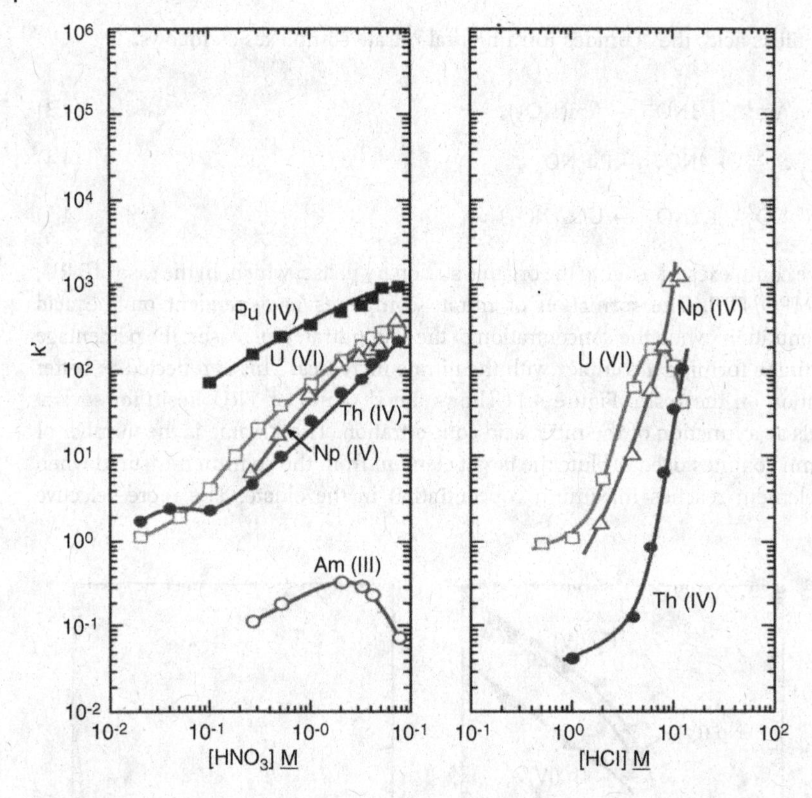

Figure 4.12 k' values for different metals for UTEVA column as a function of the nitrate and chloride concentrations. k' value is the number of column volumes at maximum metal concentration in the eluate when eluting the column with the acid of specified concentration. (Horwitz, E.P., Chiarizia, R., Dietz, M.L., and Diamond, H. (1993) Separation and Preconcentration of Actinides from Acidic Media by Extraction Chromatography. *Anal. Chim. Acta*, **281**, 361).

the resin for the element the higher is the k' value, which is proportional to the distribution ratio in solvent extraction. The same information is given for UTEVA and TEVA resins in Figures 4.12 and 4.13.

In comparison with earlier methods (e.g., anion exchange), extraction chromatography offers two notable advantages:

- The amounts of strong acids are less and reduce waste disposal problems.
- The times required for the analysis are sometimes noticeably shortened.

As an example of the application of extraction chromatography, the use of UTEVA, TEVA, and TRU Resins in the determination of Pu and Am in sediment (Figure 4.14) may be considered. The sediment is first leached in strong nitric acid and hydrochloric acids. After oxalate precipitation to remove iron, the sample is dissolved in 3 M HNO_3, plutonium is reduced with $Fe(NH_2SO_3)_2$ and ascorbic acid to Pu^{3+}, and

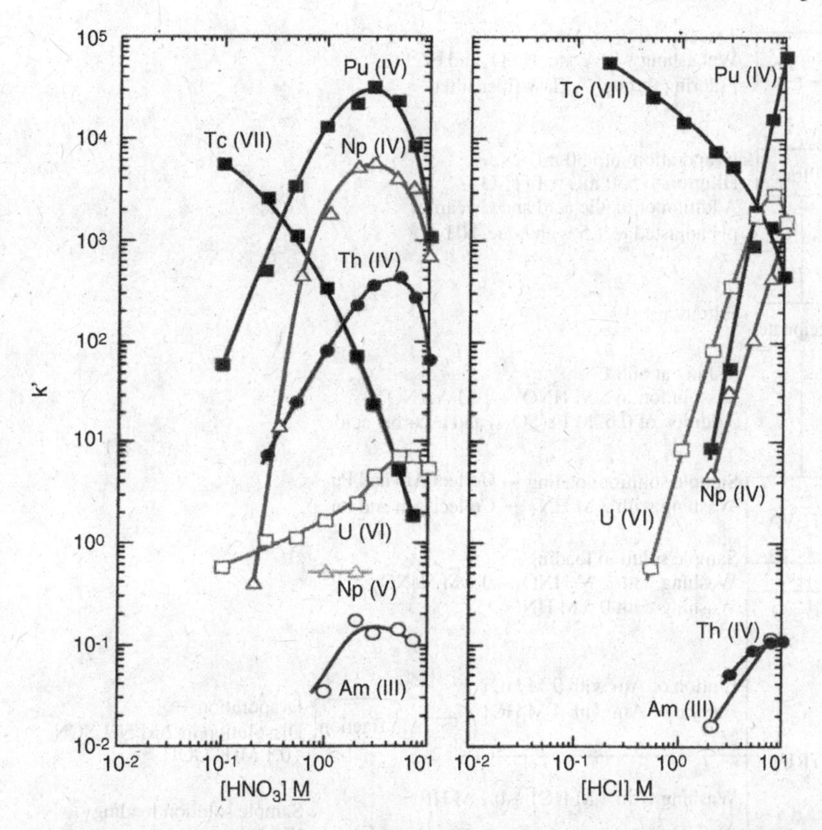

Figure 4.13 k' values for different metals for TEVA column as a function of the nitrate and chloride concentrations. k' value is the number of column volumes at maximum metal concentration in the eluate when eluting the column with the acid of specified concentration. (Horwitz, E.P., Dietz, M.L., Chiarizia, R., Diamond, H., Maxwell, S.L. III, and Nelson, M. (1995) Separation and preconcentration of actinides by extraction chromatography using a supported liquid anion exchanger: Application to the characterization of high-level nuclear waste solutions. *Anal. Chim. Acta*, **310**, 78).

the solution is poured into a UTEVA column. Because the UTEVA does not bind actinides in oxidation states three, Pu^{3+} and Am^{3+} pass through while U(VI) and Th (IV) are retained. The solution containing americium and plutonium is poured into the TRU column and the plutonium is oxidized to oxidation state four with $NaNO_2$ so that it binds more strongly to the column and is not eluted from it along with americium in 4 M HCl, after which the plutonium is reduced to oxidation state three with $TiCl_3$ and eluted with 4 M HCl. Americium fraction is further purified from lanthanides with a TEVA column in 2 M thiocyanate solution, where americium forming a negative complex with thiocyanate is retained in the column while lanthanides pass the column. Finally Am is eluted out of the column with 2 M HCl.

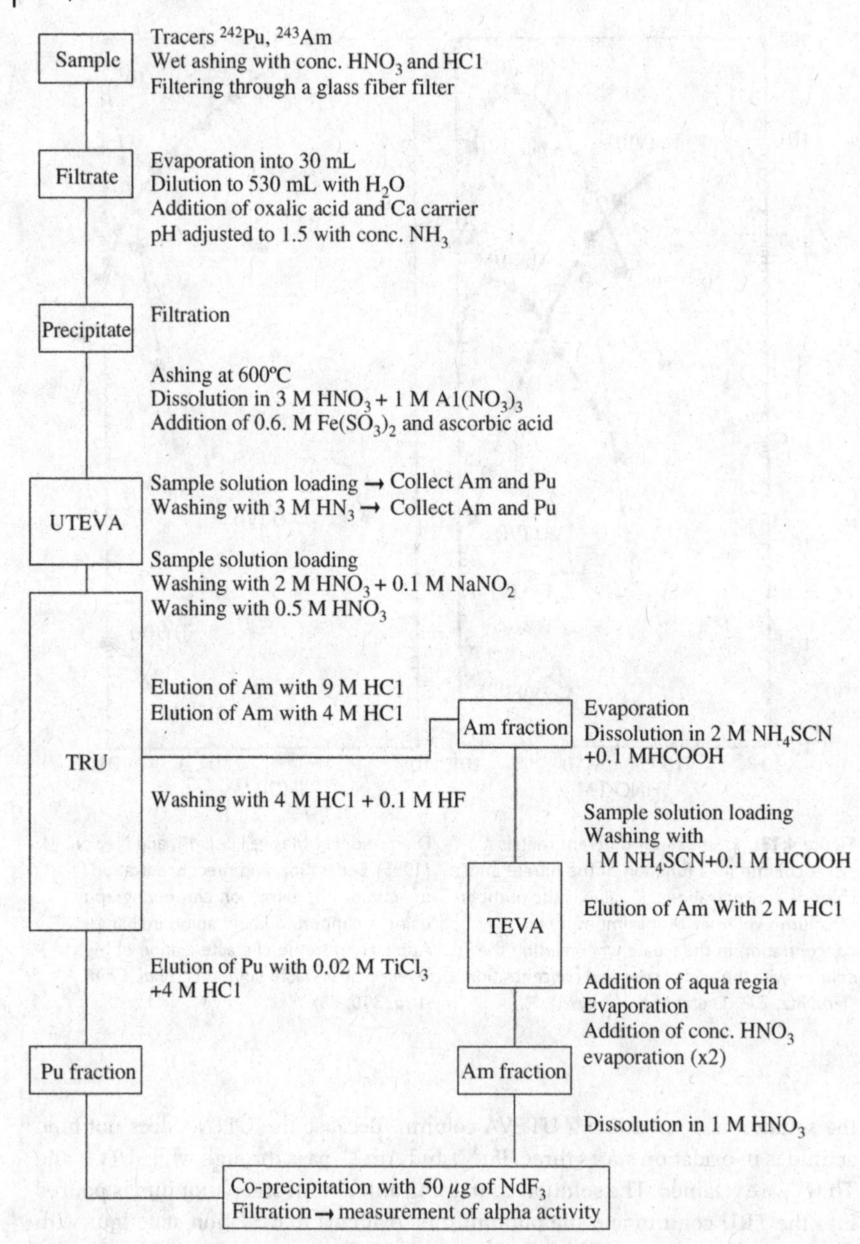

Figure 4.14 Separation of plutonium and americium from sediment by extraction chromatography resins UTEVA, TEVA and TRU resins (Lusa, M., Lehto, J., Leskinen, A., Hölttä, P., Salminen, S., and Jaakkola, T. (2009) ^{137}Cs, ^{239,240}Pu and ^{241}Am in bottom sediments and surface water of Lake Päijänne. *J. Environ. Radioact.*, **100**, 468).

5
Yield Determinations and Counting Source Preparation

At the end of radiochemical separation procedures there are two important steps. First, the chemical yield needs to be determined in order to calculate the activity of the radionuclide in Th the initial sample. Second, the sample, purified from interfering radionuclides and matrix elements, needs to be transformed into a suitable form – a source for activity measurement.

5.1
The Determination of Chemical Yield in Radiochemical Analyses

The final result in a radiochemical analysis is the counting rate (cps) obtained through the transformation of the observed energy of the particles or rays in the source into electric pulses in the detector. When the counting efficiency of the instrumentation is known, the counting rate can be converted into the activity (Bq) of the sample. However, during the chemical separation, not all of the analyte can be separated and recovered in the final sample used for the measurement; this means that some portion of the analyte is lost in the separation process. To find out the proportion of the analyte recovered, the chemical yield has to be measured. The chemical yield is determined with the aid of a stable carrier or a radioactive tracer. Both carriers and radioactive tracers may be isotopic or nonisotopic – the nonisotopic forms being used when suitable isotopic carriers and radioactive tracers of the same element as the analyte are not available.

5.1.1
Use of Stable Isotopic Carriers in Yield Determinations

The chemical yield can be determined by adding a known amount of a stable isotopic carrier to the original sample: before the activity is measured in the final sample the amount of the carrier is measured and compared to the amount added to get chemical yield. In the measurement of ^{90}Sr, for example, a known amount of stable strontium,

Chemistry and Analysis of Radionuclides. Jukka Lehto and Xiaolin Hou

ISBN: 978-3-527-32658-7

typically tens of milligrams, is added to the original sample. As the last step of the analysis, $SrCO_3$ is precipitated and its mass weighed. From this, the recovery of strontium can be calculated. The strontium carbonate is dissolved in a weak acid and liquid scintillation cocktail is added to the solution. The counting rate (e.g., 100 cps) obtained from the liquid scintillation counter is corrected with the counting efficiency (e.g., 70%) obtained from the quenching curve; the activity in the measured sample is then calculated as $100\,\text{cps}/0.7 = 143\,\text{Bq}$. A correction is then made for the chemical yield (e.g., 80%) and the activity in the original sample is obtained: $143\,\text{Bq}/0.8 = 179\,\text{Bq}$. The amount of stable carrier remaining in the sample after the separation, needed for the determination of the yield, is determined gravimetrically or by measuring the concentration of the carrier in the solution by AAS or ICP-MS. Since the stable isotopes of the target radionuclide may also exist in the original sample, the amount of carrier added has to be high enough to enable the contribution from the original sample to be neglected, or the concentration of stable isotopes of radionuclide in the original sample is measured and the contribution to the total amount of the stable isotope is taken into account. The stable isotope in the original sample can sometimes be used as a yield determinant tracer if its concentration in the original sample is high enough. In this case, the chemical yield is determined by the measured amount of stable isotope in the initial sample and in the separated samples.

The use of stable strontium in the measurement of ^{90}Sr is an example of the use of an isotopic carrier in the determination of the chemical yield. A nonisotopic carrier is used for radionuclides that do not have stable isotopes or any radioactive isotope suitable as tracer. Examples are the radium isotopes ^{226}Ra and ^{228}Ra, for which barium is used as a nonisotopic carrier, since radium does not have any stable isotopes or a suitable radioactive tracer. The measurements of ^{99}Tc can be made with the chemically similar element rhenium used as a nonisotopic carrier; there are also isotopic radioactive tracers (^{95m}Tc, ^{99m}Tc) for technetium.

5.1.2
Use of Radioactive Tracers in Yield Determinations

Radioactive tracers can be utilized in yield determinations in several ways. In the measurements of beta-emitting radionuclides, a known amount of gamma-emitting isotope of the same element can be added to the original sample; after the separations, the amount of gamma-emitting isotope that remains in the sample is measured with a gamma counter. This method is used in the measurements of ^{90}Sr, for example, where the gamma-emitting ^{85}Sr isotope is used as a tracer. The gamma activities are measured before the beta counting. ^{85}Sr decays by electron capture and does not emit beta particles. Thus, adding it to the system does not disturb the measurement of the ^{90}Sr beta spectrum.

For chemical yield determination of the alpha emitting isotope of radium, ^{226}Ra, a suitable alternative to stable barium is the gamma-emitter ^{133}Ba, which decays by electron capture and does not interfere with the measurement alpha radiation of ^{226}Ra by LSC.

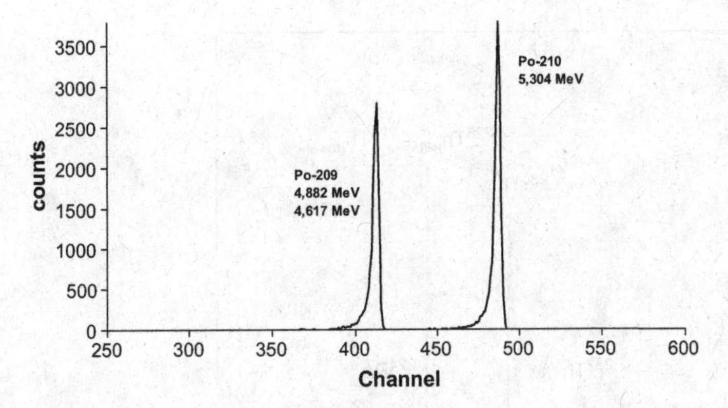

Figure 5.1 Alpha spectrum of polonium.

For alpha emitters the yield can be determined with the aid of another alpha-emitting isotope of the same element. An example is the measurement of ^{210}Po, in which ^{209}Po is used as the tracer. Both isotopes have just one alpha peak (the alpha energy of ^{210}Po is 5.407 MeV, and that of ^{209}Po is 4.979 MeV), and they are clearly distinguishable in the alpha spectrum (Figure 5.1). The ratio of the areas of the two peaks gives the concentration ratio of the two isotopes in the original sample, and since the amount of ^{209}Po added to the sample in the beginning of the analysis is known, the activity of ^{210}Po in the original sample is easily calculated. Thus, measuring the activity of an unknown sample does not require determination of the chemical yield at all. Knowledge of the yield is nevertheless essential in evaluating an analytical method since the larger the yield, the better is the accuracy of the measurement. Determination of the yield in this case also requires that we know the efficiency of the measuring system, that is, how large a proportion of the alpha particles emitted by the counted sample the system is capable of being observe as electric pulses. In alpha spectrometry, as measured with standards of known activity, the counting efficiency is typically in the range of 20–40%. ^{226}Ra is an exception among alpha-emitting radionuclides: there is no suitable radium alpha isotope available to be used as a tracer, and nonisotopic methods have therefore to be used.

Activities of the isotopes of thorium, ^{228}Th, ^{230}Th, and ^{232}Th, can be counted by adding the artificial isotope ^{229}Th as the tracer. If this is not available, ^{228}Th can be used instead. This is a nuclide that is present in all natural samples. The amount of natural ^{228}Th in the sample is measured by carrying out two parallel analyses of the sample, one with the added ^{228}Th tracer and one without. Subtracting the value for the sample without the tracer from that for the sample with the tracer gives the pulses due to the added ^{228}Th. Comparing these pulses to the ^{228}Th activity added to the sample in turn gives the yield, which can be used to calculate activities of all three isotopes (Figure 5.2). Table 5.1 summarizes the yield determination methods used for most radionuclides discussed in this book.

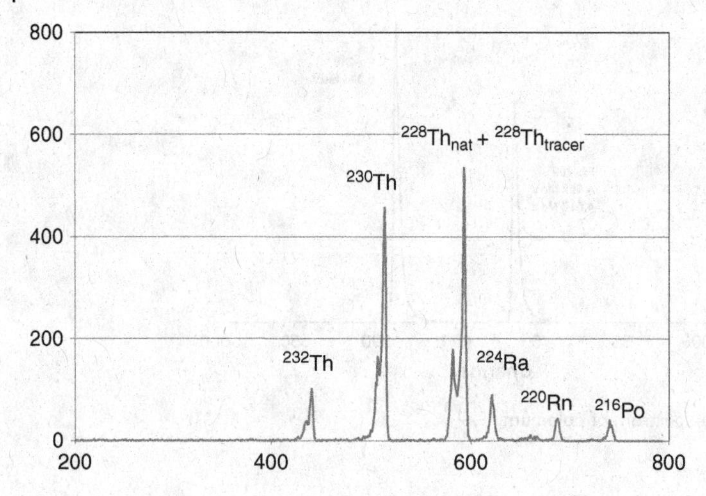

Figure 5.2 Alpha spectrum of thorium, including daughter nuclides.

Table 5.1 Methods used in yield determinations of important radionuclides.

Radionuclide	Decay mode	Yield determination method
^{3}H	beta	100% recovery in general
^{14}C	beta	100% recovery in general
^{36}Cl	beta	stable chlorine
^{41}Ca	EC	stable calcium
^{55}Fe	EC	stable iron
^{63}Ni	beta	stable nickel
^{79}Se	beta	stable selenium
^{90}Sr	beta	stable strontium or gamma-active ^{85}Sr
^{99}Tc	beta	stable rhenium or gamma-active ^{99m}Tc
^{93}Mo	beta	stable molybdenum
^{129}I	beta	^{125}I or stable iodine
^{210}Pb	beta	stable lead
^{210}Po	alpha	alpha-active ^{208}Po and ^{209}Po
^{226}Ra	alpha	stable barium or gamma-active ^{133}Ba
^{228,230,232}Th	alpha,beta	^{229}Th or ^{228}Th
^{234,235,238}U	alpha	alpha-active ^{232}U and ^{236}U
^{237}Np	alpha	^{239}Np and ^{236}Np
^{238,239,240}Pu	alpha	alpha-active ^{236}Pu and ^{242}Pu
^{241}Pu	beta	alpha-active ^{236}Pu and ^{242}Pu
^{241}Am	alfa	alpha-active ^{243}Am
^{242}Cm	alfa	alpha-active ^{244}Cm or ^{243}Am

5.2 Preparation of Sources for Activity Counting

After the interfering radionuclides and matrix elements have been removed, the sample is prepared for counting. Different methods are used for gamma emitters and radionuclides emitting alpha and beta particles. Activities of gamma-emitting radionuclides can usually be measured directly without radiochemical separation. Radionuclides decaying by beta and alpha emissions and emitting no measurable gamma radiation will require radiochemical separations (see Chapter 4) before the measurement of the activity. The activities of alpha-emitting radionuclides are measured with a semiconductor detector or liquid scintillation counter. The resolution of a semiconductor in alpha counting is about 20–30 keV, whereas the resolution of the liquid scintillation counter is ten times larger. The activities of beta-emitting radionuclides are usually determined with liquid scintillation counters, but they can also be determined with gas ionization counters, especially proportional counters.

5.2.1 Preparation of Source for Gamma Emitters

Gamma-emitting radionuclides are measured with a semiconductor detector – usually a Ge detector, or a solid scintillation detector – usually an Na(I) detector. The resolution of the latter is clearly worse: the energy resolution of semiconductor detectors is under 2 keV, whereas that of Na(I) detectors is several tens of keV. Usually, radiochemical separations are not required for gamma-emitting radionuclides, and with the semiconductor detector tens of radionuclides can be measured at the same time. For gamma spectrometry, the sample is packed into a counting vial and usually placed above the detector. The volume and the mass of the samples are determined prior to measurement. The measurement of aqueous samples is most straightforward – a given volume of solution is placed in a counting vial.

Prior to the gamma measurement, the counting system must be calibrated for energy and efficiency. The energy calibration is carried out with a known mixture of radionuclides of varying gamma energy. The efficiency calibration is a somewhat more demanding task. First, the system has to be calibrated with respect to geometry: the efficiency for standard radionuclides of known activity is determined for each type of sample vial and various sample volumes. In addition, the efficiencies are determined for various distances between the sample and the detector. This is typically enough for aqueous samples. The situation in the case of solid samples is, however, more complicated. The efficiency calibration for solid samples has to take into account variable self-absorption of gamma rays in the sample. If solid standards of known activity and with identical or at least similar composition are available, the efficiency calibration can be carried out with the aid of these standards. However, this is usually not the case, and the self-absorption correction to counting efficiency is done by the density of the sample or by a calculation using the Monte Carlo simulation. The latter is much more demanding and requires knowledge of the chemical composition of the sample. Self-absorption is especially marked for gamma rays with energies below 100 keV.

The counting efficiency of conventional semiconductor detectors falls off sharply at energies below 100 keV, and low-energy gamma rays cannot be measured, particularly when activities are low. However, low-energy or broad-energy gamma detectors, which have thinner windows, effectively detect low-energy radiation as well; sometimes they can be utilized to measure radionuclides that otherwise would require radiochemical separations. One example of a situation where radiochemical separation can be dispensed with is the measurement of ^{210}Pb for age dating of layers in a sediment. Generally, radiochemical separation is required for this task, and ^{210}Pb is measured by alpha spectrometric measurement of its alpha-emitting daughter ^{210}Po assuming there is a radiochemical equilibrium between these two radionuclides. Much time and effort can be saved by the direct measurement of the low-energy (46.5 keV) photons emitted by ^{210}Pb using a semiconductor detector capable of measuring low-energy radiation. Since the intensity of the gamma radiation of ^{210}Pb is only 4%, the detection limit of gamma measurement is not as low as that in alpha spectrometry (intensity of alpha emissions is 100%). Another example is ^{241}Am, which, in addition to alpha particles, emits low-energy (59 keV) gamma radiation of 36% intensity. If the ^{241}Am activity is high enough, it can be measured directly by gamma spectrometry. Deposition from nuclear testing is the main source of ^{241}Am in the environment; however, it is generally present in high enough concentrations to be measured by gamma spectrometry only at nuclear test sites and in other heavily contaminated areas.

An additional problem in gamma spectrometric measurement is the inhomogeneity of the sample. Solid samples are, whenever possible, ground to fine grains or to powder and homogenized by blending. The radioactivity in the sample may, however, be inhomogeneously distributed in spite of grinding and homogenization if only a few larger radioactive particles are present, and this results in irreproducibility of the results. This may be especially the case when air filters are measured. For aqueous samples, two factors can result in inhomogeneity. First, the radionuclides can adsorb on the counting vials' surfaces; this can be avoided by the acidification of the samples. Second, radionuclide-bearing particles dispersed in the sample may deposit on the counting vial bottom during measurement, especially if the counting time is long. This can be avoided by filtration of the sample and measuring both the filtered fraction and the particle fraction – the latter preferably after dissolution.

5.2.2
Sample Preparation for LSC

The liquid scintillation counter measures liquid samples, as is evident from the name. After radiochemical separations of interfering radionuclides, scintillation cocktail is added to the solution containing the target nuclide for activity measurement. The scintillation cocktail is a mixture of organic solvents and organic scintillation molecules, which transform the energy of the beta particles into light. If the radionuclide is in an organic solution, which is seldom the case, it is readily soluble in the cocktail. Most frequently, the measured radionuclide is in an aqueous solution which is not completely soluble in the organic cocktail; however, mixtures up to about 50% of water can be obtained.

The use of a scintillation cocktail can be avoided if the beta energy is high enough to result in the formation of Cherenkov light radiation, which is detected by the counter's photomultiplier tubes. The minimum beta energy to create Cherenkov radiation in water is 263 keV; however, to become efficient compared to a liquid scintillation counter, Cherenkov counting requires maximum beta energies clearly higher than one MeV. By using Cherenkov counting, the generation of organic waste can be avoided. In addition, the scintillation vials, usually of 20 mL volume, can be filled up since no scintillation cocktail is needed.

LSC can also be used to measure solid samples: chromatographic papers or chromatographic masses containing radionuclides can be directly introduced into scintillation vials with cocktails. If the radionuclide is dissolved from the paper or the mass in the scintillation cocktail, the situation is more or less identical with an ordinary counting situation and the paper piece or the mass, if not of high quantity, do not essentially interfere with counting. If the radionuclide is not dissolved from the mass or other solid samples are measured, the grain size of the solid sample should be as low as possible and the grains should be uniformly distributed in the cocktail which can be obtained by the use of gelatinous agents. The main problem in measuring solid samples is the self-absorption, which lowers the counting efficiency. Measuring solid samples also results in the irreproducibility in count rate response and should be avoided when possible.

In the determination of tritium and radiocarbon from organic matter (see Chapter 13) two methods are typically used: solubilization or combustion. In solubilization, organic compounds are made soluble in the scintillation cocktail by the addition of appropriate reagents such as hydrogen peroxide. A more effective sample preparation is combustion, followed by trapping of tritium as water and radiocarbon as carbon dioxide and mixing them with the scintillation cocktail for activity measurement. The latter method results in the efficient removal of matrix elements, which could interfere with LSC. In addition, it allows the measurement of both nuclides separately which improves the counting accuracy.

Typically, the main challenge in LSC is the determination of counting efficiency. Depending on the composition of the sample solution, the efficiency varies – sometimes in a wide range. Therefore, the counting efficiency must be determined for each individual sample. There are several methods used for this purpose, including the use of internal standards, the use of external standards, and the use of three photomultiplier tubes, which are briefly described in Chapter 1.

5.2.3
Source Preparation for Alpha Spectrometry with Semiconductor Detectors and for Beta Counting with Proportional Counters

For measurements with the semiconductor detector and the proportional counter, the preparation for counting is the same: the sample is prepared in solid form and the

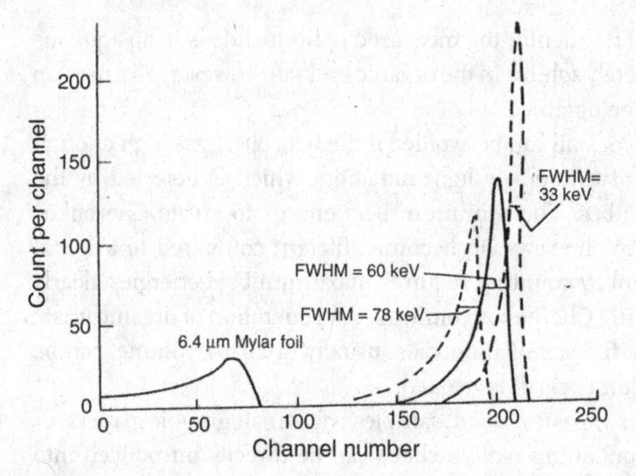

Figure 5.3 Effect of sample mass on the shape (resolution) of peaks in an alpha spectrum (Holm, E. (2001) Source preparations for alpha and beta measurements, Report NKS-40).

final sample amount should be as small as possible to minimize self-absorption. If the quantity of sample to be measured is too large, self-absorption increases, the intensity of the alpha peaks weakens, and the peaks move to lower channels and broaden. An example of this can be seen in Figure 5.3.

Three widely employed methods of preparing sources that will be measured with semiconductors and proportional counters are described below. These methods are electrodeposition, micro-coprecipitation and spontaneous deposition.

5.2.3.1 **Electrodeposition**

After the removal of interfering radionuclides and matrix elements, the solution containing the target radionuclide is poured into an electrodeposition vessel and mixed with ammonium, sulfate, chloride, oxalate, hydroxide, or formate as the electrolyte and the solution is made slightly acidic. A metal disk – usually made of polished steel or sometimes platinum – is tightly mounted to the lower part of the electrodeposition vessel. A platinum thread is put in the vessel and a constant current (10–150 mA/cm^2) is set up between the platinum thread and the metal disk so that the platinum thread operates as the anode and the metal disk as the cathode (Figure 5.4). The current causes reduction of the metals in the solution and their deposition in metallic form or as hydroxides on the surface of the steel plate. The electrodeposition takes place over one to four hours. Several electrodeposition vessels can be connected in series in the same system. The electrodeposition vessel normally needs to be cooled with a water bath during the electrodeposition, especially when a higher current is applied, for example for Pu and Am, to prevent evaporation loss of the solution at high temperatures.

5.2.3.2 **Micro-coprecipitation**

Counting sources of actinides can be prepared by the coprecipitation of the actinides with very small amounts of hydroxide, fluoride, or sulfate. Typically, the coprecipitation is carried out with lanthanide fluorides: 10–50 µg La, Ce, or Nd is added to the solution, and the fluoride (LaF_3, CeF_3, NdF_3) is precipitated through the addition of

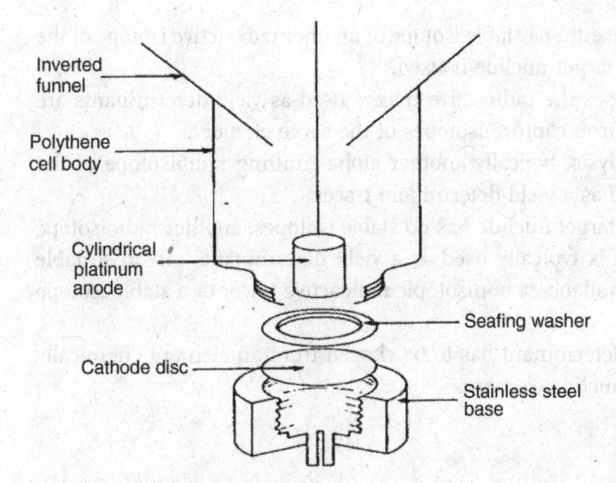

Figure 5.4 Electrodeposition equipment (Holm, E. (2001) Source preparations for alpha and beta measurements, Report NKS-40).

HF. Since actinides will only coprecipitate with lanthanide fluoride if they are in An(III) or An(IV) form, UO_2^{2+}, for example, must first be reduced to U^{4+}, and plutonium in the higher valence state (PuO_2^{+}/PuO_2^{2+}) must be reduced to a lower valence state (Pu^{4+}/Pu^{3+}). After precipitation, the precipitate is collected on a membrane filter, dried, and mounted on the measurement plate with glue for alpha counting. The sample to be measured for radium can be prepared by micro-coprecipitation with barium sulphate. Because the mass, and so the self-absorption of the sample, is larger, the resolution obtained after micro-coprecipitation will be somewhat poorer than that after electrodeposition. Micro-coprecipitation is a distinctly more rapid technique, however, and the resolution is usually adequate.

5.2.3.3 **Spontaneous Deposition**

The counting source for ^{210}Po is normally prepared by spontaneous deposition onto a nickel or silver disk. Polonium is reduced on these metals because it is a more noble metal. The deposition system is similar to ordinary electrodeposition, except that no external voltage is required. In addition, instead of cooling, heating the solution and stirring during the deposition are needed to improve the deposition efficiency.

5.3
Essentials in Chemical Yield Determination and in Counting Source Preparation

5.3.1
Yield Determination

- In radiochemical separations the chemical yield is determined by adding a known amount of yield determinant in the beginning of the analysis and calculating the yield from the recovery of the yield determinant at the end of the analysis.

- As a yield determinant either a stable isotope or another radioactive isotope of the same element as the target nuclide is used.
- In beta nuclide analyses the radioactive tracers used as yield determinants are gamma-emitting electron capture isotopes of the same element.
- In alpha nuclide analysis, typically another alpha-emitting radioisotope of the same element is used as a yield determinant tracer.
- If the element of the target nuclide has no stable isotopes, another radioisotope of the same element is typically used as a yield determinant – if no suitable radioactive tracer is available, a nonisotopic radioactive tracer or a stable isotope has to be used.
- A nonisotopic yield determinant has to be chosen from an element chemically closest to the target nuclide element.

5.3.2
Counting Source Preparation

- Gamma-emitting radionuclides are usually measured directly from samples without radiochemical separations.
- For beta and alpha measurement with LSC, the sample needs to be transformed into a liquid, which is then mixed with liquid scintillation cocktail and measured for the activity. In some cases solid samples can be measured as well.
- For the measurement of beta emitters with proportional counting and alpha emitters with a semiconductor detector the sample is prepared by electrodeposition or spontaneous deposition as a thin metallic or oxidic layer on a metal surface. The other choice is to coprecipitate the target nuclide with a small amount of precipitate and collect the precipitate on a membrane filter for the measurement.

6
Radiochemistry of the Alkali Metals

6.1
Most Important Radionuclides of the Alkali Metals

In terms of radiochemistry, radioactivity in the environment, and human health, the most important radionuclides of the alkali metals are ^{40}K and the cesium isotopes ^{137}Cs and ^{134}Cs (Table 6.1). The isotope ^{40}K is an extremely long-lived naturally occurring radionuclide forming 0.0118% of natural potassium. ^{137}Cs is a long-lived fission product of uranium and plutonium with a high fission yield, and is certainly the main source of radioactive pollution passing to humans in food. ^{134}Cs in turn, is generated in nuclear fuel and reactor materials through neutron activation of stable cesium. All three radionuclides emit high-energy gamma radiation, allowing their radioactivity to be measured easily, nearly always without prior purification. ^{135}Cs, a fission product of uranium and plutonium, is another radioisotope of cesium and is a pure beta emitter. It could in principle be measured by its daughter's gamma rays, but the very large excess of ^{137}Cs makes this impossible. Thus, the radiochemical separation of ^{135}Cs has to be completed and its measurement has to be done in a non-radiometric way. ^{87}Rb, the extremely long-lived radioisotope of rubidium, is used to measure the age of rock up to billions of years old from the ratio of its concentration to that of its daughter ^{87}Sr. Measurements are done by mass spectrometry.

6.2
Chemical Properties of the Alkali Metals

The alkali metals constitute Group 1 of the periodic table. All have one s-orbital electron in their outer shell and form compounds with the oxidation state +I. In aqueous solution they exist as hydrated M^{+1} ions. With a small charge and large size, they are strongly electropositive, becoming more electropositive as the atomic number increases, as follows: Li < Na < K < Rb < Cs < Fr. Because they are electropositive, the alkali metals mostly form ionic bonds, and their common compounds, such as the halides, are readily soluble. The ionic radius of the alkali metals increases with the atomic number. Like other metal ions, in aqueous solution they appear in

Chemistry and Analysis of Radionuclides. Jukka Lehto and Xiaolin Hou

ISBN: 978-3-527-32658-7

Table 6.1 Important radionuclides of the alkali metals.

Element	Important radionuclides	Half-life	Decay mode	Gamma emission	Source/Use
Lithium	^{8}Li	0.8 s	β^-		
Sodium	^{22}Na	2.6 y	β^+	yes	tracer
	^{24}Na	15 h	β^-	yes	activation product
Potassium	^{40}K	1.3×10^9 y	β^-/EC	yes	natural nuclide
	^{42}K	12 h	β^-	yes	tracer
Rubidium	^{86}Rb	19 d	β^-	yes	tracer
	^{87}Rb	4.7×10^{10} y	β^-		natural nuclide
Cesium	^{137}Cs	30 y	β^-	yes	fission product
	^{134}Cs	2.1 y	β^-/EC	yes	activation product, tracer
	^{135}Cs	2.3×10^6 y	β^-	no	fission product
Francium	^{223}Fr	22 m	β^-	yes	natural nuclide

aqua complex forms, as $M(H_2O)_x^+$. Because of their smaller size, the lightest alkali metals more strongly attract water molecules, and the size of the hydrated ions decreases from lithium to cesium. The alkali metals hydrolyze, that is, form hydroxyl complexes (MOH), only sparingly and at a very high pH. Their hydroxides are highly soluble. Alkali metal ions do not form complexes with very many organic compounds because there are no free vacancies in their electron shells allowing coordination bond formation. Consequently, they cannot be usually separated by solvent extraction with organic complexing agents. Some important physical properties of the alkali metals are listed in Table 6.2.

6.3 Separation Needs of Alkali Metal Radionuclides

Most radionuclides of the alkali metals are gamma emitters and do not require chemical separation prior to measurement. The only exception is ^{135}Cs, which is a

Table 6.2 Physical properties of selected alkali metals.

Element	Electronegativity (Pauling)	Ionic radius in crystal (pm)	Hydrated ionic radius (pm)	Boiling point (°C)
Lithium	0.98	86	340	1342
Sodium	0.93	112	276	883
Potassium	0.82	144	232	759
Rubidium	0.82	158	228	688
Cesium	0.79	184	228	671
Francium	about 0.7	about 190		

pure beta emitter and must be separated before measurement. Because of the low concentration of ^{137}Cs in natural waters, however, some pre-concentration is normally necessary before its activity measurement. Most gamma-emitting alkali metal radionuclides, especially ^{40}K, ^{134}Cs, and ^{137}Cs, also emit beta particles and/or conversion electrons which interfere with the radiometric measurement of other beta-emitting radionuclides. Hence, if a target nuclide is a pure beta-emitter or decays by electron capture, they need to be separated before the activity of the target nuclide is measured. Alkali metals can be removed from radionuclides that hydrolyze in aqueous solution by precipitation of their hydroxides, leaving the alkali metals in solution. Alkali metals can be separated from alkaline earth metals, ^{90}Sr for example, by precipitation of carbonate, oxalate, and phosphate. The alkaline earth metals are removed, while the alkali metals remain in solution.

6.4
Potassium – ^{40}K

^{40}K is a primordial radionuclide, which has existed since the creation of the Earth about five billion years ago. With its exceedingly long half-life of 1.3×10^9 years, it is still present in the environment. Potassium is the eighth most common element in the Earth's crust (1.8%), and the isotope ^{40}K comprises 0.0118% of natural potassium, which is present in common rock-forming minerals such as feldspar and muscovite, and in seawater. It is an essential trace element for humans and animals and is present in all living cells at a constant concentration. Accordingly, radioactive ^{40}K is also present in cells at a fixed concentration. The concentration of ^{40}K in the human body is about 70 Bq per kilo and it is responsible for 5% of the total dose to humans. ^{40}K emits high-energy gamma radiation (1.46 MeV) (Figure 6.1) that is easily detected by gamma spectrometry, and the activity of a source can be determined

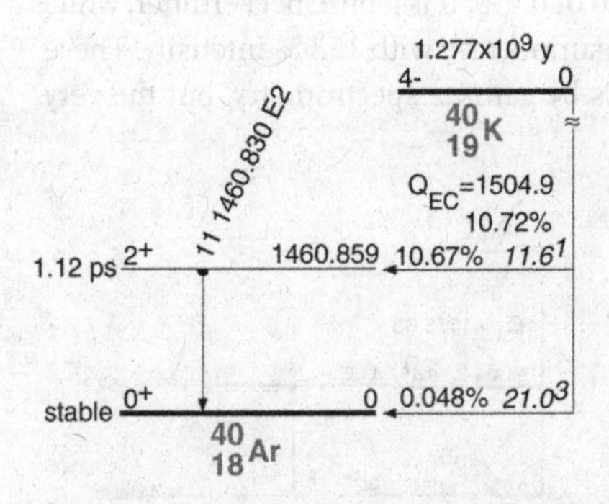

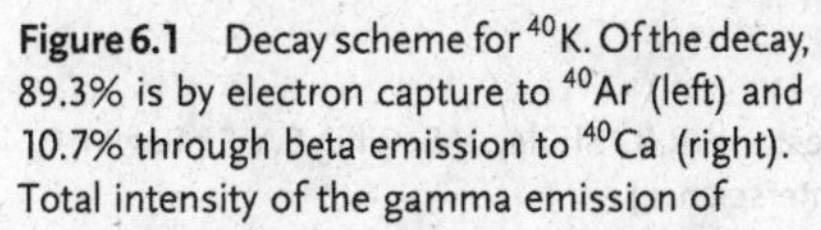

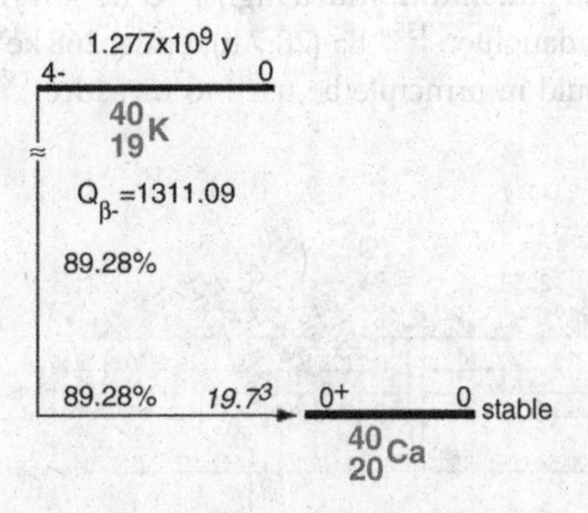

Figure 6.1 Decay scheme for ^{40}K. Of the decay, 89.3% is by electron capture to ^{40}Ar (left) and 10.7% through beta emission to ^{40}Ca (right). Total intensity of the gamma emission of 1.46 MeV is 10.67% (Firestone, R.B., Shirley, V. S., Chu, S.Y.F., Baglin, C.M., and Zipkin, J. (1996) *Table of Isotopes*, Wiley-Interscience).

without prior radiochemical separation. ^{40}K is also used for finding the age of rocks by measuring their ^{40}K concentration compared to that of its daughter nuclide ^{40}Ar.

6.5
Cesium – ^{134}Cs, ^{135}Cs, and ^{137}Cs

6.5.1
Sources and Nuclear Characteristics

Radionuclides ^{137}Cs and ^{90}Sr are the two most important long-lived fission products of uranium and plutonium, each with a half-life of about 30 years and a ^{235}U fission yield of 6.3%. ^{137}Cs is widely present in the environment as a consequence of fallout from atmospheric nuclear weapons testing in the fifties, sixties, and seventies, and the Chernobyl accident in 1986. ^{134}Cs is produced from the stable isotope ^{133}Cs by neutron activation. While ^{134}Cs was not present in the fallout from nuclear weapons testing, it comprised a significant portion of the fallout from the Chernobyl accident because it had accumulated during long-term irradiation of the fuel and construction materials in the reactor. Since the half-life of ^{134}Cs is just 2.1 years, the pollution from ^{134}Cs in the Chernobyl fallout has now disappeared from the environment.

Both ^{134}Cs and ^{137}Cs emit high-energy gamma radiation (Figure 6.2), which in most cases can be measured by gamma spectrometry directly from environmental and nuclear waste samples without the need for a purification step. ^{137}Cs has only one observable gamma transition, at 662 keV, while ^{134}Cs has several, the most intensive ones being at 604 keV and at 795 keV. Because cesium resembles potassium, it follows potassium into human and animal tissues, and the radionuclides of cesium in fallout have clearly caused larger radiation doses to people than have other man-made radionuclides. Relative to the doses caused by natural nuclides they are still fairly low, however.

^{135}Cs has a very long half-life of 2.3×10^6 y. It too is a fission product of uranium and plutonium, with a high ^{235}U fission yield of 6.5%. It is a pure beta emitter, while its daughter, ^{135m}Ba (28.7 h), emits 268 keV gamma rays with 15.5% intensity. These could in principle be used to measure ^{135}Cs by gamma spectrometry, but the very

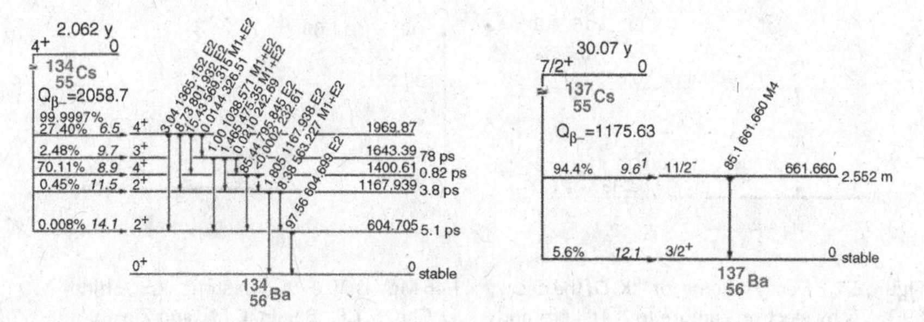

Figure 6.2 Decay schemes of ^{134}Cs and ^{137}Cs (Firestone, R.B., Shirley, V.S., Chu, S.Y.F., Baglin, C. M., and Zipkin, J. (1996) *Table of Isotopes*, Wiley-Interscience).

high excess of coexisting ^{137}Cs makes this measurement impossible. The 268 keV gamma rays are indistinguishable from the Compton background of the ^{137}Cs 662 keV gamma rays. The masses of ^{135}Cs and ^{137}Cs in irradiated uranium and plutonium are approximately the same, but the activity of the former is only about one part in 80 000 of the latter. Thus, radiochemical separation has to be carried out to remove the interference of matrix elements, and the measurement of ^{135}Cs has to be done using techniques other than radiometry, such as neutron activation analysis and mass spectrometry.

The average concentration of stable cesium (^{133}Cs) in the Earth's crust is 2.6 ppm (2 $\mu mol\,kg^{-1}$). Stable cesium is also present in natural waters. In seawater the average concentration is 0.3 ppb, or about 2 $nmol\,L^{-1}$. Radioactive cesium studies, therefore, must always take account of the presence of the natural carrier in the system.

The boiling point of cesium is very low, 671 °C. Thus, cesium released in nuclear explosions and Chernobyl-type accidents completely evaporates, and, as the cloud cools, the cesium binds with atmospheric aerosol particles. Cesium is not found in large particles (i.e., hot particles), which are mostly fuel fragments or fragments of nuclear weapons material, because it breaks away from these during evaporation. In the fallout from the Chernobyl accident, the ^{137}Cs was mostly bound to small particles, 0.1–1 μm in size. The small particle size and the solubility of cesium compounds mean that any cesium deposited on the ground easily migrates into the environment. Though cesium in general is very soluble, it is tightly bound to clay minerals in soil, sediments, and water.

In nuclear waste, most of the radioactive cesium will be present in the spent nuclear fuel. Since the half-life of ^{137}Cs in only 30 years it cannot be considered an essential risk in the final disposal of spent nuclear fuel. ^{135}Cs, however, has such a long half-life that it will be the prevailing fission product after one million years. The radioactivity levels at that time, however, will be very low. Because of its low boiling point and high solubility, cesium will be rather readily released from the fuel into the reactor coolant if cracks appear in the fuel cladding during reactor operation. Cesium is mostly removed from the coolant into the ion exchange resins used to purify it, and finally ends up in the repositories for low and intermediate active power plant waste. The Sellafield nuclear fuel reprocessing plant in the UK released large amounts of ^{137}Cs into the Irish Sea until 1985, when the SIXEP effluent purification system was put into operation. These releases have been distributed over a long path from the Irish Sea via the North Sea coast up to the Arctic Sea.

6.5.2
Preconcentration of Cesium Nuclides from Natural Waters

Sometimes ^{137}Cs is present in the environment in such small amounts that it cannot be determined within a reasonable time frame directly from samples. Water samples, in particular, require enrichment of ^{137}Cs before counting. Lake and stream water can be evaporated to reduce the volume and increase the concentration. Alternatively, all cations, including Cs^+, can be separated from the water onto a strong organic cation exchange resin. An ion exchange capacity of a few hundred milliliters is sufficient for the separation of all cations from a few tens of liters of fresh water. Seawater presents a different problem, as it contains such large concentrations of salts that the capacity of a cation exchanger is rapidly exceeded, rendering the method ineffective. However, ^{137}Cs can be separated from seawater by coprecipitation, for example, with ammonium phosphomolybdate, $(NH_4)_3PMo_{12}O_{40}\cdot 3H_2O$, or potassium cobalt hexacyanoferrate, $K_2CoFe(CN)_6\cdot H_2O$. Both are highly selective agents for cesium precipitation. The latter can also be prepared in granular form and used in column separations. Before the availability of gamma spectroscopy in the early 1960s, these reagents were also used in radiochemical separations. Cesium was precipitated by the above-mentioned reagents, and activity measurements were done by counting the beta radiation of ^{137}Cs with a Geiger–Muller counter. Other cesium-selective precipitation agents, which form insoluble salts with cesium, were also used. These included hexanitrocobaltate, phosphotungstate, and hexachloroplatinate. In addition, a filtration method is also used for the separation of radioactive cesium from seawater, especially *in-situ* separation. In this case, $K_2CoFe(CN)_6$ or $K_2CuFe(CN)_6$ is impregnated onto a cotton filter cartridge fitted to a filtration system for a large water sample (500–2000 L).

6.5.3 Determination of ^{135}Cs

As already mentioned, radiometric measurement of ^{135}Cs is not possible. Two alternative methods for its measurements are neutron activation analysis and mass spectrometry.

6.5.3.1 Determination of ^{135}Cs by Neutron Activation Analysis

In the determination of ^{135}Cs by neutron activation analysis (NAA), cesium is chemically separated, and the ^{135}Cs is activated in a neutron flux to form gamma-emitting ^{136}Cs. The thermal neutron capture cross section of ^{135}Cs is reasonably high at 9 barns. The resulting ^{136}Cs has a half-life of 13 days and emits high-energy gamma rays of 818, 1048 and 1235 keV. The chemical separation of cesium is accomplished by adding ammonium molybdophosphate (AMP) powder into a cesium-bearing solution. AMP, which takes up cesium quantitatively, is then dissolved in 0.06 M NaOH, and the cesium is collected in a strongly acidic cation exchange resin column. After irradiation of the cesium-containing resin, the cesium is removed from the resin with 5 M HNO_3, and interfering radionuclides are removed by precipitating the cesium as AMP. For final purification, AMP is dissolved in 0.06 M NaOH and reprecipitated in 0.1 M HCl for gamma spectrometric measurement of the cesium radioisotopes ^{134}Cs, ^{136}Cs, and ^{137}Cs. The most important drawback in the neutron activation method for ^{135}Cs determination is the formation of ^{134}Cs from the stable cesium isotope ^{133}Cs. ^{134}Cs has gamma rays in the same range as ^{136}Cs, resulting in the formation of a high Compton background to the gamma spectrum. This makes the detection of ^{136}Cs peaks difficult and results in a rather high detection limit, too high for environmental samples. (Chao, Jiunn-Hsing and Tseng, Chia-Lian

(1996) Determination of ^{135}Cs by neutron activation analysis. *Nucl. Instr. Meth. Phys. Res.* A, **372**, 275).

6.5.3.2 Determination of ^{135}Cs by Mass Spectrometry

Mass spectrometry offers a much lower detection limit than that of neutron activation analysis. In the measurement of ^{135}Cs with mass spectrometry the main challenges, causing interference, are the tailings due to the stable cesium isotope ^{133}Cs and the isobaric interference due to ^{135}Ba. In environmental samples, the ratio of ^{135}Cs to ^{133}Cs is typically 10^{-9}, and thus very high mass separation efficiency is needed. Several mass spectrometric techniques have been applied to ^{135}Cs measurement, including thermal ionization mass spectrometry (TIMS), inductively coupled plasma mass spectrometry (ICP-MS), and resonance ionization mass spectrometry (RIMS). This last technique shows the most promising performance, since it can more or less completely remove the isobaric interference due to ^{135}Ba. They all require chemical separation of cesium from the matrix elements, especially from barium. Typically, separation is carried out by AMP. An example of such a separation for TIMS measurement is given in Figure 6.3.

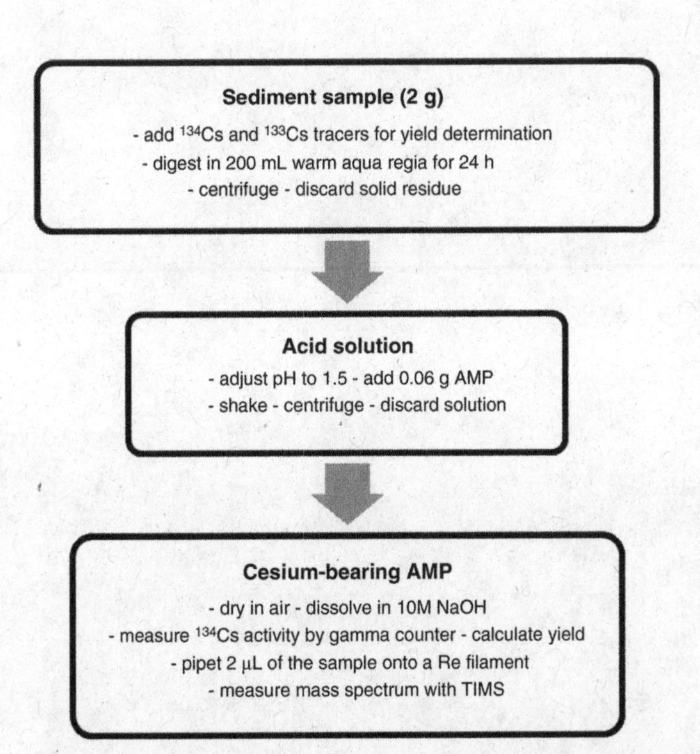

Figure 6.3 Sample preparation for ^{135}Cs measurement with thermal ionization mass spectrometry (TIMS). (Karam, L.R., Pibida, L., and McMahon, C.A. (2002) Use of resonance ionization mass spectrometry for determination of Cs ratios in solid samples. *Appl. Radiat. Isotopes*, **56**, 369).

6.6
Essentials in the Radiochemistry of the Alkali Metals

- The most important radionuclides of alkali metals are the naturally occurring very long-lived ^{40}K, the fission product ^{137}Cs, and an activation product ^{134}Cs.
- All these radionuclides emit high-energy gamma rays, which can be readily measured for the determination of their activities – no radiochemical separation is usually needed for these activity measurements.
- The determination of ^{137}Cs in natural waters requires preconcentration prior to activity measurement – this is accomplished by evaporation, ion exchange, or coprecipitation.
- ^{135}Cs is a pure beta emitter and requires radiochemical separation before the activity measurement. ^{135}Cs cannot be measured radiometrically because of the interference of ^{137}Cs, which is in large excess in activity. ^{135}Cs can be measured by neutron activation analysis and by mass spectrometry, the latter being a much more sensitive method.

7
Radiochemistry of the Alkaline Earth Metals

7.1
Most Important Radionuclides of the Alkaline Earth Metals

From the point of view of radiochemistry and radioactivity in the environment, the most important radionuclides of the alkaline earth metals are ^{90}Sr, originating in nuclear weapons testing, the Chernobyl accident, and nuclear waste, and the ^{226}Ra and ^{228}Ra isotopes generated in the natural radioactive decay series. ^{90}Sr is a pure beta emitter, while the isotopes of radium are alpha and beta emitters. Consequently, they all require radiochemical separations before counting. The third radionuclide of interest in radiochemical separations is the very long-lived isotope of calcium, ^{41}Ca, formed as an activation product in nuclear waste. The important radionuclides of the alkaline earth metals are listed in Table 7.1 and are discussed below.

7.2
Chemical Properties of the Alkaline Earth Metals

The alkaline earth metals belong to group 2 of the periodic table and have 2 s-orbital electrons in their outer shell. They form compounds in oxidation state +II and appear in aqueous solutions as M^{2+} ions. Alkaline earth metals are moderately electropositive, but less electropositive than the alkali metals since their charge is twice as large. The electropositivity of the alkaline earth metals increases with the atomic number as follows: Be < Mg < Ca < Sr < Ba < Ra (Table 7.2). Because they are electropositive, the alkaline earth metals, like the alkali metals, mostly form ionic bonds, and most of their common compounds, like halides, are highly soluble, though quite a number are sparingly soluble. Beryllium is exceptional among the alkaline earth metals in its very small ion size (ionic radius 31 pm), and therefore it forms compounds with covalent bonds. In their ion size the alkaline earth metals follow the same trend as the alkali metals: the ion size increases with atomic number

Chemistry and Analysis of Radionuclides. Jukka Lehto and Xiaolin Hou

ISBN: 978-3-527-32658-7

Table 7.1 Important radionuclides of the alkaline earth metals.

	Important radionuclides	Half-life	Decay mode	Gamma emission	Source or use
Beryllium	^{7}Be	53 d	EC[a)]	yes	cosmogenic
	^{10}Be	1.6×10^{6} y	β^-	no	cosmogenic
Magnesium	^{28}Mg	21 h	β^-	yes	tracer
Calcium	^{45}Ca	163 d	β^-	no	activation product, tracer
	^{41}Ca	1.03×10^{5} y	EC[a)]	no	activation product
Strontium	^{85}Sr	65 d	β^-	yes	tracer
	^{89}Sr	50 d	β^-	no	fission product
	^{90}Sr	29 y	β^-	no	fission product
Barium	^{133}Ba	10.7 y	EC[a)]	yes	tracer, activation product
	^{140}Ba	13 d	β^-	yes	fission product
Radium	^{226}Ra	1600 y	α	yes	natural nuclide
	^{228}Ra	5.75 y	β^-	no	natural nuclide

a) Electron capture

(Table 7.2) while the hydrated ion size increases in the reverse order. Alkaline earth metals hydrolyze more readily than alkali metals. Beryllium ions form a hydroxide at a pH as low as 5–6, and this hydroxide is amphoteric. Magnesium forms a hydroxide at pH 9–10, but the hydroxides of the others alkaline earth metals are highly soluble. However, calcium can also be precipitated as a hydroxide in highly alkaline solutions, while the heavier alkaline earth metals do not precipitate as hydroxides. The alkaline earth metals form sparingly soluble carbonates in alkaline solution, though magnesium carbonate is clearly more soluble than the carbonates of the heavier alkaline earth metals. Beryllium and magnesium hydroxides are less soluble than their carbonates.

Separations of radionuclides of the alkaline earth metals $^{41,45}Ca$, $^{89,90}Sr$, and $^{226,228}Ra$ are typically carried out with hydroxides and carbonates or phosphates. Hydroxide precipitations are used to remove hydrolyzable metals (see Tables 4.1 and 4.2).

Table 7.2 Electronegativities and ionic radii of alkaline earth metals.

	Electronegativity (Pauling)	Crystal ionic radius (pm)
Beryllium	1.5	31
Magnesium	1.2	78
Calcium	1.0	106
Strontium	1.0	127
Barium	0.9	143
Radium	0.9	157

Table 7.3 Solubility products of common compounds of alkaline earth metals. Radium behaves much like barium.

	Magnesium	Calcium	Strontium	Barium
Carbonate – MCO_3	3.5×10^{-8}	3.8×10^{-9}	1.1×10^{-10}	5.1×10^{-9}
Hydroxide – $M(OH)_2$	1.8×10^{-11}	5.5×10^{-6}	soluble	soluble
Oxalate – MC_2O_4	7×10^{-7}	2.0×10^{-8}	4×10^{-7}	2.3×10^{-8}
Sulfate – MSO_4	soluble	9.1×10^{-6}	3.2×10^{-7}	1.1×10^{-10}
Chromate – $MCrO_4$	soluble	7.1×10^{-4}	2.2×10^{-5}	2.2×10^{-10}
Phosphate – $M_3(PO_4)_2$	1×10^{-25}	1×10^{-26}	1×10^{-31}	3×10^{-23}

Most metal ions precipitate as hydroxides, while $^{41,45}Ca$, $^{89,90}Sr$, and $^{226,228}Ra$ remain in solution together with alkali metals. To remove alkali metals and anionic components, carbonate or phosphate precipitation is carried out: $^{41,45}Ca$, $^{89,90}Sr$ and $^{226,228}Ra$ precipitate as carbonates, but alkali metals and anionic components do not. A few special applications of precipitation used in radiochemical analyses of $^{41,45}Ca$, $^{89,90}Sr$, and $^{226,228}Ra$ also exist. First, barium and radium can be separated from calcium and strontium by chromate precipitation: barium and radium chromates are far less soluble than those of calcium and strontium, and this precipitation, when carried out at pH 4, can achieve a nearly complete separation. Second, a special precipitation makes use of the different solubilities of calcium and strontium nitrates in 70% nitric acid (see Table 7.6): since strontium is much less soluble in these circumstances it can be removed from calcium by repeating the precipitation two to three times. The third special precipitation is that of sulfates in strong sulfuric acid: barium and radium sulfates are much less soluble than those of calcium and strontium. The fourth special precipitation makes use of the precipitation of calcium as hydroxide in highly alkaline solutions, the heavier alkaline earth metals, strontium, barium, and radium being removed since their hydroxides are completely soluble. Table 7.3 gives the solubility products of common compounds of alkaline earth metals.

Like the alkali metals, the alkaline earth metals do not readily form coordination complexes with organic compounds because no vacancies occur in their electron shells with which to form coordination bonds. Thus, solvent extraction is seldom used for chemical separation. Crown ethers, which form a strong complex with strontium and a relatively strong complex with radium, are an exception. Conversely, solvent extraction is used to remove interfering compounds from samples being measured for strontium. For example, all tri- and tetravalent ions can be removed by extraction with HDEHP (di-2-ethylhexylphosphoric acid). As these move into the organic phase, strontium and other di- and monovalent metals remain in the aqueous phase. Solvent extraction is also used in indirect measurement of ^{90}Sr. When the accumulation of ^{90}Y, the daughter nuclide of ^{90}Sr, reaches equilibrium with ^{90}Sr, the ^{90}Y is extracted into tributylphosphate or HDEHP and its activity is measured.

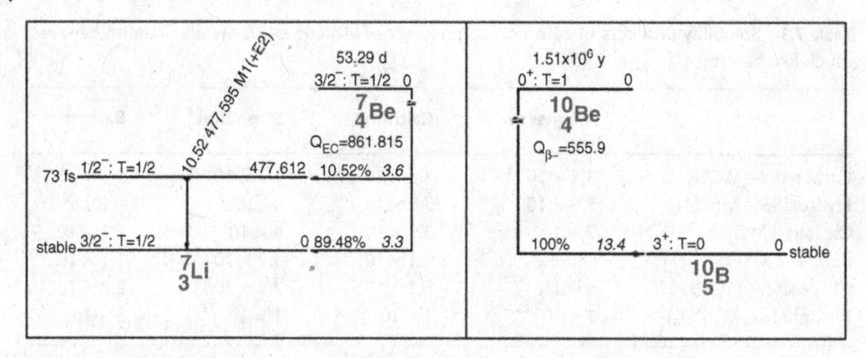

Figure 7.1 Decay schemes of ^{7}Be and ^{10}Be (Firestone, R.B., Shirley, V.S., Chu, S.Y.F., Baglin, C.M., and Zipkin, J. (1996) *Table of Isotopes*, Wiley-Interscience).

7.3
Beryllium – ^{7}Be and ^{10}Be

^{7}Be and ^{10}Be (Figure 7.1) are radionuclides of cosmogenic origin, which arrive at the Earth's surface adsorbed on aerosol particles. The half-life of ^{7}Be is 53 days, and this radionuclide can be measured, for example, on air filters with a gamma spectrometer (gamma energy 478 keV) without radiochemical separations. Long-lived ^{10}Be ($t_{1/2} = 1.6 \times 10^6$ y), which is a pure beta emitter, is more difficult to detect, and an accelerator mass spectrometer (AMS) is often used to measure it.

7.4
Calcium – ^{41}Ca and ^{45}Ca

7.4.1
Nuclear Characteristics and Measurement

^{45}Ca is a pure beta emitter, with a maximum energy of 0.257 MeV and a half-life of 163 days. It is widely used as a calcium tracer in biomedical studies. ^{41}Ca, a long-lived radionuclide ($t_{1/2} = 1.03 \times 10^5$ y), decays by electron capture without any gamma emission. Both ^{41}Ca and ^{45}Ca appear in nuclear waste, especially in concrete used as biological shielding in reactor buildings, as neutron activation products of stable ^{40}Ca and ^{44}Ca isotopes. Because of the short half-life of ^{45}Ca, its concentration in reactor concrete is low during decommissioning after some years cooling of the reactor materials. The ^{41}Ca level in concrete is of major interest because of its long half-life and relatively high mobility in the environment. The specific activity of ^{41}Ca is, however, very low. A further difficulty in measuring it radiometrically is that it decays by electron capture, and emits only low energy X-rays ($k_{\alpha 1,2}$ 3.3 keV, total intensity 11.4%) and Auger electrons (3.0 keV, intensity 77%), which are not easily detected by LSC. ^{41}Ca can also be measured by mass spectrometry, for example, by AMS. The decay schemes of ^{41}Ca and ^{45}Ca are shown in Figure 7.2.

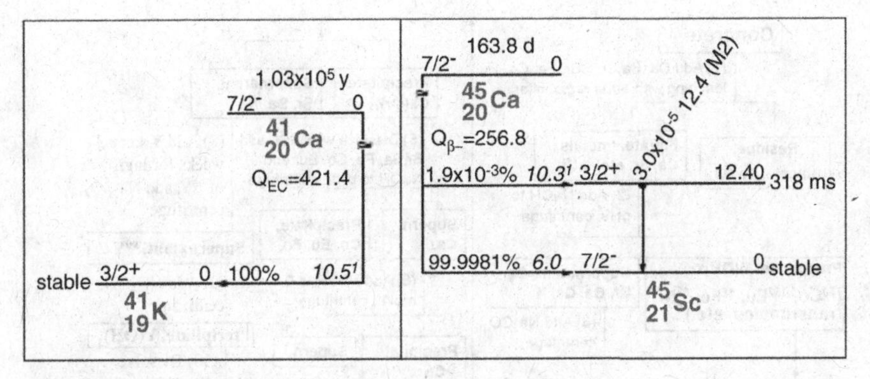

Figure 7.2 Decay schemes of ^{41}Ca and ^{45}Ca (Firestone, R.B., Shirley, V.S., Chu, S.Y.F., Baglin, C.M., and Zipkin, J. (1996) *Table of Isotopes*, Wiley-Interscience).

7.4.2
Determination of ^{45}Ca and ^{41}Ca in Concrete

To remove interfering radionuclides and matrix elements in the determination of calcium isotopes, several successive precipitations are typically used. As previously mentioned, carbonate or phosphate precipitations are used to remove alkali metals and anionic components (calcium is precipitated), hydroxide precipitation at pH 6–9 to remove hydrolyzable metals (calcium remains in solution), and hydroxide precipitation in a highly alkaline solution to remove strontium, barium, and radium (calcium is precipitated). In the following two methods, calcium isotope separation from concrete is described.

One method of separating Ca from concrete for the determination of ^{41}Ca is based on the final precipitation as $Ca(OH)_2$ in NaOH solution, shown in Figure 7.3. Two alternative methods to release Ca from a concrete sample can be used, alkali fusion and acid leaching: alkali fusion dissolves all the calcium while leaching with *aqua regia* releases more than 95% from the sample. Calcium is separated from interfering radionuclides in three main steps. First, Ca is separated from the transition metals, such as Co, Eu, Fe, Ni, and transuranics by hydroxide precipitation at pH 9 by adding NaOH. Interfering radioisotopes of these elements are precipitated, while Ca remains in the solution with Sr, Ba, Ra, and the alkali metals. This step is repeated three times for a good separation. Ca in the supernatant is then precipitated as carbonate by adding Na_2CO_3. Alkali metals and non-metallic elements remain in the supernatant, whereas Sr, Ba, and Ra pass to the precipitate. The carbonate precipitate is then dissolved in a dilute acid, and Ca is finally separated from Sr, Ba, and Ra by precipitation as $Ca(OH)_2$ in a 0.5 M NaOH solution. In these circumstances, calcium forms a hydroxide, but strontium, barium, and radium hydroxides are completely soluble. This step is repeated and the separated $Ca(OH)_2$ is dissolved in HCl for the activity measurement using liquid scintillation counting (LSC) after neutralizing to pH 6–8. The chemical yield of ^{41}Ca is monitored by the measurement of Ca before and after chemical separation using ICP-OES or ICP-MS. The measured chemical

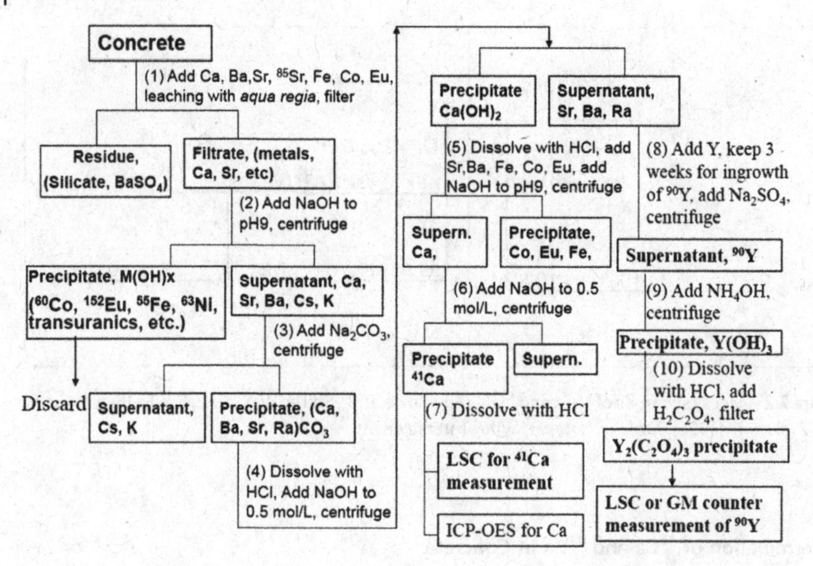

Figure 7.3 Separation procedure for the determination of ^{41}Ca in concrete (Hou, X. (2005) Radiochemical determination of ^{41}Ca in nuclear reactor waste. *Radiochim. Acta*, **93**, 611).

yields for ^{41}Ca is on average 85%. The decontamination factors for the interfering radionuclides such as ^{60}Co, ^{152}Eu, ^{133}Ba, ^{85}Sr, ^{137}Cs, ^{55}Fe, and ^{63}Ni are higher than 10^5. This method could naturally not separate ^{45}Ca from ^{41}Ca. If both isotopes exist in the sample, interference of the ^{45}Ca on the ^{41}Ca beta spectrum has to be corrected. This can be accomplished by determining the contribution of ^{45}Ca pulses on the ^{41}Ca window at lower channels where the pulses of the Auger electrons of ^{41}Ca are recorded.

Another method for the determination of ^{45}Ca and the estimation ^{41}Ca activity in reactor concrete is given in Figure 7.4. The concrete is extracted with 8 M HCl, and calcium, including ^{45}Ca, partially dissolves out of the sample. Not all the calcium needs to be dissolved since the total concentration of calcium in the concrete is separately determined by X-ray fluorescence spectroscopy. The concentration of calcium in the acid leachate is measured by AAS or ICP-MS to obtain the percentage of dissolved calcium. Scavenging with ferric hydroxide is carried out to remove most of the hydrolyzable metals, including their radionuclides (e.g., ^{60}Co, ^{152}Eu, ^{63}Ni and ^{59}Fe), that are present in the concrete. In the subsequent carbonate precipitation, alkali metals which do not form sparingly soluble carbonates and remain in solution are removed. In this step, cesium isotopes ^{134}Cs and ^{137}Cs are removed. The carbonate precipitate is dissolved in 8 M HCl, and anion exchange separation is carried out on a strongly basic anion exchange column. Calcium does not adsorb to the resin, while ^{60}Co, ^{59}Fe, ^{51}Cr, and ^{65}Zn, are adsorbed and retained in the column. The effluent is evaporated to dryness and dissolved in 8 M HNO_3, after which a further anion exchange is carried out to remove the lanthanides, such as ^{141}Ce. A final purification of the calcium-bearing effluent is done with ferric hydroxide as a

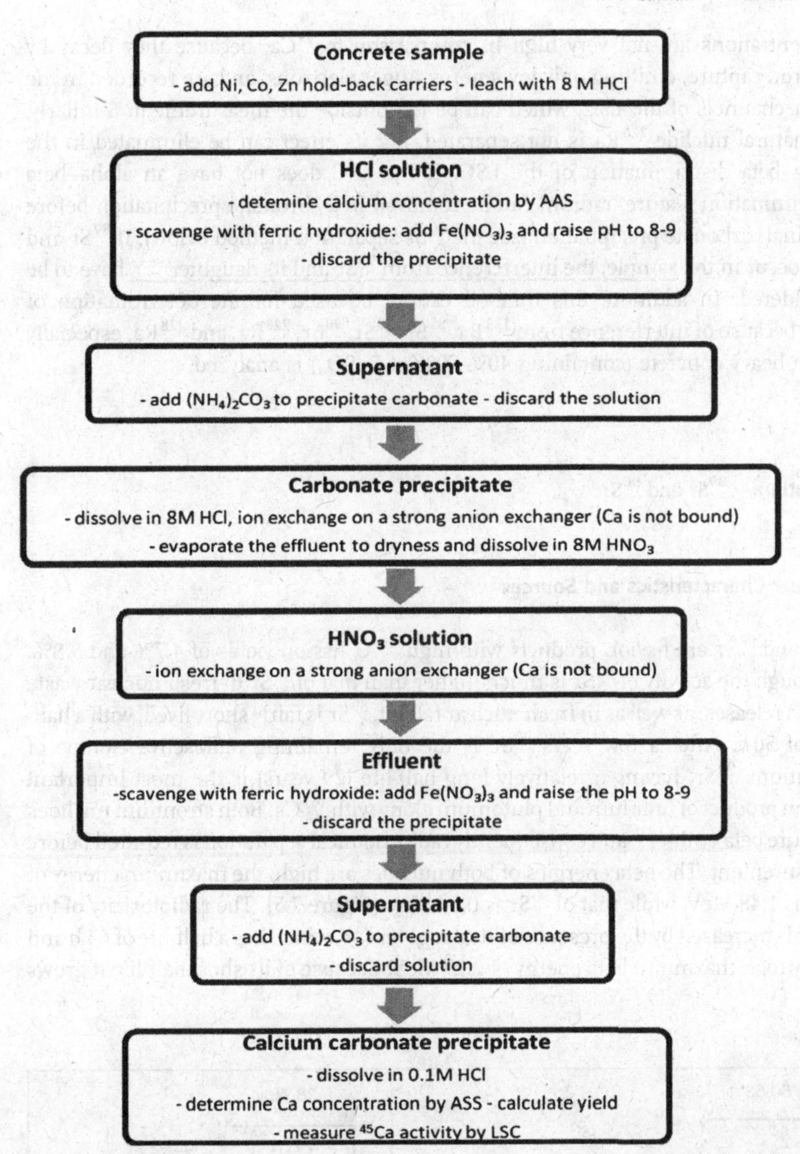

Figure 7.4 Determination of ^{45}Ca in concrete (Ervanne, H., Hakanen, M., Lehto, J., Kvarnström, R., and Eurajoki, T. (2009) Determination of ^{45}Ca and gamma-emitting radionuclides in concrete from a nuclear power plant. *Radiochim. Acta*, **97**, 631).

scavenger, and the calcium that is retrieved is precipitated as carbonate. The carbonate precipitate is dissolved in a weak acid solution, which is used to measure the activity of ^{45}Ca with an LSC and to determine stable calcium by AAS for yield determination. Neither ^{85}Sr nor ^{133}Ba is removed by this separation method, but neither of these isotopes interferes with the measurement of ^{45}Ca activity when their concentrations are not very high in comparison to ^{45}Ca, because they decay by electron capture, emitting only low-energy Auger electrons, and are recorded in the lower channels of the LSC, which can be left outside the measurement. Similarly, the natural nuclide ^{226}Ra is not separated, but its effect can be eliminated in the alpha–beta discrimination of the LSC. If the LSC does not have an alpha–beta discrimination feature, radium can be removed in a chromate precipitation before the final carbonate precipitation (see the ^{90}Sr separation method below). If ^{89}Sr and ^{90}Sr occur in the sample, the interference from ^{90}Sr and its daughter ^{90}Y have to be considered. In addition, this method cannot be used for the determination of ^{41}Ca because of interference from ^{133}Ba, ^{85}Sr, ^{89}Sr, ^{90}Sr, ^{226}Ra, and ^{228}Ra, especially when heavy concrete (containing 40%-70% of $BaSO_4$) is analyzed.

7.5 Strontium – ^{89}Sr and ^{90}Sr

7.5.1 Nuclear Characteristics and Sources

^{89}Sr and ^{90}Sr are fission products with high ^{235}U fission yields of 4.7% and 5.8%. Although the activity of ^{89}Sr is much higher than that of ^{90}Sr in fresh nuclear waste and in releases, as well as in fresh nuclear fallout, ^{89}Sr is fairly short-lived, with a half-life of 50 d. After a few years ^{90}Sr is the only remaining radioactive isotope of strontium. ^{90}Sr, having a relatively long half-life (29 years) is the most important fission product of uranium and plutonium along with ^{137}Cs. Both strontium nuclides are pure beta emitters, and consequently radiochemical separation is required before measurement. The beta energies of both nuclides are high: the maximum energy of ^{89}Sr is 1.48 MeV, while that of ^{90}Sr is 0.54 MeV (Figure 7.5). The radiotoxicity of the latter is increased by the presence of its daughter ^{90}Y, which has a half-life of 64 h and a very high maximum beta energy of 2.27 MeV. Because of its short half-life it grows

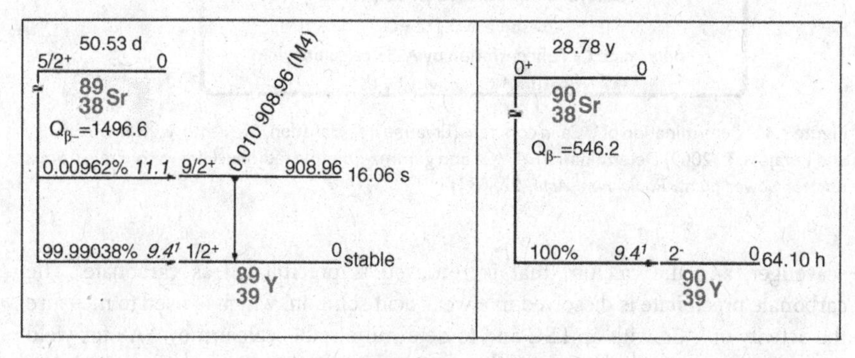

Figure 7.5 Decay schemes of ^{89}Sr and ^{90}Sr (Firestone, R.B., Shirley, V.S., Chu, S.Y.F., Baglin, C.M., and Zipkin, J. (1996) *Table of Isotopes*, Wiley-Interscience).

Table 7.4 Nuclear characteristics of important radioisotopes of strontium and yttrium.

	Decay mode	Half-life	Radiation energy (MeV)	^{235}U fission yield (%)
^{90}Sr	beta	29 y	0.54 (E_{max})	5.8
^{89}Sr	beta	50 d	1.48 (E_{max})	4.8
^{85}Sr	EC[a)]	65 d	0.514 γ (intensity 96%)	(tracer)
^{90}Y	beta	64 h	2.27 (E_{max})	5.5

a) Electron capture

into the ^{90}Sr-bearing sample in a fairly short time; complete ingrowth essentially takes less than a month. Table 7.4 presents nuclear characteristics of major radioisotopes of strontium and of ^{90}Y.

In nuclear fuel, ^{90}Sr occurs at approximately the same level as ^{137}Cs, and together they provide the great majority of the total radioactivity in the first few hundred years after fuel removal from a reactor. They decay away in less than one thousand years, and thus form no major radiation hazard in the final disposal of spent nuclear fuel, since the technical barriers surrounding the spent fuel will prevent their release from the fuel matrix. Nuclear power plant waste, that is, spent organic resins used to purify the primary coolant water, waste effluents, and other low and medium active waste, contains a small percentage of ^{90}Sr compared to ^{137}Cs since the extent of release from the fuel of the former is much lower because of its lower volatility and solubility. In global fallout from nuclear weapons tests, the ^{90}Sr deposition was somewhat lower (about 30%) than that of ^{137}Cs, while in the Chernobyl fallout the fraction of ^{90}Sr was much lower, roughly in the range 1–10%. In total the amount of ^{90}Sr in the Chernobyl release was only a little more than 1% of the global weapons test fallout.

7.5.2 Measurement of Strontium Isotopes

Strontium isotopes are measured by either LSC or a gas ionization detector. The beta energies of both strontium isotopes are so high that their measurement by the latter method, not to mention the former, does not cause a big problem due to self-absorption in the sample itself nor in the window of the gas ionization detector. Even though the counting efficiency obtained by measuring the activity with a gas ionization detector is clearly lower, it enables measurement of lower activities due to lower background.

7.5.2.1 Measurement of ^{90}Sr Activity

Several alternatives exist for measuring ^{90}Sr activity:

- Measurement of the ^{90}Sr beta spectrum after the separation procedure. The measurement should be carried out immediately after separation to prevent the

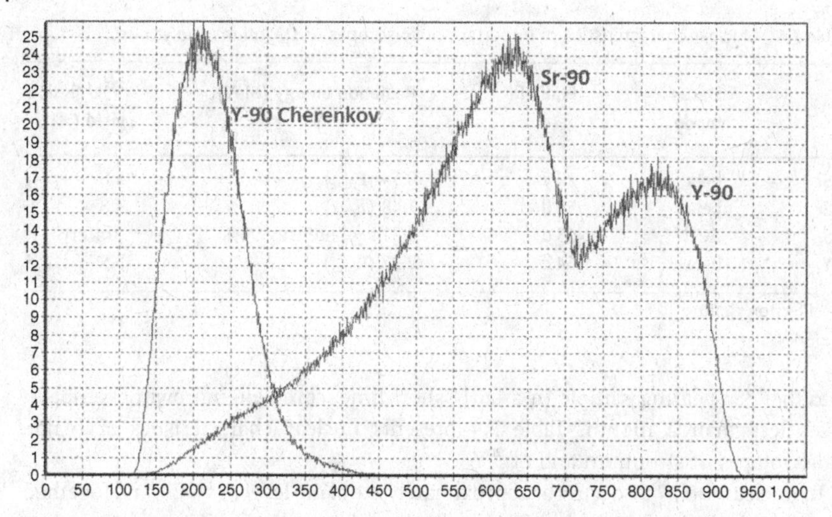

Figure 7.6 Spectra of ^{90}Sr, and ^{90}Y measured by LSC and Cherenkov counting.

ingrowth of ^{90}Y. If several samples need to be analyzed, immediate measurement is probably not feasible because of the rapid ingrowth of ^{90}Y. This method is only applicable for single samples and in the case of a need for rapid information on ^{90}Sr activity levels.

- Measurement of the total beta spectrum of ^{90}Sr and ^{90}Y after ingrowth of ^{90}Y in three weeks (Figure 7.6) during which time the ^{90}Y ingrowth percentage has reached 99.6%. This method has the advantage of an improved counting efficiency, since pulses from both ^{90}Sr and ^{90}Y are recorded (Table 7.5).
- Measurement of the ^{90}Y beta spectrum after ingrowth of ^{90}Y. This method requires separation of ^{90}Y from ^{90}Sr after the ingrowth, accomplished by the removal of ^{90}Sr as a sulfate precipitate. It is not necessary to wait for the total ingrowth of ^{90}Y. The ingrowth percentage can be calculated from the time elapsed since the removal of ^{90}Y from ^{90}Sr.
- Measurement of ^{90}Y by Cherenkov counting after ingrowth of ^{90}Y (Figure 7.6). In this method the removal of ^{90}Sr from the counting sample is not

Table 7.5 Counting efficiencies for ^{90}Sr and ^{90}Y measured individually and together by LSC and Cherenkov counting.

Nuclides measured	Counting efficiency, %	
	LSC	**Cherenkov counting**
^{90}Sr	77	1
$^{90}Sr + ^{90}Y$	168	57
^{90}Y	91	57

necessary since the Cherenkov counting efficiency is only 1.4% for ^{90}Sr whereas it is as high as 57% for ^{90}Y due to its high beta energy. Thus the fraction of the extra pulses caused by the presence of ^{90}Sr in the counting sample is only 2%. As in the previous method it is not necessary to wait for the total ingrowth.

7.5.2.2 Simultaneous Determination of ^{89}Sr and ^{90}Sr

If a nuclear fallout or nuclear waste sample is fresh, and short-lived ^{89}Sr is still present, the activity can be measured in two ways. First, the total activity of strontium ($^{90}Sr + ^{89}Sr$) is measured by LSC immediately after the separation procedure. ^{90}Y is then allowed to grow in, and ^{90}Sr and ^{89}Sr are removed by sulfate precipitation, allowing the ^{90}Y activity to be measured by LSC or Cherenkov counting. Since the activity of ^{90}Y in equilibrium is the same as that of ^{90}Sr it can be subtracted from the total activity of radioactive strontium measured in the first step, so obtaining the activity of ^{89}Sr. The second method uses the difference of the Cherenkov counting efficiencies between ^{89}Sr and ^{90}Sr. Because of the high beta energy (1.49 MeV), the former has a rather high Cherenkov counting efficiency of 40%, while the efficiency for the latter is low, 1.4%. Thus the ^{89}Sr activity can be measured by Cherenkov counting after yttrium removal, with a nominal error caused by ^{90}Sr presence. It is important, of course, that the measurement is carried out as soon as possible after yttrium removal so that ingrowth of ^{90}Y does not produce an error in the result. After Cherenkov counting of ^{89}Sr, ^{90}Y is allowed to grow in and the ^{90}Sr activity is calculated with the aid of ingrown ^{90}Y.

7.5.3 Radiochemical Separations of ^{90}Sr and ^{89}Sr

Below, we present two types of radiochemical methods typical for the separation of radioactive strontium. The first method is based on several precipitations, and the second involves the separation of strontium by extraction chromatography using a crown ether-containing extraction agent. The latter method has increasingly replaced the former. In all these separations the most challenging task is the removal of calcium, the closest chemical element to strontium.

In general, strontium can be removed from hydrolyzing metals by hydroxide precipitation (strontium remains in solution) and from alkali metals and anionic components by carbonate, oxalate, and phosphate precipitations (strontium is precipitated). An additional major challenge is separating it from calcium. Calcium, an abundant element in all sample types, is chemically the closest element to strontium. Despite the fact that calcium does not usually have any radionuclides that need to be removed, the presence of stable calcium interferes with the separation of strontium. Large amounts of calcium are most noticeably present in milk. Calcium interferes with the yield determination of strontium and activity measurement of ^{90}Sr by LSC. It also interferes with the sorption of strontium to the extraction chromatographic resin. Calcium is removed from strontium in two ways:

- By nitrate precipitation in 70% nitric acid, in which strontium precipitates as $Sr(NO_3)_2$ almost completely whereas calcium only partly. To obtain a good separation the precipitation has to be repeated, depending on the calcium content of the sample.
- By scavenging calcium by $Ca(OH)_2$ precipitation in 0.2 M NaOH, in which strontium is soluble.

In the extraction chromatographic separation of ^{90}Sr, the main interfering element is potassium, which can be removed by carbonate, oxalate, and phosphate precipitations leaving potassium in the solution.

7.5.3.1 Determination of Chemical Yield in Radiostrontium Separations

Two alternative methods of yield determination for the measurement of strontium activity can be used. In the first method, a known amount of stable strontium carrier is added at the beginning, and the yield is determined as the amount of Sr present in the final solution after the separation procedure by AAS or ICP-MS or by gravimetry after precipitation of strontium as carbonate. If strontium is to be determined in soil or sediment, which contain considerable amounts of stable strontium, this stable strontium must be measured before radiochemical separation, after dissolution of the sample and before addition of the strontium carrier. The second method is to add a known amount of the gamma-emitting tracer isotope ^{85}Sr ($t_{1/2} = 65$ d, 514 keV, intensity 96%) at the start of the analysis and to measure the activity remaining in the solution after the final purification. The measurement is usually done by gamma counting with an Na(I) detector.

7.5.3.2 Separation of Radiostrontium by the Nitrate Precipitation Method

The classical method, developed in the 1940s and still is in use, of separating strontium is based on several precipitations, the most essential being the separation of strontium from calcium in a nitrate precipitation. The method, called the nitrate method, is reliable and repeatable but very laborious and time consuming. It is also somewhat dangerous because it calls for the use of fuming nitric acid. Figure 7.7 presents a scheme for the separation of ^{90}Sr from lichen that has been decomposed by ashing in an oven. The ash is dissolved in strong nitric acid, and stable strontium and barium are added as carrier and hold-back carrier. After filtration to remove the undissolved residue, fuming (100%) nitric acid is added until the concentration reaches 70%. At this nitrate concentration, nitrates of other metals are soluble, while those of Sr, Ba, and Ra are insoluble, and consequently precipitate. Strontium nitrate is far less soluble than calcium nitrate, but part of the calcium also precipitates with the $Sr(NO_3)_2$, making one precipitation insufficient to remove calcium, and a further precipitation or two must be performed. Table 7.6 shows the percentages of Ca, Sr, and Ba precipitated in different concentrations of nitric acid. It can be seen that the optimal concentration of nitric acid is 70%. At this level almost all of the Sr precipitates, while 90% of the calcium is retained.

The precipitate containing strontium and barium nitrates is dissolved in water, and a carbonate precipitation is carried out by adding $(NH_4)_2CO_3$ to reduce the volume.

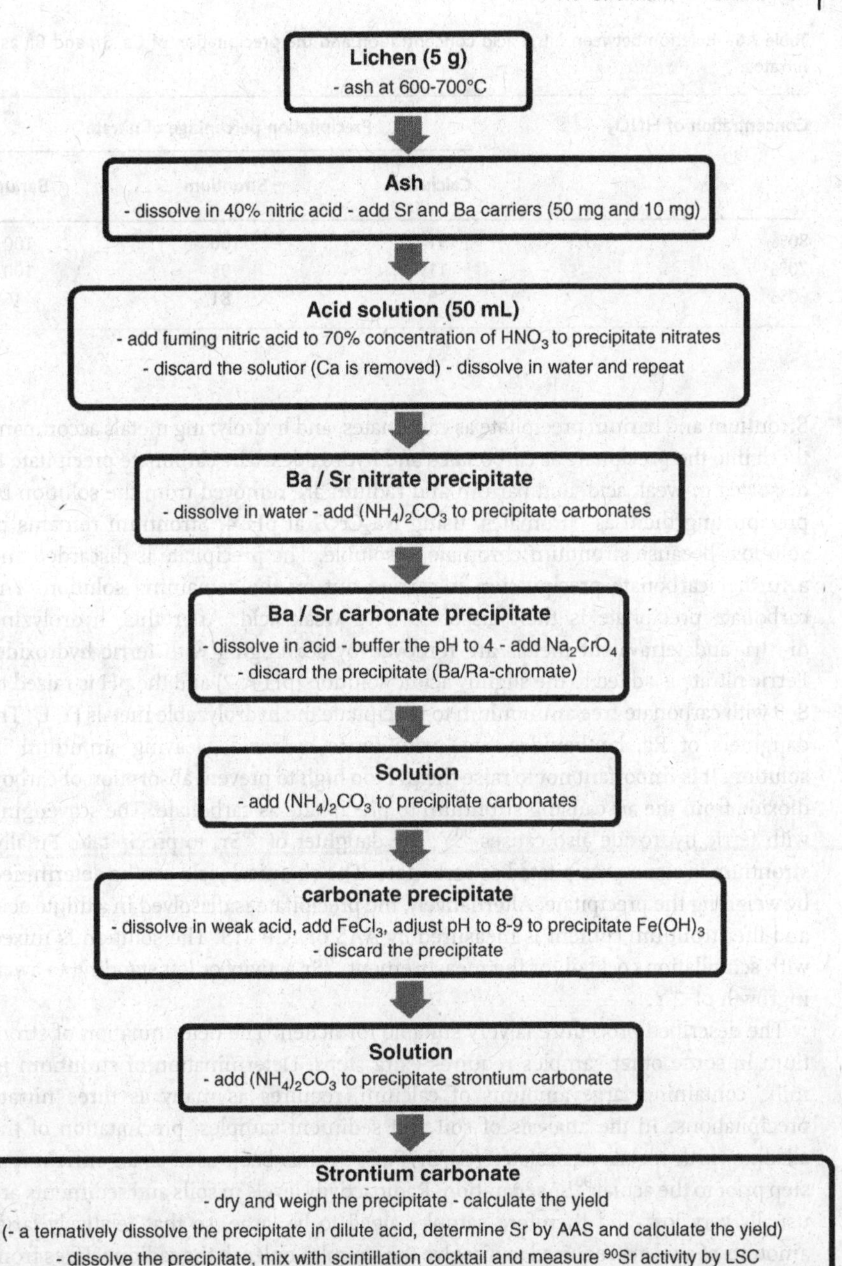

Figure 7.7 Determination of ^{90}Sr in lichen by the nitrate precipitation method.

Table 7.6 Relation between nitric acid concentration and the precipitation of Ca, Sr, and Ba as nitrates.

Concentration of HNO_3	Precipitation percentage of nitrate		
	Calcium	Strontium	Barium
80%	51	100	100
70%	11	98	100
60%	3	81	86

Strontium and barium precipitate as carbonates, and hydrolyzing metals accompany them into the precipitate as carbonates and hydroxides. The carbonate precipitate is dissolved in weak acid, and barium and radium are removed from the solution by precipitating them as chromates, using Na_2CrO_4 at pH 4. Strontium remains in solution, because strontium chromate is soluble. The precipitate is discarded and a further carbonate precipitation is carried out on the remaining solution. The carbonate precipitate is then dissolved in a weak acid. After this, hydrolyzing di-, tri- and tetravalent metals are removed by scavenging with ferric hydroxide. Ferric nitrate is added to the slightly acidic solution ($pH < 2$) and the pH is raised to 8–9 with carbonate-free ammonium to precipitate the hydrolyzable metals (Y, U, Th, daughters of Ra, lanthanides, etc.) with ferric hydroxide, leaving strontium in solution. It is important not to raise the pH too high to prevent absorption of carbon dioxide from the air causing strontium to precipitate as carbonate. The scavenging with ferric hydroxide also causes ^{90}Y, the daughter of ^{90}Sr, to precipitate. Finally, strontium is again precipitated as carbonate. The chemical yield can be determined by weighing the precipitate. Alternatively, the precipitate is dissolved in a dilute acid and the strontium content is measured by AAS or ICP-MS. The solution is mixed with scintillation cocktail for the measurement ^{90}Sr activity or left standing to await ingrowth of ^{90}Y.

The described procedure is very suitable for lichen. The determination of strontium in some other samples requires extra steps. Determination of strontium in milk, containing large amounts of calcium, requires as many as three nitrate precipitations. In the analysis of soil and sediment samples, precipitation of the alkaline earth metals as oxalates ($Ca/Sr/BaC_2O_4$) has been used as an enrichment step prior to the actual ^{90}Sr separation. Radioactivity levels in soils and sediments are usually very low, and therefore samples need to be large, so that relatively large amounts of acid are required to dissolve the samples or leach the radionuclides from them. Hence, enrichment of radionuclides from these large acid volumes is necessary. Oxalate precipitation is carried out by adding oxalic acid and ammonium acetate and adjusting the pH to 1.5. Besides concentrating the alkaline earth metals, the oxalate precipitation also removes iron, always found in considerable amounts in soils and sediments. Trivalent iron remains in solution because it does not form a sparingly soluble oxalate, and the pH is so low that $Fe(OH)_3$ does not precipitate either. The remaining iron, not removed in this step, is finally removed in the ferric hydroxide scavenging.

Determinations of strontium in waters are simpler than those in solid samples, such as soil and sediments, because interfering metals are less abundant. Again, the most challenging interfering metal is calcium, which is always present in natural waters, and this is removed with two successive nitrate precipitations. The difficulty with determinations in aqueous solutions is that the activity concentrations of ^{90}Sr are very low, and to obtain enough ^{90}Sr to exceed to detection limit the sample must be preconcentrated. Environmental water samples must always have a volume of at least 10 L, and a preconcentration is needed as the first step. This can be done by evaporation or carbonate precipitation.

7.5.3.3 Separation of Radiostrontium by a $Ca(OH)_2$ Precipitation Method

Because of the risk of operating with fuming nitric acid and the high cost of this reagent, as well as the difficulties associated with the precipitation of chromate due to the need for critical control of the pH, an alternative precipitation method has been developed. Several types of precipitations are applied in this method, but here it is called the calcium hydroxide separation method since the separation of strontium from calcium is accomplished by calcium hydroxide precipitation. In this procedure, Ca is precipitated in a NaOH solution at a concentration of 0.2–0.5 M. At this concentration, more than 96% of Ca is precipitated as $Ca(OH)_2$, while more than 99% of Sr remains in the solution. This procedure is very suitable for the determination of ^{90}Sr in high calcium-bearing samples, such as seawater, milk, and soil. Figure 7.8 shows the separation procedure for determination of ^{90}Sr in a sample comprising a large volume of seawater. An ^{85}Sr tracer and Sr carrier are

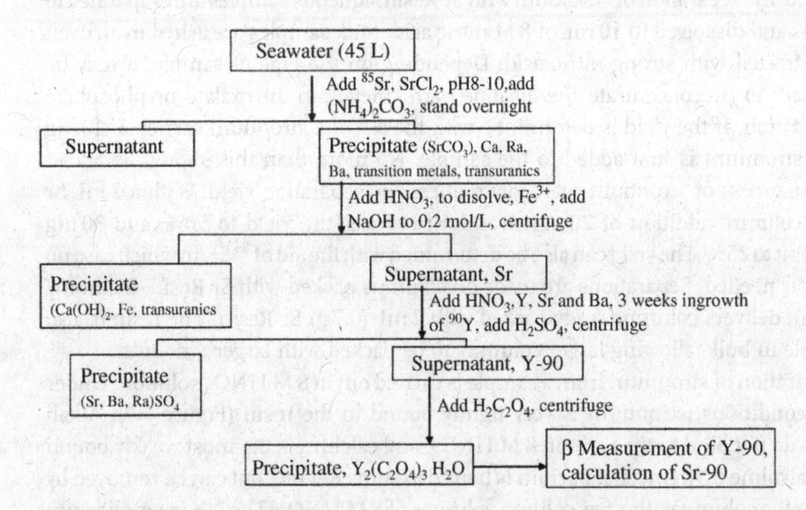

Figure 7.8 Separation procedure for the determination of ^{90}Sr in sea water (Chen, Q.J., Hou, X.L., Yu, Y.X., Dahlgaard, H., and Nielsen, S.P. (2002) Separation of Sr from Ca, Ba and Ra by means of $Ca(OH)_2$ and $Ba(Ra)Cl_2$ or $Ba(Ra)SO_4$ precipitation for the determination of radiostrontium. *Anal. Chim. Acta*, **466**, 109–116).

first added to the water sample, and the pH is adjusted to 8–9. The water is then heated to boiling, $(NH_4)_2CO_3$ is added, the solution is allowed to stand for 5–12 h, and the carbonate precipitate is separated by siphoning off the supernatant. The separated precipitate is dissolved in dilute HNO_3, and $FeCl_3$ is added to the solution. Dilute NaOH is then added to adjust the pH to 8–9, after which 6 M NaOH is added until an NaOH concentration of 0.2 M is reached. The resulting hydroxide precipitate containing calcium and transition metals is separated by centrifugation. To the supernatant, Na_2CO_3 is added, and the solution is heated and stirred for 1 h. After cooling to room temperature, the carbonate precipitate is separated by centrifugation. The precipitate is dissolved in HNO_3, and the chemical yield of Sr is measured by counting ^{85}Sr with a gamma counter. A Y^{3+} carrier is then added to the solution. After three weeks of ^{90}Y ingrowth, H_2SO_4 is added to precipitate Ba, Sr, and Ra sulfates, which are then separated by centrifugation. To the supernatant, $H_2C_2O_4$ is added to precipitate $Y_2(C_2O_4)_3$, and the solution is again stirred and heated. The precipitate is separated by filtration on a filter paper. The chemical yield of ^{90}Y is measured by a gravimetric method, and the activity of ^{90}Y is measured by beta counting. Finally, ^{90}Sr activity is calculated from the measured ^{90}Y activity, the chemical yields of Sr and Y, and the ingrowth time after separation of ^{90}Y and the decay time of ^{90}Y after separation of ^{90}Sr.

7.5.3.4 Separation of Radiostrontium by Extraction Chromatography

As described in Chapter 4 (Separation Methods), extraction chromatography is carried out with organic extraction reagents fixed to a porous organic polymer or silica that act as inert supports. In Eichrom/Triskem Sr Resin, polymer pores are filled with 18-crown-6 ether dissolved in 1-octanol (see Figure 4.9).

Before the separation of strontium with Sr Resin, aqueous samples are evaporated to dryness and dissolved in 10 mL of 8 M nitric acid. Milk samples are ashed in an oven and extracted with strong nitric acid. Depending on the type of sample, it may be necessary to preconcentrate the alkaline earth metals in an oxalate or phosphate precipitation. If the yield is determined with the aid of a strontium carrier, 5–10 mg stable strontium is first added to the sample. No more than this should be added because excess of strontium markedly reduces the separation yield. With a 2 mL Sr Resin column, addition of 20 mg strontium reduces the yield to 50%, and 30 mg reduces it to 25%. The yield can also be determined with the aid of ^{85}Sr, in which case no carrier is needed. Separations are made on columns packed with Sr Resin. Eichrom/Triskem delivers columns ready packed with 2 mL (0.7 g) Sr Resin. The resin is also available in bulk, allowing larger columns to be packed with larger amounts.

Separation of strontium from a sample is carried out in 8 M HNO_3 solution. Under these conditions, strontium is very tightly bound to the resin (Figure 7.9). Alkali metals do not bind to the resin in 8 M HNO_3, and calcium is the most weakly bound of the alkaline earth metals. Barium is bound to some degree, but can be removed by rinsing the column with a few column volumes of 8 M HNO_3. The ^{90}Y in equilibrium with ^{90}Sr is also eluted with 8 M HNO_3, and the time of the elution should be noted. ^{90}Sr is eluted from the column using a low concentration of acid or water. The activity of ^{90}Sr in the eluate is then measured by LSC, and the chemical yield of Sr is

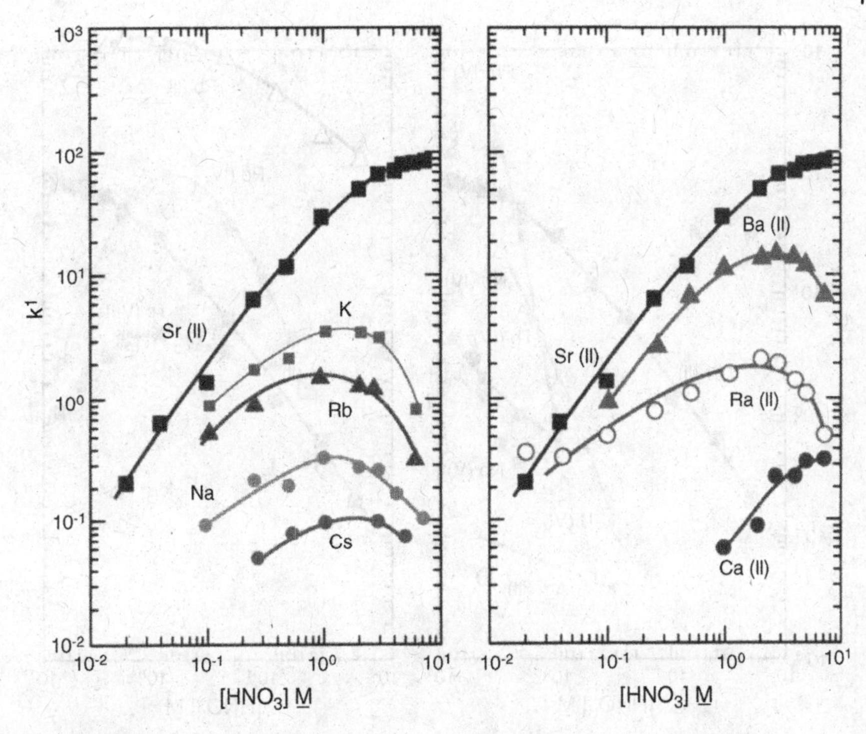

Figure 7.9 Binding of alkali and alkaline earth metals to Sr Resin as a function of nitric acid concentration. *k'* values are column volumes in which the metal achieves maximum value in the eluate when the column is eluted with acid of specified concentration (Horwitz, E.P., Chiarazia, R., and Sietz, M.L. (1993) A novel strontium-selective extraction chromatographic resin. *Solvent Extr. Ion. Exc.*, **10**, 313).

determined. If the amount of strontium is measured gravimetrically, strontium should first be precipitated as carbonate. The carbonate is then dissolved in a weak acid for activity measurement by LSC.

In addition to ^{90}Sr, 18-crown-6 ether binds ^{210}Pb, forming an even a stronger complex with lead (Figure 7.10). ^{210}Pb does not, however, cause a problem in ^{90}Sr analysis since it is not eluted from the column with dilute acid or water. If the sample contains plutonium and neptunium in the oxidation state + IV, these will be strongly bound to the Sr Resin in the 8 M HNO_3 solution (Figure 7.10). Plutonium and neptunium are removed by rinsing the column with 3 M HNO_3–0.05 M oxalic acid. Under these conditions they form an oxalate complex and are eluted from the column.

After stable strontium, the major source of interference in ^{90}Sr separations is potassium. Potassium concentrations greater than 0.01 M reduce the uptake of ^{90}Sr on the Sr Resin dramatically. Moreover, the beta-active ^{40}K in potassium interferes with strontium determinations if it gets into the sample to be measured for ^{90}Sr activity. If potassium is present in large amounts, the alkaline earth metals should be preconcentrated in an oxalate, carbonate, or phosphate precipitation. Sodium and calcium interfere in a minor way, beginning to reduce the uptake of ^{90}Sr on the Sr

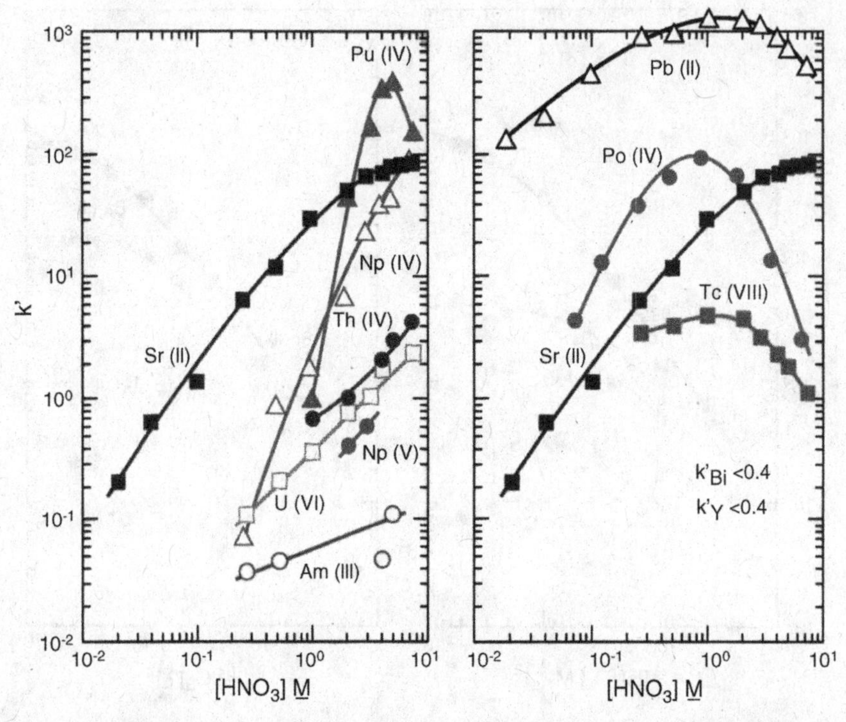

Figure 7.10 Sorption of Sr, Pb, Po, Tc and actinides to Sr Resin as a function of nitric acid concentration (Horwitz, E.P., Chiarazia, R., and Sietz, M.L. (1993) A novel strontium-selective extraction chromatographic resin, *Solvent Extr. Ion. Exc.*, **10**, 313).

Resin only if they are present in concentrations ten times greater than that of potassium. For the samples with large amounts of calcium, such as large seawater samples, soil, sediment, and milk, however, the preseparation of Ca and K from Sr is necessary before separation using Sr Resin.

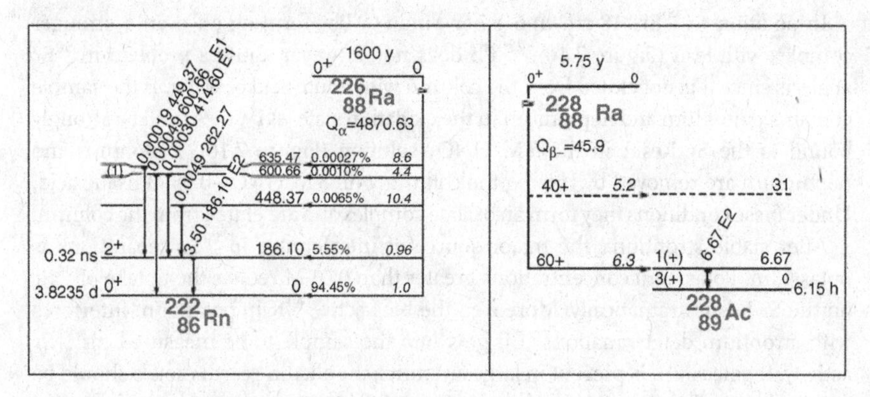

Figure 7.11 Decay schemes of ^{226}Ra and ^{228}Ra (Firestone, R.B., Shirley, V.S., Chu, S.Y.F., Baglin, C.M., and Zipkin, J. (1996) *Table of Isotopes*, Wiley-Interscience).

7.6 Radium – ^{226}Ra and ^{228}Ra

Radium holds a special place in the history of radiochemistry. Marie and Pierre Curie painstakingly separated a minute amount of radium from pitchblende in 1898, just a few months after the discovery of polonium. Radium was to become quite a fashionable substance in the early years of the twentieth century.

7.6.1 Nuclear Characteristics of Radium Isotopes

Radium exists in nature as four isotopes, formed in the decay of uranium and thorium (Table 7.7). In the decay series starting from ^{238}U, the most important radionuclide of radium is ^{226}Ra, which has a half-life of 1600 years. ^{226}Ra is formed in the alpha decay of ^{230}Th and further decays by alpha decay to ^{222}Rn, mostly to its ground state (Figure 7.11). Gamma rays are also emitted in the decay, but the intensity of these is so low (5.6%) that gamma radiation cannot be utilized for direct measurement of ^{226}Ra in low active samples. A further difficulty in direct gamma measurement of ^{226}Ra is that the energy (186 keV) of the strongest gamma transition is the same as the almost equally intense peak of ^{235}U. ^{226}Ra is therefore usually determined either by measuring its alpha radiation or indirectly by measuring the alpha radiation of its progeny by LSC as described later. Measurement through its progeny requires the radium to be in radioactive equilibrium with its daughters. Another relatively long-lived isotope of radium, ^{228}Ra, is a daughter nuclide of ^{232}Th and has a half-life of 5.8 years. It is a pure beta emitter, the beta energy being very low, only 7.2 keV on average, making it difficult to measure.

7.6.2 Measurement of the Activity of Radium Isotopes

The activity of radium can be measured in two ways. In the first method, after radiochemical separation, radium is coprecipitated with a small amount of barium sulfate by adding 40% sodium sulfate and acetic acid to a solution containing 0.1 mg of barium. The Ba(Ra)SO_4 precipitate is collected on a filter, washed, dried,

Table 7.7 Natural isotopes of radium.

Isotope	Half-life	Decay mode	Radiation energy (MeV)	Decay series
^{223}Ra	11 d	alpha	5.716 (51.6%), 5.606 (25.2%)	actinium
^{224}Ra	3.6 d	alpha	5.685 (94.9%), 5.448 (5.1%)	thorium
^{226}Ra	1600 y	alpha	4.784 (94.45%), 4.601 (5.55%)	uranium
^{228}Ra	5.8 y	beta	46 keV (E_{max})	thorium

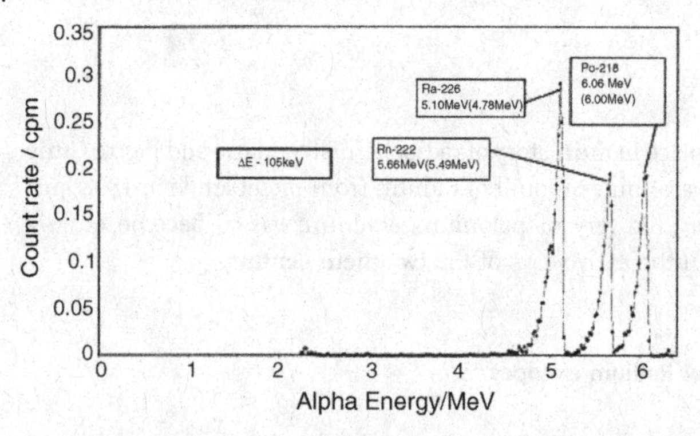

Figure 7.12 Alpha spectrum of ^{226}Ra and its daughters ^{222}Rn and ^{218}Po.

and glued onto a sample disk for measurement of the alpha spectrum of ^{226}Ra (Figure 7.12).

In the second method, the activity of ^{226}Ra is measured by counting the activity of its progeny radionuclide ^{214}Po or the sum of ^{226}Ra with all its alpha progeny (^{222}Rn, ^{218}Po and ^{214}Po) at equilibrium by LSC. The solution obtained from ^{226}Ra separation is tightly sealed in a Teflon-lined scintillation vial to prevent radon escape, and the ingrowth of ^{222}Rn and its progeny is awaited. One progeny nuclide, ^{214}Po, has a much higher alpha energy (7.687 MeV) than those of ^{218}Po, ^{222}Rn and ^{226}Ra, and it can be distinguished in the liquid scintillation spectrum (Figure 7.13). The measurement can commence after about one month, upon achievement of equilibrium. In addition, the activity of ^{133}Ba, as a tracer, can also be measured in the beta window of the LSC with an alpha-beta discrimination feature. ^{133}Ba, a gamma-emitting radionuclide, typically used as a yield determinant tracer in ^{226}Ra separation, also

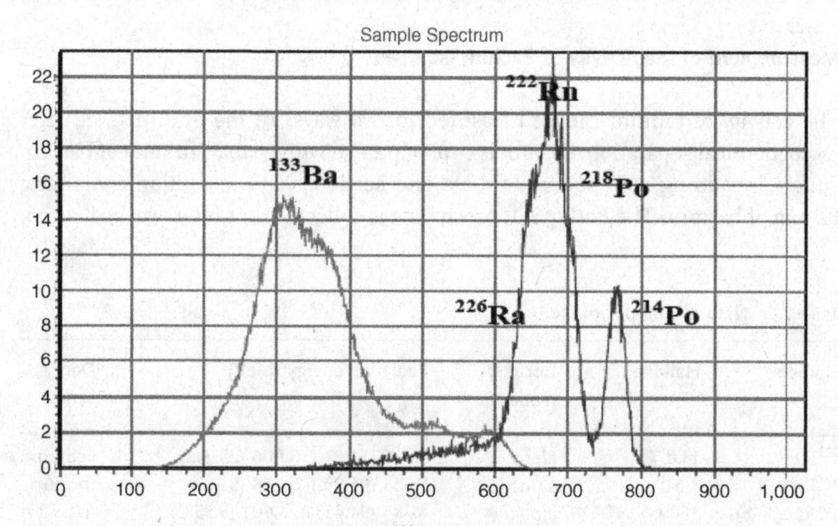

Figure 7.13 Liquid scintillation spectrum of ^{226}Ra and its daughters together with the conversion electron spectrum of ^{133}Ba used as a yield determination tracer.

emits conversion electrons, the spectrum of which can be determined with LSC (Figure 7.13).

Since the beta energy of ^{228}Ra is very low (7.2 keV on average), the measurement of activity by counting these beta emissions is very difficult. ^{228}Ra decays to short-lived ($t_{1/2} = 6$ h) ^{228}Ac, which is also a beta emitter but has a much higher average beta energy (380 keV). Measurement of this beta radiation by LSC and a gas ionization detector is the typical way to measure ^{228}Ra activity. A method to separate ^{228}Ac, in equilibrium with ^{228}Ra, is described later.

7.6.3 Need for Determining the Activity of Radium Isotopes

The need for radium separation mostly concerns geological samples: rocks and groundwater. In order to understand the transport and sorption of radionuclides in bedrock and soil, the tracing of radium routes into the human food chain and investigation of the disequilibria of radionuclides in the uranium and thorium decay series are important research objectives. All parent nuclides of radium isotopes are thorium isotopes. Radium is much more soluble than thorium. In almost all compounds of thorium, the thorium appears in the oxidation state +IV, and these are very insoluble. Radium always appears in oxidation state +II, and its compounds are relatively soluble. At the rock–water interface, the energy of recoil received by radium atoms during alpha decay may help release the radium into the water phase. Once in the water phase, as a divalent positive ion, it is not totally soluble, however, but adsorbs on rock surfaces through ion exchange or precipitation. The concentrations of radium in groundwater are so low that the solubility products of its compounds are not exceeded, but it can be coprecipitated with chemically similar elements, such as calcium and barium.

7.6.4 Radiochemical Separations of Radium

Coprecipitation, ion exchange, and solvent extraction are used in radiochemical separations of radium. The radium forms sparingly soluble carbonate, phosphate, oxalate, and sulfate salts, the sulfate salt being the most widely used in separations. Nevertheless, radium is only present in very small amounts in nature and cannot be precipitated alone, therefore requiring the addition of a carrier prior to separation. Because radium has no stable isotopes, its nearest chemical analog, barium, is normally used as a carrier. Also, no suitable radioactive tracer of radium could be used in yield determinations. Stable barium or the gamma-emitting isotope ^{133}Ba ($t_{1/2} = 10.5$ a, γ 356 keV, intensity 62%) is used in yield determinations of radium. As seen in Figures 4.5 and 4.6, radium does not form anionic complexes in nitric or hydrochloric acid and it cannot be isolated by anion exchange. Instead, other radionuclides in the sample, such as ^{210}Po and isotopes of uranium and thorium, can be removed by anion exchange. Moreover, radium mostly appears in solution as Ra^{2+} ions, and this can be removed with cation exchangers, to which it binds from

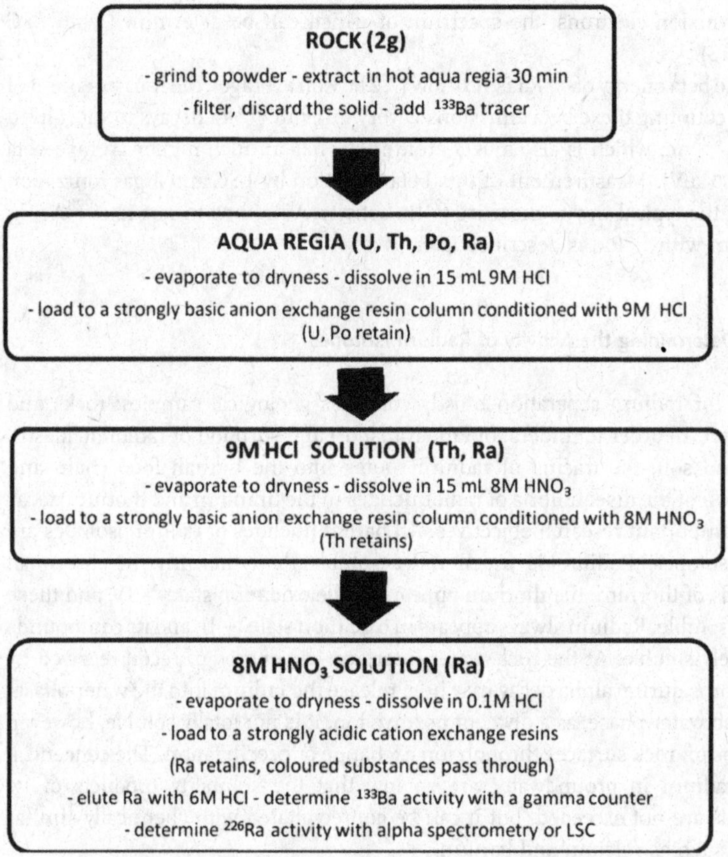

ROCK (2g)
- grind to powder - extract in hot aqua regia 30 min
- filter, discard the solid - add ^{133}Ba tracer

AQUA REGIA (U, Th, Po, Ra)
- evaporate to dryness - dissolve in 15 mL 9M HCl
- load to a strongly basic anion exchange resin column conditioned with 9M HCl (U, Po retain)

9M HCl SOLUTION (Th, Ra)
- evaporate to dryness - dissolve in 15 mL 8M HNO_3
- load to a strongly basic anion exchange resin column conditioned with 8M HNO_3 (Th retains)

8M HNO_3 SOLUTION (Ra)
- evaporate to dryness - dissolve in 0.1M HCl
- load to a strongly acidic cation exchange resins (Ra retains, coloured substances pass through)
- elute Ra with 6M HCl - determine ^{133}Ba activity with a gamma counter
- determine ^{226}Ra activity with alpha spectrometry or LSC

Figure 7.14 Separation of radium from rock (Juntunen, P., Ruutu, A., and Suksi, J. (2001) Determination of ^{226}Ra from Rock Samples using LSC, Proceedings of the International Conference on Advances in Liquid Scintillation Spectrometry, Karlsruhe, Germany, May 7–11, pp. 299–302).

solutions ranging from weakly acidic to basic. Since radium does not readily form organic complexes, solvent extraction with organic complexing agents is not an option either. Instead, as in anion exchange, solvent extraction is used to remove other metals from the solution.

Two methods for the separation of radium are presented below, one suitable for the measurement in rock samples (Figure 7.14), and the other for measurements in water.

7.6.4.1 Separation of ^{226}Ra in Rock Samples with Use of Ion Exchange

This method was briefly described in Chapter 4 as an example of ion exchange separation. Here it is in more detail.

In this method, the rock is first ground to a fine powder and leached with hot concentrated hydrochloric acid, and a ^{133}Ba tracer is added to the solution for yield determination. The solid material is removed on a filter, and the filtrate is evaporated to dryness. The evaporation residue is dissolved in 9 M HCl, and the solution is applied to an ion exchange column packed with a strongly basic anion exchange resin. The uranium and ^{210}Po (which follows radium in the uranium decay series) form negatively charged chloride complexes and are bound in the column. Thorium and radium do not form chloride complexes and pass through the column. The effluent is evaporated to dryness and the residue is dissolved in 8 M HNO_3. This solution is similarly applied to a column containing a strongly basic anion exchange resin. The negatively charged nitrate complex of thorium is retained in the column while the radium passes through it. The effluent is evaporated to dryness and dissolved in dilute hydrochloric acid, and a further purification is carried out by applying it to a cation exchange column in dilute HCl (pH 1.5). The Ra^{2+} ions sorb in the resin, but some interfering colored substances do not. Radium, along with ^{133}Ba, is eluted with strong hydrochloric acid. The activity of ^{133}Ba is counted with a gamma counter and compared with the added activity for the determination of yield. Finally, ^{226}Ra activity is measured either by alpha spectrometry or by LSC.

7.6.4.2 Determination of ^{226}Ra and ^{228}Ra in Water by Extraction Chromatography

In water samples, both the alpha-active isotope ^{226}Ra and the beta-active isotope ^{228}Ra are measurable with extraction chromatography. Nitric acid is added to about one liter of water sample, containing ^{133}Ba for the yield determination, to adjust the pH to 2. Radium nuclides are removed by pouring the solution onto a column filled with a strongly acidic cation exchange resin. Radium, barium, and any metals absorbed on the column are eluted with 8 M HNO_3, and the eluate is evaporated to dryness. The residue is dissolved in 10 mL of 0.1 M HNO_3. The solution is poured into an Eichrom/Triskem Ln Resin column, and ^{228}Ac, the daughter of ^{228}Ra is absorbed on the column, while ^{226}Ra and ^{228}Ra pass through. The Ln Resin contains the organic extraction agent, HDEHP, in the pores of an organic support polymer. In 0.1 M HNO_3, this extractant binds actinium but not radium or barium. (http://www.eichrom.com/radiochem/methods/eichrom/index.cfm).

Radium and barium are present in the effluent from the Ln Resin column. The activity of the tracer, ^{133}Ba, in the effluent is counted with a gamma counter for determination of the yield. After this, the radium and barium are micro-coprecipitated from the solution as barium sulfate for the measurement of the activity. A barium carrier (0.075 mg) is added to the effluent, and barium sulfate is precipitated by the addition of 40% sodium sulfate and acetic acid. The precipitate is collected onto a 0.1-μm filter, washed, dried, and glued onto the sample disk for alpha counting of ^{226}Ra.

The activity of ^{228}Ra is measured as the activity of the ^{228}Ac trapped in the Ln Resin column. The ^{228}Ac is eluted from the column with 10 mL of 0.35 M HNO_3 and is precipitated by micro-coprecipitation with CeF_3 for measurement of its activity. For this, a 0.3 mg Ce carrier and 1 mL of strong hydrofluoric acid are added to the eluate, whereupon Ac is coprecipitated with CeF_3. The precipitate is collected onto a 0.1-μm filter, washed, dried, and glued onto the sample disk for beta counting with a proportional counter. If a proportional counter is not available, the activity of ^{228}Ac can be determined by LSC. In determining the activity of ^{228}Ac, it is important to remember that since the half-life of ^{228}Ac is only 6.1 h, the activity decreases fairly rapidly after the separation of ^{228}Ac from ^{228}Ra in the column.

7.7 Essentials in the Radiochemistry of the Alkaline Earth Metals

- ^{41}Ca is a very long-lived ($t_{1/2} = 1.03 \times 10^5$ y) beta emitter produced by neutron activation in reactor-shielding concrete, relevant in the decommissioning of nuclear facilities.
- ^{90}Sr is a fairly long-lived ($t_{1/2} = 29$ y) beta-emitting fission product, present in spent nuclear fuel, other nuclear waste, releases from nuclear facilities, and fallout from nuclear explosions.
- ^{226}Ra is a long-lived ($t_{1/2} = 1600$ y) naturally occurring alpha-emitting radionuclide and a member of the ^{238}U decay chain occurring in the ground; another relatively long-lived radium isotope is ^{228}Ra ($t_{1/2} = 5.8$ y), which is pure beta emitter.
- Calcium exists as a divalent cation, Ca^{2+}. It is not readily hydrolyzable and forms carbonate, phosphate, and oxalate. ^{41}Ca is separated by a series of hydroxide and carbonate precipitations and measured either by LSC or AMS.
- Strontium is chemically very similar to calcium. Removal of calcium is often the most challenging task in ^{90}Sr determinations. ^{90}Sr is separated by a series of carbonate, hydroxide, and nitrate precipitations, or by extraction chromatography using crown ether attached on a chromatographic support and measured by LSC or gas ionization detector.
- Radium separations are typically needed on geological samples, rocks, and groundwater. ^{228}Ra can be separated from other radionuclides in the uranium and thorium decay series (Po, U, and Th) by anion exchange and is measured by either alpha spectrometry or LSC. The activity of ^{228}Ra is determined by measuring the beta activity of its short-lived daughter ^{228}Ac.

8
Radiochemistry of the 3d-Transition Metals

8.1
The Most Important Radionuclides of the 3d-Transition Metals

Elements in the fourth period (or row) of the periodic table, from scandium to zinc, include several important radionuclides beginning with chromium (Table 8.1). Scandium, titanium, and vanadium have no radionuclides of interest and are thus not discussed in this book. All the radionuclides listed in Table 8.1 are activation products generated in reactors by neutron activation of the elements in steel and other materials. Water in the primary circuit causes steel to corrode, releasing activation products which then move to the primary circuit and waste water purification systems, especially ion exchange resin beds. These radionuclides are also formed in the concrete structures surrounding reactors.

Most 3d-radionuclides are gamma emitters and can be measured by their emitted gamma rays. Three members of the group, ^{55}Fe, ^{59}Ni and ^{63}Ni, cannot, however, be measured without prior separation. Most of the 3d-radionuclides are relatively short-lived with half-lives of less than one year, and their radiological significance in the final disposal of nuclear waste is minor. The longer-lived nuclides are ^{55}Fe (2.7 y), ^{60}Co (5.3 y), and especially ^{63}Ni (100 y) and ^{59}Ni (7.6×10^4 y). ^{60}Co, which emits high-energy gamma rays (1173 keV and 1332 keV), is generated in relatively large amounts in nuclear power plants and is responsible for larger radiation doses to plant workers than any other radionuclide. With regard to final waste disposal, the longer-lived ^{63}Ni is the most important of them as, after a few tens of years, its activity will be the strongest in low- and medium-activity wastes and in wastes from decommissioned reactors. ^{59}Ni and ^{63}Ni are also present in spent nuclear fuel if disposed of without reprocessing, since they is formed by neutron-induced reaction in zircaloy fuel cladding. Because of the very long half-life of ^{59}Ni, it will be present in the final disposal repository for hundreds of thousands of years. ^{55}Fe was also generated in significant amounts during the atmospheric weapons tests through the neutron activation of steel bomb casings. In the ensuing fallout during the 1950s and 1960s, ^{55}Fe was deposited in considerable amounts around the world but has since decayed. ^{55}Fe is not particularly radiotoxic because it decays by electron capture and emits only low-energy X-ray rays and Auger electrons.

Chemistry and Analysis of Radionuclides. Jukka Lehto and Xiaolin Hou

ISBN: 978-3-527-32658-7

Table 8.1 Important radionuclides of the 3d-transition metals. All radionuclides are activation products.

	Most important radionuclides	Half-life	Decay mode	Gamma emission
Chromium	^{51}Cr	28 d	β^-	yes
Manganese	^{54}Mn	313 d	EC	yes
Iron	^{55}Fe	2.7 y	EC	no
Cobalt	^{58}Co	271 d	EC	yes
	^{60}Co	5.3 y	β^-	yes
Nickel	^{63}Ni	100 y	β^-	no
	^{59}Ni	7.6×10^4 y	EC	no
Zinc	^{65}Zn	244 d	EC	yes

The major challenges that the 3d-metals present for radiochemistry are the radiochemical separations needed to measure ^{63}Ni, ^{59}Ni, and ^{55}Fe activities. Other nuclides must be removed because they emit not only gamma rays but also beta particles, which interfere with the measurement of ^{63}Ni, ^{59}Ni, and ^{55}Fe. Another important task for radiochemistry is the separation of stable iron as it almost always interferes with the analyses of other radionuclides. Iron is the fourth most common element in nature and is idely present in the environment and in the steel constructions in nuclear power plants.

8.2 Chemical Properties of the 3d-Transition Metals

Elements in the fourth period have full 3s and 3p shells, and as the atomic number increases, the 3d shell is increasingly filled (Table 8.2). There are also two electrons in

Table 8.2 Electronic structures and valence states of the 3d-transition metals that have important radionuclides.

	Electronic structure	Typical oxidation states	Typical forms in solution
Chromium	$[Ar]3d^54s^1$	+III	Cr^{3+}
		+VI	CrO_4^{2-}
Manganese	$[Ar]3d^54s^2$	+II	Mn^{2+}
		+IV	Mn^{4+}, MnO_2
		+VII	MnO_4^-
Iron	$[Ar]3d^64s^2$	+II	Fe^{2+} (ferro)
		+III	Fe^{3+} (ferri)
Cobalt	$[Ar]3d^74s^2$	+II	Co^{2+}
Nickel	$[Ar]3d^84s^2$	+II	Ni^{2+}
Copper	$[Ar]3d^{10}4s^1$	+II	Cu^{2+}
Zinc	$[Ar]3d^{10}4s^2$	+II	Zn^{2+}

the 4s shell, except in chromium and copper, where there is only one. As a result of the filling of the 3d shell, these elements, along with those that fill their 4d shells, are called transition elements. Table 6.2 shows chemical properties of the metals of the fourth period. The most soluble are the divalent metal ions; the solubility decreases with oxidation state, being lowest for the tetravalent form, and both Ti(IV) and Mn(IV) form oxides that are very sparingly soluble. At higher oxidation state levels than +IV, the solubility increases, and the metals at higher oxidation states, Cr(VI) and Mn(VII), form soluble anionic species CrO_4^{2-} and MnO_4^- in neutral and alkaline solutions. Tri- and tetravalent metal ions hydrolyze strongly, while divalent metal ions hydrolyze much less. All 3d-transition metals in oxidation states +II and +III readily form complexes. Metals in oxidation state +II form sparingly soluble hydroxides, carbonates, oxalates, and sulfides; the least soluble are the sulfides, the solubilities of the other compounds varying with the metal. Fe(III) and Cr(III) also form sparingly soluble hydroxides but not carbonates or oxalates.

8.3 Iron – ^{55}Fe

Iron has only one radionuclide, ^{55}Fe, requiring radiochemical separations before measurement; it is neither very long-lived nor particularly significant in terms of radiation protection since it emits exceedingly low-energy radiation. Stable iron, in contrast, is highly significant in radiochemical separations because it may be present in large amounts and so interferes with the separation of radioisotopes of other elements. It is thus very important to eliminate iron in radiochemical analysis. Iron, however, does have its uses: coprecipitation with ferric hydroxide is a very common step in the radiochemical analysis of many radionuclides.

8.3.1 Nuclear Characteristics and Measurement of ^{55}Fe

^{55}Fe is generated in nuclear reactors and nuclear weapons when stable iron is activated in neutron flux via a neutron capture reaction $^{54}Fe(n,\gamma)^{55}Fe$. It decays through electron capture to ^{55}Mn (Figure 8.1). As a consequence of the electron capture, the daughter nuclide emits X-rays (5.9 keV, intensity 24%) and Auger electrons (5.2 keV, intensity 60%), and the measurement of these radiations by semiconductor X-ray detector or liquid scintillation counter (LSC) allows the determination of ^{55}Fe. Because the energies are very low, the sample to be measured needs to be radiochemically very pure.

8.3.2 Chemistry of Iron

Iron comprises 6.2% of the Earth's crust, being the fourth most common element. It appears in two oxidation states: Fe(II) or ferrous iron and Fe(III) or ferric iron.

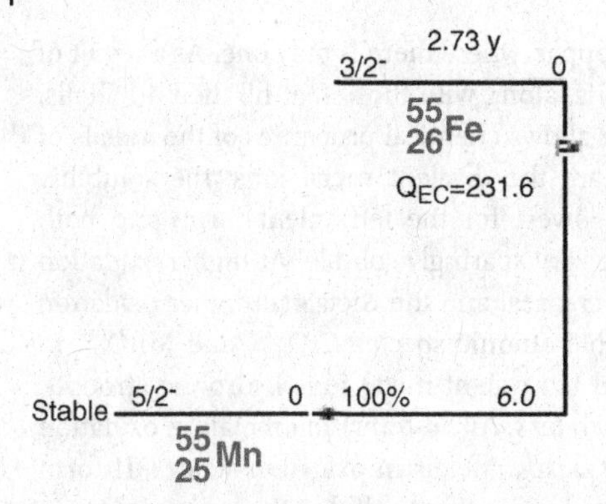

Figure 8.1 Decay scheme of ^{55}Fe (Firestone, R.B., Shirley, V.S., Chu, S.Y.F., Baglin, C.M., and Zipkin, J. (1996) *Table of Isotopes*, Wiley-Interscience).

Figure 8.2 presents the Eh-pH diagram of iron, which shows that ferrous iron is only stable in reducing and acidic conditions. Most of the iron in the Earth's crust appears in the form of oxides, carbonates, and sulfides. In oxidizing conditions, iron appears as hematite (Fe_2O_3) and goethite (FeOOH), in which the iron is in ferric form. In somewhat reducing conditions, iron appears as Fe_3O_4, where it is partially reduced to ferrous form. The most common ferrous mineral in the Earth's crust is FeS_2 (pyrite or marcasite), which forms in anoxic conditions where iron

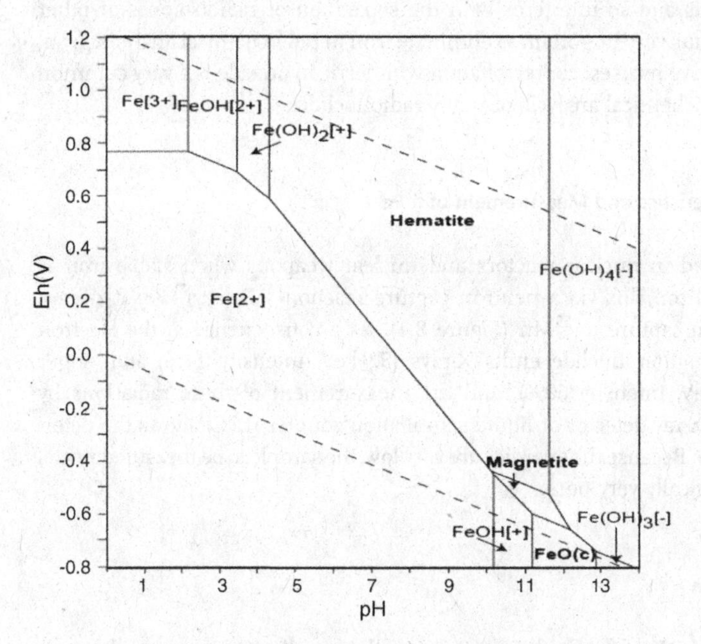

Figure 8.2 Eh-pH diagram of iron. Hematite: Fe_sO_3, magnetite: Fe_3O_4 (Atlas of Eh-pH diagrams, Geological Survey of Japan Open File Report No. 419, 2005).

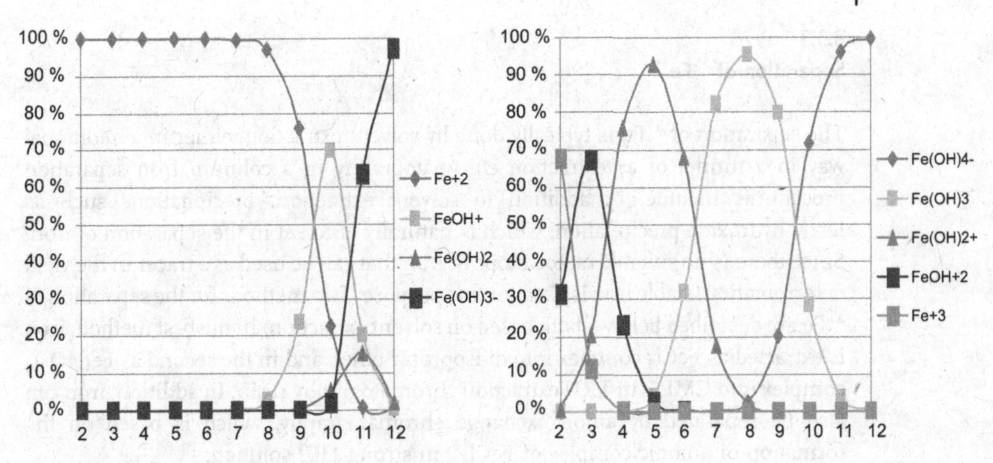

Figure 8.3 Hydrolysis of Fe(II) and Fe(III) in 10^{-6} M solution.

is reduced to oxidation state II and sulfate to sulfide. The oxidation state of iron in the overburden and bedrock is of great significance for the oxidation states of other elements. If, for example, oxidic water bearing soluble uranium (UO_2^{2+}) encounters a pyrite front, the uranium will be reduced to insoluble UO_2 and can form uranium ore. When present in natural waters, iron can determine the behavior of other elements in many ways. For instance, when oxic water sinks toward lake bottoms as the weather warms in the spring, Fe(II) in the bottom water oxidizes to Fe(III) and precipitates as hydroxide, which, in turn, co-precipitates other metals. As conditions become more anoxic, iron reduces to Fe(II) and partially goes into solution.

Fe(III) hydrolyzes much more readily than divalent iron (Figure 8.3). Trivalent iron begins to hydrolyze just above pH 2, forming hydrolysis species $FeOH^{2+}$, Fe $(OH)^{2+}$, and $Fe(OH)_3$(aq). The solubility minimum of Fe(III) is at pH 8, where the maximum amount of iron that can be in solution is about 10^{-11} M. In alkaline solution, iron partially dissolves as $Fe(OH)^{4-}$. In more concentrated iron solutions, ferric ions not only form monomeric hydrolysis species but also polymeric species, of which, in 0.1 M solution, $Fe_3(OH)_4^{5+}$ is predominant between pH 3 and pH 6. Fe(II) begins to hydrolyze only at pH 7, and its solubility minimum, about 10^{-7} M, is at pH 10.5.

In addition to soluble hydroxides, both Fe(II) and Fe(III) form sparingly soluble sulfides. Fe(II) also forms sparingly soluble oxalate and carbonate, which Fe(III) does not form. Oxalate precipitation is used in the separation of iron from many other trivalent metals, in particular lanthanides and actinides: iron is oxidized to Fe(III) and remains in solution, while lanthanides and actinides precipitate as oxalates. Carbonate precipitation cannot be used in the same way because Fe(III) would precipitate as hydroxide along with the carbonates. The oxalate precipitation is carried out in acidic solution, in which Fe(III) does not precipitate as hydroxide. Iron ions, especially Fe(III) ions, readily form complexes.

8.3.3
Separation of ^{55}Fe

The separation of ^{55}Fe is typically done by solvent extraction, either in a traditional way in a funnel or as extraction chromatography in a column. Iron separation procedures include, in addition to solvent extraction, precipitations such as ferric hydroxide precipitation, which is naturally efficient in the separation of iron. Since there is no suitable radioisotope of iron that can be used as a tracer in the yield determination, stable iron is used for this purpose. Two methods for the separation of ^{55}Fe are described below – both based on solvent extraction. In the first method, iron is extracted as $FeCl_3$ complex into di-isopropyl ether, and in the second as $Fe(NO_3)_3$ complex into CMPO in TRU extraction chromatography resin. In addition, iron can also be separated by anion exchange chromatography, which is based on the formation of anionic complex of $FeCl_4^-$ in strong HCl solution.

8.3.3.1 Separation of ^{55}Fe by Solvent Extraction

Figure 8.4 shows the separation of ^{55}Fe from the diverse group of radionuclides present in a spent ion exchange resin used to purify the primary circuit coolant water in a nuclear power plant. In this method, ^{55}Fe is first eluted from the resin with strong acid. Many other nuclides, such as ^{60}Co, are eluted at the same time. Nuclides are sorbed on the resin both as exchangeable ions and as hydroxide particles, which originate in the corrosion of the reactor pressure vessel and adsorb to the resin bed by filtration. Stable iron as carrier is added to the effluent, after which the iron is precipitated as hydroxide by the addition of ammonium hydroxide. In this step, the alkali and alkaline earth metals, such as ^{137}Cs, remain in the solution and are separated. The precipitate is dissolved in 8 M hydrochloric acid, and iron is separated from the aqueous solution by extraction with di-isopropyl ether. Iron is extracted as neutral $FeCl_3$ complex. The extraction removes most of the interfering radionuclides but not ^{125}Sb or $^{103,106}Ru$, which are present as anionic species. Iron, the ^{125}Sb, and the $^{103,106}Ru$ are back-extracted into the aqueous phase by water alone, whereupon the $FeCl_3$ complex breaks down. Iron is again precipitated as iron hydroxide, the precipitate is dissolved, and the ion exchange is carried out in 0.2 M HCl on a strongly acidic cation exchange resin. The iron remains in the column (see Figure 4.5), while ^{125}Sb and $^{103,106}Ru$ are washed out of the column as anions. Iron is then eluted from the column with weak oxalic acid, which forms an oxalate complex with iron and removes it from the exchanger. A drop of hydrogen peroxide is added to the eluate to ensure that the iron is in oxidation state +III. Iron is again precipitated as hydroxide, and this is dissolved in a small amount of strong hydrofluoric acid, which also forms a colorless fluoride complex with iron. The complexation of iron is necessary because Fe(III) is colored and quenches the liquid scintillation spectra. Water (1 mL) is added to the dissolved iron solution, and a subsample is taken to determine the chemical yield of iron by AAS or ICP-MS. Scintillation cocktail is added to the remaining solution, and the activity of ^{55}Fe is determined by liquid scintillation counting.

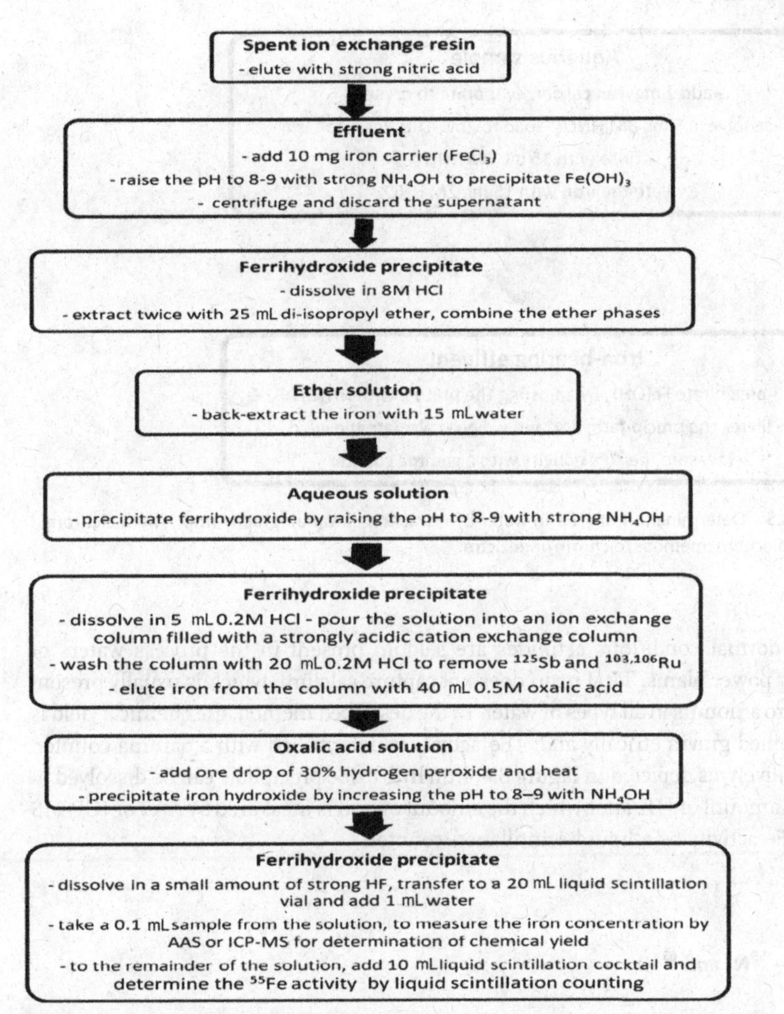

Figure 8.4 Chemical procedure for determining ^{55}Fe in spent ion exchange resin used to purify primary coolant water at a nuclear power plant.

8.3.3.2 Separation of ^{55}Fe by Extraction Chromatography

Eichrom reports a simple method for the determination of ^{55}Fe in aqueous samples. The method utilizes TRU resin, in which the extraction reagent is carbamoylphosphine oxide (CMPO) (see Figure 4.10). The separation method is presented in Figure 8.5.

First, the iron carrier is added to the aqueous sample, which is evaporated to dryness; the residue is then dissolved in 8 M HNO_3. The iron is separated using a TRU column. Figure 4.11 shows the selectivity of TRU resin for selected metals in nitric acid solution. As can be seen, Fe(III) sorbs efficiently to the resin in 8 M HNO_3 and desorbs well in 2 M HNO_3. Actinides also sorb efficiently in 8 M HNO_3, although

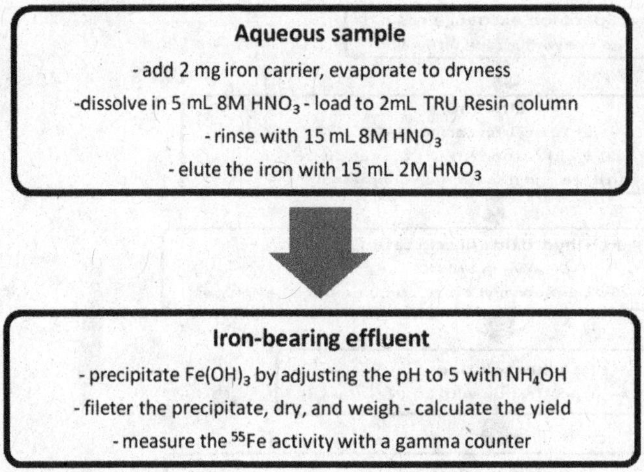

Aqueous sample
- add 2 mg iron carrier, evaporate to dryness
- dissolve in 5 mL 8M HNO_3 - load to 2mL TRU Resin column
- rinse with 15 mL 8M HNO_3
- elute the iron with 15 mL 2M HNO_3

Iron-bearing effluent
- precipitate $Fe(OH)_3$ by adjusting the pH to 5 with NH_4OH
- fileter the precipitate, dry, and weigh - calculate the yield
- measure the ^{55}Fe activity with a gamma counter

Figure 8.5 Determination of ^{55}Fe in water by extraction chromatography http://www.eichrom.com/radiochem/methods/eichrom/index.cfm.

under normal conditions actinides are seldom present in the process waters of nuclear power plants. TRU resin does not capture calcium, which is usually present in macro amounts in all types of water. In the described method, the chemical yield is determined gravimetrically and ^{55}Fe activity is determined with a gamma counter. Alternatively, as depicted in Figure 8.4, iron hydroxide precipitate can be dissolved in a small amount of HF, after which the amount of iron is measured by AAS or ICP-MS and ^{55}Fe activity by a liquid scintillation counter.

8.4 Nickel – ^{59}Ni and ^{63}Ni

There are two significant radionuclides of nickel, ^{59}Ni and ^{63}Ni, present in nuclear power plant waste and in the metallic parts surrounding spent nuclear fuel. ^{63}Ni is a pure beta emitter, and ^{59}Ni decays by electron capture, which means that radiochemical separations are required for their measurement.

8.4.1 Nuclear Characteristics and Measurement of ^{59}Ni and ^{63}Ni

^{59}Ni and ^{63}Ni are produced by neutron activation reactions of stable Ni and Cu ($^{58}Ni(n,\gamma)^{59}Ni$, $^{62}Ni(n,\gamma)^{63}Ni$; $^{63}Cu(n,p)^{63}Ni$) in the iron pressure vessel of nuclear reactors and other materials in the reactor. They are also formed in the same neutron-induced reactions in the Zircaloy cladding of the nuclear fuel. Through the primary circuit cleaning system, in particular the ion exchange resins, ^{63}Ni finds its way into the waste of the nuclear power plant. ^{63}Ni, a relatively long-lived

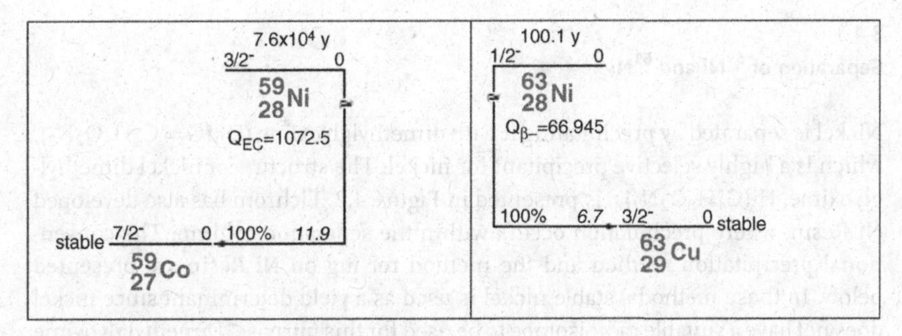

Figure 8.6 Decay schemes of ^{59}Ni and ^{63}Ni (Firestone, R.B., Shirley, V.S., Chu, S.Y.F., Baglin, C.M., and Zipkin, J. (1996) *Table of Isotopes*, Wiley-Interscience).

radioisotope of Ni with a half-life of 100 years, decays by pure beta emission to stable ^{63}Cu (Figure 8.6). The maximum energy of the beta particles of ^{63}Ni is 67 keV. A gas ionization counter, beta spectrometry with semiconductor detectors, and LSC have been used for its measurement. The other radioisotope of nickel in nuclear waste, ^{59}Ni, has a very long half-life of 76 000 years. ^{59}Ni decays by electron capture to ^{59}Co, emitting low-energy X-rays (6.9 keV, intensity 30.8%) and Auger electrons (6.1 keV, intensity 54%) (Figure 8.6).

Chemical separations do not separate ^{63}Ni from ^{59}Ni. Both isotopes are produced by thermal neutron reactions with stable nickel ^{62}Ni (3.65%) and ^{58}Ni (68.1%), and the initial activity ratio of ^{63}Ni/^{59}Ni is around 100 or higher. Because of this high activity ratio and the fact that the beta spectrum of ^{63}Ni ovelaps with the X-ray and Auger electron spectra of ^{59}Ni, only ^{63}Ni is measurable by LSC. The signal of ^{59}Ni in LSC occurs in the low-energy part of the ^{63}Ni spectrum, which makes the interference of ^{59}Ni to the ^{63}Ni spectrum very small and can be completely avoided by excluding ^{63}Ni measurement in the lowest channels. For a high-level ^{59}Ni sample (>0.1 Bq/g), X-ray spectrometry can be used, but a thin source has to be prepared to prevent self-absorption of the low energy X-rays. Electroplating the separated Ni on stainless steel is a good choice for this purpose. To determine very low levels of ^{59}Ni, accelerator mass spectrometry (AMS) must be used. In this case, the separation of Ni from Cu is a key issue because of the serious isobaric interference of stable ^{59}Co (abundance of 100%) to the AMS measurement of ^{59}Ni.

8.4.2 Chemistry of Nickel

The chemistry of nickel closely resembles that of Fe(II); in normal conditions, the sole oxidation state of nickel is +II and it is present in solution as Ni^{2+} ion. Like Fe(II), nickel forms a sparingly soluble carbonate, oxalate, hydroxide, and sulfide, whose solubility decreases in this order. Nickel hydrolyzes similarly to Fe(II); hydrolysis begins at a pH of about 7, and the solubility has a minimum, 10^{-8} mol/L, at pH 10.

8.4.3 Separation of ^{59}Ni and ^{63}Ni

Nickel is separated by precipitating it with dimethylglyoxime (DMG = $C_4H_8O_2N_2$), which is a highly selective precipitant for nickel. The structure of nickel dimethylglyoxime, $Ni(C_4H_8O_2N_2)_2$, is presented in Figure 4.2. Eichrom has also developed Ni Resin, where precipitation occurs within the separation column. The conventional precipitation method and the method relying on Ni Resin are presented below. In these methods, stable nickel is used as a yield determinant since nickel does not have a suitable radioisotope to be used for this purpose. Dimethylglyoxime precipitation is not, however, efficient enough to prepare pure nickel samples for AMS measurement, for which a method based on $Ni(CO)_4$ formation is also presented.

8.4.3.1 Separation of Nickel by the DMG Precipitation Method

Figure 8.7 shows the conventional method of separating ^{63}Ni by the DMG precipitation method from the ion exchange resin used in the purification of the primary circuit water of nuclear power plants.

In the method outlined in Figure 8.7, the radionuclides are first eluted from the ion exchange resin with strong nitric acid. Ni, Co, Cr, Mn, and Zn carriers are added to the effluent. Other carriers besides nickel are added because beta active $^{58,60}Co$, ^{51}Cr, ^{54}Mn, and ^{65}Zn are present in the ion exchange resin and must be removed before measuring the nickel. The solution is evaporated to near dryness and the residue is dissolved in 4M nitric acid, after which precipitation as hydroxide is carried out by increasing the pH to 8–9 with NaOH. Most of the interfering radionuclides follow the nickel into the precipitate, while the radionuclides of alkali metals and most alkaline earth metals, like $^{134,137}Cs$ and ^{90}Sr are removed. Simultaneously, anionic radionuclides are removed. The hydroxide precipitate is dissolved in strong hydrochloric acid and evaporated to near dryness; the evaporation residue is dissolved in 8M HCl. The acid solution is passed through an ion exchange column containing a strongly basic anion exchange resin. The nickel does not sorb to the ion exchanger under these conditions but passes through (see Figure 4.6). Trivalent iron, cobalt, and zinc are sorbed to the column. Ammonium citrate is added to the effluent to form complexes with other metals (Mn, Eu, etc.) to prevent hydroxide precipitate forming after raising the pH and to prevent their sorption to the dimethylglyoxime precipitate. The pH is adjusted to pH 8–9 with ammonium hydroxide, and DMG solution is added to form a bright red Ni-DMG precipitate. All remaining interfering radionuclides are removed at this stage. To confirm the separation, the Ni-DMG precipitation is carried out once again. The Ni-DMG precipitate is dissolved in strong nitric acid, heated to decompose organic material and evaporated to near dryness. The residue is dissolved in 0.1 M HCl. A subsample is taken from this solution and measured for Ni concentration by AAS or ICP-MS to determine the chemical yield. The remaining solution is measured for ^{63}Ni activity by liquid scintillation counting.

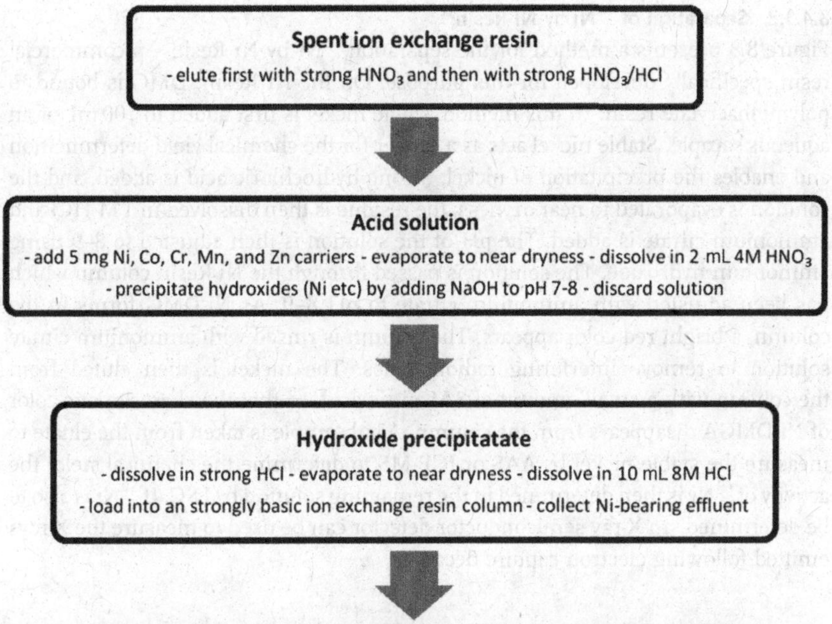

Spent ion exchange resin
- elute first with strong HNO_3 and then with strong HNO_3/HCl

Acid solution
- add 5 mg Ni, Co, Cr, Mn, and Zn carriers - evaporate to near dryness - dissolve in 2 mL 4M HNO_3
- precipitate hydroxides (Ni etc) by adding NaOH to pH 7-8 - discard solution

Hydroxide precipitatate
- dissolve in strong HCl - evaporate to near dryness - dissolve in 10 mL 8M HCl
- load into an strongly basic ion exchange resin column - collect Ni-bearing effluent

Effluent
- add 2 g ammonium citrate - adjust pH to 9-10 with strong NH_4OH
- precipitate Ni-DMG by adding 10 mL 1% DMG solutions - discard solution

Ni-DMG precipitate
- dissolve in 25 mL 8M HCl - repeat precipitation with DMG
- dissolve Ni-DMG in 15 mL conc. HNO_3 - evaporate to near dryness - dissolve in 0.1M HCl

Ni solution
- measure stable Ni by AAS or ICP-MS - calculate yield
- mix with scintillation cocktail - determine ^{63}Ni activity by LSC

Figure 8.7 Determination of ^{63}Ni in ion exchange resin used to purify the primary circuit water of a nuclear reactor (Puukko, E. and Jaakkola, T. (1992) Actinides and beta emitters in the process water and ion exchange resin samples from the Loviisa power plant, Report YJT-92-22, Nuclear Waste Commission of Finnish Power Companies).

8.4.3.2 Separation of ^{63}Ni by Ni Resin

Figure 8.8 presents a method for the separation ^{63}Ni by Ni Resin – a commercial resin specifically developed for this purpose. On the Ni Resin, DMG is bound to polymethacrylate resin. In this method, stable nickel is first added to 100 mL of an aqueous sample. Stable nickel acts as a carrier for the chemical yield determination and enables the precipitation of nickel. Strong hydrochloric acid is added, and the solution is evaporated to near dryness; the residue is then dissolved in 1 M HCl and ammonium citrate is added. The pH of the solution is then adjusted to 8–9 using ammonium hydroxide. The solution is passed through the Ni Resin column which has been adjusted with ammonium citrate to pH 8–9. As Ni-DMG forms in the column, a bright red color appears. The column is rinsed with ammonium citrate solution to remove interfering radionuclides. The nickel is then eluted from the column with a small amount of 3 M nitric acid, so that the characteristic color of Ni-DMGA disappears from the column. A subsample is taken from the eluate to measure the stable nickel by AAS or ICP-MS to determine the chemical yield; the activity of ^{63}Ni is then determined in the remaining solution by LSC. If ^{59}Ni is also to be determined, an X-ray semiconductor detector can be used to measure the X-rays emitted following electron capture decay.

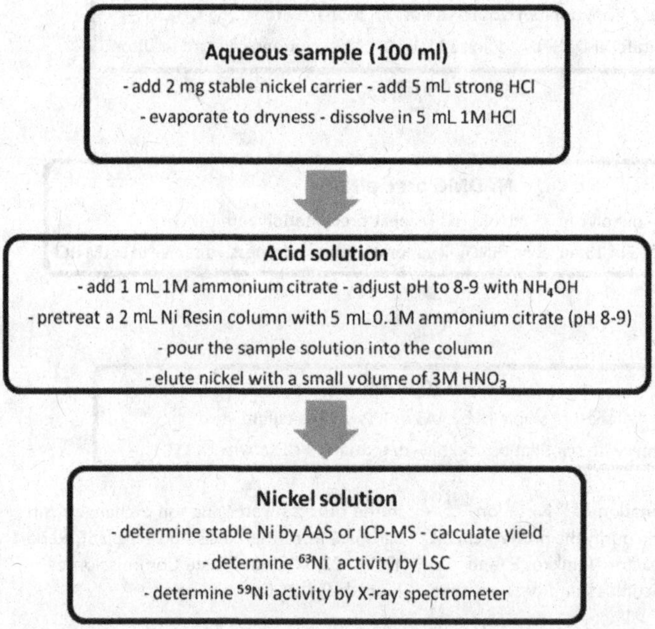

Aqueous sample (100 ml)
- add 2 mg stable nickel carrier - add 5 mL strong HCl
- evaporate to dryness - dissolve in 5 mL 1M HCl

Acid solution
- add 1 mL 1M ammonium citrate - adjust pH to 8-9 with NH_4OH
- pretreat a 2 mL Ni Resin column with 5 mL 0.1M ammonium citrate (pH 8-9)
- pour the sample solution into the column
- elute nickel with a small volume of 3M HNO_3

Nickel solution
- determine stable Ni by AAS or ICP-MS - calculate yield
- determine ^{63}Ni activity by LSC
- determine ^{59}Ni activity by X-ray spectrometer

Figure 8.8 Radiochemical procedure for the determination of $^{59,63}Ni$ in aqueous solution using Ni Resin column (Horwitz, E.P., Chirizia, R., Dietz, M.L., Diamond, H., and Nelson, D.M. (1993) Separation and preconcentration of actinides from acidic media by extraction chromatography. *Anal. Chim. Acta*, **281**, 361).

Table 8.3 Decontamination factors achieved on Ni Resin for interfering radionuclides in aqueous samples (http://www.eichrom.com/products/info/ni_resin.cfm).

Nuclide	Decontamination factor	Nuclide	Decontamination factor
^{51}Cr	>37 000	^{54}Mn	270 000
$^{58,60}Co$	110 000	^{95}Nb	14 000
^{134}Cs	>9000	^{137}Cs	58 000

This method gives very good decontamination factors for interfering radionuclides in the final solution (Table 8.3). Compared to conventional Ni-DMG precipitation, the total analysis time is reduced from 12 h to 4 h.

8.4.3.3 Separation of Nickel for the Measurement of Nickel Isotopes with AMS

In AMS measurements, the major interference comes from stable isobaric isotopes, ^{59}Co for ^{59}Ni, and ^{63}Cu for ^{63}Ni. It is a difficult task to separate nickel from samples, such as activated metals from nuclear reactors, as pure forms 'free' of cobalt and copper. Even very small amounts of cobalt and copper make the AMS measurement impossible since the amounts of nickel isotopes are at very low levels. Applying conventional DMG precipitation for nickel separation does not allow the measurement of the low levels of nickel radioisotopes typically found in nuclear facilities. Separating nickel from copper and cobalt can be considerably improved by separating nickel as the volatile nickel tetracarbonyl – a method typically used to produce pure nickel metal. To produce $Ni(CO)_4$, the sample is dissolved in acids and placed in a flask. The pH is raised to 10–11 with ammonium hydroxide: the formation of nickel ammonium complex prevents its precipitation as hydroxide. $NaBH_4$ is added to the solution, and a mixture of CO and He is bubbled through the Ni solution to produce volatile $Ni(CO)_4$, which is collected in a liquid nitrogen-cooled trap. The $Ni(CO)_4$ collected in the trap is heated and transferred to the AMS holder by He flow, where it is thermally decomposed to Ni. By this process, a Co/Ni ratio of $<2 \times 10^{-8}$ in the sample is obtained resulting in a very low detection limit of 2×10^{-11} for $^{59}Ni/Ni$ ratio or 0.05 mBq of ^{59}Ni. (McAninch, J.E., Hainsworth, L.J., Marchetti, A.A., Leivers, M.R., Jones, P.R., Dunlop, A.E., Mauthe, R., Vogel, S., Proctor, I.D., and Straume, T. (1997) Measurement of ^{63}Ni and ^{59}Ni by accelerator mass spectrometry using characteristic projectile X-rays. *Nucl. Instr. Methods B*, **123**, 137).

8.4.3.4 Simultaneous Determination of ^{55}Fe and ^{63}Ni

In many cases, both ^{55}Fe and ^{63}Ni are to be determined. A radiochemical procedure for the simultaneous separation of Fe and Ni from interfering radionuclides and from each other is given in Figure 8.9.

In this procedure, stable Ni and Fe carriers and holdback carriers Co, Eu, Mn, Cu, Sr, Cr, and Zn are added to the solution obtained by decomposing by digestion or leaching of contaminated graphite, concrete, and metals from a research reactor.

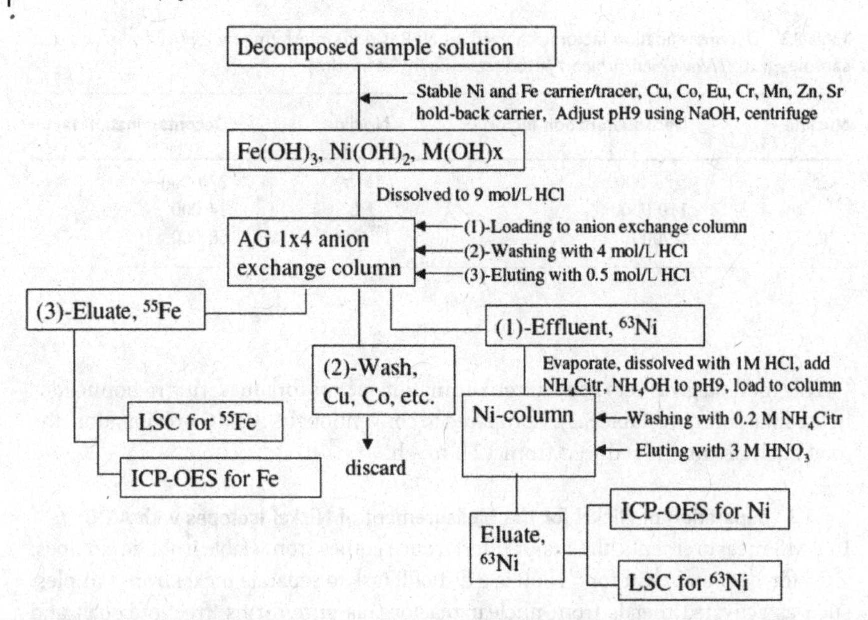

Figure 8.9 Combined radiochemical procedure for the separation of ^{63}Ni and ^{55}Fe (Hou, X.L., Østergaard, L.F., and Nielsen, S.P. (2005) Radiochemical determination of ^{55}Fe and ^{63}Ni in nuclear waste samples. *Anal. Chim. Acta.*, **535**, 297).

NaOH solution is then added to adjust the pH to 8–9, which results in the formation of hydroxide precipitate containing Fe and Ni and other hydrolyzable metals. Hydroxide precipitate is separated by centrifugation and dissolved in 9 M HCl. The solution is loaded to an anion exchange column preconditioned with 9 M HCl. To remove the nickel, which does not absorb to anion exchange resin in these conditions, the column is rinsed with of 9 M HCl (30 mL), and the rinsing solution is combined with the effluent. Co, Cu, and so on, are then removed by eluting the column with 4 M HCl (30 mL); finally, the Fe in the column is eluted with 0.5 M HCl (30 mL). The Fe eluate is evaporated to near dryness, and the residue is dissolved in 1 M H_3PO_4 (2 mL): the solution becomes colorless in a few minutes when iron forms a phosphate complex. The solution is transferred to a LSC vial, and 0.1 mL of the solution is taken for the ICP-OES measurement of stable Fe to determine the chemical yield. The remaining solution is used for LSC the counting of ^{55}Fe. The Ni fraction is evaporated to dryness and the residue is dissolved with 1 M HCl (2 mL). 1 M ammonium citrate solution (2 mL) is added to the solution, and the pH adjusted to 8–9 using ammonium hydroxide solution: the solution becomes green. This solution is loaded to a 2-mL Ni Resin column, which is preconditioned with 0.2 M ammonium citrate (10 mL) (pH 8). The column is washed with 0.2 M ammonium citrate (20 mL), and the Ni in the column is finally eluted with 3 M HNO_3 (7 mL). The eluate containing Ni is evaporated to 0.2–0.5 mL and transferred to an LSC vial; the beaker is washed with water, and the wash is combined with eluate in the LSC vial.

The total weight of the Ni solution is measured, and 0.1 g of the solution is taken for ICP-OES measurement of Ni for determination of the chemical yield.

The remaining solution is used for the measurement of ^{63}Ni. When evaporating Ni-bearing solutions in nitric acid, they should not be evaporated to dryness since nickel can be lost because of the volatility of $Ni(NO_3)_2$.

8.5 Essentials in 3-d Transition Metals Radiochemistry

- Iron has one radioisotope that requires radiochemical separations: ^{55}Fe. It is mainly formed in a neutron-induced activation reaction from stable iron in metallic constructions in nuclear reactors. It is fairly short-lived ($t_{1/2} = 2.7$ y) and decays by electron capture. Because of the short half-life and emission of only low-energy X-rays and Auger electrons, it cannot be regarded as a major radionuclide in nuclear waste from the point of view of radiation protection. X-rays and Auger electrons are used to measure the activity of ^{55}Fe by LSC or X-ray spectrometry, and solvent extraction is the main method to separate ^{55}Fe.
- Stable iron is a major interfering element in the determination of radionuclides from most solid samples, such as soil and sediment, which contain considerable amounts of iron that needs to be removed. Stable iron is also important in controlling oxidation state and solubility of many radionuclides in natural systems.
- Nickel has two important isotopes in nuclear waste: ^{59}Ni ($t_{1/2} = 76\,000$ y) and ^{63}Ni ($t_{1/2} = 100$ y). The former is an EC isotope emitting only low-energy X-rays and Auger electrons, while the latter is a beta emitter ($E_{max} = 67$ keV). They are formed in neutron-induced activation reactions in stable nickel and copper in metal constructions in nuclear power reactors, including metallic parts surrounding nuclear fuel. The shorter-lived isotope ^{63}Ni is a major radionuclide in low and intermediate nuclear power plant waste. A standard method in the separation of ^{63}Ni is its precipitation with dimethylglyoxime – a very specific precipitation agent for nickel. While ^{63}Ni activity is typically measured by LSC, gas ionization detectors can also be used. ^{59}Ni cannot be usually measured radiometrically because of the large excess of ^{63}Ni always present in same sample. AMS is the only choice for the measurement of low levels of ^{59}Ni. The major interference in AMS comes from the stable ^{59}Co isotope, and thus a careful removal of cobalt is needed before proceeding with the AMS measurement of ^{59}Ni.

9
Radiochemistry of the 4d-Transition Metals

9.1
Important Radionuclides of the 4d-Transition Metals

The elements in the fifth period of the periodic table include several fission products (Table 9.1), with mass numbers that place them in the lower mass peak in the fission product yield curve. Most of these nuclides (^{95}Zr, ^{95}Nb, ^{99}Mo, ^{103}Ru, ^{105}Rh) are fairly short-lived, with half-lives from a few days to a few months, and all can be measured by their clearly observable gamma emissions. In fresh nuclear fallout, such as that originating in the Chernobyl accident, these nuclides are present in large amounts but decay within about a year. ^{106}Ru has a half-life of 374 days, considerably longer than the nuclides mentioned above, and for years after the Chernobyl incident it was still present in the environment in measurable amounts. ^{106}Ru is a pure beta emitter but decays to short-lived ^{106}Rh; the emitted gamma rays of ^{106}Rh can be used to determine the equilibrium amount of ^{106}Ru.

The most important radionuclide of the 4d-elements with regard to environmental radioactivity, and nuclear waste is ^{99}Tc, which has a half-life of 211 000 years. ^{99}Tc is a pure beta emitter, and its activity cannot be measured without radiochemical separation. Zirconium also has a long-lived radionuclide (^{93}Zr) produced by the fission of uranium and plutonium. ^{93}Zr is also formed in a neutron activation reaction from stable ^{92}Zr in the Zircaloy fuel cladding of the nuclear fuel. Niobium and molybdenum also have long-lived radionuclides (^{93}Mo, ^{94}Nb) produced by the neutron activation of stable molybdenum and niobium (^{92}Mo, ^{93}Nb) in the Zircaloy fuel cladding and other metallic parts surrounding nuclear fuel. These radionuclides are important in evaluating long-term behavior of spent nuclear fuel and high-level nuclear waste in the final disposal conditions. Their chemistry is challenging since they are present as neutral and anionic oxospecies in groundwater conditions and thus are expected to be rather mobile. Of these three nuclides, only ^{94}Nb emits gamma rays that can be used in its activity measurements. ^{93}Zr is a pure beta emitter, and ^{93}Mo decays by electron capture. Thus, the determination of these nuclides requires radiochemical separations.

Chemistry and Analysis of Radionuclides. Jukka Lehto and Xiaolin Hou

ISBN: 978-3-527-32658-7

Table 9.1 The most important radionuclides of the 4d-transition metals.

Element	Most important radionuclides	Half life	Decay mode	Gamma emissions	Fission yield for ^{235}U (%)
Zirconium	^{95}Zr	64 d	β^-	yes	6.4
	^{93}Zr	$1{,}53 \times 10^6$ y	β^-	no	6.4
Niobium	^{95}Nb	35 d	β^-	yes	6.4
	^{94}Nb	20 300 y	β^-	yes	—
Molybdenum	^{99}Mo	65 d	β^-	yes	5.9
	^{93}Mo	4000 y	EC	no	—
Technetium	^{99}Tc	$2{,}11 \times 10^5$ y	β^-	no	5.9
Ruthenium	^{103}Ru	39 d	β^-	yes	3.2
	^{106}Ru	374 d	β^-	no	5.3
Rhodium	^{105}Rh	35 h	β^-	yes	1.2
	^{106}Rh	30 s	β^-	yes	5.3

9.2 Chemistry of the 4d-Transition Metals

In the 4d-transition metal period, the 4d orbitals are filled, as the name indicates. Some chemical properties of these elements are given in Table 9.2, and the chemistry of each element is described in more detail later. The prevailing oxidation state increases systematically from +IV for zirconium to +VII for technetium. The increase in the oxidation state results in the increasing formation of oxoanions. Zirconium, a tetravalent metal, is the least soluble of them – solubility increases with the atomic number; molybdenum and especially technetium are very soluble in neutral and alkaline solutions.

9.3 Technetium – ^{99}Tc

Technetium has one important, long-lived radioisotope, ^{99}Tc, which is present in large amounts in nuclear waste. Besides promethium, technetium is the only

Table 9.2 Some chemical properties of Zr, Nb, Mo, and Tc.

Element	Concentration in earth's crust	Prevailing oxidation states	Electron configuration	Typical form in solution
Zirconium	162 ppm	+IV	$[Kr]4d^2 5s^2$	$Zr(OH)_4$
Niobium	20 ppm	+V	$[Kr]4d^4 5s^1$	$Nb(OH)_5, NbO_3^-$
Molybdenum	1.2 ppm	+VI	$[Kr]4d^5 5s^1$	MoO_4^{2-}
Technetium	none	+VII	$[Kr]4d^5 5s^2$	TcO_4^-

element lighter than polonium that has no stable isotopes. Technetium is virtually a fully artificial element – it is found in the Earth's crust only in exceedingly small amounts as a product of spontaneous uranium fission. The main sources of ^{99}Tc in the environment are fallout from nuclear weapons tests and the releases from the Chernobyl accident. Considerable amounts of ^{99}Tc have been released to the marine environment from nuclear fuel reprocessing plants, especially from the Sellafield plant in the UK. In the final disposal of high-level nuclear wastes, Tc will constitute the most important component 100 000 years after sealing the repositories.

9.3.1
Chemistry of Technetium

The electron configuration of technetium is $[Kr]4d^5 5s^2$. Thus, the 4d shells are half-filled and the 5s shell is full. The chemistry of technetium is similar to that of manganese, which lies above it in the periodic table with the Group 7 elements, though it more resembles rhenium, the element below. Like Mn and Re, technetium forms oxides with the oxidation states +VII and +IV, that is, Tc_2O_7 and TcO_2. The former is very soluble, forming pertechnetic acid $HTcO_4$ in water solutions. The solubility of the latter oxide is very low, and in reducing conditions such as those found in deep underground repositories for spent nuclear fuel, TcO_2 will be the solubility-limiting solid phase. Unlike manganese, Tc does not form stable divalent ions in solution; rather, its only form in solution is the pertechnetate ion TcO_4^-. Tetravalent technetium hydrolyzes strongly and forms sparingly soluble TcO_2 – the prevailing species in the reducing conditions expected in deep geological repositories for spent fuel (Figure 9.1). Unlike the permanganate ion, the pertechnetate ion is not a strong oxidizing agent, which reflects its higher stability. Pertechnetate can, however, be reduced readily to Tc(IV) by common reducing agents. The pertechnetate ion does not readily form complexes. Moreover, as a large monovalent anion it does not easily adsorb onto surfaces and so is readily soluble and mobile in natural waters. The boiling point of technetium metal is very high, 4567 °C; however, in the presence of oxygen, technetium forms the easily volatile heptaoxide Tc_2O_7, which has a boiling point of just 311 °C. Since technetium has no stable isotopes, nonisotopic rhenium is used as a carrier.

9.3.2
Nuclear Characteristics and Measurement of ^{99}Tc

^{99}Tc has a half-life of 211 000 years and is an almost pure beta emitter, with maximum beta energy of 294 keV. ^{99}Tc decays to stable ^{99}Ru (Figure 9.2). When ^{99}Tc decays it also emits 90 keV gamma rays; however, as their intensity is only 0.0006%, they cannot be used for ^{99}Tc measurement. The ^{235}U fission yield of ^{99}Tc is high (6%). In addition, ^{99}Tc appears in the short-lived isomeric form ^{99m}Tc, which is widely used in radiopharmaceuticals for organ imaging. It is a gamma emitter, with a gamma energy of 140 keV, and is also used as a tracer in ^{99}Tc

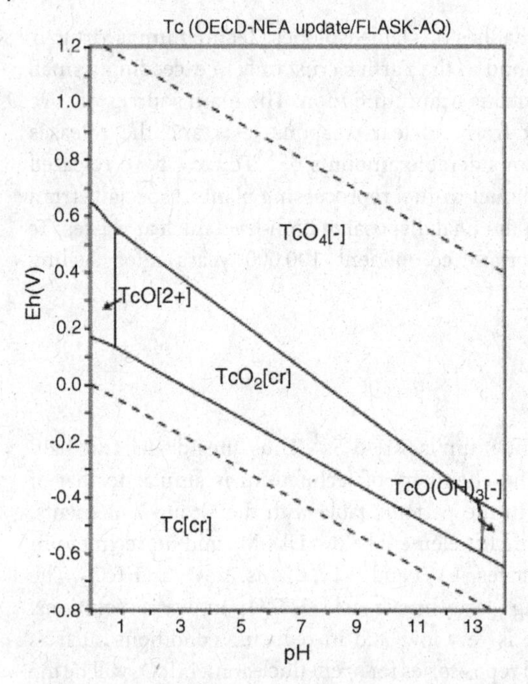

Figure 9.1 Eh-pH diagram of technetium (Atlas of Eh-pH diagrams, Geological Survey of Japan Open File Report No. 419, 2005).

investigations. Its short half-life (6 h) imposes some limitations, however. After radiochemical separation, the determination of ^{99}Tc activity is carried out by liquid scintillation counting or with a gas ionization detector. For measurement with a gas ionization detector, ^{99}Tc is electrodeposited on a stainless steel disk. ICP-MS,

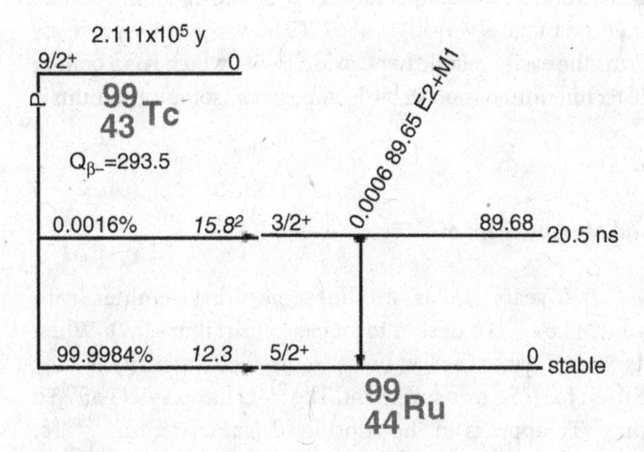

Figure 9.2 Decay scheme for ^{99}Tc (Firestone, R.B., Shirley, V.S., Chu, S.Y.F., Baglin, C.M., and Zipkin, J. (1996) *Table of Isotopes*, Wiley-Interscience).

which offers much lower detection limits, has recently become popular for measuring technetium concentrations in solution. With liquid scintillation and gas ionization, determination limits are in the range of 1–100 mBq/sample depending on the separation method: lower detection limits are observed with gas ionization than with liquid scintillation. ICP-MS, in turn, may allow measurements even as low as 0.001 mBq activities in a sample.

9.3.3 Separation of ^{99}Tc

Because of its high solubility in oxidizing conditions, ^{99}Tc can be found in natural waters, and thus the separation of technetium from water samples has been an important task. Concentrations of technetium in seawater and its accumulation in seaweed have been widely investigated in the Irish Sea and North Sea, for example. Another important research area concerns the final disposal of nuclear waste, especially the behavior of spent nuclear fuel and reprocessing waste in geological repositories.

For the separation of ^{99}Tc, various methods have been used, including solvent extraction, ion exchange, and extraction chromatography. Pertechnetate forms a complex with a number of organic ligands, such as triisooctyl amine (TIOA) and tributylphosphate (TBP), which are used in the analytical separations of ^{99}Tc. Pertechnetate ion is strongly bound to anion exchange resins in dilute acids, and, by using anion exchange in these conditions, Tc can be removed from actinides which are not retained in the anion exchange resins in dilute acids. The same principle is used for technetium separation by extraction chromatography with TEVA Resin; this has the same functionality as strongly basic anion exchange resins. Since the concentration of technetium in natural waters is very low, a preconcentration step is needed. This is usually based on the reduction of pertechnetate to tetravalent technetium and coprecipitation with iron hydroxide. Since iron is also reduced, ferrous hydroxide, $Fe(OH)_2$, is formed. Iron can be removed from technetium by dissolving the precipitate and oxidizing technetium to pertechnetate and iron to ferric iron, after which the latter is precipitated as hydroxide while technetium remains in solution as TcO_4^-.

9.3.3.1 Yield Determination in ^{99}Tc Analyses

Technetium has one suitable gamma-emitting isotope, ^{99m}Tc, with a half-life of 6 h, which is used as a tracer in determining chemical yield. ^{99m}Tc is obtained from a generator in which ^{99}Mo is adsorbed in an aluminum oxide column as MoO_4^{2-}. ^{99}Mo is obtained by neutron irradiation of stable ^{98}Mo or ^{235}U in a reactor. In the column, ^{99}Mo decays by beta emission to ^{99m}Tc, which can be eluted with sodium chloride solution as TcO_4^- while molybdenum remains. Since ^{99m}Tc also emits conversion electrons which overlap with the beta spectrum of ^{99}Tc, it must decay before measuring the beta activity of ^{99}Tc: first, the ^{99m}Tc activity is measured with a gamma counter; and then after several days, the ^{99}Tc activity is measured with LSC or with gas ionization counting. Stable rhenium, the chemically closest element, can also be used in chemical yield determinations if isomer ^{99m}Tc is not available as a tracer or if its half-life (6 h) is too short for the task. In addition, gamma-emitting technetium isotopes ^{95m}Tc ($t_{1/2}=61$ d) and ^{96}Tc ($t_{1/2}=4.3$ d) can be used as tracers, but these are not readily available.

A further alternative way to determine the yield is by parallel analyses of the same sample: one with the addition of a known amount of ^{99}Tc, that is, the same isotope as the analyte, and another without this addition. The chemical yield is then calculated as the difference in count rates between the sample with added ^{99}Tc and the sample without added ^{99}Tc. The difference in count rates of these samples divided by the added activity gives the chemical yield.

9.3.3.2 Enrichment of ^{99}Tc for Water Analyzes

Concentrations of ^{99}Tc in natural water are very low. In the North Sea, for example, it is a few bequerels per cubic meter, which means that samples need to be as large as 50–200 liters. The first step in radiochemical separations is to enrich the technetium in seawater into a smaller volume. Two methods are used: ion exchange and precipitation.

In ion exchange, filtered seawater is directly loaded to a strongly basic anion exchange column, in which anionic TcO_4^- is strongly sorbed, and the interfering ruthenium is captured as $RuCl_4^-$ complex. Ruthenium is removed from the column by oxidizing it from Ru^{3+} to Ru^{4+} with NaClO and eluting it with EDTA. Ru^{4+} forms a strong neutral complex which does not remain in the column. The TcO_4^- remaining in the column is eluted with concentrated nitric acid (200–600 mL).

In the precipitation method, iron hydroxide is used as a coprecipitant to precipitate technetium as TcO_2. The TcO_4^- is reduced to tetravalent Tc^{4+} at pH 4 with potassium pyrosulfite ($K_2S_2O_5$). Ferric iron is also added, which is reduced to Fe^{2+} by the action of pyrosulfite. The pH is then adjusted to 9 with sodium hydroxide, which results in the coprecipitation of TcO_2 with the forming $Fe(OH)_2$. The precipitate is dissolved in hydrochloric acid, after which technetium is oxidized to TcO_4^- and iron to Fe^{3+} with hydrogen peroxide. The iron is then removed by precipitating it at pH 9 as $Fe(OH)_3$, while TcO_4^- remains in solution.

9.3.3.3 Separation of ^{99}Tc from Water by Precipitation and Solvent Extraction

After the enrichment of ^{99}Tc by anion exchange and the separation of ruthenium by the NaClO/EDTA method (see above), concentrated samples are treated as described in Figure 9.3.

After the addition of NaClO and H_2SO_4, the solution is heated to remove the remaining Ru (^{106}Ru) as volatile RuO_4. Silver ($AgNO_3$) as a hold-back carrier is added for the removal of ^{110m}Ag as AgCl precipitate. Iron is then added, and iron hydroxide is precipitated by raising the pH to 9 to remove U, Th, Pu, Am, Po, Np, and Ag by coprecipitation with $Fe(OH)_3$. The precipitate containing both $Fe(OH)_3$ and AgCl is removed by centrifugation and discarded. The solution is acidified with sulfuric acid, after which potassium persulfate is added, and TcO_4^- is extracted with triisooctyl amine. The rest of the U and Th remain in the water phase. Technetium

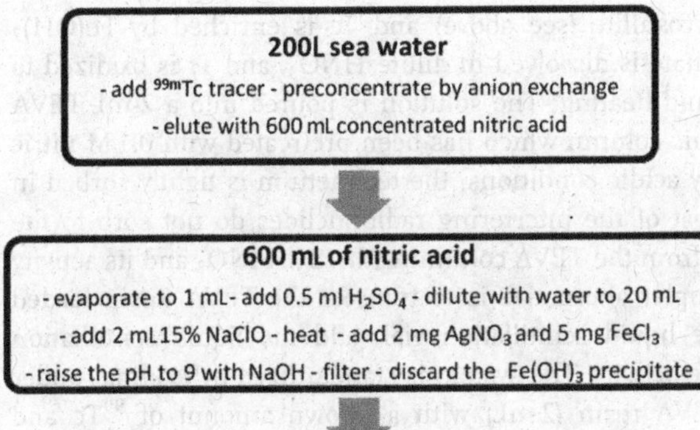

200L sea water

- add ^{99m}Tc tracer - preconcentrate by anion exchange

- elute with 600 mL concentrated nitric acid

600 mL of nitric acid

- evaporate to 1 mL - add 0.5 ml H_2SO_4 - dilute with water to 20 mL

- add 2 mL 15% NaClO - heat - add 2 mg $AgNO_3$ and 5 mg $FeCl_3$

- raise the pH to 9 with NaOH - filter - discard the $Fe(OH)_3$ precipitate

Water solution

- add 5 mL 10M H_2SO_4 and 1 g $K_2S_2O_8$

- extract with 10% tri-iso-octylamine (TIOA) in xylene

TIOA/xylene solution

- back-extract with 2M NaOH

NaOH solution

- measure ^{99m}Tc with a gamma counter - calculate yield

- electrodeposit Tc on stainless steel disk - wait a week

- measure ^{99}Tc activity with a gas ionization detector

Figure 9.3 Determination of ^{99}Tc in seawater (Chen, Q., Aarkrog, A., Nielsen, S.P., Dahlgaard, H., Lind, B., Kolstad, A.K., and Yu, Y. (2001) Procedures for determination of ^{239,240}Pu, ^{241}Am, ^{237}Np, ^{234,238}U, ^{228,230,232}Th, ^{99}Tc and ^{210}Pb-^{210}Po in environmental samples, Risø-R-1263(EN) Report, Risö National Laboratory, Roskilde, Denmark,).

is back-extracted into sodium hydroxide solution, and ^{99m}Tc activity is measured with a gamma counter to determine the chemical yield. Technetium is then electrodeposited on a steel disk and its activity measured after seven days with a gas ionization detector after ^{99m}Tc has decayed. In addition to TIOA, tributyl phosphate (TBP) has been used as a liquid extraction reagent in the separation of technetium.

9.3.3.4 Separation of ^{99}Tc by Extraction Chromatography

In the separation of ^{99}Tc by extraction chromatography, a ^{99m}Tc tracer is added to the seawater to determine the chemical yield. Thereafter, TcO_4^- is reduced to

Tc^{4+} with potassium pyrosulfite (see above) and Tc is enriched by $Fe(OH)_2$ precipitation. The precipitate is dissolved in dilute HNO_3, and Tc is oxidized to TcO_4^- by adding H_2O_2 and heating. The solution is poured into a 2-mL TEVA extraction chromatographic column which has been pretreated with 0.1 M nitric acid. Under these slightly acidic conditions, the technetium is tightly sorbed in the column, whereas most of the interfering radionuclides do not sorb to the resin. ^{99}Tc can be eluted from the TEVA column with 12 M HNO_3 and its activity measured by LSC. A simpler alternative is to transfer the TEVA Resin loaded with ^{95}Tc directly to the liquid scintillation vial, add the liquid scintillation solution, and measure the activity. The system is standardized by loading, under the same conditions, TEVA resin (2 mL) with a known amount of ^{99}Tc and comparing the counting rates obtained from the unknown samples with those of the spiked samples to clarify the quenching caused by measuring the activity directly from the resin. (http://www.eichrom.com/docs/methods/pdf/tcw01-16_tc-water.pdf).

The functional group in TEVA resin is an aliphatic tertiary amine $R_3NCH_3^+NO_3^-$(or Cl^-). This is the same functional group as that in strongly basic anion exchange resins. The amine in TEVA resin is in solution and forms more flexible and stronger complexes than the same group when attached to a polymer. Tetravalent actinides are strongly bound to TEVA resin when in acid solution – especially in 2–4 M nitric acid where they form negatively charged nitrate complexes (see Figure 4.13). Technetium, in turn, binds strongly to TEVA in weakly acidic, neutral, and even alkaline solutions, where the actinides are not bound and they and other cationic radionuclides are removed.

9.3.3.5 Separation of ^{99}Tc by Distillation

TcO_4^- can be separated by distillation in acids as Tc_2O_7. The best yields are achieved in perchloric acid medium, $HClO_4$, in which 75% of the technetium is distilled when 20% of the perchloric acid has been distilled. Relatively good yields are also obtained in sulfuric acid if an oxidizing agent, such as $Na_2S_2O_8$ or KIO_4, is added as well. However, yields are low in nitric acid medium.

9.4 Zirconium – ^{93}Zr

In nuclear explosions and in the Chernobyl accident, considerable amounts of gamma-emitting ^{95}Zr were released to the environment. Because of its short half-life of only 64 days, it disappeared in two years after deposition. The main interest in zirconium comes from the fact that a longer lived zirconium radioisotope ^{93}Zr is formed in nuclear fuel by fission and in the Zircaloy cladding surrounding the fuel by neutron activation. Thus, ^{93}Zr is important for evaluating the long-term safety of the final disposal of spent nuclear fuel, and is the reason why its radioactivity levels and solubility are currently being studied.

9.4.1 Chemistry of Zirconium

Zirconium is located in the fourth group in the periodic table, with the electron configuration of $[Kr]4d^25s^2$. The only important oxidation state of zirconium is +IV, and the corresponding oxide is ZrO_2, which is very difficult to dissolve in acids but dissolves as an anionic hydroxide complex ($Zr(OH)_5^-$) in alkaline solutions. Because of its good properties such as corrosion resistance, hardness, high melting point (1852 °C), and low neutron absorption cross section (0.18 b), zirconium metal is widely used as a cladding material for nuclear fuel.

The solution chemistry of Zr is very complicated and not very well understood. This is mainly because zirconium readily forms colloids and because zirconium ions undergo extensive hydrolysis and polymerization, which strongly depend on the pH and concentration of Zr. At zirconium concentrations above 10^{-9} M and in acidic solutions, zirconium mainly exists as Zr^{4+} ion, which is complexed in high acid concentrations with the anion of the acid, forming complexes, for example, $ZrCl_6^{2-}$ in hydrochloric acid and $Zr(SO_4)_3^{2-}$ in sulfuric acid. Because of the high charge of zirconium, hydrolysis starts in acidic solutions, and already at pH 0 about one tenth of zirconium is as monohydroxide complex $Zr(OH)^{3+}$. With increasing pH, higher hydroxide complexes ($Zr(OH)_2^{2+}$, $Zr(OH)_3^+$) appear, and already at pH 3 the neutral species $Zr(OH)_4$ prevails (Figure 9.4). Colloid formation and precipitation of zirconium take place at very low zirconium concentrations since the solubility of $Zr(OH)_4$ is very low, around 10^{-8} M. Zirconium hydroxide is not a stoichiometric compound but a hydrous oxide which, after time, crystallizes toward ZrO_2.

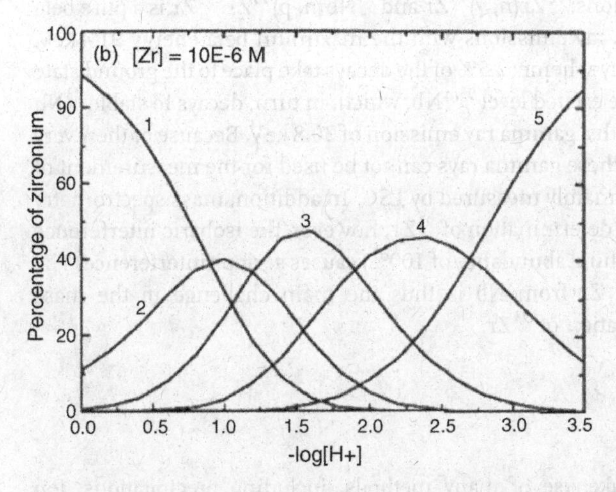

Figure 9.4 Hydrolysis of zirconium at a total concentration of 10^{-6} M in 1 M (Na,H)ClO$_4$ medium, 1: Zr^{4+}; 2: $ZrOH^{3+}$; 3: $Zr(OH)_2^{2+}$; 4: $Zr(OH)_3^+$; 5: $Zr(OH)_4$. (Bombard, A. (2005) Determination of the long-lived radionuclides ^{93}Zr, ^{93}Mo and ^{94}Nb in samples originally from the nuclear industry, PhD thesis, Université de Nantes).

Zirconium hydroxide/oxide will be the solubility limiting solid phase in deep geological repositories for spent nuclear fuel.

The nitrate, chloride, bromide, and sulfate of zirconium are soluble in acid solutions. Phosphate, iodate, phenyl arsonate, cupferrate, oxalate, and hydroxide of zirconium, and barium hexafluorozirconate are insoluble, the latter being widely used to separate zirconium from matrix elements and from other transition metals. Trace amounts of zirconium are strongly coprecipitated with most precipitates in acid solution if no complex forming ions, such as fluoride, are present. An example of such a coprecipitating compound is $CePO_4$.

Zirconium forms complexes with many ions such as Cl^-, NO_3^-, HSO_4^-, oxalate, and F^-, of which the fluoride complex is the most stable. Zirconium also forms chelate complexes with a number of organic compounds such as TTA (thenoyltrifluoroacetone), TOA (trioctylamine), PMBP(1-phenyl-3-methyl-4-benzoyl-5-pyrazolone), BPHA (*N*-benzoyl-*N*-phenylhydroxylamine) and TBP (tri-*n*-butylphosphate). These agents are used for the separation of zirconium by solvent extraction.

9.4.2
Nuclear Characteristics and Measurement of ^{93}Zr

Zirconium has five stable isotopes and more than 25 radioisotopes, most being short-lived (half-life <1 day). The most important radioisotope, considering the final disposal of spent nuclear fuel and high-level nuclear waste, is ^{93}Zr, a fission product of uranium and plutonium, since it has a very long half-life (1.5×10^6y) and a high fission yield (6.3%). It is also produced by neutron activation reaction from stable ^{92}Zr (17.15% natural abundance) and ^{93}Nb (100% natural abundance) via neutron activation reactions: $^{92}Zr(n, \gamma)^{93}Zr$ and $^{93}Nb(n, p)^{93}Zr$. ^{93}Zr is a pure beta emitter without gamma ray emissions with the maximum beta energy 91.4 keV. Figure 9.5 shows its decay scheme: 2.5% of the decays take place to the ground state of ^{93}Nb and 97.5% to the excited level ^{93m}Nb, which, in turn, decays to stable ^{93}Nb by internal transition with a gamma ray emission of 30.8 keV. Because of their very low intensity (0.006%), these gamma rays cannot be used for the measurement of ^{93}Zr. Therefore, ^{93}Zr is mainly measured by LSC. In addition, mass spectrometry can be also used for the determination of ^{93}Zr; however, the isobaric interference from the stable ^{93}Nb (natural abundance of 100%) causes serious interference. The complete separation of Zr from Nb is thus the main challenge in the mass spectrometric determination of ^{93}Zr.

9.4.3
Separation of ^{93}Zr

Separations of ^{93}Zr make use of many methods, including precipitations, ion exchange, and solvent extraction. Important materials which are being analyzed for ^{93}Zr are spent nuclear fuel and its Zircaloy cladding. As mentioned earlier, the measurement of zirconium is done either by LSC or by mass spectrometry. If the

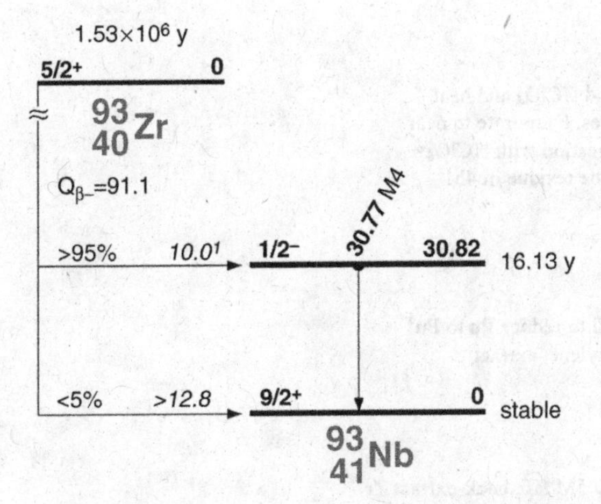

Figure 9.5 Decay scheme of ^{93}Zr (Firestone, R.B., Shirley, V.S., Chu, S.Y.F., Baglin, C.M., and Zipkin, J. (1996) *Table of Isotopes*, Wiley-Interscience).

measurement is done by LSC, a suitable tracer for the chemical yield determination is the short-lived zirconium isotope ^{95}Zr ($t_{1/2} = 64$ d) emitting gamma rays with energies of 368 keV (intensity 54.5%) and 401 keV (intensity 44.3%).

9.4.3.1 Determination of ^{93}Zr by TTA Extraction and Measurement by LSC

A method for the determination of ^{93}Zr from nuclear waste and environmental samples using TTA extraction and activity measurement by LSC is presented in Figure 9.6. In this method, the sample is first digested with HF, HNO_3 and $HClO_4$. The ensuing solution is evaporated to near dryness, and the residue is digested with HF-$HClO_4$ to ensure the complete dissolution of Zr. The final residue is dissolved in HCl- $AlCl_3$ solution and $NH_2OH.HCl$ is added to reduce the plutonium to Pu^{3+} to avoid its extraction to TTA. Thereafter, TTA-xylene solution is added to extract Zr to the organic phase as $ZrCl_4$. In these conditions, ^{55}Fe as trivalent iron is coextracted to the organic phase. The Zr in the organic phase is then back-extracted to the aqueous phase using HCl-HF, whereby ^{55}Fe remains in the organic phase. The activity of ^{93}Zr in the aqueous phase is finally measured by LSC. This method results in reasonably high decontamination factors for all interfering radionuclides such as: ^{55}Fe, ^{60}Co, ^{63}Ni, ^{65}Zn, ^{90}Sr, ^{94}Nb, ^{133}Ba, ^{137}Cs, ^{152}Eu, and Pu and in high chemical yields of more than 80%.

9.4.3.2 Separation of ^{93}Zr by Coprecipitation and Solvent Extraction for the Zr Measurement by ICP-MS

Another separation procedure using coprecipitation of $BaZrF_6$, the extraction of Zr with cupferron, and anion exchange chromatography for purification of ^{93}Zr from ^{55}Fe has also been reported for the determination of ^{93}Zr in reactor materials (irradiated Zr cladding material) (Figure 9.7). The irradiated Zircaloy material is first dissolved with HNO_3 and HF, and the Zr in the solution is then coprecipitated as $BaZrF_6$ by

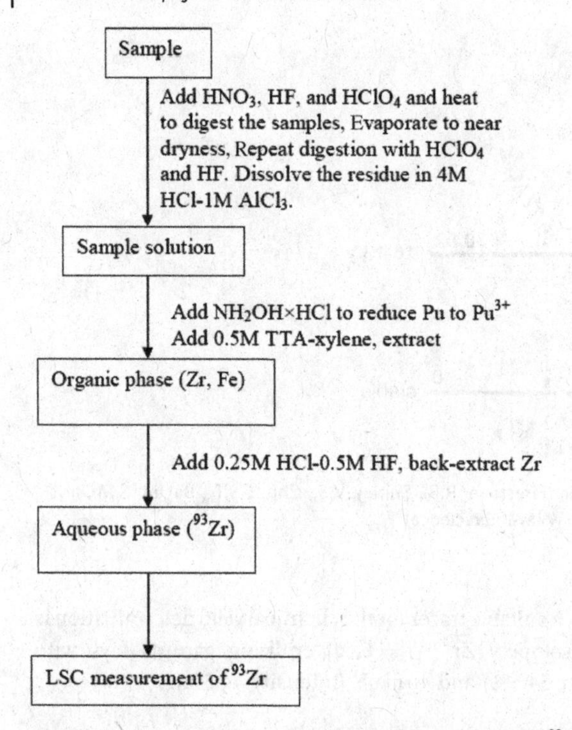

Figure 9.6 Separation procedure for the determination of ^{93}Zr in nuclear waste and environmental samples (Espartero, A.G., Suarez, J.A., Rodriguez, M., Pina, G. (2002) Radiochemical analysis of ^{93}Zr. *Appl. Radiat. Isot.*, **56**, 41).

adding solid $Ba_3(AlF_6)_2$ to the solution. After the separation of the precipitate by centrifugation, the coprecipitated Zr is dissolved with concentrated HCl-HNO_3. The daughter nuclide ^{93m}Nb of ^{93}Zr in the solution is then removed by extraction using BPHA in $CHCl_3$. The remaining Zr in the aqueous phase is then extracted with 5% cupferron in $CHCl_3$ and then back-extracted with 6 M HNO_3. A further coprecipitation step is followed by adding Al^{3+} and adjusting the pH to neutral range to form $Al(OH)_3$, which results in the removal of the matrix components and some of the interfering radionuclides that are not precipitated with aluminum hydroxide. Al $(OH)_3$ is dissolved with 4 M HCl-0.15 M $H_2C_2O_4$, and Zr is purified by extraction using 1% BPHA in $CHCl_3$ after the destruction of $H_2C_2O_4$ in the aqueous phase with $KMnO_4$ and acidification to 6–8 M HCl. The Zr in the organic phase is back-extracted with 0.025 M $H_2C_2O_4$. The solvent extraction step is repeated to remove most of the ^{93m}Nb and ^{93}Mo interferences. An anion exchange chromatographic separation follows to remove the ^{55}Fe. In this step, HCl is first added to the solution to form $FeCl_4^-$ anion, and the solution is then loaded to the anion exchange column; the $FeCl_4^-$ is adsorbed on the column, while Zr as $Zr(C_2O_4)_2$ complex passes through the column and exists in the effluent; this solution is then used for the measurement of ^{93}Zr using ICP-MS or LSC.

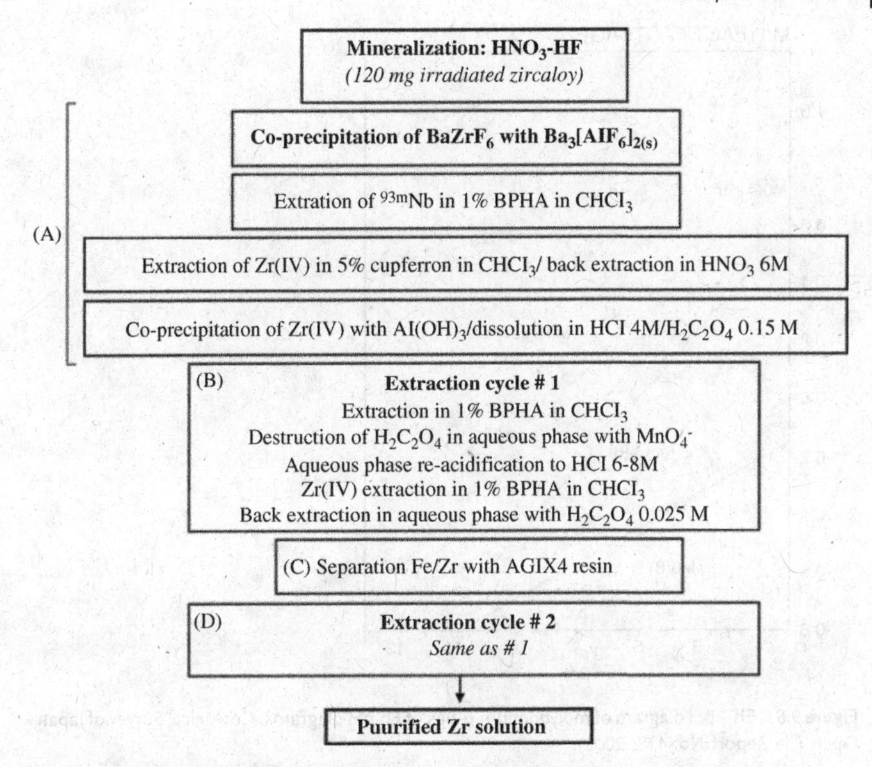

Figure 9.7 A scheme for the separation of Zr from irradiated cladding materials of nuclear fuel to determine ^{93}Zr using ICP-MS (Chartier, F., Isnard, H., Degros, J.P., Faurce, A.L., and Frechou, C. (2008) Application of the isotope dilution technique for ^{93}Zr determination in an irradiated cladding material by multiple collector ICP-MS. *Int. J. Mass Spectrom.*, **270**, 133).

9.5 Molybdenum – ^{93}Mo

9.5.1 Chemistry of Molybdenum

Molybdenum is in the sixth group in the periodic table, with the electron configuration [Kr]$4d^5 5s^1$. It can exist in oxidation states from 0 to +VI, the most stable oxidation states in aqueous solution being +VI and +IV, the former prevailing in alkaline media and in oxidizing conditions, while the latter is more stable in acidic solutions and in reducing conditions (Figure 9.8). In the chemical conditions prevailing in deep geological repositories for spent nuclear fuel, the most probable solid phase, limiting the solubility of molybdenum is the tetravalent molybdenum oxide MoO_2. Molybdenum is a hard metal with a melting point of 2620 °C. The metal is soluble in dilute nitric acid, *aqua regia*, and concentrated sulfuric acid. MoO_3 is the

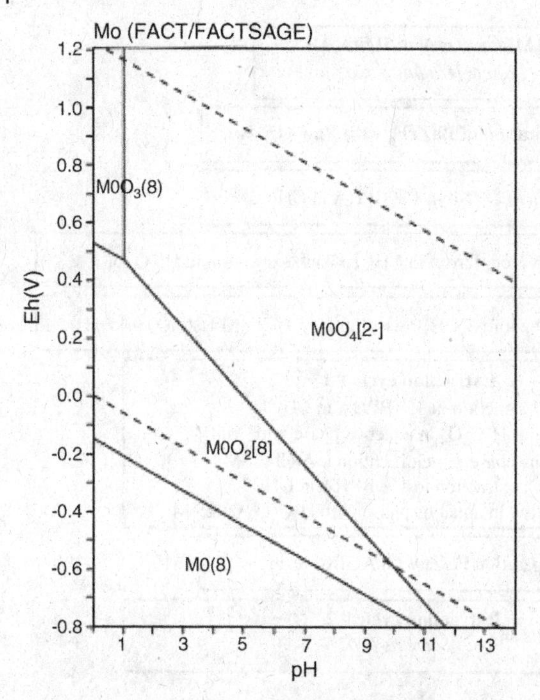

Figure 9.8 Eh – pH diagram of molybdenum (Atlas of Eh-pH diagrams, Geological Survey of Japan Open File Report No. 419, 2005).

other stable oxide of molybdenum, with a relatively low melting point (791 °C). It is slightly soluble in water and readily soluble in alkaline solutions, forming molybdate ions (MoO_4^{2-}). MoO_3 can also be dissolved in HF and concentrated H_2SO_4. Most molybdates, including ammonium, sodium, potassium, and magnesium salts, are water soluble while $PbMoO_4$, Hg_2MoO_4, $CaMoO_4$, Ag_2MoO_4, and $Th(MoO_4)_2$ are water insoluble. In addition, cupferron, 8-hydroxyquinoline, and α-benzoinoxime form solid compounds with molybdenum, and these are often used in the gravimetric determination of Mo. For trace amounts of molybdenum, MnO_2, 8-hydroxyquinolinolates of indium or aluminum, and Bi_2S_3 have been used for the coprecipitation. Mo(VI) is easily reduced by the standard reducing reagents such as Zn, $SnCl_2$, SO_2, sulfite, and hydrazine to Mo(III), Mo(IV), or Mo(V), depending on the conditions in solution.

Molybdenum chemistry is very complex and not thoroughly understood. In slightly acidic to alkaline solutions, Mo mainly exists as molybdate ions (MoO_4^{2-}), while in acidic solution, molybdate ions condense to form polymolybdate ions, and in strong acids such as concentrated HNO_3 and HCl, molybdic acid (H_2MoO_4) forms. Molybdenum – a typical transition metal – has a strong tendency to form complex ions: both Mo(VI) and Mo(V) form complexes with F^-, Cl^-, Br^-, CN^- and SCN^-, oxalate, citrate, thioglycolic acid, phenylhydrazine, catechol, and EDTA. Chelate compounds of Mo with cupferron, 8-hydroxyquinoline, and α-benzoinoxime (ABO), dissolved in organic solvent, can be used to extract Mo for separation and analysis. Acetylacetone ($C_5H_7O_2$), toluene-3, 4-dithiol, and EDTA also form chelate complexes with Mo.

Anion exchange chromatography is often used to separate Mo from other metals, since MoO_4^{2-}, the most typical form of molybdenum in solution, is strongly adsorbed on strongly basic anion exchange resins in nitric acid solutions (see Figure 4.5). Mo is also strongly adsorbed on alumina, which is widely used to separate Mo from uranium and fission products and to prepare a ^{99}Mo-^{99m}Tc generator.

9.5.2
Nuclear Characteristics and Measurement of ^{93}Mo

Molybdenum has six stable isotopes (^{92}Mo, ^{100}Mo and ^{94}Mo-^{98}Mo), and more than 25 radioisotopes. Most of radioisotopes are short-lived ($t_{1/2} < 7$h). A somewhat longer-lived radioisotope of molybdenum is ^{99}Mo (65.9 h), a fission product often used for the determination of the burn-up of nuclear fuel. Another application of this nuclide is its use in ^{99m}Tc generators: ^{98}Mo is activated to ^{99}Mo in a neutron flux or produced as a fission product by irradiating ^{235}U in a reactor and adsorbed as MoO_4^{2-} on an aluminum oxide column. Eluting the column with NaCl releases the shorter-lived daughter ^{99m}Tc (6 h), from the column, while ^{99}Mo remains. $^{99m}TcO_4^-$ is one of the most applied radiopharmaceuticals used in nuclear medicine. In nuclear waste, the only important molybdenum isotope is ^{93}Mo, which has a long half-life of 4000 years. ^{93}Mo is an activation product (^{92}Mo(n,γ)^{93}Mo) of stable ^{92}Mo (14.84%), and is found as an impurity in construction materials in nuclear reactor systems. Because of its long half-life, ^{93}Mo is among the important radionuclides in the final disposal of spent nuclear fuel and of waste in the decommissioning of nuclear power plants.

^{93}Mo decays to ^{93m}Nb (88%) and ^{93}Nb (12%) by electron capture (Figure 9.9). ^{93m}Nb decays to stable ^{93}Nb by internal transition, mainly by internal conversion, and only a small fraction decays by emission of gamma rays. The use of these low-energy gamma rays (30.8 keV) for the activity measurement of ^{93}Mo is more or less impossible because of their very low intensity (0.0006%). In the decay of ^{93}Mo, X-rays are, however, emitted with energies of 16.5 keV (21.4%), 16.6 keV (40.9%), and 18.6 keV (9.4%). By measuring these X-rays using X-ray spectrometry and the Auger electrons using LSC, the activity measurement of ^{93}Mo can be accomplished. In this case, complete chemical separation of Mo from other radionuclides is necessary because of due their interference in the low-energy window of LSC for ^{93}Mo. The need to remove interfering radionuclides is also important in X-ray spectrometry – ^{93m}Nb, in particular, which has X-rays in the same energy range as ^{93}Mo, needs to be carefully removed. In addition to radiometry, mass spectrometry can be used to measure ^{93}Mo; however, the elements causing isobaric interference, especially stable ^{93}Nb (100%) and ^{93}Zr, have to be completely removed before measurement. The interference of polyatomic ions, such as ^{92}MoH and ^{92}ZrH, and the tailing of ^{92}Mo and ^{94}Mo also cause some problems when determining ^{93}Mo by mass spectrometry.

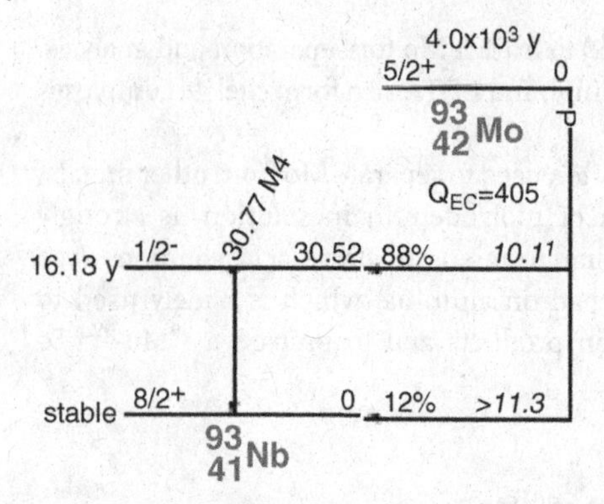

Figure 9.9 Decay scheme of ^{93}Mo (Firestone, R.B., Shirley, V.S., Chu, S.Y.F., Baglin, C.M., and Zipkin, J. (1996) *Table of Isotopes*, Wiley-Interscience).

9.5.3 Separation of ^{93}Mo

9.5.3.1 Separation of Radioactive Molybdenum by Aluminum Oxide

The often used chemical procedure for the separation of Mo from uranium and fission products is alumina column chromatography. In this method, the sample (irradiated uranium) is dissolved in 6 M HNO_3 catalyzed by $Hg(NO_3)_2$; the solution obtained is then passed through an alumina column, which selectively absorbs Mo. Uranium and fission products which are not adsorbed on the column are removed from the alumina column by washing with 1 M HNO_3, H_2O, and 0.01 M NH_4OH. The Mo is then eluted from the column using 1 M NH_4OH. The separated Mo in the 1 M NH_4OH solution is loaded to an anion exchange column, and the impurities that remain are removed by washing the column with NH_4OH and water. The Mo on the column is finally eluted by 1.2 M HCl (Boyd, R.E. (1982) Molybdenum-99: Technetium-99m generator. *Radiochim. Acta*, **30**, 123).

9.5.3.2 Separation of ^{93}Mo by Solvent Extraction

A radiochemical procedure for the determination of ^{93}Mo in nuclear waste is given in Figure 9.10. In this method, the nuclear waste sample is first dissolved in 1.5 M HF/1.2 M HCl/0.3 M HNO_3, and stable Mo is added as a carrier and for the chemical yield determination. To remove the radioactive cesium isotopes ^{137}Cs and ^{134}Cs, potassium cobalt hexacyanoferrate is added to the solution and mixed for 10 min, after which the hexacyanoferrate powder is separated by centrifugation. The supernatant is then loaded onto a strongly acidic cation exchange column preconditioned with 1 M NH_4Cl, and most of the interfering radionuclides, such as ^{60}Co, ^{90}Sr, ^{63}Ni, ^{55}Fe, ^{152}Eu, and so on, are removed by absorption on the column. The effluent, containing ^{93}Mo as molybdate ion and other anions, is evaporated to dryness; the residue is dissolved in 4 M HCl, and $AlCl_3$ is added up to a concentration of 1 M. Hydroxylamine hydrochloride is added to the solution to reduce plutonium to Pu

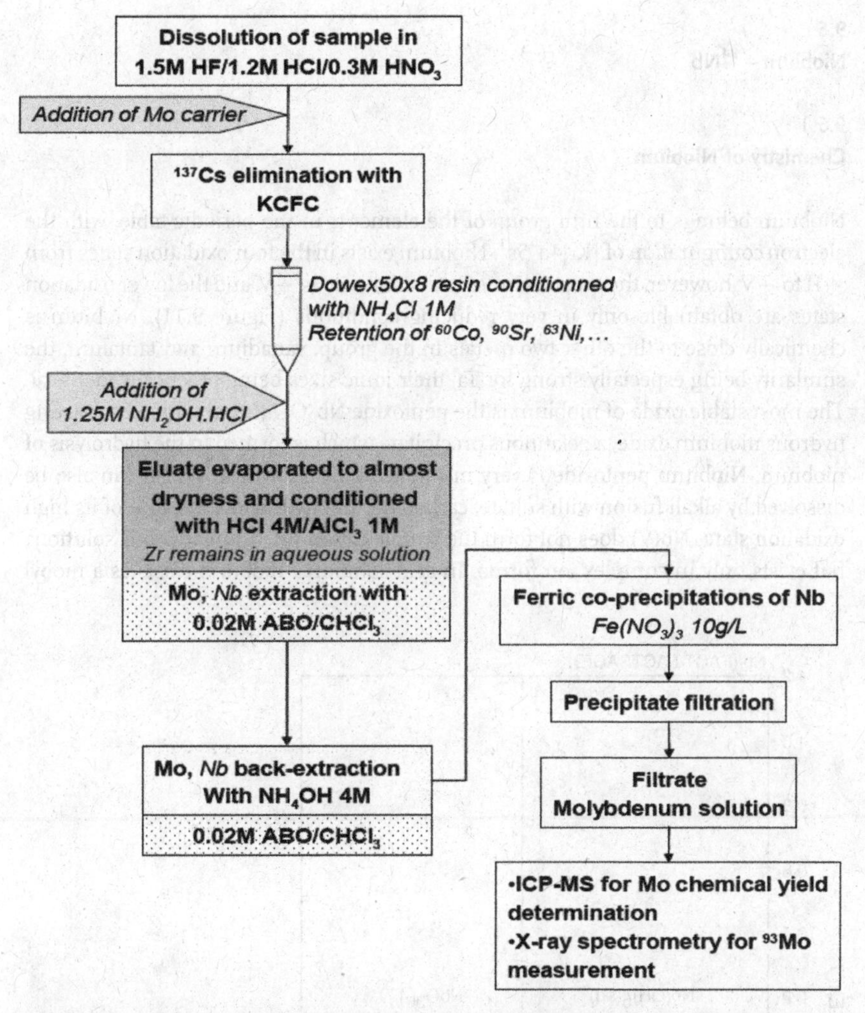

Figure 9.10 Chemical separation procedure for the determination of ^{93}Mo in nuclear waste (Bombard, A. (2005) Determination of the long-lived radionuclides ^{93}Zr, ^{93}Mo and ^{94}Nb in samples originally from the nuclear industry, PhD thesis, Université de Nantes).

(III) to prevent its extraction during the next step. 0.02 M ABO (α-benzoinoxime)/$CHCl_3$ is mixed with the solution to extract ^{93}Mo into the organic phase. Niobium is co-extracted along with the molybdenum. The Mo, in the organic phase, is then back-extracted from the organic phase using 4 M NH_4OH solution. Niobium, still present in the solution, is removed by coprecipitation with $Fe(OH)_3$. After the removal of the $Fe(OH)_3$ precipitate by filtration, an aliquot of the solution is taken for ICP-MS measurement to determine the chemical yield of the ^{93}Mo separation. The remaining solution is used to measure ^{93}Mo activity by X-ray spectrometry or LSC.

9.6
Niobium – ^{94}Nb

9.6.1
Chemistry of Niobium

Niobium belongs to the fifth group of the elements in the periodic table, with the electron configuration of $[Kr]4d^4 5s^1$. Niobium exists in the four oxidation states from +II to +V; however, the most common oxidation sate is +V, and the lower oxidation states are obtainable only in very reducing conditions (Figure 9.11). Niobium is chemically close to the other two metals in the group, vanadium and tantalum, the similarity being especially strong for Ta, their ionic sizes being practically identical. The most stable oxide of niobium is the pentoxide Nb_2O_5. It is obtained by roasting hydrous niobium oxide, a gelatinous precipitate which is formed in the hydrolysis of niobium. Niobium pentoxide is very insoluble, but dissolves in HF. It can also be dissolved by alkali fusion with sulfate, carbonate, and hydroxide. Because of its high oxidation state, Nb(V) does not form the simple cation (M^{5+}) in aqueous solutions but exists only in complex ion forms. In acid solutions, niobium exists as a niobyl

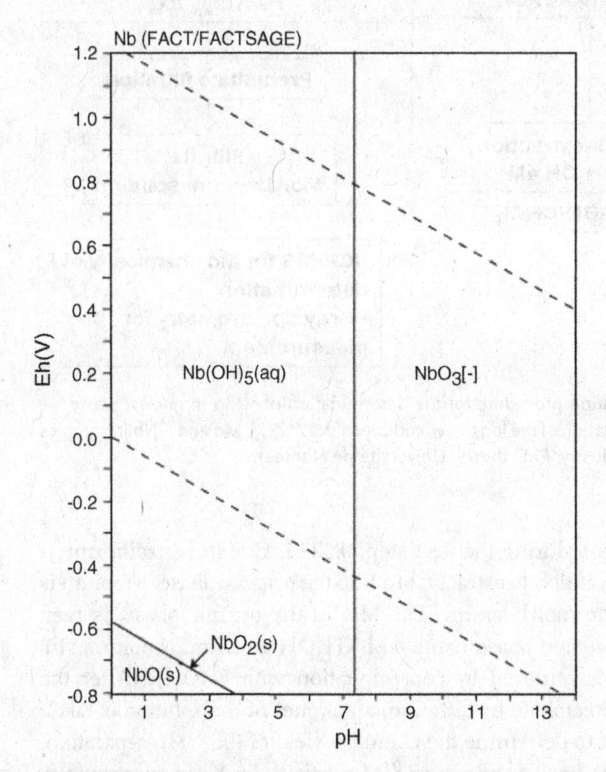

Figure 9.11 Eh – pH diagram of niobium (Atlas of Eh-pH diagrams, Geological Survey of Japan Open File Report No. 419, 2005).

oxocation NbO^{3+} and forms complexes with the anions of the acid, such as $NbOF_5^{2-}$ in hydrofluoric acid. Niobium forms nonoxic complexes MF_6^- and NbF_7^{2-} only in concentrated hydrofluoric acid. Niobium is readily hydrolysable, forming insoluble hydrous oxide with a nonstoichiometric composition. The soluble hydrolysis product in slightly acidic to neutral pH is niobic acid $HNbO_3$ (or $Nb(OH)_5$), which turns into niobate NbO_3^- (or $Nb(OH)_6^-$) at a pH above 7. Since the solubility of niobium is very low, these species exist in solution in only very low concentrations. Nb has a strong tendency to form colloids in solution. To ensure a true solution, strong complexing agents such as F^- and $C_2O_4^{2-}$ are added to the solution. Trace amounts of niobium are unstable in acid solution without complex-forming anions and tend to adsorb on both glassware and on many precipitates such as sulfides, hydroxides, zirconium phosphate, and silica gel.

9.6.2 Nuclear Characteristics and Measurement of Niobium Radionuclides

Niobium has only one stable isotope, ^{93}Nb, and more than 30 radioisotopes. Of these radioisotopes ^{93m}Nb (16.1 y) and ^{94}Nb (2×10^4 y) are the most important. Although they are both fission products of uranium and plutonium, the fission yield of ^{94}Nb is very low at only one millionth of a percent. ^{93m}Nb is a daughter nuclide of the long-lived ^{93}Zr and ^{93}Mo, discussed earlier in this chapter. ^{94}Nb and ^{93m}Nb are also produced by neutron activation reactions of stable niobium, $^{93}Nb(n, \gamma)^{94}Nb$ and $^{93}Nb(n, n)^{93m}Nb$. Neutron activation is the major route in the formation of ^{94}Nb. Stable niobium exists as an impurity in the structural components of nuclear reactor pressure vessels and the cladding of nuclear fuel. The long half-lives of these radionuclides, especially the very long half-life of ^{94}Nb, make them important radionuclides in the decommissioning of nuclear reactors and in the final disposal of spent nuclear fuel. ^{94}Nb decays to the excited state of ^{94}Mo by the emission of beta particles with a maximum energy of 471 keV. This excited state, in turn, decays to stable ^{94}Mo by the emission of gamma rays with energies of 702.6 keV (97.9%) and 871.1 keV (100%) (Figure 9.12). ^{94}Nb could thus be measured by gamma spectrometry. However, the concentration of ^{94}Nb in the nuclear waste is normally very low compared with other radionuclides such as ^{137}Cs and ^{60}Co. Major gamma emitting radionuclides cause a high Compton background, which makes the direct detection of ^{94}Nb gamma rays very difficult, and thus chemical separation is needed before its measurement.

9.6.3 Separation of ^{94}Nb

In the following, three methods for the radiochemical separation of ^{94}Nb are given. The first two make use, among other methods, of the precipitation of sparingly soluble Nb_2O_5 oxide; the third method applies solvent extraction for the separation of ^{93}Zr, ^{93}Mo, and ^{94}Nb. For the chemical yield determination, both stable niobium and a gamma-emitting ^{95}Nb radioisotope can be used.

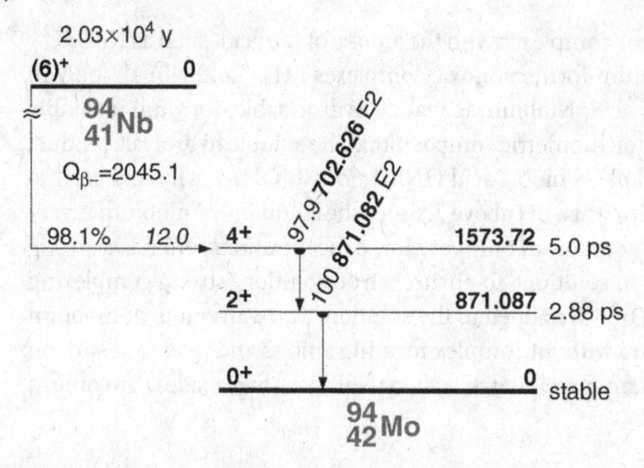

Figure 9.12 Decay scheme of ^{94}Nb (Firestone, R.B., Shirley, V.S., Chu, S.Y.F., Baglin, C.M., and Zipkin, J., (1996) *Table of Isotopes*, Wiley-Interscience).

9.6.3.1 **Separation of ^{94}Nb by Precipitation as Nb_2O_5**

The first method employs the selective precipitation of niobium for the determination of ^{94}Nb in nuclear waste, spent ion exchange resins, and evaporator concentrate. Samples of the resins and the concentrate are digested with H_2SO_4 and H_2O_2 to decompose the organics, after which the solution is evaporated to dryness and the residue is dissolved by a mixture of HNO_3, $HClO_4$, HCl, and HF acids. This solution is evaporated to form moist salts, and the residue is dissolved with HCl and HF acids. Stable Nb carrier and hold-back carriers are added to this solution, which is heated to evaporate it slowly to dryness, and the solid residue is treated with 4 M HCl. In these conditions, Nb(V) remains insoluble in the form of the pentoxide. These steps are repeated twice in order to get adequate decontamination of the pentoxide precipitate. At the end of the radiochemical separation process, the pentoxide precipitate is dissolved in HF acid to form the complex $NbOF_5^{2-}$, which is stable in aqueous solution. The ^{94}Nb in the solution is measured by gamma spectrometry; the chemical yield of the process is obtained by the measurement of the stable Nb carrier in the final solution using the spectrophotometric method. The decontamination factors for most interfering radionuclides, such as ^{137}Cs, ^{125}Sb, ^{65}Zn, ^{60}Co, and ^{54}Mn, are greater than 10^4, and the chemical yield is 50–60% (Espartero, A.G., Suarez, J.A., and Rodriguez, M. (1998) Determination of ^{93m}Nb and ^{94}Nb in medium and low level radioactive wastes, *Appl. Radiat. Isot.*, **49**, 1277).

9.6.3.2 **Separation of ^{94}Nb by Precipitation as Nb_2O_5 and by Anion Exchange**

Another method for the separation of ^{94}Nb from nuclear waste uses the precipitation of Nb_2O_5 and anion exchange chromatography. In this method, ^{95}Nb tracer and 10 mg Nb carrier are first added to the sample, which is then destroyed by digestion with 65% HNO_3. The residue, containing niobium as Nb_2O_5, is taken up in 1 M HCl. The solution is filtered, and the Nb_2O_5 precipitate on the filter is rinsed with 1M HCl, which is then dissolved in 22.5 M HF, diluted with distilled water, and filtered. The filtrate is loaded to a strongly basic anion exchange column, and the column is rinsed with 22.5 M HF. Niobium absorbed in the column as a negative fluoride complex is then eluted with 7 M HNO_3. The strip solution is gently evaporated in a Teflon beaker and evaporated in a glass beaker twice with 65% HNO_3 (10 mL), which results in the formation of Nb_2O_5 precipitate. This is taken up in 1 M HCl (20 mL), filtered, and rinsed with 1 M HCl. After drying, the ^{94}Nb as well as the ^{95}Nb tracer activities are determined by gamma spectrometry (Osvath, S., Vajda, N., and Molnar, Z. (2008) Determination of long-lived Nb isotopes in nuclear power plant wastes. *Appl. Radiat. Isot.*, **66**, 24).

9.6.3.3 Separation of ^{94}Nb by Solvent Extraction

The scheme of a combined procedure for the determination of ^{93}Zr, ^{93}Mo, and ^{94}Nb in nuclear waste is shown in Figure 9.13. In this method, ^{93}Zr, ^{93}Mo, and ^{94}Nb can be separated from interfering radionuclides and matrix components, as well as from each other. The separated nuclides are then measured using different techniques including LSC, ICP-MS, X-ray spectrometry, and γ-spectrometry. In this combined procedure, the sample is first decomposed with a mixture of acids, and stable Zr, Nb, and Mo are added to be used as carriers as well as yield tracers. The ^{137}Cs in the solution is first removed by adding potassium cobalt hexacyanoferrate to the solution to specifically adsorb ^{137}Cs, which is then separated by centrifuge. The supernatant is loaded to a cation exchange column: since Zr, Mo, and Nb exist as complex anions they are not retained in the column, unlike most transition elements including Co, Ni, Fe and other elements such as ^{90}Sr that exist as cations and are retained in the column. The effluent containing Zr, Mo, and Nb is then evaporated to dryness, the residue is dissolved in 4 M HCl, and $AlCl_3$ is added to the solution. The Zr is then separated by solvent extraction using HTTA/Xylene after adding $NH_2OH{\cdot}HCl$ to reduce Pu to Pu^{3+} to keep it in the aqueous phase during extraction. The extracted Zr is back-extracted using HF-HNO_3, and from this solution the ^{93}Zr is measured by ICP-MS or LSC. The Mo and Nb remaining in the aqueous phase during extraction are extracted with 0.02 M ABO-$CHCl_3$ and back extracted using 4 M NH_4OH. The Nb in the 4 M NH_4OH solution is separated by coprecipitation with $Fe(OH)_3$, and the ^{94}Nb is measured using gamma spectrometry. The ^{93}Mo remaining in the 4 M NH_4OH solution is measured by LSC or ICP-MS.

9.7 Essentials in the Radiochemistry of 4-d Transition Metals

- 4-d transition elements have several important long-lived radionuclides which require radiochemical separations ^{99}Tc, ^{93}Zr, ^{93}Mo, and ^{94}Nb. Of these, ^{99}Tc and ^{93}Zr are fission products and ^{93}Mo and ^{94}Nb neutron activation products formed in the metallic structures in nuclear power plants. In addition to fission, ^{93}Zr is also formed by neutron activation in the Zircaloy cladding of the nuclear fuel. The only radionuclide of these four found in the environment is ^{99}Tc, a very soluble radionuclide. The other three are only present in nuclear waste, in spent nuclear

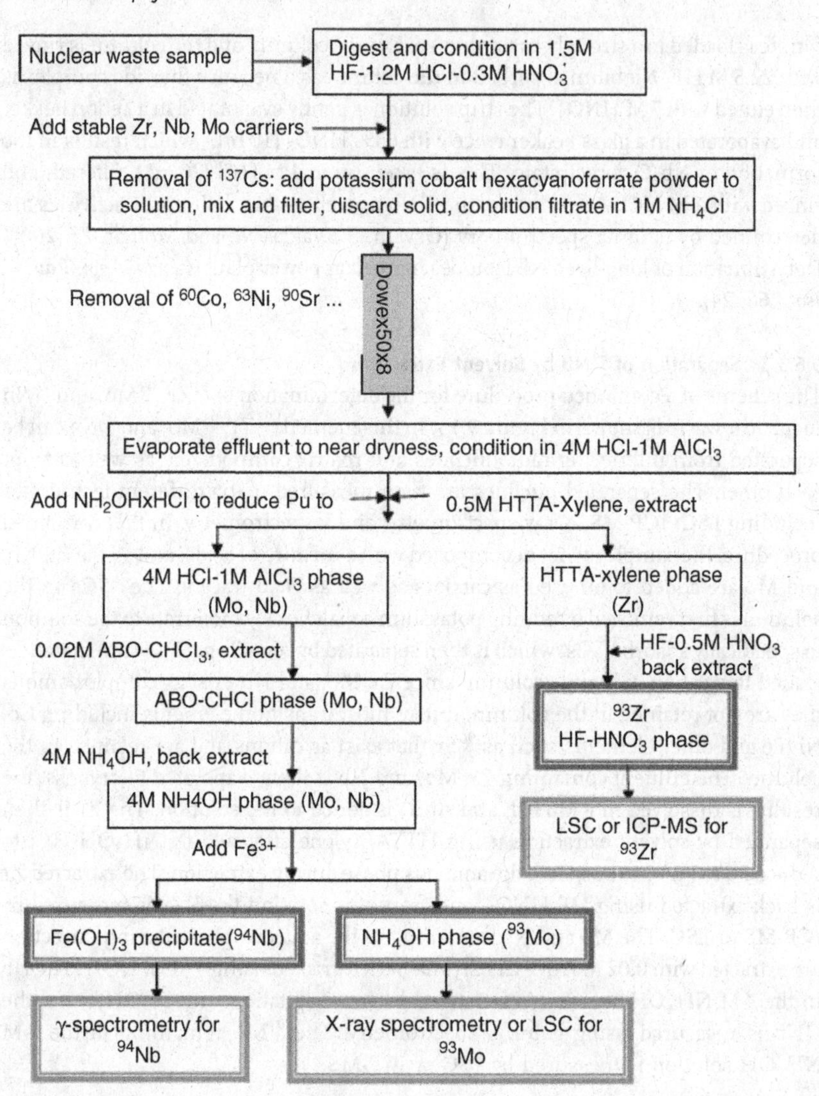

Figure 9.13 Combined analytical procedure for the determination of ^{93}Zr, ^{93}Mo and ^{94}Nb in nuclear waste (Bombard, A. (2005) Determination of the long-lived radionuclides ^{93}Zr, ^{93}Mo and ^{94}Nb in samples originally from the nuclear industry, PhD thesis, Université de Nantes).

fuel, and in reactor constructions, and are mainly relevant in connection with the final disposal of nuclear waste.

- ^{99}Tc is a pure beta emitter ($E_{max} = 294$ keV) with a very long half-life of 2.1×10^5 years. It can be measured by LSC, gas ionization detectors, and ICP-MS. The most prevailing oxidation state of technetium is +VII, and in aqueous solutions technetium is present as the pertechnetate ion, TcO_4^-, which is very soluble and mobile. In reducing conditions it is reduced to tetravalent technetium, which forms a sparingly soluble oxide, TcO_2. In natural waters, ^{99}Tc concentrations are so low that preconcentration is needed. Typical methods to separate technetium for ^{99}Tc determinations are solvent extraction, anion exchange, and extraction chromatography.

- ^{93}Zr is a pure beta emitter ($E_{max} = 91.4$ keV) with a very long half-life of 1.5×10^6 years. It can be measured by LSC, gas ionization detectors, and ICP-MS. The only important oxidation state of zirconium is +IV. It is a strongly hydrolyzing element and forms an insoluble $Zr(OH)_4$ even in slightly acidic solutions, and readily forms colloids in aqueous solutions. Precipitations, ion exchange, and solvent extraction are typical methods used in ^{93}Zr separations.
- ^{93}Mo decays by electron capture, and has a long half-life of 4000 years. It can be measured by detecting Auger electrons by LSC, by X-rays with an X-ray spectrometer, or by ICP-MS. The most typical oxidation state of molybdenum is +VI, and the typical chemical form in aqueous solution is the molybdate ion, MoO_4^{2-}. Molybdenum can be reduced to lower oxidation states, and, in reducing conditions, tetravalent molybdenum forms a sparingly soluble MoO_2 oxide. Typical methods used in ^{93}Mo separations are ion exchange and solvent extraction.
- ^{94}Nb is a beta emitter ($E_{max} = 471$ keV) with a long half-life of 20 000 years. It also emits high-energy gamma rays with high intensities. However, these gamma rays cannot be used to directly measure ^{94}Nb from nuclear waste samples because of the large presence of other gamma-emitting radionuclides. After chemical separation, ^{94}Nb can be measured by gamma spectrometry, LSC, and ICP-MS. The most typical oxidation state of niobium is +V, and it forms a sparingly soluble pentoxide Nb_2O_5 even in acidic solutions. The typical form in the solution phase is the niobate NbO_3^- ion (or $Nb(OH)_6^-$). Typical methods used in ^{94}Nb separations are ion exchange, solvent extraction, and precipitation as Nb_2O_5.

10
Radiochemistry of the Lanthanides

10.1
Important Lanthanide Radionuclides

Although the lanthanides include several fission-product radionuclides, their half-lives are generally relatively short. Table 10.1 lists fission-product radionuclides of lanthanides with half-lives of more than one day and fission yields greater than 1%. The most important of these in fresh fallout is ^{144}Ce, which has a half-life of almost a year and emits 133 keV (11.1%) gamma rays, forming a strong source of external radiation. Owing to its high fission yield and the high intensity of the gamma rays, the determination of ^{144}Ce is rather simple. The only radionuclides without observable gamma emission are ^{151}Sm and ^{147}Pm, and, for these, radiochemical separations are required before counting. With technetium, promethium is one of the two elements lighter than polonium that has no stable isotopes. ^{143}Pr has also no gamma emissions, but because of its short half-life it is not discussed further.

10.2
Chemical Properties of the Lanthanides

Lanthanides are conventionally known as rare-earth metals, but this is misleading since they are not particularly rare. The lanthanides are classed in the side group 3 of the periodic table. Scandium, yttrium, and particularly lanthanum are chemically similar to the lanthanides, and we therefore include lanthanum in this discussion of the lanthanides. Chemically, the lanthanides all behave similarly. All have full K, L, and M shells, and full 4s, 4p, 4d, 5s, and 5p orbitals. The lanthanide series is characterized by the successive addition of an electron to the seven 4f orbitals, so altogether, there are 14 lanthanides, or 15 with lanthanum included. The full $6p^2$ and $5p^6$ orbitals effectively shield the 4f electrons, which therefore behave like inner-shell electrons and do not participate in the formation of chemical bonds. An exception is cerium, the first in the group, after lanthanum, whose one f-electron takes part in the

Chemistry and Analysis of Radionuclides. Jukka Lehto and Xiaolin Hou

ISBN: 978-3-527-32658-7

Table 10.1 Important fission product lanthanides. All nuclides decay by β^- decay.

Element	Fission product nuclide	Half-life	Fission yield for ^{235}U (%)	Gamma emission
Lanthanum	^{140}La	1.7 d	6.5	yes
Cerium	^{141}Ce	33 d	5.8	yes
	^{144}Ce	285 d	4.4	yes
Praseodymium	^{143}Pr	14 d	6.0	no
Promethium	^{147}Pm	2.6 y	2.3	no
Samarium	^{151}Sm	90 y	0.4	no
Neodymium	^{147}Nd	11 d	2.3	yes

formation of chemical bonds. Indeed, cerium is the only lanthanide with a stable oxidation state of +IV. Except for Ce and Eu, the lanthanides typically appear in oxidation state +III and are present in aqueous solution as Ln^{3+} ions. Ce also appears as Ce^{4+} and europium as Eu^{2+}. The Ln^{3+} ions are formed through the loss of 5s and 5p electrons. Because the 4f orbitals are shielded, the lanthanides rarely form complexes with organic ligands. Complexes are only formed with strongly complexing chelates. Most lanthanide compounds are ion bonded and therefore relatively soluble. The ion sizes of the lanthanides decrease systematically from lanthanum to lutetium (Figure 10.1). Since the 4f orbitals are shielded, the addition of electrons does not expand the electron cloud, but the growing nuclear charge increases the electrostatic attraction on the electrons, thereby decreasing the ion size. This decrease in size is known as the lanthanide contraction.

The ion size affects both the solubility of the lanthanides and the formation of complexes. The smaller the ion, the stronger is the electric interaction with an oppositely charged ion and the poorer is the solubility. Similarly, as the atomic number increases, complex formation, including hydrolysis, strengthens. Because the lanthanides are chemically so similar, their separation from each other is fairly difficult. The separation processes depend on factors that reflect their differences in size, for example, their complex-forming ability.

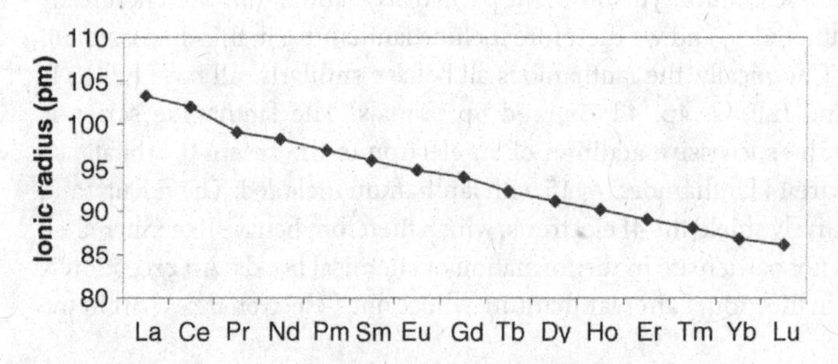

Figure 10.1 Ionic radii of lanthanides in six-coordination.

10.3
Separation of Lanthanides from Actinides

A major challenge for radiochemists wishing to determine the activity of the actinides is separating them from the stable isotopes of the lanthanides. Environmental samples naturally contain relatively large amounts of stable isotopes of lanthanides, and these amounts are much greater than the amounts of actinides that appear in the environment as pollutants, such as Pu, Np, Am, and Cm. The lanthanides behave very similarly to the trivalent actinides (Ac^{3+}, Pu^{3+}, Am^{3+}, and Cm^{3+}). Lanthanides interfere with the purification of actinides, especially in the preparation of sources for alpha counting involving coprecipitation with LnF_3. The lanthanides precipitate at the same time and increase the self-absorption in proportion to the mass of the source. In the case of plutonium, oxidation to higher oxidation states can be exploited in the removal of lanthanides. This, however, is not possible for actinium, americium, and curium, because, in practice, their only oxidation state is +III. Plutonium can be oxidized to the oxidation state +IV, and in 7–8 molar nitric acid it is trapped to a strong anion exchange resin. Trivalent lanthanides and actinides are not retained in the resin and pass the column. Plutonium can also be separated from trivalent lanthanides on a TRU extraction chromatographic resin (see Figure 4.11). Trivalent actinides, the most important of which is ^{241}Am, are separated from the lanthanides on either anion exchange resin or TEVA extraction chromatographic resin. In both methods, use is made of thiocyanate, SCN^-, which forms a complex with americium but not with the lanthanides. The sample, in 1 M HNO_3, is loaded into the anion exchange resin, and the lanthanides are removed from the column by rinsing with 0.5 M NH_4SCN, while americium (and curium) is retained in the column as a thiocyanate complex. In the case of TEVA resin, the column is rinsed with a stronger solution, 2 M NH_4SCN. Again, the lanthanides are removed, while americium (and curium) remains in the resin. These methods are presented in more detail in Chapter 15, in which the determination of americium and curium is discussed.

10.4
Lanthanides as Actinide Analogs

Because trivalent lanthanides behave in much the same way as actinides in the same oxidation state, they can be used as analogs for actinides. Figure 10.2 expresses the similarity very well. When trivalent actinides and lanthanides are bound on a strong cation exchange resin and eluted with ammonium-α-hydroxyisobutyrate, they elute from the column as mated pairs, at the same elution positions, in the order of their atomic numbers. The order is systematic, the larger the atomic number of the lanthanide or actinide, the faster it elutes from the column. In both series, the size of the trivalent ion decreases as the atomic number increases, and so the complex with ammonium-α-hydroxyisobutyrate is also stronger.

The use of lanthanides as analogs is particularly helpful in investigations requiring macro amounts of compounds. Actinides are rarely used in macro amounts, except

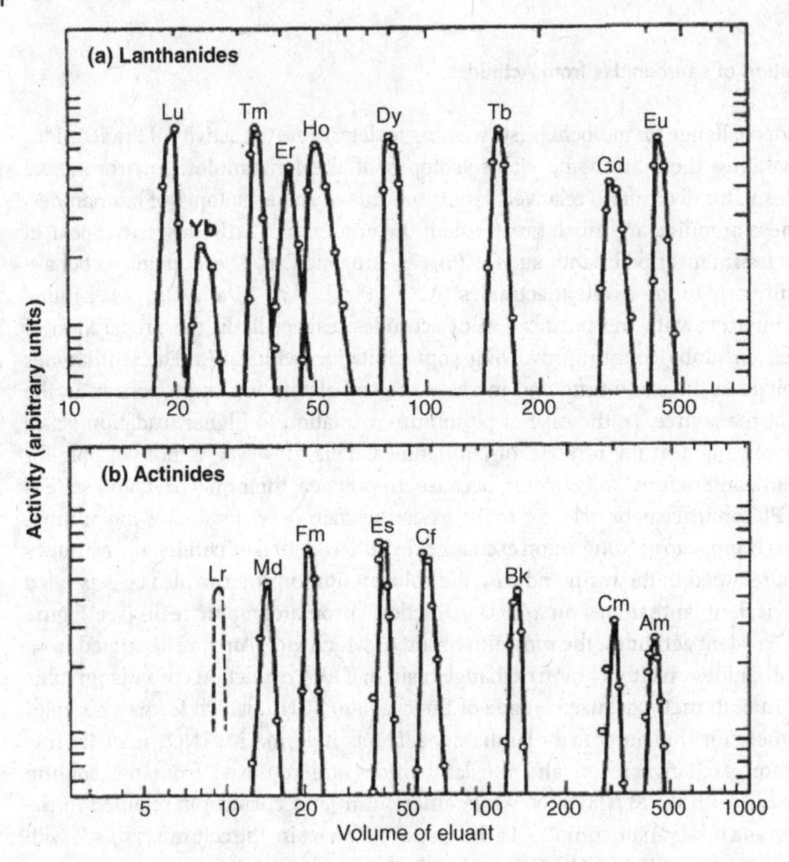

Figure 10.2 Elution of trivalent lanthanides and actinides from strong cation exchange resin with ammonium-α-hydroxyisobutyrate (Katz, J.J. and Seaborg, G.T. (1957) *The Chemistry of the Actinide Elements*, John Wiley & Sons, Inc., New York).

Th and U. Even if they were available, the activity levels would be so high that the radiation risk would be unacceptable or the treatment would require 'hot boxes' with remote handling devices. Corresponding species in the lanthanide series can be used to represent trivalent actinides: samarium for Pu, europium for Am, and gadolinium for Cm. Frequently, europium is used to represent all trivalent actinides. Moreover, europium has a useful radioisotope, ^{152}Eu, and its gamma emissions are more easily measured than the alpha activity of actinides. Lanthanides are also useful analogs in the spectroscopic investigation of chemical forms of actinides. Actinides are typically present in solution in such low concentrations that they cannot be observed, for example, by fluorescence spectroscopy. Laser-induced fluorescence spectroscopy (LIFS) is considerably more sensitive than normal UV/VIS spectroscopy, but even this technique is not sufficiently sensitive for the study of actinides (except uranium) in environmental samples. Model experiments, where chemical species in solution or on mineral surfaces are studied under precisely defined conditions, can be carried out with lanthanides instead. Indirect conclusions about the behavior of trivalent actinides under corresponding conditions can be drawn from the results.

10.5 ^{147}Pm and ^{151}Sm

10.5.1 Nuclear Characteristics and Measurement of ^{147}Pm and ^{151}Sm

In the decommissioning of nuclear facilities and in studies on environmental radioactivity, the most important rare earth radionuclides requiring radiochemical separations are ^{147}Pm and ^{151}Sm because of their fairly long half-lives of 2.6 y and 90 y and high fission yields of 2.3% and 0.4% for ^{235}U. Promethium and samarium are light rare-earth elements, promethium having no stable isotopes; that is, Pm is entirely artificial. ^{151}Sm is also produced by the neutron activation reaction, ^{150}Sm (n,γ)^{151}Sm, of stable ^{150}Sm, which is an impurity in the materials of construction of nuclear reactors with a very high neutron activation cross section (15 200 b).

^{147}Pm decays to ^{147}Sm ($t_{1/2} = 10^{11}$ y) through emission of beta particles at a maximum energy of 224 keV, very low-intensity 121.2 keV γ-rays (0.0028%), and X-rays (40.1 keV, 0.00115%) (Figure 10.3). ^{147}Pm is normally measured by detecting the beta particles with a liquid scintillation counter. Mass spectrometry can also be used for the determination of ^{147}Pm, but, because of the isobaric interference from stable ^{147}Sm and polyatomic ions, a very good chemical separation of Pm from Sm, as well as other elements, is required.

^{151}Sm decays to stable ^{151}Eu by emission of low-energy beta particles with a maximum energy of 76.7 keV and very low-intensity and low-energy γ-rays (21.5 keV, 0.03%) and X-rays (5.85 keV, 0.026%). ^{151}Sm is normally measured by LSC. ^{151}Sm can also be measured by mass spectrometry, but the isobaric interference of the stable isotope ^{151}Eu and polyatomic ions requires an extremely efficient separation of Sm from Eu. In addition, the tailing effect of ^{150}Sm and interference due to polyatomic ions, such as ^{150}SmH and ^{150}NdH, also need to be taken into account.

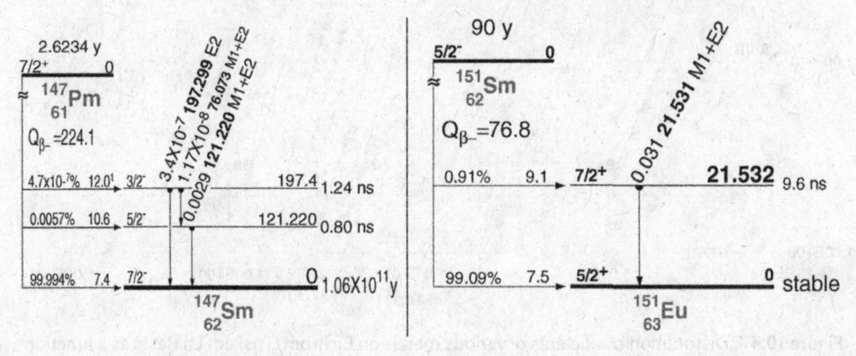

Figure 10.3 Decay schemes of ^{147}Pm and ^{151}Sm (Firestone, R.B., Shirley, V.S., Chu, S.Y.F., Baglin, C.M., and Zipkin, J. (1996) *Table of Isotopes*, Wiley-Interscience).

10.5.2
Separation of ^{147}Pm and ^{151}Sm

Analytical methods have been reported for the determination of ^{147}Pm and ^{151}Sm in nuclear waste and in environmental and biological samples. Coprecipitation, solvent extraction, ion exchange chromatography, extraction chromatography, and high-performance liquid chromatography have been used for the separation of ^{147}Sm and ^{151}Sm from matrix components, transuranics, other fission products, and other rare earth elements. ^{147}Pm and ^{151}Sm measurements have been done by LSC and ICP-MS.

10.5.2.1 Separation with Ln Resin

A specific extraction chromatography resin lanthanides, Ln Resin, is supplied by Eichrom/Triskem. In Ln Resin, the extractant, di(2-ethylhexyl)orthophosphoric acid (HDEHP), is grafted on a hydrophobic support. The distribution coefficients (K_d) versus nitric acid concentrations for various metal ions on the Ln Resin are shown in Figure 10.4. The affinity of rare earth elements to Ln Resin increases with the mass number of the element. Using this resin, ^{151}Sm can be separated from other rare

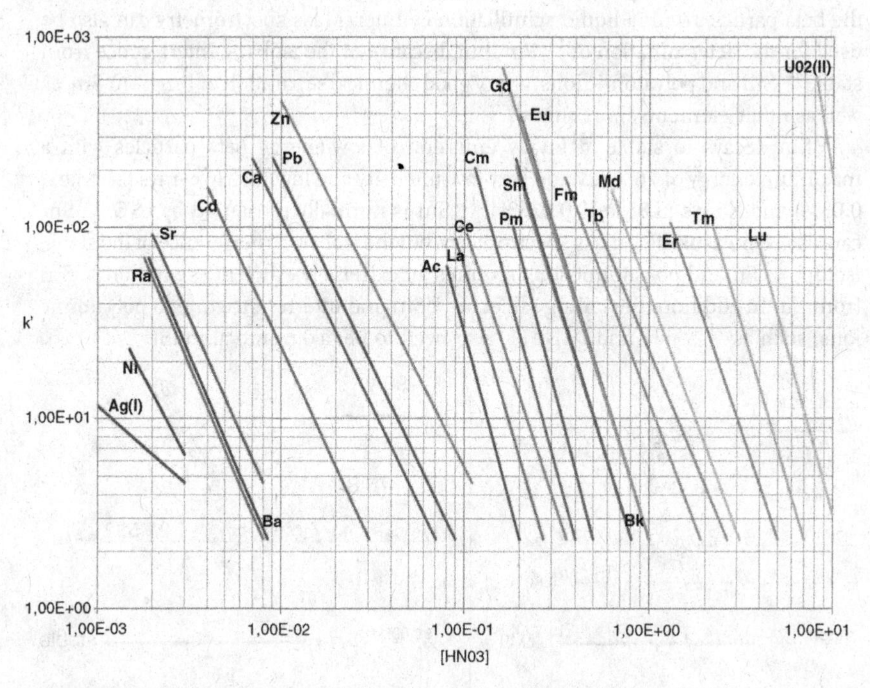

Figure 10.4 Distribution coefficients of various metals on Eichrom/Triskem Ln Resin as a function of HNO_3 concentration (Horwitz, E.P. *et al.* (1975) Chemical separations for super-heavy element searches in irradiated uranium targets. *J. Inorg. Nucl. Chem.*, **37**, 425).

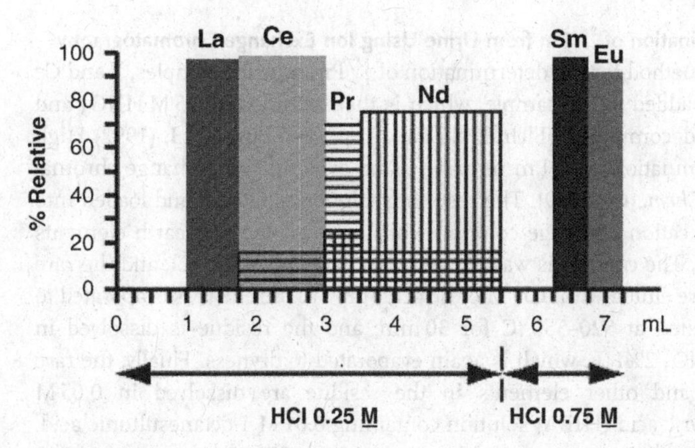

Figure 10.5 Elution of La, Ce, Pr, Nd, Sm, and Eu from Eichrom/Triskem Ln Resin column (Pin, Christian *et al.* (1996) Sequential separation of rare-earth elements, thorium and uranium by miniaturized extraction chromatography: application to isotopic analyses of silicate rocks. *Anal. Chim. Acta.*, **339**, 79).

earth elements. The sample is first prepared in 0.05 M HNO_3 and then loaded to a Ln resin column (0.3 g, 50–100 μm), and different elements are then eluted from the column with HCl solutions. Figure 10.5 shows the elution of La, Ce, Pr, Nd, Sm, and Eu. A total of 5.5 mL of 0.25 M HCl is used to strip the La, Ce, Pr, and Nd, with no detectable Sm. Then, after passing 0.75 M HCl (0.75 mL), Sm appears in the next 0.5 mL, and finally Eu is collected in the next 0.5 mL of 0.75 M HNO_3.

Ln Resin has been successfully used in the analysis of ^{147}Pm in environmental and nuclear waste samples. As an example, a method for the separation of ^{147}Pm from water is given below (Cable, Peter *et al.* (1997) Analysis of Promethium in Aqueous Samples. 43rd Annual Conference on Bioassay, Analytical and Environmental Radiochemistry, Charleston, SC, November, 1997). The samples are first preconcentrated, either by evaporation or by calcium phosphate coprecipitation, and then prepared in 0.2 M HNO_3 with ascorbic acid. The ascorbic acid reduces Fe(III) to Fe (II), which passes through the column without significant adsorption. Interfering americium, potentially present, is rinsed from the column along with strontium by 0.2 M HNO_3. Promethium is retained in the column along with bismuth, yttrium, and lanthanides. The promethium, along with other lanthanides, is eluted with 1 M HNO_3 (5 mL). Excellent decontamination is achieved from ^{60}Co, ^{134,137}Cs, and ^{89,90}Sr. The separated solution can be used for measurement of ^{147}Pm. This procedure, however, cannot separate Sm, Gd, and other lanthanides from ^{147}Pm. Radioisotopes of these elements may interfere with the measurement of ^{147}Pm if their concentrations are at the same level or higher than that of ^{147}Pm. For example, a high ^{151}Sm concentration in the sample interferes with the measurement of ^{147}Pm by LSC, and high concentrations of stable ^{147}Sm interfere with the measurement of ^{147}Pm by ICP-MS. In these cases, further purification of Pm with respect to these elements is needed.

10.5.2.2 Determination of ^{147}Pm from Urine Using Ion Exchange Chromatography

In an analytical method for the determination of ^{147}Pm in urine samples, Y and Ce carriers are first added to the sample, which is then acidified to 0.5 M HNO_3 and boiled for 1 h to decompose it (Elchuk, S., Lucy, C.A., and Burns, K.I. (1992) High resolution determination of ^{147}Pm in urine using dynamic ion exchange chromatography. *Anal. Chem.*, **64**, 2339). The resulting solution is filtered and loaded into a strongly acidic cation exchange column, which absorbs the rare earth elements including ^{147}Pm. The column is washed with H_2O and 1.0 M NH_4Cl, and the rare earth elements are eluted with 0.03 M citric acid (pH 5). The eluate is evaporated to dryness and ignited at 520–550 °C for 30 min, and the residue is dissolved in concentrated HNO_3-2%HF, which is again evaporated to dryness. Finally, the rare earth elements and other elements in the residue are dissolved in 0.05 M α-hydroxyisobutyric acid (HIBA) solution containing 0.01 M 1-octanesulfonic acid. From this solution, ^{147}Pm is separated by employing an HPLC system with a C18 reversed-phase separation column and mobile phase of 0.05–0.25 M HIBA/0.01 M 1-octanesulfonate.

Reversed-phase HPLC (RP-HPLC or RPC) has a nonpolar stationary phase and an aqueous, moderately polar mobile phase. One common stationary phase is RMe_2-SiCl-treated silica, where R is a straight-chain alkyl group such as $C_{18}H_{37}$ or C_8H_{17}. If the $C_{18}H_{37}$ alkyl group in the silica is in the stationary phase, the column is called a C18 column. With these stationary phases, retention time is longer for molecules which are more nonpolar, while polar molecules elute more readily. The retention time of an analyte can be increased by adding more water to the mobile phase, thereby making the affinity of the hydrophobic analyte for the hydrophobic stationary phase stronger relative to the now more hydrophilic mobile phase. Similarly, the retention time of the analyte can be reduced by adding more organic solvent to the eluent. The affinity to the C18 column or nonpolarity of the rare earth elements decreases from La to Sm, and their retention time therefore decreases from La to Sm. (Figure 10.6). For effective elution of elements from the column, especially the high-affinity elements, a gradient elution with an increased concentration of hydroxyisobutyric acid is used. ^{147}Pm separated from other radionuclides and interfering elements is measured by liquid scintillation counting. From Figure 10.6, it can be seen that the reported method can also be used for the separation and determination of ^{151}Sm.

10.5.2.3 Separation of ^{147}Pm from Irradiated Fuel by Ion Exchange Chromatography

A method for the determination of ^{147}Pm in irradiated nuclear fuel by employing anion exchange chromatography and HPLC with a cation exchange column has been reported (Brennetot, R., Stadelmann, G., Gaussignac, C., Gombert, C., Fouque, M., and Lamouroux, C. (2009) A new approach to determine ^{147}Pm in irradiated fuel solution. *Talanta*, **78**, 681). In this method the sample is first dissolved and prepared in a 10 M HCl solution, the solution is then loaded into a strongly basic anion exchange column, and the column is washed with 10 M HCl. The column does not adsorb ^{147}Pm and other rare earth elements that do not form anionic complexes with Cl^-. These are collected in the effluent and a 10 M HCl rinse solution. Pu, U, and most of the transition metal fission and neutron activation products, such as ^{55}Fe, ^{60}Co, ^{99}Tc,

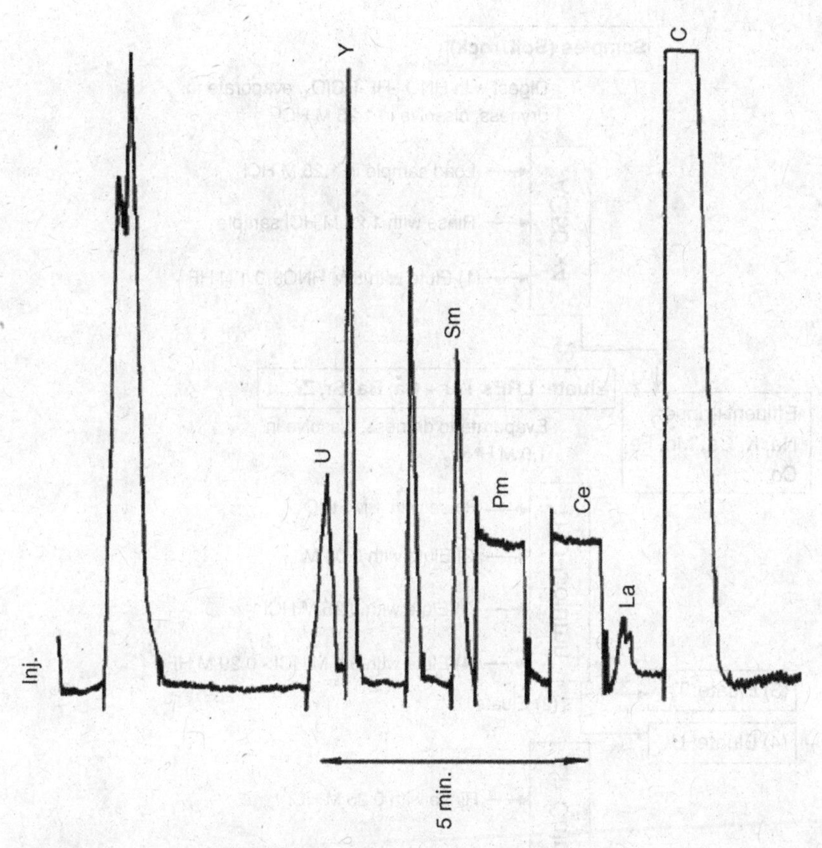

Figure 10.6 Separation spectrum of elements for Pm and Sm using HPLC (C18 reverse column) (Elchuk, S., Lucy, C.A., and Burns, K.I. (1992) High resolution determination of ^{147}Pm in urine using dynamic ion exchange chromatography. *Anal. Chem.*, **64**, 2339).

and ^{99}Mo, which exist as anions in 10 M HCl, are absorbed in the column, separating them from ^{147}Pm. The effluent solution with ^{147}Pm is evaporated to dryness and then dissolved in 0.2 M HNO_3 solution. This solution is injected into an HPLC system with a cation exchange column (NUCLEOSIL SA), and 2-hydroxy-methylbutyric acid (HMB) is used as eluent. In this step, the rare earth elements and other elements not forming anions in 10 M HCl and not absorbed on the anion exchange column in the previous anion exchange step (^{63}Ni, ^{137}Cs, ^{90}Sr, etc.) are absorbed on the cation exchange column. Depending on the different affinities of these cations, they can be eluted using a complex agent, HMB, with different retention times. The ^{147}Pm is therefore separated from other rare earth elements and other cations. The separated ^{147}Pm is measured using LSC.

10.5.2.4 Determination of ^{147}Pm and ^{151}Sm in Rocks

The separation of Pm and Sm from the matrix components, U, Th, and other rare earth elements in rocks utilizes an analytical method based on ion exchange and

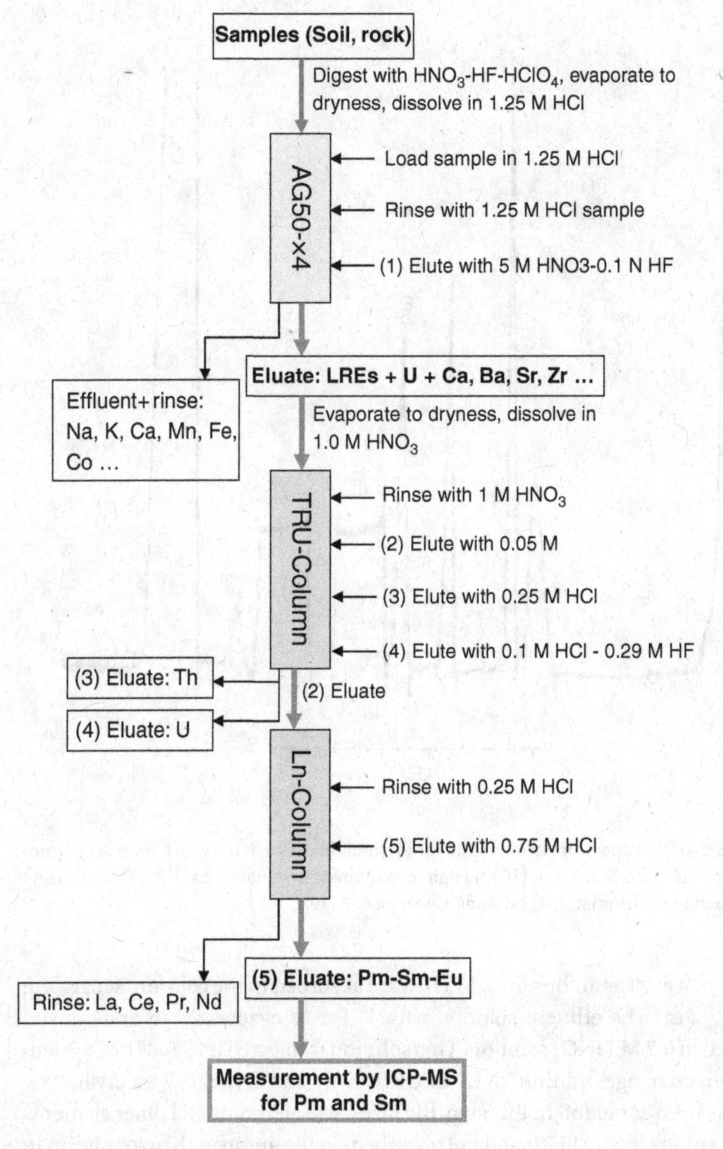

Figure 10.7 Separation procedure for the determination of ^{147}Pm and ^{151}Sm in rock samples (Pin, C. and Zalduegui, J.F.S. (1997) Sequential separation of light rare earth elements, thorium and uranium by miniature extraction chromatography: Application to isotopic analyses of silicate rocks, *Anal. Chim. Acta*, **339**, 79).

extraction chromatography (Figure 10.7). In this method, the sample is first digested with HF-HNO_3-$HClO_4$, and the solution is then evaporated to dryness. The evaporation residue is dissolved in 1.25 M HCl. This solution is loaded into a strongly acidic cation exchange resin column. Most matrix elements and many interfering radionuclides such as ^{137}Cs, ^{40}K, ^{55}Fe, and ^{60}Co pass through the column without retention. The light rare-earth elements (LREs), including ^{151}Sm and ^{147}Pm, retained in the column are then eluted along with U, Th, Ca, Ba, Sr, and Zr using 5 M HNO_3-0.1 M HF. The eluate is evaporated to dryness, and the residue is dissolved in 1 M HNO_3 solution, which is then loaded into an Eichrom/Triskem TRU Resin column. After rinsing with 1 M HNO_3, the LREs are eluted with 0.05 M HNO_3 and the eluate is loaded into an Ln Resin column. The TRU Resin column is treated with 0.25 M HCl to remove Th and with 0.1 M HCl – 0.29 M HF to remove U. The LREs in the Ln Resin column are removed by a two-step elution with 0.25 M and 0.75 M HCl (see Figure 10.5). Both ^{147}Pm and ^{151}Sm are separated from other rare earths and interfering elements by collecting the corresponding fractions, and finally measured by ICP-MS or LSC. For LSC measurement ^{147}Pm and ^{151}Sm still need to be separated from each other.

10.6
Essentials of Lanthanide Radiochemistry

- Lanthanides have several fission product radionuclides with high fission yields. Most of them emit readily detectable gamma rays and thus do not need radiochemical separations. Two lanthanide radionuclides, however, ^{147}Pm and ^{151}Sm, have fairly long half-lives, 2.6 years and 90 years, and are pure beta emitters with no gamma emissions. The maximum beta energies of ^{147}Pm and ^{151}Sm are 224 keV and 77 keV, typically measured with LSC after radiochemical separation.
- Lanthanides are a group of fourteen elements in which the 4f-shell is progressively filled. With only a few exceptions lanthanides occur exclusively at an oxidation state of +III and exist as Ln^{3+} ions in aqueous solutions. Their bonding in compounds is ionic in nature, and thus the solubility of most compounds is high. Furthermore, lanthanides do not hydrolyze strongly. The ionic size of trivalent lanthanide ions decreases with increasing atomic number, and consequently the tendency to hydrolyze and form complexes also increases.
- Trivalent lanthanides closely resemble trivalent actinides (Ac^{3+}, Pu^{3+}, Am^{3+}, Cm^{3+}) and are therefore used as analogs in actinide studies where macro amounts of elements are needed. When determining actinides in solid samples, such as soil and sediments, separation of lanthanides is an important task to avoid excessive masses in alpha counting sources. Lanthanide separations from actinides are typically accomplished by anion exchange or extraction chromatography, making use of their different capabilities of forming complexes with thiocyanate.
- Separation of ^{147}Pm and ^{151}Sm is usually done by ion exchange and extraction chromatography. Eichrom/Triskem manufactures an extraction chromatography resin for lanthanides, Ln Resin, which has di(2-ethylhexyl)orthophosphoric acid (HDEHP) as an extraction agent in the pores of an inert support.

11
Radiochemistry of the Halogens

11.1
Important Halogen Radionuclides

Of the halogens, only iodine has significant fission product nuclides: ^{129}I, ^{131}I, ^{132}I, ^{133}I, ^{134}I, and ^{135}I (Table 11.1). In the long term the most important of these is ^{129}I because of its very long half-life of 1.6×10^7 years. The others are short-lived, the longest half-life among them being that of ^{131}I: only 8 days. In fresh nuclear fallout the radiation dose to humans due to, especially ^{131}I, is around half of the total radiation dose received during the first few weeks after the fallout. Not only are these short-lived iodine isotopes responsible for an external dose, they also are transported into the body through the respiratory tract, as well as via food, especially milk from cows that have fed on polluted grass and hay. All the short-lived iodine isotopes emit detectable gamma rays and do not require radiochemical separation. The long-lived iodine isotope also emits gamma rays, but the energy and intensity of these are rather low, and since the specific activity of ^{129}I is very low its determination requires radiochemical separation. Other sources of long-lived ^{129}I in the environment, besides nuclear weapons tests and the Chernobyl accident, are water and air discharges from fuel reprocessing plants. Because of its long life, activity concentrations of ^{129}I in the environment are very low. However, because of its very long half-life and high mobility in the geosphere, ^{129}I is of long-term concern in the final depository of nuclear waste.

The most important radioactive isotope of fluorine is ^{18}F. It is a short-lived positron emitter and has an important application in positron emission tomography for medical purposes. Bromine has no radiochemically significant radionuclides. Chlorine, in turn, has one long-lived radionuclide, ^{36}Cl, which is generated in nuclear reactors by neutron activation reaction of existing stable ^{35}Cl (natural abundance 75.77%) in the nuclear fuel and construction materials in the nuclear reactor. In addition, nuclear reactions induced by cosmic radiation produce ^{36}Cl in the environment. Because of the very high mobility of Cl in the geosphere, ^{36}Cl is also an important radionuclide in the decommissioning of nuclear facilities and deposition

Chemistry and Analysis of Radionuclides. Jukka Lehto and Xiaolin Hou

ISBN: 978-3-527-32658-7

Table 11.1 Important halogen radionuclides.

Element	Radionuclide	Half-life	Fission yield (%)	Decay mode	Gamma emission	Source or use
Fluorine	^{18}F	1.8 h		β^+	yes	tracer
Chlorine	^{36}Cl	301 000 y		β^-	no	activation product, cosmogenic
Bromine	^{82}Br	35 h		β^-	yes	activation product
Iodine	^{129}I	1.6×10^7 y	0.66%	β^-	yes	fission product, cosmogenic
	^{131}I	8.0 d	2.9%	β^-	yes	fission product
	^{132}I	2.3 h	4.3%	β^-	yes	fission product
	^{133}I	20.8 h	6.7%	β^-	yes	fission product
	^{134}I	52.5 min	7.8%	β^-	yes	fission product
	^{135}I	6.57 h	6.3%	β^-	yes	fission product
Astatine	$^{215-219}At$	0.1 ms-56 s		α/β^-	no	natural nuclides

of nuclear waste. In sum, just two important halogen radionuclides, ^{36}Cl and ^{129}I, require radiochemical separations before measurement.

11.2
Physical and Chemical Properties of the Halogens

Halogens are nonmetals and are located in group 17 of the periodic table with the electron structure $[Ng]ns^2np^5$. There are seven electrons in the outermost shell, that is they have the octet structure of the noble gases but with one electron missing. As a result, they readily form negative ions with a single charge and have the oxidation state of −I (F^-, Cl^-, Br^-, I^-, At^-). In the case of fluorine, −I is, in practice, the only oxidation state. The halogens are extremely electronegative, fluorine being the most electronegative element of all. Salts formed with these negative ions of single charge, the halides, are highly soluble, and solubility increases with the atomic number. In higher oxidation states, the halogens also form oxoacids. For example, in the oxidation state +V, iodine forms iodic acid, whose anion is iodate (IO_3^-), and in the oxidation state +VII it forms periodic acid, whose anion is periodate (IO_4^-). As seen from Figure 11.1, iodide is the predominant chemical form. Iodate is present in oxidizing conditions in alkaline solutions. Formation of periodate requires oxidizing conditions more extreme than any found in natural systems. Chlorine most commonly appears in solution as the chloride ion, Cl^-, and the perchlorate ion, ClO_4^-, can only be found in very oxidizing conditions, again not found in natural systems (Figure 11.1). The halogens can also be in the form of gases (F_2, Cl_2, Br_2, I_2, At_2), and as such are moderately soluble in water.

Most chlorides and iodides are very soluble, excluding those of silver, mercury, lead, cuprous copper, thallium, and palladium. Silver chloride and iodide are typically used in separations and as targets in accelerator mass spectrometry (AMS)

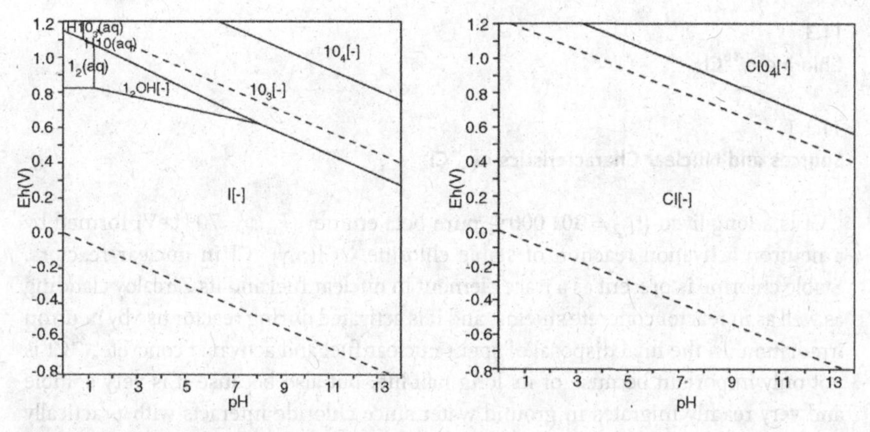

Figure 11.1 Eh–pH diagrams of iodine and chlorine (Atlas of Eh–pH diagrams, Geological Survey of Japan, Open File Report, No. 419, 2005).

measurements. Palladium iodide, PdI_2, is another compound used in separations. It can be used to separate iodine from chlorine, since chlorine does not form an insoluble compound with palladium. Most iodates are sparingly soluble, except alkali metal and ammonium iodates, which are very soluble.

Boiling points of the halogens are very low (Table 11.2). An important consequence is that large amounts of iodine are released in nuclear explosions and accidents such as that at Chernobyl in 1986. For example, 20% of the iodine in the Chernobyl reactor was released when it exploded, compared with 13% of the ^{137}Cs and just 4% of the ^{90}Sr. Iodine in the explosion cloud was either bound to aerosol particles or present in a gaseous form as elemental iodine (I_2) or methyl iodide. At the moment of the explosion, most of the iodine released from the reactor was in gaseous form. After the explosion, I_2, which is chemically reactive, was converted into other forms: iodide and iodate, which were associated with aerosol particles. In addition, tellurium isotopes (especially ^{131m}Te and ^{134}Te, which have relatively long half-lives) bound to aerosols decayed to short-lived iodine isotopes of corresponding mass number, and consequently the proportion of iodine in gaseous form was lower in the explosion cloud and fallout than at the moment of the explosion.

Table 11.2 Physical characteristics of the halogens.

Element	Ionization energy (kJ/mol)	Boiling point (°C)	Ionic radius X^- (pm)	Electronegativity (Pauling)
Fluorine	1680	−188	133	3.98
Chlorine	1256	−34	184	3.16
Bromine	1143	60	196	2.96
Iodine	1009	185	220	2.66
Astatine	about 926			2.20

11.3
Chlorine – ^{36}Cl

11.3.1
Sources and Nuclear Characteristics of ^{36}Cl

^{36}Cl is a long-lived ($t_{1/2}$ = 301 000 y) pure beta emitter (E_{max} = 709 keV) formed by a neutron activation reaction of stable chlorine $^{35}Cl(n,\gamma)^{36}Cl$ in nuclear reactors. Stable chlorine is present as a trace element in nuclear fuel and its Zircaloy cladding as well as in reactor concrete shields, and it is activated during reactor use by neutron irradiation. In the final disposal of spent nuclear fuel and activated concrete, ^{36}Cl is not only important because of its long half-life, but also because it is very soluble and very readily migrates in ground water since chloride interacts with practically no mineral surfaces in the bedrock and soil. Less important sources, compared to nuclear reactors, are nuclear reactions induced by cosmic radiation in the atmosphere, water, and bedrock. Because of its low specific activity, ^{36}Cl cannot be considered a substantial danger to humans. However, it makes a useful tracer in interpreting environmental processes. The decay scheme of ^{36}Cl is presented in Figure 11.2.

^{36}Cl can also be used in geological age determinations. ^{36}Cl is generated in a spallation reaction of argon induced by cosmic radiation. The natural ratio of ^{36}Cl to stable chlorine ($^{36}Cl/Cl$) on the earth's surface is 7×10^{-13}. In deeper soil, sediment, and ice layers with no contact with the atmosphere, this ratio decreases with time. AMS measurement is powerful enough to reveal very low ratios in deep sediments, old groundwater, and thick ice layers, and the ages of the various layers can be calculated. ^{36}Cl dating is a good complement to ^{14}C dating, since much older samples can be dated with ^{36}Cl.

11.3.2
Determination of ^{36}Cl

For the determination of ^{36}Cl in soil, concrete, and other activated solid materials, chlorine is separated from the sample by vaporizing the chlorine to Cl_2 gas by either

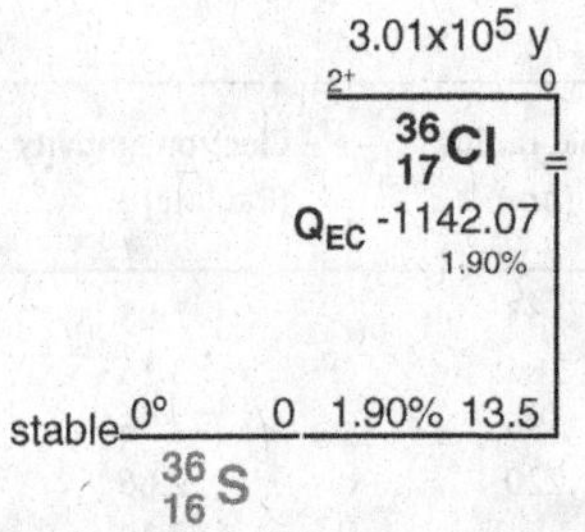

Figure 11.2 Decay scheme of ^{36}Cl (Firestone, R.B., Shirley, V.S., Chu, S.Y.F., Baglin, C.M., and Zipkin, J. (1996) *Table of Isotopes*, Wiley-Interscience).

heating the sample or boiling it in strong oxidizing acids. The evaporated chlorine is bound into an activated carbon column cooled with liquid nitrogen, and chlorine from the boiling vapors is captured in an NaOH trap. The sample can also be dissolved so that the chlorine remains in solution as chloride ions. In all three cases, a known amount of chlorine carrier, in the form of NaCl, for example, is added to the sample. Chlorine gas in the activated carbon column or in the NaOH trap is reduced with $NaNO_2$ to chloride. The chloride is then precipitated as silver chloride (AgCl) by adding $AgNO_3$ to a weakly acidic solution. Iodine precipitates along with the silver chloride, which means that the long-lived radionuclide ^{129}I will also be present in the precipitate. Iodine can be removed by extracting the precipitate with ammonia solution, in which AgCl dissolves, but AgI does not. The AgI precipitate is discarded, and the chloride in the ammonia solution is reprecipitated by acidification to $pH < 2$. At the end of the analysis, the chemical yield of the separation is determined by measuring the amount of chlorine with ICP-MS or ion chromatography.

Finally, after dissolving the precipitate and adding scintillation cocktail, the ^{36}Cl in the purified AgCl precipitate is measured by liquid scintillation counting (LSC) or directly from the precipitate by AMS. Detection limits are much lower in AMS than in LSC. In LSC they are about 10 mBq activity at best. In AMS a $^{36}Cl/Cl_{tot}$ ratio of 10^{-15} can be measured, corresponding to 1 nBq if 1 mg ^{35}Cl carrier is used. The relatively high ^{36}Cl activities in nuclear waste can be measured by LSC, but the very low ^{36}Cl concentrations in the environment can be determined only by AMS.

Chlorine has no suitable radioactive tracer for use to monitor the chemical yield in ^{36}Cl analyses. For this purpose, an addition of stable chlorine is used. Since the samples also contain stable chlorine, the amount of the added chlorine has to be clearly higher than what was present initially. When determining the ^{36}Cl by AMS, stable chlorine is added before sample decomposition and chemical separation. The $^{36}Cl/^{35}Cl$ ratio is then measured by AMS, and the concentration of ^{36}Cl in the sample can be calculated from the measured $^{36}Cl/^{35}Cl$ ratio and the amount of ^{35}Cl (stable Cl) added. If the initial $^{36}Cl/^{35}Cl$ ratio in the sample is needed, the concentration of stable chlorine needs to be measured in it by an independent method, by ICP-MS, for example.

Most ^{36}Cl-bearing samples also contain ^{129}I. When measuring these nuclides by radiometry they have to be separated from each other. Since they are chemically very similar, they follow each other in most separation steps. In the following methods given for the determinations of both ^{36}Cl and ^{129}I, their mutual separation is accomplished by solvent extraction, making use of their different chemical species at various oxidation states. In the first step, both elements are reduced to the −I oxidation state (chloride/iodide) with sodium bisulfite, which is followed by oxidation of iodine to I_2 with hydrogen peroxide or nitrite leaving chlorine as chloride ions. After these treatments, iodine (I_2) can be extracted with CCl_4, whereas chlorine (Cl^-) remains in the aqueous phase.

11.3.2.1 Determination of ^{36}Cl from Steel, Graphite, and Concrete by Solvent Extraction and Ion Exchange

A method for the determination of ^{36}Cl and ^{129}I in nuclear waste samples, such as graphite, concrete, and various metals is shown in Figure 11.3. In this method, the

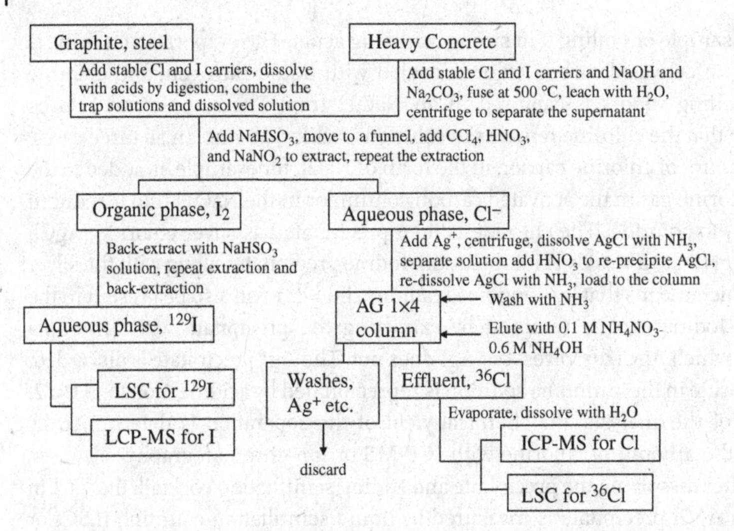

Figure 11.3 Analytical procedure for the determination of ^{129}I and ^{36}Cl in nuclear waste samples (Hou, X.L., Østergaard, L.F., and Nielsen, S.P. (2007) Determination of ^{36}Cl in nuclear waste from reactor decommissioning. *Anal. Chem.*, **79**, 3126).

samples are first decomposed. Graphite is decomposed by acid digestion with a mixture of acids (H_2SO_4:HNO_3:$HClO_4$ = 15 : 4 : 1). The metals lead, aluminum, and stainless steel are dissolved with concentrated HNO_3, H_2SO_4, and H_2SO_4-H_3PO_4, respectively. The acid solutions are then heated and bubbled with nitrogen gas, and the released gases are trapped in three solutions connected in series: the first one is H_2O and the two others 0.4 M NaOH. The concrete sample is decomposed using alkaline fusion by mixing the sample with NaOH-Na_2CO_3 and heating at 500 °C for 3–4 h. The fused cake is dissolved with H_2O, and the leachate is used for the determination of ^{36}Cl. Before sample decomposition, a stable chloride carrier and hold-back carriers are added.

The decomposed sample solution is then transferred to a separation funnel and the pH is adjusted to below 2. $NaHSO_3$ solution is added to convert all the iodine to iodide and chlorine to chloride. After addition of CCl_4, $NaNO_2$ solution is added to oxidize iodine to I_2, while chlorine remains as Cl^- ions. I_2 is then extracted to the CCl_4 phase by shaking. ^{129}I (and ^{131}I, if present) is now in the organic phase, which can be used for the determination of ^{129}I. After removal of iodine, $AgNO_3$ solution to give an Ag^+/Cl^- weight ratio >5 is added to the aqueous phase, and the solution is stirred to aggregate the AgCl precipitate. This precipitate is separated by centrifugation and washed with 1M HNO_3 to remove interfering metal radionuclides. To further purify the precipitate it is dissolved with 25% NH_3 and reprecipitated by adding HNO_3 to lower the pH below 2. The precipitate is separated by centrifugation and the supernatant containing interfering metal radionuclides is discarded. The AgCl precipitate is then dissolved in 25% NH_3 solution, and the solution is loaded into an anion exchange column converted into the OH^- form with 25% NH_4OH. The column is washed with 25% NH_3 until the wash effluent is free of Ag^+. Chloride on the column is then eluted with 0.2 M NH_4NO_3 – 0.6 M NH_4OH and the eluate is evaporated to dryness on a hotplate. The residue is dissolved in water (2 mL) and transferred to an LSC vial. To determine stable chlorine by ICP-MS for chemical yield calculation, 0.1 mL of the solution is taken. Scintillation cocktail (15 mL) is added to the remaining solution and ^{36}Cl activity is measured by LSC. The decontamination factors for all interfering radionuclides are higher than 10^6, and the recovery of Cl is higher than 90%. Since the final solution of ^{36}Cl is in just 2 mL of water, the quenching level is very low and the counting efficiency is high (>98%).

11.4 Iodine – ^{129}I

11.4.1 Sources and Nuclear Characteristics of ^{129}I

The short-lived iodine radioisotopes $^{131-135}$I are only relevant in a fresh nuclear fallout, and since they all emit easily detectable gamma rays they are not further discussed in this book. In the long term the only important iodine radioisotope is the very long-lived ^{129}I (1.6×10^7 y), a fission product with a ^{235}U fission yield of 0.66%. ^{129}I decays by beta emission to the excited state of xenon isotope ^{129}Xe (Figure 11.4), and the maximum energy of the beta particles is 154 keV. The decay of the excited state ^{129m}Xe takes place mostly by internal conversion, and the intensity of the 39.6 keV gamma ray emissions is only 7.5%. X-rays, including 29.5 keV (20.4%) and 29.8 keV (37.7%), are also emitted. Conversion and Auger electrons are also emitted with energies of 3–30 keV and intensities of 2–79%. As a tracer in ^{129}I analyses ^{125}I can be used. It is an EC nuclide with a half-life of 59 days. It emits 35.5 keV gamma rays with the intensity of 6.7% and several X-rays in the energy range of 27–31 keV with intensities of 4–74%.

^{129}I, naturally present in the environment, is generated in the atmosphere from xenon by cosmic radiation-induced nuclear reactions and in the soil from uranium through spontaneous fission and fission due to thermal neutrons. The concentration

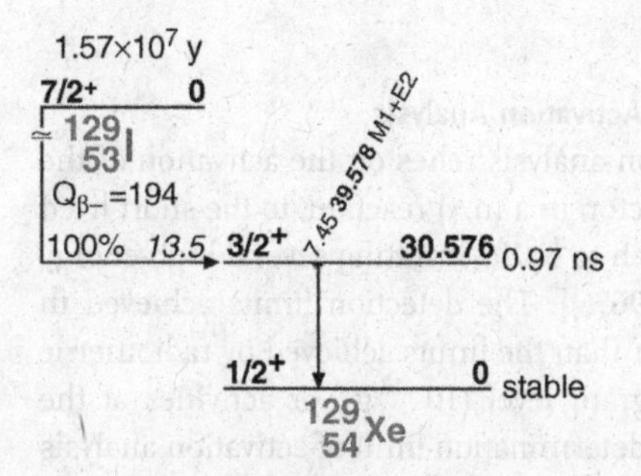

Figure 11.4 Decay scheme of ^{129}I (Firestone, R.B., Shirley, V.S., Chu, S.Y.F., Baglin, C.M., and Zipkin, J. (1996) *Table of Isotopes*, Wiley-Interscience).

of natural ^{129}I in the environment is so low that it is not usually expressed as an activity concentration, but in proportion to the only stable isotope of iodine, ^{127}I. Before the nuclear age, the $^{129}I/^{127}I$ ratio was about 10^{-12} in the sea and somewhat higher in groundwater. Now, as a result of nuclear weapons tests, nuclear accidents, and emissions from nuclear facilities, the level of ^{129}I in the environment has risen, and the $^{129}I/^{127}I$ ratio has increased by several orders of magnitude, up to as high as 10^{-4}. The most extensive releases of ^{129}I to the environment have occurred at the spent nuclear fuel reprocessing facilities at Sellafield (UK) and La Hague (France) since the early 1990s. Most of the ^{129}I generated in the nuclear power plants is present in the spent nuclear fuel. Because of its very long half-life it will be a prevailing radionuclide in the final repositories in the distant future. Because of the high mobility, in the geosphere and soil, of iodide (I^-), the typical form of iodine, it is an essential task in the safety analysis of the final disposal of spent nuclear fuel to evaluate the behavior of this radionuclide in these environments.

11.4.2 Measurement of ^{129}I

The very long half-life of ^{129}I means that its specific activity is very low (6.5×10^6 Bq g^{-1}). Thus, radiometric methods are only suitable for samples in which the activity of the isotope is fairly high. Such samples are found in nuclear power plants and nuclear fuel reprocessing facilities, or environmental samples around the reprocessing plants. Radiometric measurements can be accomplished by gamma and X-ray spectrometers and by LSC, the latter method being more sensitive because of the high intensities of the beta, conversion, and Auger electron emissions, which are all detected in LSC. Radiometric methods are not sensitive enough for measuring the very low activity concentrations of ^{129}I in the environment. More sensitive methods are activation analysis and especially AMS. ^{129}I can also be measured by other mass spectrometric methods. ICP-MS allows determinations at more or less the same sensitivity as radiometric methods. By applying dynamic reaction cell and iodine introduction to the plasma as gas, the sensitivity of ICP-MS can be improved, but it is still difficult to use for the determination of ^{129}I in low-level environmental samples in which $^{129}I/^{127}I$ ratio is below 10^{-7}. Table 11.3 presents the detection limits for various measurement methods.

11.4.2.1 Determination of ^{129}I by Neutron Activation Analysis

Determination of ^{129}I by neutron activation analysis relies on the activation of the long-lived ^{129}I, in the neutron flux of a reactor, in a (n,γ) reaction, to the short-lived ^{130}I. The ^{130}I decays with a half-life of 12.3 h to ^{130}Xe, emitting several high-energy gamma rays (536 keV (99%), 668.5 keV (96%)). The detection limits achieved in neutron activation analysis are much lower than the limits achieved by radiometric methods. Amounts even below the picogram level (10^{-13} g), or activities at the microbequerel level, can be detected. The determination limit of activation analysis for the $^{129}I/^{127}I$ ratio is about 10^{-10}, which means that iodine isotope ratios two orders of magnitude above the natural background level are measurable. In many

Table 11.3 Detection limits of radiometric and mass spectrometric methods for the measurement of ^{129}I (Hou, X. and Roos, P. (2008) Critical comparison of radiometric and mass spectrometric methods for the determination of radionuclides in environmental, biological and nuclear waste samples. *Anal. Chim. Acta.*, **608**, 105).

Detection method	Target preparation	Detection limit	
		Bq	**^{129}I/^{127}I ratio**
X-γ spectrometry	Direct measurement	100–200 mBq	10^{-4}–10^{-5}
X-γ spectrometry	Separated iodine (AgI)	20 mBq	10^{-5}–10^{-6}
LSC	Separated iodine	10 mBq	10^{-5}–10^{-6}
RNAA	Separated MgI_2/I_2 absorbed on charcoal	1 µBq	10^{-10}
ICP-MS	Direct water measurement	40–100 µBq/ml	10^{-5}–10^{-6}
ICP-MS	Gaseous iodine	2.5 µBq/g	10^{-7}
AMS	AgI	10^{-9} Bq	10^{-13}

samples, however, the ratio is much lower than this, and AMS is required. In neutron activation, stable (^{127}I) iodine is activated in an (n,γ) reaction to ^{128}I ($t_{1/2} = 25$ min) and in an (n,2n) reaction to ^{126}I ($t_{1/2} = 13$ d). The ^{127}I in the sample can thus be determined with the aid of these nuclides, and also the ^{129}I/^{127}I ratio can be calculated. This requires, however, that no iodine carrier is added to the sample before separation. Figure 11.5 shows a gamma spectrum of iodine obtained using radiochemical neutron activation analysis. ^{129}I is determined by measuring the 546 keV gamma rays of its thermal neutron activation product, ^{130}I, and ^{127}I is

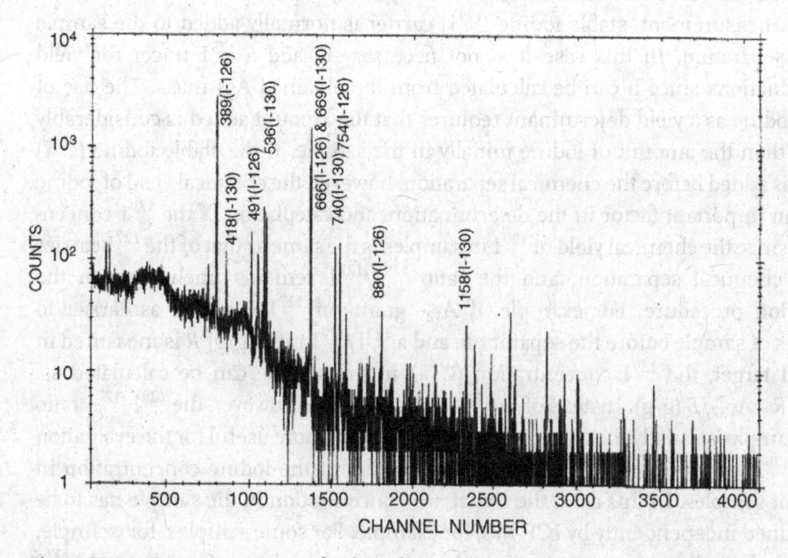

Figure 11.5 Gamma spectrum of iodine obtained by radiochemical neutron activation analysis.

determined by measuring the 491 keV gamma rays of its fast neutron activation product, ^{126}I. For the activation analysis, iodine is separated from the sample using methods described later, and at the end of the procedure iodine is converted into a solid iodide compound, such as PbI_2 or MgI_2, which is then sealed in a quartz ampoule for the neutron irradiation in a reactor. After irradiation the irradiated iodide is taken out of the ampoule, dissolved in acid solutions, and purified by solvent extraction from interfering radionuclides to decrease the Compton background in the gamma spectrum, thus increasing the measurement sensitivity.

11.4.2.2 Determination of ^{129}I by Accelerator Mass Spectrometry

AMS is unquestionably the most sensitive method for the determination of ^{129}I. The detection limit is several orders of magnitude lower than that obtainable with neutron activation analysis, or by other mass spectrometric methods. AMS allows determination of very low values of the $^{129}I/^{127}I$ ratio, and is the only method that allows the determination of natural background values of the ratio (10^{-12}) and even values as low as 10^{-14}. For iodine determinations, a solid AgI sample of 1–5 mg is prepared. The same methods as those used for neutron activation analysis and the radiometric method can be used for the separation of iodine from the samples. The chemical yield during separation is monitored by the gamma-emitting ^{125}I tracer, which is added before separation and measured after separation. In the solution, separated iodine is present as iodide, and, after adding $AgNO_3$, it is finally precipitated as AgI. The AgI is separated by centrifugation and dried at a low temperature ($<100\,°C$). The dried AgI is mixed with silver or niobium powder and pressed in a metal holder for AMS measurement.

AMS is a relative method, which only gives the $^{129}I/^{127}I$ ratio values from the samples but not the absolute amounts or concentrations. To enable the ^{129}I separation and the AMS measurement to be performed, that is, to get enough of AgI for measurement, stable iodine (^{127}I) carrier is normally added to the sample before separation. In this case it is not necessary to add a ^{125}I tracer for yield determinations since it can be calculated from the obtained AgI mass. The use of stable iodine as a yield determinant requires that the amount added is considerably higher than the amount of iodine initially in the sample. If the stable iodine (^{127}I) carrier is added before the chemical separation, however, the chemical yield of iodine is not an important factor in the determination, and calculation of the ^{129}I concentration since the chemical yield of ^{129}I in samples is the same as that of the ^{127}I carrier in the chemical separation, and the ratio $^{129}I/^{127}I$ remains unchanged in the separation procedure. For example, if A_{127} grams of ^{127}I is added as carrier to B grams of sample before the separation, and a $^{129}I/^{127}I$ ratio (g/g) R is measured in the AgI target, the ^{129}I concentration (C_{129}) in the sample can be calculated as: $C_{129} = R^* A_{127}/B$ (g/g). Instead of the ^{129}I concentration, however, the $^{129}I/^{127}I$ ratio in the sample is usually measured because this value is more useful for the evaluation of the ^{129}I level in the investigated area regardless of the iodine concentration in different samples. In this case, the initial ^{127}I concentration in the sample has to be determined independently by ICP-MS, for example. For some samples, for example, seaweed, the iodine concentration is high enough and it is not necessary to add ^{127}I

as carrier to obtain enough AgI for AMS measurement. In this case, the measured $^{129}I/^{127}I$ ratio by AMS is also the ratio of $^{129}I/^{127}I$ in the sample. For obtaining the ^{129}I concentration in the sample, ^{127}I concentration has to be measured independently.

11.4.3
Radiochemical Separations of ^{129}I

For all measurement techniques, radiometry, neutron activation analysis, and AMS, the same radiochemical separation methods can be used. Differences arise only at the end of the separation procedure. For gamma and X-ray spectrometry, a solid sample is finally prepared, for LSC the aqueous sample is mixed with a liquid scintillation cocktail, for neutron activation analysis a solid PbI_2 or MgI_2 source is prepared, and for AMS measurement a solid AgI source is prepared. Neutron activation analysis has one more difference: the irradiated PbI_2/MgI_2 source is further dissolved and purified from interfering radionuclides.

Typically, iodine separations are done by solvent extraction of molecular iodine (I_2), typically with CCl_4 or $CHCl_3$. For extraction into CCl_4, iodine is oxidized to I_2 which is readily soluble in organic solvents but only very slightly in water (0.33 g L^{-1}). The solutions with molecular iodine have a violet color. For back extraction into an aqueous phase, iodine is reduced back to the highly water soluble I^-. Removal of the most chemically similar radionuclide, ^{36}Cl, is accomplished by the oxidation of I^- to I_2 with hydrogen peroxide or nitrite, neither of which oxidize Cl^- ions, leaving them in the aqueous phase while iodine transforms as I_2 into an organic phase.

For the chemical yield determination either a ^{125}I tracer or stable iodine (^{127}I) is used. ^{125}I is determined by gamma spectrometry, measuring its 35.5 keV gamma rays, and stable iodine is measured by gravimetry. Stable iodine is added either as iodide or iodate. Before extraction, all iodine is turned into iodide (I^-) by addition of bisulfite solution and acidification to $pH < 2$.

In the radiochemical separations of ^{129}I, oxidation and reduction of iodine is required. Typically oxidation to molecular iodine for extraction and reduction to iodide for precipitation are needed. A list of the most relevant oxidation/reduction reactions and agents used to obtain the desired oxidation state is given below:

- $I^- \rightarrow I_2$ oxidation with hydrogen peroxide or sodium nitrite in acid
- $I^- \rightarrow IO_3^-$ oxidation with NaClO in base or $KMnO_4$ in acid
- $I_2 \rightarrow I^-$ reduction with bisulfite or sulfite or sulfur dioxide
- $IO_3^- \rightarrow I_2$ reduction with $NH_2OH{\cdot}HCl$ in acid.

The redox reactions between I^- and I_2 are fast since only electron transfer is involved in these reactions. Reactions between I^- or I_2 and IO_3^- are slow since these also involve the breaking or forming of a covalent bond between iodine and oxygen.

11.4.3.1 Separation of ^{129}I by Solvent Extraction
An example of solvent extraction separation of iodine is given in Figure 11.6, which depicts a procedure to separate ^{129}I from spent ion exchange resin used in the

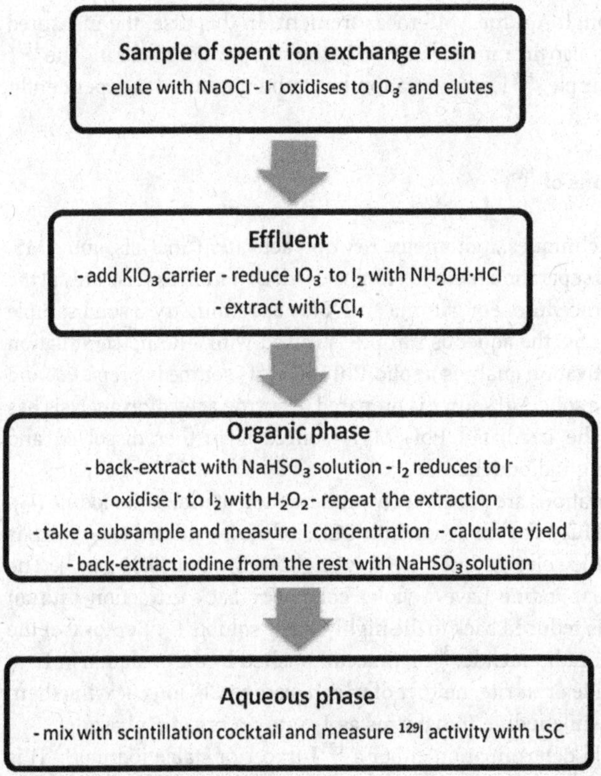

Sample of spent ion exchange resin

- elute with NaOCl - I^- oxidises to IO_3^- and elutes

Effluent

- add KIO_3 carrier - reduce IO_3^- to I_2 with $NH_2OH{\cdot}HCl$
- extract with CCl_4

Organic phase

- back-extract with $NaHSO_3$ solution - I_2 reduces to I^-
- oxidise I^- to I_2 with H_2O_2 - repeat the extraction
- take a subsample and measure I concentration - calculate yield
- back-extract iodine from the rest with $NaHSO_3$ solution

Aqueous phase

- mix with scintillation cocktail and measure ^{129}I activity with LSC

Figure 11.6 Separation of ^{129}I from the spent ion exchange resin from a nuclear power plant by solvent extraction for the measurement of ^{129}I activity by LSC (Puukko, E. and Jaakkola, T., Actinides and Beta Emitters in the Process Water and Ion Exchange Resin Samples from the Loviisa Power Plant. Report YJT-92-22).

purification of the primary circuit of nuclear power plants. In this method the iodine in the resin is eluted with sodium hypochlorite (NaOCl), which oxidizes iodide adsorbed in the exchange resin to iodate (IO_3^-), which has no affinity to the resin and elutes from the resin. After this, KIO_3 is added to the solution as a carrier, and the iodine (IO_3^-) is reduced to I_2 with hydroxylamine hydrochloride, $NH_2OH{\cdot}HCl$. The reduction of iodine to I_2 is performed at a pH below 2, since I_2 is only stable in acidic solutions (Figure 11.1). The I_2 is extracted into carbon tetrachloride (CCl_4) and back-extracted into water with dilute sodium bisulfite ($NaHSO_3$) solution, which reduces I_2 to iodide (I^-). The iodide is then oxidized to I_2 with H_2O_2 in acidic conditions, and the extraction step is repeated with toluene. At this step ^{36}Cl is removed since it is not oxidized and stays in the solution as Cl^- ions. A subsample is taken from the toluene solution, and the absorbance is recorded by spectrophotometry for determination of the chemical yield of iodine. The iodine in the remaining part of the toluene solution is back-extracted into an aqueous solution with $NaHSO_3$

solution. A scintillation cocktail is added to the aqueous solution and the activity of ^{129}I is measured by LSC.

For measuring lower ^{129}I activities by neutron activation analysis (NAA), this method can be modified by precipitating iodine obtained from the final aqueous solution as lead or magnesium iodide (PbI_2, MgI_2), instead of mixing it with a scintillation cocktail. Lead iodide is then sealed into a quartz ampoule and irradiated in a reactor. After irradiation (5–10 h) and cooling (1–3 h), the lead/magnesium iodide is dissolved in nitric acid, and the iodide is oxidized with hydrogen peroxide to iodine, extracted into CCl_4 or toluene, and back-extracted with sodium bisulfite into an aqueous phase. This extraction is done to remove interfering radionuclides generated in the neutron irradiation, such as ^{82}Br and ^{24}Na, which interfere with the measurement of ^{130}I and ^{126}I by raising the Compton background of the gamma spectra. Finally, the activities of ^{130}I and ^{126}I are measured with a gamma spectrometer.

An example of ^{129}I separation for measurement with NAA is given in Figure 11.7. For the neutron irradiation the purified iodine, as iodide, is converted to MgI_2 by adding MgO to the back-extracted solution and evaporating the solution to dryness at <150 °C. The source for gamma spectrometric measurement is PdI_2, which is formed by adding $PdCl_2$ to the final purified iodine solution. Iodide forms black PdI_2 precipitate, which is separated by filtration on a membrane filter for gamma

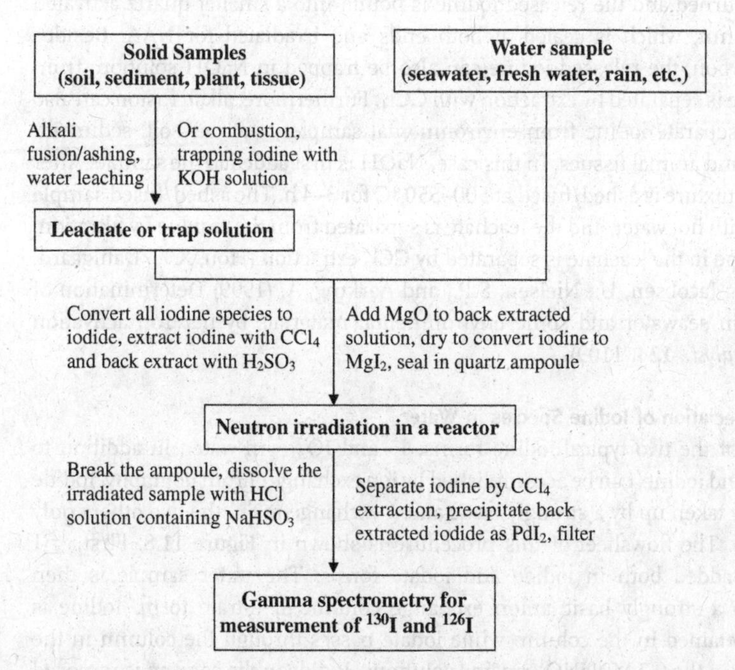

Figure 11.7 Diagram of the analytical procedure for the determination of ^{129}I by neutron activation analysis (Hou, X.L., Dahlgaard, H., and Nielsen, S.P. (2002) Level and origin of iodine-129 in the Baltic Sea. *J. Environ. Radioactiv.*, **61**, 331).

spectrometric measurement. In this method the chemical yield of iodine is determined by adding a known amount of ^{125}I tracer to the sample before the separation. The isotope ^{125}I emits 35 keV gamma rays, and the activity can be measured in the final sample together with ^{130}I and ^{126}I. ^{131}I cannot be used as a tracer for this purpose, because ^{131}I solutions always contain some amount of ^{129}I impurity, the activity of which may be higher than the ^{129}I levels in the low-activity samples.

11.4.3.2 Pretreatment of Samples for ^{129}I Analyses

The samples requiring ^{129}I analyses vary widely in type and in activity concentration. Figure 11.6 shows how iodine was removed from a spent organic resin from a nuclear power plant. It was oxidized to iodate, which has no affinity, and elutes out of the resin. Figure 11.7 shows methods for various liquid and solid samples. The activity concentration of ^{129}I in most water samples is so low that a preconcentration is needed. This can be accomplished by anion exchange. Iodine in 30–50 liters of seawater, for example, is first reduced to iodide using bisulfite solution at $pH < 2$ and is loaded into a strongly basic anion exchange column in NO_3^- form. The absorbed iodide is then eluted with 2 M $NaNO_3$ solution and separated by CCl_4 extraction. If ^{129}I is to be determined in a solid sample, from soil for example, iodine is separated by heating the sample and then binding the released iodine into an activated carbon column cooled with liquid nitrogen. To reduce the volume, the activated carbon column is burned and the released iodine is bound into a smaller quartz activated carbon column, which is sealed at both ends and irradiated for NAA. Besides activated carbon, the released iodine can also be trapped in NaOH solution, from which iodine is separated by extraction with CCl_4. Furthermore, alkali fusion can also be used to separate iodine from environmental samples, such as soil, sediment, vegetation, and animal tissues. In this case, NaOH is first added to the sample. After drying, the mixture is ashed/fused at 500–550 °C for 3–4 h. The ashed/fused sample is leached with hot water, and the leachate is separated from the residue by filtration. Finally, iodine in the leachate is separated by CCl_4 extraction (Hou, X.L., Dahlgaard, H., Rietz, B., Jacobsen, U., Nielsen, S.P., and Aarkrog, A. (1999) Determination of iodine-129 in seawater and some environmental materials by neutron activation analysis. *Analyst.*, **124**, 1109).

11.4.3.3 Speciation of Iodine Species in Water

Separation of the two typical iodine forms, I^- and IO_3–, in water, in addition to organic-bound iodine, can be accomplished by ion exchange chromatography. Iodide is effectively taken up by a strongly basic anion exchange resin, the two others only very slightly. The flowsheet of this procedure is shown in Figure 11.8. First, ^{125}I tracers are added both in iodide and iodate forms. The water sample is then poured into a strongly basic anion exchange column in nitrate form. Iodide is effectively retained in the column while iodate passes through the column in the effluent and in the 0.2 M $NaNO_3$ rinsing solutions. Iodide in the column is removed with 2 M $NaNO_3$ solution. From this solution iodine is separated by solvent extraction with CCl_4, and ^{129}I is measured by either AMS or NAA and stable iodine by ICP-MS

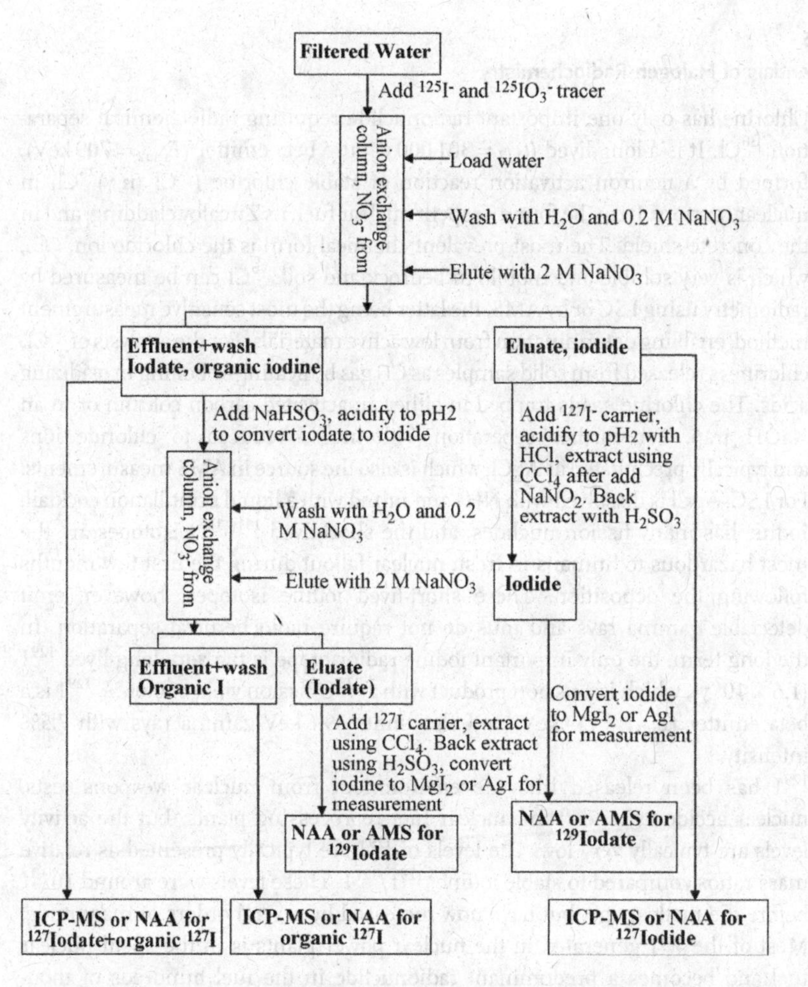

Figure 11.8 Separation of various iodine species (iodide, iodate, organic-bound) from natural waters (Hou, X., Hansen, V., Aldahan, A., Possnert, G., Lind, O.C., and Lujaniene, G. (2009) A review on speciation of iodine-129 in the environmental and biological samples. *Anal. Chim. Acta.*, **632**, 181).

or NAA. Iodine in the iodate fraction is reduced by bisulfite solution to iodide, and another ion exchange separation is carried out. Iodide in this solution is taken up by the resin, while the organic-bound iodine passes through the column. Iodide is then removed from the column with 2 M $NaNO_3$ solution, and again the solvent extraction separation is utilized to separate iodine. In case of iodide fraction, ^{129}I is measured by either AMS or NAA and stable iodine by ICP-MS or NAA. For organic-bound iodine only stable iodine can be determined. In natural waters the speciation of iodine is strongly dependent by two factors, pH and redox potential. A high pH and oxic waters favor the iodate form over iodide, while iodide is prevalent in more acidic and nonoxic waters.

11.5 Essentials of Halogen Radiochemistry

- Chlorine has only one important radionuclide requiring radiochemical separation: ^{36}Cl. It is a long-lived ($t_{1/2}$ = 301 000 y) pure beta emitter (E_{max} = 709 keV), formed by a neutron activation reaction of stable chlorine (^{35}Cl (n,γ)^{36}Cl) in nuclear reactors. It can be found in spent nuclear fuel, its Zircaloy cladding, and in the concrete shield. The most prevalent chemical form is the chloride ion, Cl^-, which is very soluble and mobile in bedrock and soil. ^{36}Cl can be measured by radiometry using LSC or by AMS, the latter being the most sensitive measurement method, enabling determination from low active materials. For the analysis of ^{36}Cl, chlorine is released from solid samples as Cl_2 gas by heating or boiling in oxidizing acids. The chlorine gas is trapped in either an activated carbon column or in an NaOH trap. For further separation, chlorine is reduced to chloride ions and typically precipitated as AgCl, which is also the source in AMS measurements. For LSC, AgCl is dissolved with NH_3 and mixed with a liquid scintillation cocktail.
- Iodine has many fission nuclides, and the short-lived $^{131-135}I$ isotopes are the most hazardous to humans in fresh nuclear fallout during the first few months following the deposition. These short-lived iodine isotopes, however, emit detectable gamma rays and thus do not require radiochemical separation. In the long term, the only important iodine radioisotope is the very long-lived ^{129}I (1.6×10^7 y), which is a fission product with a ^{235}U fission yield of 0.66%. ^{129}I is a beta emitter (E_{max} = 154 keV) and also emits 39.6 keV gamma rays with 7.5% intensity.
- ^{129}I has been released into the environment from nuclear weapons tests, nuclear accidents, and from nuclear fuel reprocessing plants, but the activity levels are typically very low. The levels of ^{129}I are typically presented as relative mass ratios compared to stable iodine, $^{129}I/^{127}I$. These levels were around 10^{-12} before the nuclear age, but have now increased by several orders of magnitude. Most of the ^{129}I generated in the nuclear power plants is in the spent nuclear fuel and becomes a predominant radionuclide in the fuel hundreds of thousands of years after closure of the repositories. Because of high mobility of iodide ions, the prevailing chemical form of iodine in aqueous solutions, ^{129}I could be one of the few radionuclides that can enter the biosphere from spent fuel repositories.
- ^{129}I can be measured by radiometric methods, neutron activation analysis, and mass spectrometry. Radiometric methods, gamma and X-ray spectrometry, and LSC, can be used to measure ^{129}I from samples with relatively high activities, typically found only in nuclear waste. Neutron activation analysis allows much lower detection limits, down to a $^{129}I/^{127}I$ ratio of 10^{-10}. The most sensitive method is AMS, capable of measuring $^{129}I/^{127}I$ ratios as low as 10^{-14}. The detection limits obtainable by ICP-MS are comparable to those of radiometric methods.
- For all measurement methods the same radiochemical separation procedures can be used. Most typically iodine is separated by solvent extraction as I_2 with CCl_4.

Iodine is stripped from the organic phase by reducing it to iodide ions. For neutron irradiation in NAA, iodine is converted into solid iodide such as MgI_2 or PbI_2. After neutron irradiation the iodide is dissolved and purified from interfering radionuclides by a further solvent extraction step. For AMS, an AgI source is prepared at the end of the iodine separation procedure.

12
Radiochemistry of the Noble Gases

12.1
Important Radionuclides of the Noble Gases

The radiochemistry of helium and neon is of no significance. Helium is generated by emission of alpha particles in alpha decay, and these become a source of helium in the environment. Argon and krypton have long-lived radionuclides of cosmogenic origin, distributed more or less homogeneously in the atmosphere. ^{85}Kr is the only long-lived fission product among the noble gases. It is generated in nuclear explosions and nuclear accidents and released into the air from nuclear reactors. The short-lived isotopes of xenon, ^{131m,133,133m,135}Xe, are fission products. They are the only noble gas nuclides that emit gamma rays (Table 12.1), and are measured in the air in monitoring activities related to the nuclear weapons test ban. The most important of the noble gas radionuclides is ^{222}Rn, which is generated in rocks and soils through alpha decay of ^{226}Ra in the uranium series. Part of the radon gas diffuses into the atmosphere before decaying. When radon-containing air, especially indoor air, is inhaled, the daughter nuclides of ^{222}Rn, radioisotopes of Bi, Po, and Pb, attached to aerosol particles may become trapped in the alveoli of the lungs. The radiation dose attributable to radon is about half of the total annual radiation dose to the average person.

12.2
Physical and Chemical Characteristics of the Noble Gases

The noble gases make up group 18 of the periodic table. Their electron structure is ns^2np^6. Since all electron shells are full, the noble gases are highly inert and seldom form compounds with other elements. Instead, they are present in nature as monoatomic gases. The few compounds that they do form are of no radiochemical significance and are not discussed here. Because the noble gases are inert, there is no associated radiochemistry. Some of them, especially radon, are nevertheless important in contributing to radioactivity in the environment and the radiation dose to humans. Their separation and measurement is therefore discussed below.

Chemistry and Analysis of Radionuclides. Jukka Lehto and Xiaolin Hou

ISBN: 978-3-527-32658-7

Table 12.1 Major radionuclides of the noble gases.

Element	Radionuclide	Half-life	Fission yield (%)	Decay mode	Gamma emission	Source
Argon	^{39}Ar	269 y		β^-	no	cosmogenic
	^{42}Ar	33 y		β^-	no	cosmogenic
Krypton	^{81}Kr	2.3×10^5 y		β^-	no	cosmogenic
	^{85}Kr	10.8 y	0.3	β^-	minor	fission product/activation
Xenon	^{131m}Xe	12 d	0.04	IT	yes	fission product
	^{133}Xe	5.2 d	6.7	β^-	yes	fission product
	^{133m}Xe	2.2 d	0.2	IT	yes	fission product
	^{135}Xe	9 h	6.5	β^-	yes	fission product
Radon	^{222}Rn	3.8 d		α	no	uranium series

12.3 Measurement of Xe Isotopes in Air

Xenon isotopes in the air are measured in monitoring activities associated with the Nuclear Weapons Test Ban Treaty and the monitoring of releases from nuclear plants. As gases, the noble elements are readily released in nuclear explosions and in the treatment of spent fuel from nuclear power plants. All Xe isotopes are short-lived, and, because of their inertness, become distributed in the atmosphere and do not constitute a risk to radiation safety. Measurement of xenon isotope ratios in the air, however, can reveal information about the source. For example, whereas the $^{133m}Xe/^{133}Xe$ ratio in releases from nuclear power plants is about 0.1, the ratio in nuclear explosions is a hundred times larger. Similarly, whereas the $^{135}Xe/^{133}Xe$ ratio in releases from nuclear power plants is 0.01, the ratio in nuclear explosions is ten thousand times larger. Whenever such high ratios are observed in the atmosphere, there is reason to suspect a nuclear explosion as the source. The activity of Xe radioisotopes is determined by passing tens of cubic meters of air through an activated carbon column cooled with liquid nitrogen. The captured Xe is separated from the other noble gases by gas chromatography. In particular need of separation are ^{85}Kr, which is also formed in nuclear explosions and nuclear plants, and ^{222}Rn, which is clearly more abundant in air than the nuclides of other noble gases. Xe radioisotopes are measured by gamma spectrometry or by leading Xe into a gas-filled proportional counter.

12.4 Determination of ^{85}Kr in Air

^{85}Kr is generated as a fission product in nuclear power plants and nuclear explosions. It can also be produced by a neutron activation reaction of stable ^{84}Kr (natural abundance 57%). Because Kr is inert, it cannot be separated out in nuclear power plants, and it is consequently released into the atmosphere, where it is distributed by

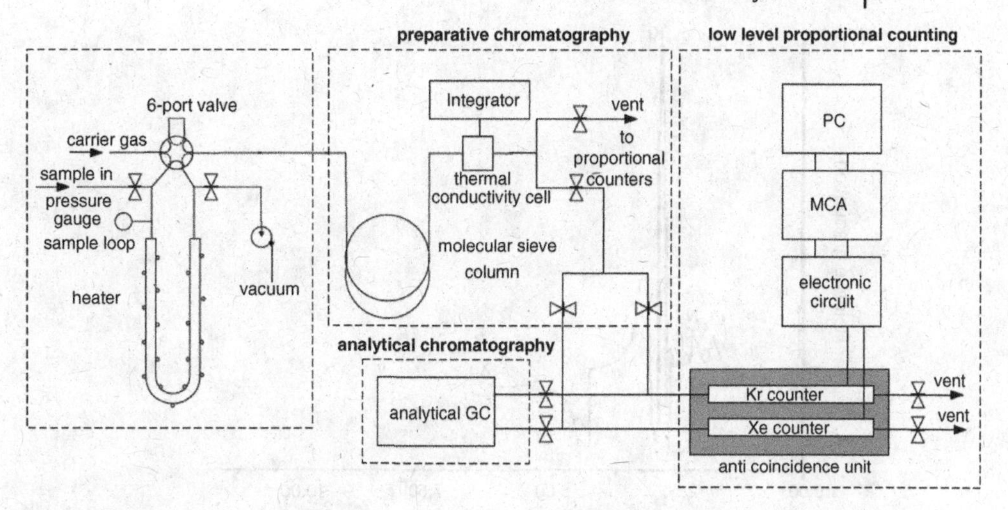

Figure 12.1 Scheme for separation and determination of ^{85}Kr and Xe isotopes. (Steinkopff, T., Dyck, W., Frank, G., Frenzel, S., and Salvamoser, J. (2004) The measurement of radioactive noble gas by DWD in the frame of the Global Atmospheric Watch Programme of WMO, *Appl. Radiat. Isot.*, **61**, 225).

air currents. ^{85}Kr is determined in the same way as xenon isotopes. It is trapped in a cooled activated carbon column and separated chromatographically from xenon and radon. ^{85}Kr is a pure beta emitter (697 keV), and its activity is measured by liquid scintillation counting (LSC) or with a gas-filled proportional counter.

Figure 12.1 shows an analytical procedure for the determination of ^{85}Kr and Xe isotopes. In this method, an air sample is first pumped through an activated carbon column cooled with liquid nitrogen for enrichment. Noble gases including Kr, Xe, and Rn are adsorbed and enriched in the activated carbon from a large air volume up to about 10 m^3. The adsorber is then warmed up to room temperature to remove N_2, O_2, and CO_2 and then to 50 °C to remove water vapor. Thereafter, noble gases are transferred into an aluminum vessel by heating the activated carbon column to 300 °C and rinsing the column with helium gas. This preconcentrated sample is pumped through a sample loop (left frame of Figure 12.1), which is filled with activated carbon and cooled with liquid nitrogen. Kr and Xe are completely adsorbed inside the sample loop. The gases are released from the loop by heating and led to a preparative gas-chromatographic system consisting of a preparative molecular sieve column to perform a baseline separation of O_2, N_2, Kr, and Xe. As seen from Figure 12.2, Kr and Xe are well separated from each other and from oxygen and nitrogen. The separated Kr and Xe fractions are routed directly into two low-level proportional counting tubes (Kr counter and Xe counter) for measuring the β-activities of ^{85}Kr and Xe radioisotopes. An aliquot of the counting gas is transferred directly into the analytical gas chromatography column for the determination of the stable Kr and Xe. In this way, radioactive ^{85}Kr and Xe radioisotope and also stable Kr and Xe can be measured.

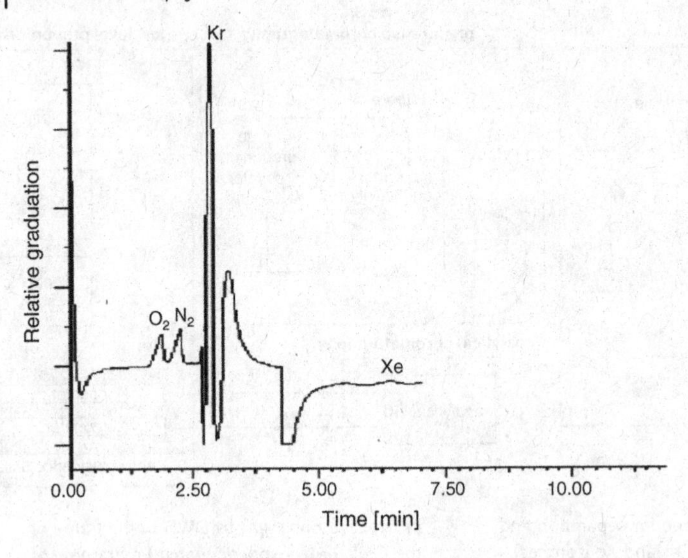

Figure 12.2 Gas chromatographic spectrum of Kr and Xe using the system shown in Figure 12.1. (Steinkopff, T., Dyck, W., Frank, G., Frenzel, S., and Salvamoser, J. (2004) The measurement of radioactive noble gas by DWD in the frame of the Global Atmospheric Watch Programme of WMO, *Appl. Radiat. Isot.*, **61**, 225.

12.5
Radon and its Determination

Three isotopes of radon form in the decay series of uranium and thorium. Of these, ^{219}Rn and ^{220}Rn have very short half-lives of just 4 s and 56 s and are therefore not significant. The half-life of ^{222}Rn is considerably longer: 3.8 days. ^{222}Rn forms in soil and bedrock during the decay of ^{226}Ra by alpha emission. Because radon is a gas, it is partly released from radium-containing minerals. The concentration of radon in soil pore spaces is 10 000–100 000 Bq m^{-3} depending on the concentration of ^{238}U and ^{226}Ra. Only a part of this diffuses into the atmosphere before it decays, however, and the concentration of radon in the air near the earth's surface is only about 10 Bq m^{-3}. Radon represents a health risk to humans when it becomes enriched in indoor air, escaping from the ground underneath, building materials, or household water. The average concentration of radon in indoor air in Finnish residences is about 120 Bq m^{-3}, but concentrations as high as 10 000 Bq m^{-3} have also been recorded. In itself, radon does not constitute a significant health risk because of its short residence time in the human body. The risk mainly comes from its short-lived alpha- and beta-active daughter nuclides, which attach themselves as solids to aerosol particles in the air. These alpha-active daughter nuclides include ^{218}Po ($t_{1/2} = 3$ min) and ^{214}Po ($t_{1/2} = 0.2$ ms) and the beta-active nuclides ^{214}Pb ($t_{1/2} = 27$ min) and ^{214}Bi ($t_{1/2} = 20$ min). When aerosol particles containing these daughter nuclides are inhaled, they attach to the lung alveoli and deliver a radiation dose.

12.5.1 Determination of Radon in Outdoor Air and Soil Pore Spaces

The concentration of radon in outdoor air can be measured by pumping an air sample into a cell coated with ZnS phosphor. When the ZnS is excited by alpha or beta particles emitted by radon or its daughter nuclides, a flash of light is emitted. The end of the cell is transparent, allowing the light to be detected by a photomultiplier tube outside the cell, which converts it into an electrical pulse. This method measures the total activity of radon and its daughters. In the case of soil pore gas, samples are collected from two to four meters depth with a probe and pumped into the ZnS cell.

The concentrations of radon and its daughter nuclides in air can also be measured separately by the two-filter method. The air is passed through a 0.1-μm filter, which traps the aerosol particle-bound radon progeny. The radon gas that passes through the filter is passed through a long chamber, where the daughter nuclides are allowed to develop to equilibrium with radon before a further filtering. Both filters are measured, for example with a scintillation counter with a ZnS phosphor. Measurement of the first filter reveals the original activity of the radon daughters and measurement of the second the activity of the radon itself.

12.5.2 Determination of Radon in Indoor Air

The most common method for determining the concentration of radon in indoor air is to use a plastic film-lined small vial (e.g., polycarbonate). The radon diffuses into the vial and reaches equilibrium with its daughter nuclides. Alpha particles emitted by radon and its daughters leave tracks on the film, which are made visible, for example, by etching. The amount of damage is directly proportional to the concentration of radon in the air. This method gives an average concentration of radon for a measuring period, typically several weeks. It is well suited for broad monitoring of radon in houses.

12.5.3 Determination of Radon in Water

If the concentration of radon in water is high enough, as, for example, in the water from a drilled well, it can be determined by measuring the daughter gamma emissions by gamma spectrometry or by adding an organic scintillation cocktail to the water and measuring the activity with a liquid scintillation counter. Both methods are simple, though care must be taken not to allow the escape of radon gas during both sampling and sample treatment. Figure 12.3 shows the liquid scintillation spectrum of a measurement of radon in water, acquired with a device that discriminates between the pulses based on alpha and beta particles. If the radon concentration is so low that direct measurement is not possible, the radon is enriched by bubbling helium gas through the sample and collecting the helium and the radon gas it picks up from the water into an activated carbon cartridge cooled with liquid

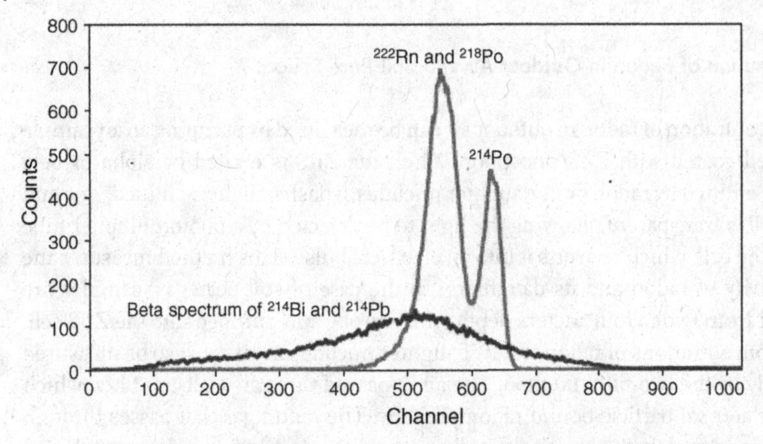

Figure 12.3 Liquid scintillation spectrum of radon and its daughter radionuclides recorded from a water sample with a liquid scintillation counter with alpha/beta discrimination capability.

nitrogen. The radon gas is released from the cartridge by heating and led to a ZnS cell for determination of the activity.

12.6
Essentials of Noble Gas Radiochemistry

- Noble gases are chemically highly inert and thus have no real radiochemistry.
- In the nuclear industry and in nuclear explosions, radioisotopes of krypton (^{85}Kr) and xenon (^{131m}Xe, ^{133}Xe, ^{133m}Xe, ^{135}Xe) are formed as fission products of uranium and plutonium. ^{85}Kr is fairly long-lived ($t_{1/2}$ = 10.8 y), but the xenon radioisotopes are short-lived ($t_{1/2}$ = 0.4–12 d). These noble gas radionuclides are more or less completely released into the atmosphere, where they become distributed in air masses. They do not cause any health hazard to humans because they are diluted into large air volumes and because of their chemical inertness. In addition, the xenon isotopes decay in a short time.
- Determination of xenon isotope ratios are used to detect undeclared nuclear weapons tests. The isotope ratios in the air after a nuclear explosion are clearly different from those generated by nuclear fuel. For the determination of xenon isotopes and ^{85}Kr, a large volume of air is passed through an activated carbon column cooled with liquid nitrogen. Krypton and xenon are separated from each other and from nitrogen and oxygen by gas chromatography and directed into a proportional counter for measuring their activities.
- ^{222}Rn is the most important source of radiation dose to humans, responsible for nearly half of the total radiation dose. ^{222}Rn is a short-lived ($t_{1/2}$ = 3.8 d) daughter of ^{226}Ra in the uranium decay series. Part of the radon diffuses from the ground into the atmosphere before its decay. In the air, radon decays to the radioisotopes of Po, Bi, and Pb, which are attached onto aerosol particles.

- In the outdoor air, radon concentration is fairly low, only a few bequerels per cubic meter. In indoor air the radon concentration is considerably higher, in some cases higher than 10 000 $Bq\,m^{-3}$. Radon in the outdoor air can be measured by pumping an air sample into a chamber which converts the radiation from radon and its daughters into light by ZnS phosphor. The light is detected by a photomultiplier tube.
- The activities of radon and its daughters can be distinguished from each other by separating the aerosol particles containing the daughter nuclides with an air filter and measuring their activities by LSC or ZnS phosphor. The radon in a volume of air that has passed through the air filter is allowed to come to equilibrium with its progeny and filtered again. This time the activity measurement gives the activity of radon initially present in the sample.
- Indoor radon concentration is typically measured by detecting the number of tracks generated by the alpha particles from radon gas in a plastic film over an exposure period of several weeks. In water, radon can be measured directly by LSC after mixing with a liquid scintillation cocktail. If the radon concentration in water is low it needs to be concentrated before the measurement. This preconcentration can be accomplished by removing radon from water by bubbling with helium gas and trapping the radon-bearing helium gas in a liquid nitrogen-cooled activated carbon column.

13
Radiochemistry of Tritium and Radiocarbon

The radioactive isotope of hydrogen, ^{3}H, called tritium, and that of carbon, ^{14}C, often called radiocarbon (though not the only radioactive isotope of carbon) are important radionuclides in radiochemistry. They are widely used in biological and life sciences as radioactive tracers with which to label organic molecules in order to study their behavior in various systems. This extensive and interesting field is, however, out of the scope of this book. Tritium and radiocarbon are produced in a variety of nuclear activities and nuclear facilities. They are found in nuclear waste and are released to the environment by nuclear weapons testing, reprocessing spent nuclear fuel, nuclear reactors, as well as nuclear accidents. They are also produced naturally in cosmic radiation-induced nuclear reactions in the atmosphere.

13.1
Tritium – ^{3}H

13.1.1
Nuclear Properties of Tritium

There are three naturally occurring isotopes of hydrogen: ^{1}H and ^{2}H, which are stable, and tritium, ^{3}H, which is radioactive. Tritium, typically having the symbol T, decays to stable ^{3}He by pure beta emission with a half-life of 12.3 years. The beta energy is low, the maximum value being only 18.6 keV and the average energy 5.7 keV (Figure 13.1).

The activity concentration of tritium is expressed in SI units as $Bq\,m^{-3}$ or $Bq\,kg^{-1}$. The concentration of tritium in water is also expressed in tritium units (TU). One TU is one tritium atom in 10^{18} hydrogen atoms ($^{3}H/^{1}H = 10^{-18}$), which is roughly equivalent to the ratio present in rain water before the nuclear age, when tritium was produced solely through the interaction of cosmic rays with atmospheric nitrogen. One TU is equivalent to a ^{3}H concentration of $0.12\,Bq\,L^{-1}$ in water.

Chemistry and Analysis of Radionuclides. Jukka Lehto and Xiaolin Hou

ISBN: 978-3-527-32658-7

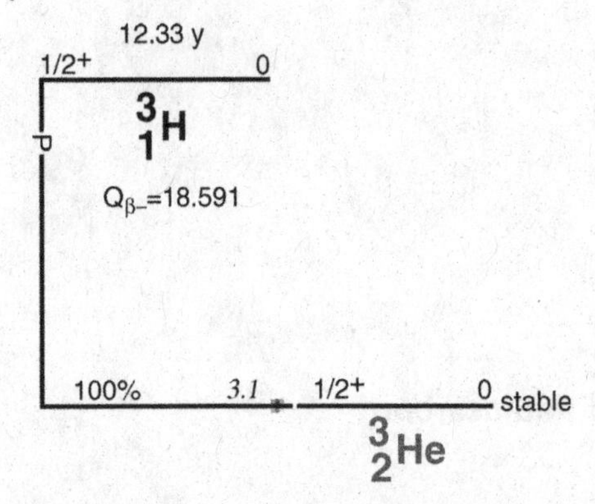

Figure 13.1 Decay scheme of tritium (Firestone, R.B., Shirley, V.S., Chu, S.Y.F., Baglin, C.M., and Zipkin, J. (1996) *Table of Isotopes*, Wiley-Interscience).

13.1.2
Environmental Sources of Tritium

Naturally occurring tritium in the environment primarily originates from nuclear reactions induced by cosmic rays in the atmosphere, especially from the reaction of neutrons with nitrogen gas $^{14}N + n \rightarrow {}^{12}C + {}^{3}H$. In the atmosphere, tritium most commonly (99%) occurs as tritiated water, HTO, but also in hydrogen gas as HT. In the biosphere, tritium can also be found in naturally occurring organic compounds such as hydrocarbons. A large amount of artificial tritium was produced and released to the environment from thermonuclear (fusion) weapons testing through the reaction $^{6}Li(n, \alpha)^{3}H$. It is also produced and released by nuclear fission weapons testing in small amounts because of its low fission yield (0.01%). Tritium is released in the reprocessing of spent nuclear fuel, and essentially all tritium ultimately finds its way into the environment as tritiated water. Atmospheric thermonuclear explosions in the 1950s and 1960s temporarily increased the amount of tritium in the atmosphere by six orders of magnitude. The amount of tritium on the earth's surface was then 100–1000 times the amount before the nuclear tests.

In nuclear reactors, tritium is also formed by the neutron activation of stable hydrogen (^{2}H), lithium, boron, and helium in the following reactions: $^{2}H(n, \gamma)^{3}H$, $^{6}Li(n, \alpha)^{3}H$, $^{10}B(n, 2\alpha)^{3}H$ and $^{3}He(n, p)^{3}H$. Of these four reactions, that of ^{6}Li is the most important because of its very high cross section (942 b). As in the case of nuclear reprocessing, tritium produced in nuclear power plants is released into the environment as tritiated water. Because tritiated water released from nuclear facilities and from nuclear explosions is strongly diluted in natural waters, it does not form any significant source for human radiation exposure. In addition to dilution, tritium's low beta energy and the fact that tritiated water does not enrich in the human body makes tritium fairly harmless as a radiation source to humans compared to most other radionuclides.

13.1.3
Determination of Tritium in Water

If the activity concentration of tritium in water is high enough, above 1 Bq L^{-1} at least, tritium can be determined by simply adding a scintillation cocktail and measuring it by liquid scintillation counting. This, however, requires that there are no other radionuclides that essentially interfere with the tritium measurement. In addition, the water should not contain any components that would quench during the measurement. Usually, liquid scintillation counting requires that tritium be separated from other radionuclides and from quenching components. Interfering radionuclides are normally separated by distillation, since most of them vaporize poorly or relatively poorly at the boiling point of water. Iodine, which could follow tritiated water in distillation as iodine gas I_2, is retained in solution by reducing it to a nonvaporizable iodide with sodium thiosulphate, ($Na_2S_2O_3$) at pH > 9. ^{14}C, another possible source of interference, could be released from the water during distillation as carbon dioxide, and it is not retained in the water being distilled. The presence of carbon dioxide in the distillate can be prevented by adding Na_2CO_3, which raises the pH high enough so that the carbon remains trapped in the initial solution as carbonate. In addition to interfering radionuclides, other components that might cause high quenching in the liquid scintillation counting are removed in the distillation step. This is important because the beta energy of tritium is very low and thus very sensitive to quenching effects. After distillation, the sample is mixed with a scintillation cocktail and the activity is measured by liquid scintillation counting (LSC). The lowest determination limits are achieved by using a low background liquid scintillation counter.

When determining tritium in the air, where it occurs as tritiated water, the water is either condensed in a cold trap or adsorbed in silica gel or a molecular sieve column. The water trapped in the columns is removed by heating and then collected by condensing it by means of cooling. The water samples thus obtained are distilled and the tritium is measured by LSC.

13.1.4
Electrolytic Enrichment of Tritium

If the activity of an aqueous sample is low (<1 Bq L^{-1}), direct counting of the distilled water using LSC is very difficult, and the tritium needs to be enriched – usually electrolytically. In electrolysis, water is decomposed into oxygen and hydrogen in an electric current. Water is placed in a cell with an anode at one end and a cathode at the other. When the electric current is switched on, the hydrogen ions move toward the cathode where they are reduced to hydrogen molecules; the hydrogen is then released as gas: $2H^+ + 2e^- \rightarrow H_2$. Correspondingly, the hydroxyl ions are oxidized to oxygen gas at the anode: $2OH^- \rightarrow O_2 + 2H^+ + 2e^-$. Thus, the dissociation of water into hydrogen and hydroxyl ions according the equation $H_2O \leftrightarrow H^+ + OH^-$ shifts to the right. Tritium is enriched in the water because the tritiated water dissociates into hydrogen and hydroxyl ions less efficiently than normal water: the dissociation constant (pK_w) at 20 °C for H_2O is 14 while for HTO it is 15.2. In the enrichment of tritium by electrolysis, a 100–500 mL sample of water is concentrated electrolytically to a volume of 3–10 mL, which results in 10- to 50-fold tritium enrichment. The detection limit for tritium is thus improved, reaching approximately 0.1–0.02 Bq L^{-1}, which is much better that the figure achieved using LSC determination directly after distillation (~ 1 Bq L^{-1}). The enrichment factor can be obtained by carrying out the electrolysis, under the same conditions, for a standard of known ^3H activity. If tritium concentrations are very low, the standard must be prepared in 'dead water', that is, tritium-free groundwater taken from deep in the bedrock. Dead water is free of tritium contamination since any cosmogenic tritium has had time to decay and thus there is no background interference in the measurement. When measuring low-level tritium samples, it is also important to use low background scintillation counters, such as the Quantulus from Perkin-Elmer, to obtain as low a detection limit as possible.

13.1.5 Determination of Tritium in Organic Material

In determining tritium from organic material, the sample, vegetation for example, is first dried by freeze-drying or at moderately elevated temperatures (60–80 °C). The water released from the sample is collected by condensing in a cold trap. Tritium is determined from this free water after distillation by LSC measurement. To obtain organic-bound tritium, the residual solid organic matter is combusted and the generated water is collected in another cold trap; tritium is then determined after distillation by LSC analysis. Perkin-Elmer has an automatic sample-preparing device (Sample Oxidizer) which combusts organic matter and collects tritium in one liquid scintillation vial and radiocarbon in another vial and adds scintillation cocktail for LSC measurement (Figure 13.2).

13.1.6 Determination of Tritium from Urine

In occupational health and other radiation protection activities, the determination of tritium in human urine is the main way to estimate the amount of tritium in the human body. Tritium can be measured directly from urine simply by adding a scintillation cocktail and measuring the activity by LSC measurement. The color of urine, however, is a problem that causes variable quenching and consequently variable counting efficiency. The counting efficiency can be determined by internal standardization, that is, by measuring two parallel subsamples from each sample: one sample with added tritium standard with a known activity and another without a standard addition. This method gives the total tritium concentration in the urine, including tritiated water and possibly co-occurring organic-bound tritium. To determine tritiated water only, a urine sample is distilled and the tritium is measured from the distillate. To avoid foaming during distillation, the urine sample is treated with activated carbon prior to distillation: urine (25 mL) is mixed with active carbon powder (1 g) in a vial and filtered into a distillation vessel.

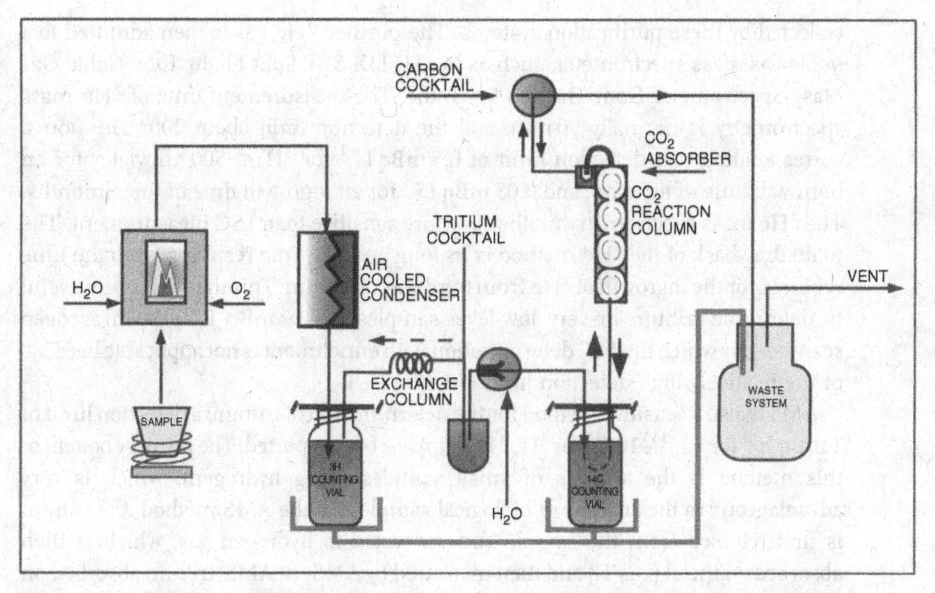

Figure 13.2 Diagram of Sample Oxidizer used for the separation of tritium and ^{14}C from nuclear waste samples by combustion.

13.1.7
Determination of Tritium after Conversion into Benzene

Tritium can be converted into benzene by a method described in the next section, in which the determination of radiocarbon is discussed. This method does not, however, essentially bring an improvement in the tritium detection levels, and, since this is considerably more laborious than distillation, it has not been much used.

13.1.8
Determination of Tritium using Mass Spectrometry

Tritium can also be measured by ^{3}He mass spectrometry. The sample is first degassed to remove the daughter nuclide ^{3}He in the sample and then kept for between one week and a few months, depending on the ^{3}H level in the sample for the ingrowth of ^{3}He from the decay of tritium. The ^{3}He produced from the tritium is removed from the sample by degassing in the following way. The water sample with the ingrown ^{3}He is heated at 90 °C for about 10 min in a stainless steel bottle equipped with vacuum valves, with the sample valve closed to promote the equilibrium partitioning of ^{3}He into the gas phase. Thereafter, the container is cooled in liquid nitrogen for about 15 min to immobilize the tritium in water. The ingrown ^{3}He in the gas phase is removed from the container by drawing off the headspace gases. The ^{3}He in these gases is purified by exposing them to Ti alloy getters at 400 °C and to activated charcoal at liquid nitrogen temperature. All gases other than He and Ne are collected by these purification systems. The purified ^{3}He gas is then admitted to a noble-gas mass spectrometer, such as the HELIX SFT Split Flight Tube Noble Gas Mass Spectrometer from Thermo Scientific. The measurement time of ^{3}He mass spectrometry is normally 90 min, and the detection limit about 200 ^{3}He atoms, corresponding to a detection limit of 0.5 mBq L^{-1} for ^{3}H in 500 mL water for an ingrowth time of ten days, and 0.05 mBq L^{-1} for an ingrowth time of three months. The ^{3}He mass spectrometry method is more sensitive than LSC measurement. The main drawback of the ^{3}He method is its long analysis time resulting from the time required for the ingrowth of ^{3}He from the decay of tritium. This method is very useful to determine tritium in very low-level samples (<0.02 mBq L^{-1}), such as deep seawater, for which the LSC determination with enrichment is not applicable because of the relatively high detection limit (50 mBq L^{-1}).

AMS is also a sensitive method for the determination of tritium: a detection limit of 1 mBq L^{-1} (or 10^{-13}–10^{-15} for ^{3}H/^{1}H ratio) has been reported. The primary benefit of this method is the analysis of small samples (2 mg hydrogen), which is very advantageous in the analysis of biological samples. In the AMS method, the tritium is first released from the sample and converted to hydrogen gas, which is then absorbed on metal (i.e., Ti) and then measured by AMS. In AMS, tritium absorbed on the metal target is inserted into a sample holder and put in the ion source. The target is sputtered with Cs^+ ions in the ion source. Tritium is then released from the target as negative T^- ions, which are selected with a deflection magnet and accelerated in a Tandem accelerator at a terminal voltage of 1.5 MeV. ^{3}He does not interfere with the determination of tritium because no stable He^- ion can be formed in the ion source. In the middle of the Tandem, T^- ions are stripped in a nitrogen gas stripper; positive T^+ ions are then formed by the removal of two electrons, and the ions are again accelerated to the ground potential. In this process, most of the HD^- and H_3^- ions are split into single atomic ions, which do not interfere with the measurement of ^{3}H at mass 3. After analysis with a switching magnet, the tritium ions are counted with a surface barrier detector. A more detailed description of the AMS principle is presented in Chapter 17. The sample preparation for AMS is more complicated than that for LSC. Because of this and the limited availability of AMS facilities, AMS is very seldom used for the determination of tritium. In a routine analysis, LSC is still the most frequent method used to determine tritium in environmental and waste samples.

13.1.9 Determination of Tritium in Nuclear Waste Samples

In heavy water reactors, large amounts of tritium are produced by the neutron activation reaction ^{2}H(n, γ)^{3}H. The produced tritium, existing as tritiated water, may be released to air in the reactor building during the purification (or other handling) of heavy water. Heavy water and the released tritium vapor may come into contact with various materials in the reactor building and contaminate them. In most construction materials in the nuclear reactor – concrete, graphite, and various metals – tritium is mainly produced by the neutron-induced reaction of stable lithium, ^{6}Li(n, α)^{3}H, and to some extent by

$^{10}B(n, 2\alpha)^{3}H$. By these reactions, tritium can be also produced in other nuclear facilities such as accelerators. Because of the normally high tritium concentration in the materials in nuclear facilities and the high mobility of tritiated water, it is one of most important radionuclides in the decommissioning of nuclear facilities.

Some methods for the determination of tritium in various samples for decommissioning nuclear facilities, including graphite, concrete, soil, sand, oil, stainless steel, paint, and silica gel, have been reported (Hou, X.L. (2005) Determination of C-14 and H-3 in reactor graphite and concrete for decommission. *App. Radiat. Isot.*, **62**, 882; Hou, X.L. (2007) Radiochemical analysis of radionuclides difficult to measure. *J. Radioanal. Nucl. Chem.*, **273**, 43). The methods are based on the combustion of materials with the aid of oxygen. In these methods, the samples are first ground to a fine powder or cut into small pieces and mixed with combustible material such as cellulose powder. The mixture is then combusted with the Perkin-Elmer Sample Oxidizer at 1000–1200 °C. Tritium is converted to tritiated water vapor which is removed from the sample with N_2 carrier gas. The tritiated water vapor is then condensed in an air cooling column and collected in a vial for LSC measurement (Figure 13.2).

Because of the very high concentration of tritium in graphite, a very small sample has to be used. The finely ground graphite powder is first mixed with cellulose powder to dilute the sample, after which a small amount is taken for analysis. To analyze metals such as stainless steel, the sample needs to be cut into small pieces before combustion with the Sample Oxidizer. Since the mechanical cutting may cause a loss of tritium in the samples, an acid digestion is applied to determine the tritium in the metal samples. In this method, the stainless steel is put into a three-necked flask, acid (HCl) is then added, and the sample is dissolved by heating. The solution is then neutralized and distilled; the distilled water is used to measure the tritium by LSC. No significant difference in tritium concentration is found in the same sample treated by acid digestion or combustion.

13.2 Radiocarbon – ^{14}C

13.2.1 Nuclear Properties of Radiocarbon

^{14}C is a long-lived ($t_{1/2} = 5730$ y) pure beta emitter. It decays to stable ^{14}N by the emission of beta particles with a maximum energy of 156 keV (Figure 13.3).

13.2.2 Sources of Radiocarbon

^{14}C is formed in the atmosphere from nitrogen in a nuclear reaction with cosmic rays $^{14}N(n, p)^{14}C$. The production rate of ^{14}C in the atmosphere is fairly constant, and,

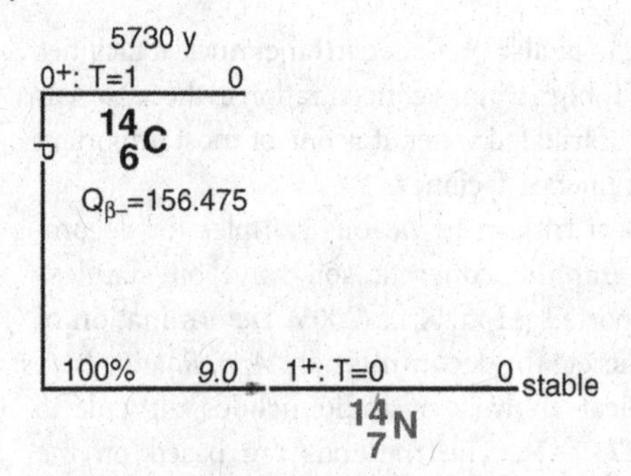

Figure 13.3 Decay scheme of ^{14}C (Firestone, R.B., Shirley, V.S., Chu, S.Y.F., Baglin, C.M., and Zipkin, J. (1996) *Table of Isotopes*, Wiley-Interscience).

because the half-life is long (5720 years), the $^{14}C/^{12}C$ ratio in the atmosphere is constant at 1.2×10^{-12}, or about 200 Bq ^{14}C per kilogram of carbon. Almost all the carbon in the atmosphere is in the form of carbon dioxide, CO_2, while less than 1% is present as methane, CH_4, and a still smaller proportion as carbon monoxide, CO. Much of the methane in the atmosphere is produced by the anaerobic decay of organic material in wetlands and sediments. The proportion of radioactive carbon in the atmosphere increased markedly in the 1950s and 1960s as a result of atmospheric nuclear weapons testing, in which radiocarbon was produced from nitrogen in the atmosphere in the same nuclear reaction with neutrons as that producing cosmogenic radiocarbon. The concentration of radiocarbon has now returned almost to the level that prevailed before nuclear testing. Increased emissions of fossil carbon dioxide (depleted in its ^{14}C concentration) in the last century have, in turn, had a diluting effect reducing the amount of radiocarbon as a fraction of the total carbon in the atmosphere.

Radiocarbon is also formed by neutron-induced nuclear reactions from nitrogen, oxygen, and carbon in the reactions $^{14}N(n, p)^{14}C$, $^{17}O(n, \alpha)^{14}C$ and $^{13}C(n, \gamma)^{14}C$. Of these, ^{14}N is the major source because of its reasonably high cross section (1.8 b) and high isotopic abundance of ^{14}N (99.6%). The contribution of the last reaction is exceedingly small. In nuclear fuel (UO_2), radiocarbon is generated from the oxygen and, in particular, the nitrogen impurities found in fuel and in fuel cladding. Most of the radiocarbon generated in this way remains in the fuel or fuel cladding and is only released during fuel reprocessing. In the coolant water of the reactor primary circuit, radiocarbon will be mostly produced from the ^{17}O isotope of water as well as from dissolved nitrogen and the nitrogen compounds added to pressurized water reactors. About 95% of the radiocarbon generated in the primary circuit water is released to the atmosphere as carbon dioxide through the ventilation system. The remainder is retained, mostly in the form of activated bicarbonate ions in the anion exchange resins used in the purification of the primary circuit. In addition to the resin, radiocarbon is contained in a variety of irradiated solid materials.

13.2.3
Chemistry of Inorganic Carbon

Carbon belongs to group 14 of the periodic table, having six electrons in configuration $1s^2 2s^2 2p^2$, and forming more compounds than any other element. Its oxidation states are +IV, 0 or −IV, and the chemical bonds are covalent. Compounds labeled with ^{14}C are widely used in organic radiochemistry; however, the important carbon compound in inorganic radiochemistry is carbon dioxide, CO_2, including its aqueous solutions. When dissolved in water, carbon dioxide forms carbonate ion, CO_3^{2-}, in alkaline solution, bicarbonate ion, HCO_3^-, in the neutral region, and carbonic acid, H_2CO_3, in acidic solution. Most of the carbon dioxide escapes from the water when the pH of carbonate-bearing solution is lowered to a value of less than 6. In a water solution system in contact with the atmosphere, the concentration of dissolved carbon dioxide in water at pH below 6 is approximately 10^{-5} mol L^{-1}. The concentration of carbon dioxide in the air and in solutions is usually expressed as a partial pressure. In the atmosphere, the partial pressure is $10^{-3.5}$ bar, while in natural waters it is in the range 10^{-4}–10^{-2} bar, with the highest concentrations found in ground waters. Carbon forms sparingly soluble carbonates with most metals, excluding alkali metals and ammonium ions. The major natural forms are calcite $CaCO_3$ and dolomite $CaMg(CO_3)_2$. In strongly reducing conditions, deep in the bedrock and in anoxic conditions in wetlands and sediments, carbon forms methane, CH_4, rather than carbon dioxide or carbonates. The speciation of carbon in different pH and Eh regions is given in Figure 13.4. The role of carbon dioxide in controlling radionuclide speciation is discussed in more detail in Chapter 3.

13.2.4
Carbon Dating of Carbonaceous Samples

The age of carbon-containing materials, or the time elapsed since their death and removal from the biosphere, can be calculated from the ratio of ^{14}C to total carbon (or ^{12}C). Plants take up carbon – in the form of carbon dioxide – from the air during photosynthesis. Because the production yield and concentration of ^{14}C in air are fairly constant, the $^{14}C/C_{tot}$ ratio in plants is also more or less constant at 0.2 Bq ^{14}C per gram of carbon. When plants die and are buried in anoxic conditions in soil or sediment so that the carbon does not escape as carbon dioxide through plant decay, the $^{14}C/C_{tot}$ ratio begins to decline in accordance with the half-life of ^{14}C. Thus, the age of trees, for example, buried in wetlands or sediment can be determined by measuring the $^{14}C/C_{tot}$ ratio. The upper limit of radiocarbon dating is approximately ten ^{14}C half-lives, or about 50 000 years. During this time, the $^{14}C/C_{tot}$ ratio in the atmosphere has not been entirely constant, but has varied, for example, with the intensity of cosmic rays. Accurate dating thus requires careful calibration and corrections on the basis of known fluctuations. The initial ^{14}C level for the age calculation of animal or vegetation samples can either be estimated or obtained by direct comparison with known year-by-year data from tree-ring data (dendrochronology) to 10 000 years of age or from cave deposits (speleothems) to about 45 000

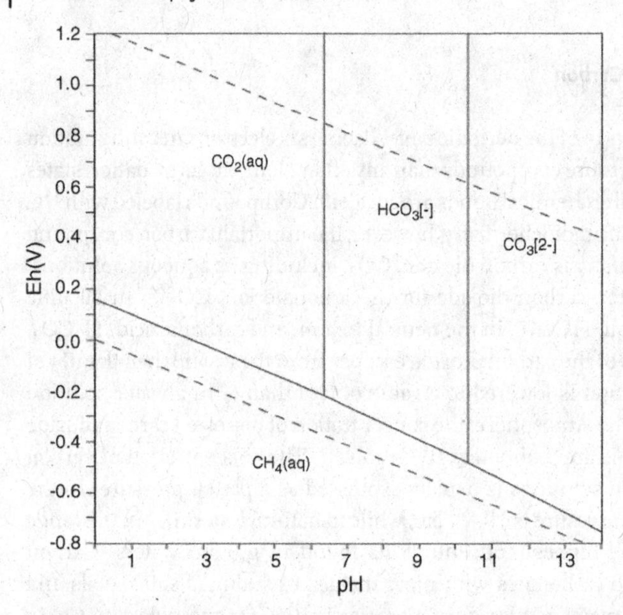

Figure 13.4 Eh–pH diagram of carbon (Atlas of Eh-pH Diagrams, Geological Survey of Japan Open File Report No. 419, 2005).

years of age. This technique has limitations within the modern industrial era because of the large atmospheric emissions during the past few centuries of fossil fuel carbon with only a little ^{14}C. Carbon dating was developed in the 1940s and 1950s. Shortly afterwards, a large amount of ^{14}C entered the atmosphere as a result of nuclear weapons testing. Because ^{14}C dating cannot, in any event, be used to measure ages of just tens or even hundreds of years, the extra input of ^{14}C through weapons testing is not a problem for age determinations.

13.2.5 Separation and Determination of ^{14}C

For age determinations and other purposes, the amount of ^{14}C is measured either radiometrically or by accelerator mass spectrometry (AMS). AMS allows the measurement of very much smaller amounts of ^{14}C and provides more accurate results. One other advantage of AMS is that the sample size can be very small – as little as a few milligrams. Radiometric methods utilize proportional counters or liquid scintillation counters. In measurements of ^{14}C activity with a gas proportional counter, the carbon is converted to carbon dioxide or methane, which is led into the detection chamber for counting. The detection limit of the liquid scintillation counter is typically about 30 mBq, sufficient for measurements of ^{14}C in nuclear waste or above-ground environmental samples. For dating purposes, however, virtually all measurements are made with AMS. With a detection limit of several orders of magnitude lower than that of the liquid scintillation counter (as low as 0.1 mBq), AMS allows dating from an extremely small sample. This is of enormous importance if the amount of material available for sampling is small, for example, if the material is an archeologically valuable artifact or if the carbon content is low, as in ice cores. ICP-MS, which is now widely used to measure long-lived radionuclides, cannot be used in ^{14}C determinations because ^{14}N is of almost the same mass and would unacceptably disturb the mass analysis of ^{14}C. Whatever the method of measurement, if the material is organic it must first be combusted, and the carbon dioxide that forms must be collected. In the case of inorganic carbon (carbonate), the carbon dioxide is separated from the sample by treating with acid so that the carbon escapes as gaseous carbon dioxide, which is then trapped for analysis or for further processing.

Tracers are not used in yield determinations for ^{14}C counting because no suitable radioactive tracers for carbon exist. Further, it is not possible to use a stable carbon carrier for yield determination because the amount of stable carbon also needs to be determined to obtain the $^{14}C/C$ ratio. Usually, almost 100% efficiency is achieved in the combustion step used to remove carbon from the sample, which means that there is almost total recovery of the carbon. The efficiency of the combustion system can be investigated by the use of standards, that is, by burning samples with a known amount of ^{14}C.

13.2.5.1 Removal of Carbon from Samples by Combustion for the Determination of ^{14}C

For measuring ^{14}C by LSC or AMS, the first step is to remove the carbon from the samples by combustion (see Figure 13.5 for a typical furnace). In combustion, carbon is transformed, by heating it in air or oxygen, into carbon dioxide, which is trapped in several ways depending on the measurement method. For LSC measurement, carbon dioxide is either trapped in an NaOH/KOH trap and converted into $CaCO_3$, or in an amine trap and converted into carbamate, or the carbon dioxide is converted into benzene. For AMS, carbon dioxide is trapped in a liquid nitrogen cooled trap and converted into graphite for measurement. All these methods are discussed in detail later.

In a combustion furnace, the sample is placed in a quartz tube on the combustion platform. The combustion gases are led with a stream of air into a tube furnace heated to 750 °C. In addition to carbon dioxide, the combustion gases contain carbon monoxide, which is oxidized to carbon dioxide in a stream of oxygen with a CuO catalyst.

13.2.5.2 Determination of ^{14}C as Calcium Carbonate by Liquid Scintillation Counting

The simplest and least accurate method to determine ^{14}C is to collect the carbon dioxide present in the combustion gases in an NaOH/KOH trap, precipitate the formed carbonate as calcium carbonate, and measure the precipitate with a liquid scintillation counter. The carbon dioxide is collected in a 2 M NaOH or KOH solution, where it dissolves as carbonate. $CaCl_2$ is added to the solution, which results in the

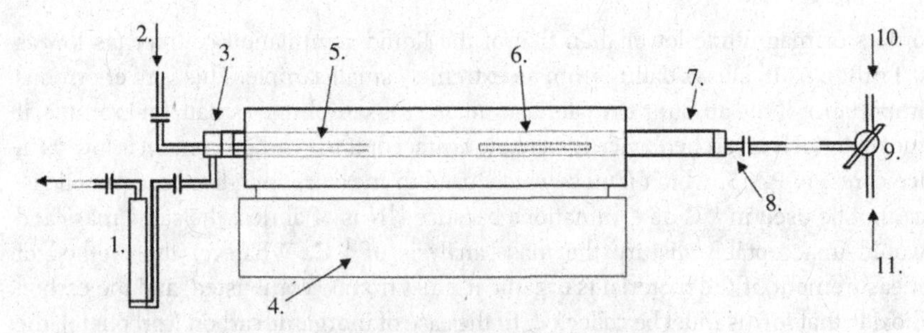

Figure 13.5 Schematic diagram and picture of combustion facility (Carbolite, UK) for ^{14}C separation. (1) Gas bubbler (filled with NaOH solution for trapping iodine); (2) Oxygen supply; (3) Exhaust gas manifold; (4) Temperature controller of combustion furnace; (5) Second furnace (for complete combustion of residue from first furnace); (6) sample boat in the first furnace; (7) Quartz working tube; (8) Gas inlet adapter; (9) Three-way valve; (10) Main oxygen supply; (11) Compressed air supply (at the beginning of the combustion, air is supplied to avoid a violent combustion under pure oxygen conditions).

precipitation of $CaCO_3$. The $CaCO_3$ is separated from the solution by filtering, then dried and weighed to determine the yield of carbon. The precipitate is transferred into a scintillation vial and a scintillation cocktail added; the beta activity of ^{14}C is then measured with a low-background liquid scintillation counter. Measuring the activity from a precipitate by LSC can be problematic. The first problem is the self-absorption or physical quench; that is, the absorption of beta radiation and the photons produced in the scintillation cocktail onto the precipitate. This lowers the counting efficiency and is significant, especially for low- and medium-energy beta emitters such as ^{14}C. Another problem is that the repeatability is not very good as the particle size in the precipitate is inhomogeneous, which causes fluctuations in the self-absorption and thus in counting efficiency.

13.2.5.3 Determination of ^{14}C by Liquid Scintillation Counting with Carbon Bound to Amine

A more advanced version of the method described above is to bind the carbon dioxide produced by combining it with an amine and not by trapping it in NaOH/KOH. The carbon dioxide forms a carbamate according to the following reaction:

$$R\text{-}NH_2 + CO_2 \rightarrow R\text{-}NHCOOH$$

The ^{14}C to be measured is obtained in soluble form, and thus the problems associated with self-absorption are avoided. Amines also dissolve well in an organic scintillator. Another advantage is that considerably more radiocarbon is collected in the amine than in the NaOH/KOH trap. A commercial amine (Carbo-Sorb E), specifically designed to trap $^{14}CO_2$ from the combustion gases, is also available.

As already mentioned, automatic sample preparation equipment for ^{14}C determinations by LSC is commercially available from Perkin-Elmer. The organic sample is burned, and the generated carbon dioxide is bound to amine, which is led directly to the liquid scintillation vial together with the liquid scintillation cocktail. A particular advantage of this equipment is that tritium can be separated, in the form of tritiated water, in another scintillation vial, so that tritium and radiocarbon can be measured from the same sample (Figure 13.2).

13.2.5.4 ^{14}C Determination by LSC in Benzene

A still more advanced way to measure ^{14}C (after combustion) is to convert the carbon into benzene, which easily mixes with the liquid scintillation cocktail. Figure 13.6 shows a scheme for the synthesis equipment used to convert carbon into benzene. The sample is burned in an external flame, and, after drying, the generated carbon dioxide is led to a reaction vessel containing molten lithium metal, where the carbon dioxide is transformed into lithium carbide, Li_2C_2. The lithium carbide is then decomposed in hot steam, and the obtained acetylene, C_2H_2, to which the carbon has transferred, is led to the reaction vessel. In the reaction vessel, the acetylene is transformed into benzene C_6H_6 by the action of chromium catalyst. Benzene is collected and mixed with scintillation cocktail to measure the ^{14}C by liquid scintillation counting. The advantage of the liquid scintillation analysis of benzene is that benzene contains over 90% carbon. As the concentration of carbon in the counting sample is high, the efficiency of the measurement is improved. The efficiency can be further improved by adding scintillation molecules rather than a liquid scintillation cocktail to the benzene, so that the benzene almost fills the liquid scintillation vial. Compared to trapping into amine, the primary benefit of benzene is that the measurement efficiency is substantially greater.

13.2.5.5 ^{14}C Determination in Graphite form by AMS

Solid samples are typically required for AMS measurement. In the case of ^{14}C, this is graphite. To convert the carbon dioxide into graphite, the sample material is first heated, and the released carbon dioxide is captured in a trap cooled with liquid nitrogen. From there, the carbon dioxide is led to the reaction chamber, where it is converted to graphite by reducing the carbon dioxide with hydrogen gas at about

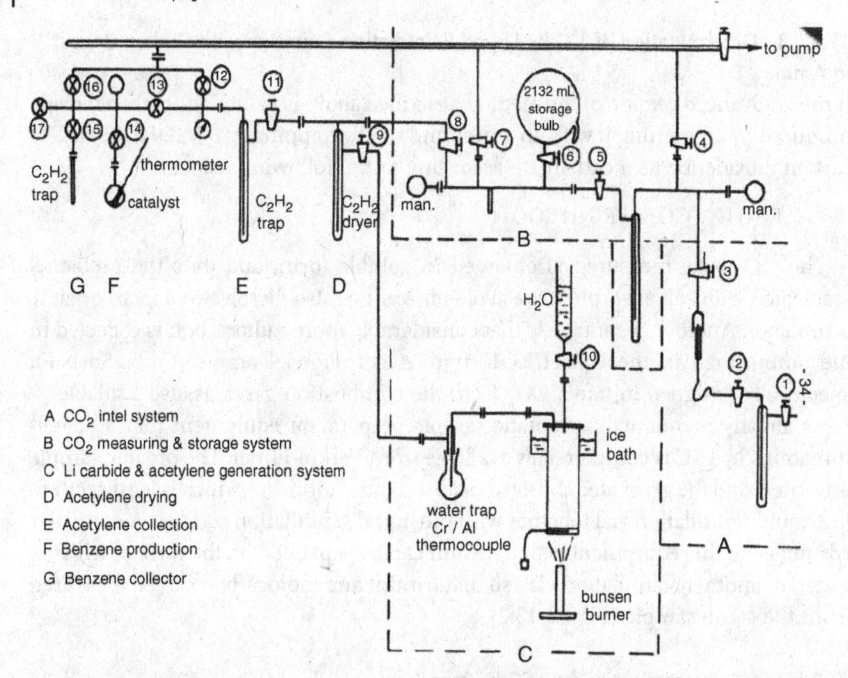

Figure 13.6 Synthesis equipment for converting the carbon in a sample into benzene.

600 °C: $CO_2 + 2H_2 \rightarrow C + 2H_2O$. The catalyst in the reaction chamber is an iron and cobalt powder. After the water has been removed from the reaction mixture, a graphite and iron/cobalt mixture is produced. This is pressed into small depressions (1-mm diameter) in an aluminum holder to be measured by AMS.

13.2.5.6 Determination of ^{14}C in Nuclear Waste

^{14}C is produced in a nuclear reactor, mainly by neutron activation of ^{14}N in the air and in construction materials, and of ^{17}O (natural abundance of 0.038%) in the UO_2 fuel and water coolant. Neutron activation of ^{13}C in the graphite reflector also makes a small contribution. Because of its relatively high concentration, long half-life, and high mobility in the environment, ^{14}C is one of most important radionuclides in the decommissioning of nuclear facilities and waste disposal.

The methods to determine ^{14}C in graphite, concrete, soil, sand, metals, oil, coolant water (including heavy water), and waste water have been reported (Hou, X.L. (2005) Determination of C-14 and H-3 in reactor graphite and concrete for decommission. *App. Radiat. Isot.*, **62**, 882; Hou, X. (2007) Radiochemical analysis of radionuclides difficult to measure for waste characterization in decommissioning of nuclear facilities, *J. Radioanal. Nucl. Chem.*, **273**, 43). To analyze graphite, concrete, soil, sand, and oil samples, the sample is first mixed with a combustible material such as cellulose powder, and the mixture is then decomposed by using a Sample Oxidizer (Figure 13.2) at 1000–1200 °C with an oxygen flow. The carbon, in the samples in organic form or as carbonate, is decomposed to CO_2, which is trapped in Carbo-Sorb

E amine solution. For the LSC measurement of ^{14}C activity, this solution is mixed with a scintillation cocktail.. Both the carbonate and the ^{14}C-labeled organic compounds are almost completely decomposed to CO_2. To treat metal samples, such as aluminum and stainless steel by a Sample Oxidizer, the samples must be cut into small pieces. By using a Sample Oxidizer, both radiocarbon and tritium can be determined simultaneously; also, each sample can be prepared in three minutes. For large metal samples, a big combustion furnace (Figure 13.5) can be used; however, it has its disadvantages in that only 1–2 samples per day can be prepared. To analyze primary coolant and waste waters, the samples are first adjusted to pH 8–9 using NaOH with 5–10 mg of Na_2CO_3 added as carrier. The solution is then evaporated to dryness to remove tritium, and the residue is dissolved with water and transferred to an LSC vial, and a scintillation cocktail is added. If there are radionuclides other than tritium present in the waste water, which is usually the case, the dissolved evaporation residue is mixed with a cellulose powder, dried by evaporation, and decomposed using a Sample Oxidizer as described above. The ^{14}C trapped in amine solution is measured by LSC. With heavy water samples, the evaporation needs to be repeated 3 to 5 times because of the extremely high tritium concentration (10^{11}–10^{12} Bq L^{-1}).

13.3 Essentials of Tritium and Radiocarbon Radiochemistry

- Tritium (^3H) is the only radioactive isotope of hydrogen. It is a pure beta emitter with a maximum energy of 18.6 keV and a half-life of 12.3 y. Tritium is formed naturally in the atmosphere in neutron-induced nuclear reactions from nitrogen atoms; 99% of tritium in the atmosphere occurs as tritiated water, HTO. Major artificial sources of tritium are atmospheric thermonuclear explosions and the nuclear industry, where the main source is the neutron-induced nuclear reaction in ^6Li.
- Tritium in water is typically determined by simply distilling the sample and measuring the tritium activity by LSC after adding scintillation cocktail to the distillate. For low-active samples, preconcentration of tritium is needed. This is done by decomposing water by electrolysis: tritium-bearing water dissociates at a slower rate than ordinary water and is thus enriched in the residue. To determine organically bound tritium, the sample is decomposed by combustion and the water formed is trapped in a cold trap and finally mixed with scintillation cocktail to measure its activity. Tritium can also be measured by mass spectrometry, the lowest detection limits being obtained with AMS. The additional advantage of AMS is that tritium can be measured from very small samples.
- ^{14}C is a long-lived ($t_{1/2} = 5730$ y) pure beta emitter with a maximum beta energy of 156 keV. It is naturally formed in the atmosphere in neutron-induced nuclear reactions from nitrogen atoms. Atmospheric nuclear weapons tests generated large additional amounts of radiocarbon into the atmosphere. Radiocarbon is produced in a nuclear reactor by neutron-induced nuclear reactions from ^{14}N,

^{17}O, and ^{13}C in various materials, such as concrete and nuclear fuel. Because of its long half-life and mobility, it is a major radionuclide in spent nuclear fuel with respect to possible radiation doses to humans in the future.

- Carbon mostly occurs in the atmosphere as carbon dioxide and, to small extent, as methane and carbon monoxide. Carbon dioxide dissolves in water and forms HCO_3^- ions in the neutral pH range and CO_3^{2-} in alkaline solutions. Most metals form sparingly soluble carbonates. $CaCO_3$ and $CaMg(CO_3)_2$ are the major forms of carbon in the geosphere.
- ^{14}C activity is measured by either LSC or AMS, the latter allowing the measurement of much lower activities. To measure its activity, carbon is released from aqueous samples by adding acid, which turns the carbonate mostly into carbon dioxide. From solid samples, carbon is released by combustion. The carbon dioxide released from the sample is trapped in various ways and converted into calcium carbonate, carbamate, or benzene to measure the ^{14}C activity by LSC. To measure by AMS, the carbon dioxide is converted into graphite.

14
Radiochemistry of Lead, Polonium, Tin, and Selenium

A large cluster of radioactive isotopes of mercury, thallium, lead, bismuth, polonium, and astatine follow after radon in the decay series of uranium and thorium (Figure 14.1, Table 14.1). With respect to radiochemistry, the most important of these radionuclides are long-lived ^{210}Pb and its short-lived daughter ^{210}Pb, which are members of the ^{238}U series. There are several important reasons for studying ^{210}Pb and ^{210}Pb. First, radon decays in the atmosphere via its short-lived daughters into longer-lived ^{210}Pb, which, together with its daughter, ^{210}Pb, are the predominant radionuclides in air, and especially in deposition. Second, the transportation of both these nuclides into humans via the respiratory and digestive tracts creates the need to determine their accumulation in humans. Third, the deposition of ^{210}Pb into sediments can be used to measure rates of sedimentation and sediment age. The deposition of ^{210}Pb from the air is relatively even, and, as new material continually accumulates on top of the old, each layer of sediment can be separately dated from the ^{210}Pb concentration. The proportion of ^{210}Pb that has decayed increases with sediment depth, and the sediment, in turn, can be dated to almost ten ^{210}Pb half-lives or about 200 years, which nicely covers the modern industrial and nuclear age. Study of the depth distribution of radionuclides of heavy metals in sediments by ^{210}Pb dating provides information on when the emissions occurred and how they have varied over time. In addition, ^{210}Po, being an alpha emitter with high radiotoxicity, makes a major contribution to the natural radiation received by humans because of its accumulation from foods, especially from fish and shellfish.

In addition to ^{210}Pb and ^{210}Po, located in the 14th and 16th group in the periodic table, another two important radionuclides in these two groups are ^{126}Sn and ^{79}Se. Because of their long half-lives (2.3×10^5 y for ^{126}Sn and 3.7×10^5 y for ^{79}Se) and high mobility, ^{126}Sn and ^{79}Se are of considerable concern in the final disposal of spent nuclear fuel.

Chemistry and Analysis of Radionuclides. Jukka Lehto and Xiaolin Hou

ISBN: 978-3-527-32658-7

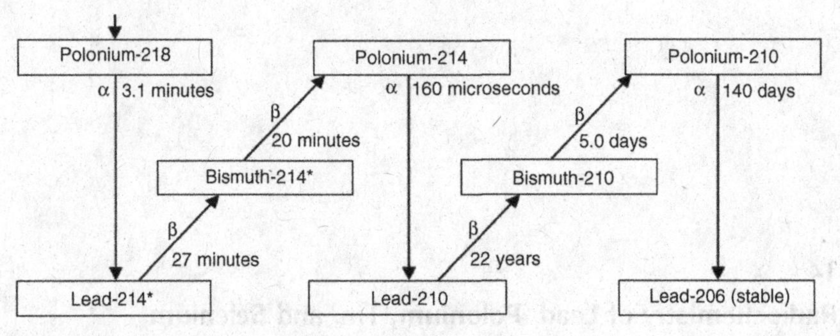

Figure 14.1 The decay series of ^{238}U after ^{222}Rn.

14.1
Polonium – ^{210}Po

14.1.1
Nuclear Characteristics of ^{210}Po

In 1898, Marie Curie separated two new elements, radium and polonium, from a few tonnes of pitchblende. Polonium was named in honor of Poland, Curie's birthplace. Because it is such a short-lived isotope, ^{210}Po was only obtained in very small invisible amounts, and conclusions about its chemical nature were made on the basis of the elements with which it coprecipitates. One ton of uranium ore contains only 0.1 mg of polonium. Although ^{210}Po is the longest-lived isotope of polonium in nature, its half-life is only 138 days. Other naturally occurring isotopes of polonium have half-lives between just 0.3 ps and 3 min. Even though the half-life of ^{210}Po is so short, it is always present in the environment because it is maintained by its parent nuclide ^{210}Pb, which decays by beta emission through ^{210}Bi to ^{210}Po. Polonium is also obtained by the bombardment of the only stable isotope of bismuth with neutrons, according to the reaction

$$^{209}\text{Bi}(\text{n},\gamma)\,^{210}\text{Bi} \xrightarrow{\beta^-, 5\,\text{d}} {}^{210}\text{Po}$$

Table 14.1 Radon progeny in the natural decay chains.

Element	Number of radioactive isotopes	Most long-lived isotope	Half-life	Decay mode	Gamma emissions
Mercury	1	^{206}Hg	8 min	β^-	no
Lead	4	^{210}Pb	22 y	β^-	yes
Bismuth	5	^{210}Bi	5 d	β^-	no
Polonium	7	^{210}Po	138 d	α	no
Astatine	3	^{219}At	1 min	α	no

Table 14.2 Half-lives, alpha energies, production, and applications of the polonium isotopes $^{208,209,210}Po$.

Isotope	Half-life	Alpha energy (MeV)	Production	Use/nature
^{208}Po	2.9 y	5.11	$^{209}Bi(d,3n)$ or $(p,2n)$	tracer
^{209}Po	102 y	4.88	$^{209}Bi(d,2n)$ or (p,n)	tracer
^{210}Po	138 d	5.31	$^{209}Bi(n,\gamma)$ or nature	heat/electricity source, natural radionuclide

Polonium can be separated from bismuth by vacuum distillation or spontaneous deposition on a silver surface. As much as a milligram of polonium can be produced in this way; however, handling this amount of polonium is extremely difficult, because polonium is highly radiotoxic and its heat production is high. Because of its high heat production, it has not been possible to study the chemistry of polonium in macro amounts. Furthermore, the intense alpha radiation causes strong radiolysis in the investigated system. For this reason, chemical tests have only been made in less than microgram amounts. In addition to ^{210}Po, two other isotopes of polonium are produced from ^{209}Bi: namely ^{208}Po and ^{209}Po, which are used as tracers in determining chemical yields in polonium analyses. Because there are no stable isotopes of polonium, only these radioactive isotopes can be used in yield determinations. Since the intensity of the gamma emission is very low, all three isotopes, ^{208}Po, ^{209}Po, and ^{210}Po, are, in practice, pure alpha emitters. Table 14.2 gives the half-lives and alpha energies of these three polonium isotopes. Figure 14.2 gives the decay chart for ^{210}Po.

14.1.2
Chemistry of Polonium

Polonium occurs in the 16th group of the periodic table. It has full 4f and 5d shells, and its outer electron shell has $6s^2$ and $6p^4$ electrons. Polonium forms compounds

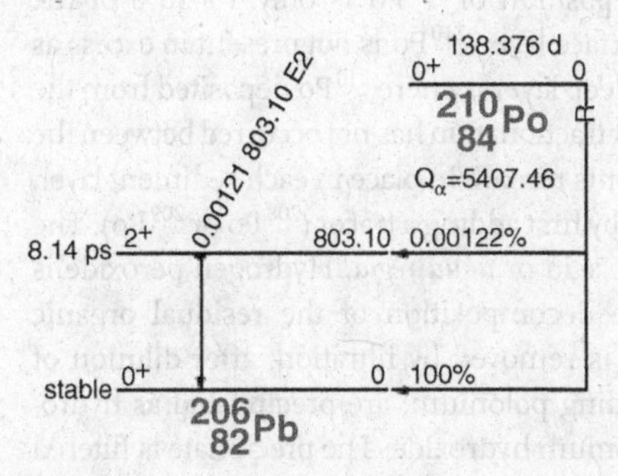

Figure 14.2 Decay chart of ^{210}Po (Firestone, R.B., Shirley, V.S., Chu, S.Y.F., Baglin, C.M., and Zipkin, J. (1996) *Table of Isotopes*, Wiley-Interscience).

in the oxidation states –II (polonides), +II, +IV, and +VI. The most common oxidation state is +IV. The oxide corresponding to oxidation state +IV are PoO_2 and the hydrous oxide $PoO(OH)_2$. In acid solution, polonium appears as the Po^{4+} ion and forms complexes with acid anions; in hydrochloric acid it forms $PoOCl_4{}^{2-}$ complexes; and in nitric acid first $PoO(NO_3)_2$ and then, as the concentration of nitric acid increases, $Po(NO_3)_5{}^{-}$ and eventually $Po(NO_3)_6{}^{2-}$. At pH values above 2.5, polonium begins to form colloids, and at pH 7 it forms $Po(OH)_4$ (solubility product about 10^{-38}). At higher pH, polonium appears as negative colloidal hydrolysis products, and when the pH rises above 12 it dissolves as the polonate $PoO_3{}^{2-}$. Because the predominant oxidation state of polonium is +IV, polonium is only sparingly soluble in natural waters, at their typical pH of about 5–9, and it tends to become attached to particulate matter. In reducing conditions, polonium can also appear in the oxidation state +II. The formation of the polonide ion requires highly reducing conditions. Because concentrations of polonium in nature are extremely low, its chemical forms cannot be determined with good accuracy. Although its boiling point is high (962 °C), polonium readily volatilizes for reasons not fully understood. Thus, ashing at high temperatures is not an option when preparing samples for polonium analysis; rather, samples must be decomposed by wet ashing.

14.1.3
Determination of ^{210}Po

The standard procedure and the only method of separation for ^{210}Po in practice is spontaneous deposition onto the surface of a silver or nickel disc. ^{210}Po deposited on the disc is then measured by alpha spectrometry. The alpha spectrum of ^{210}Po separated from sediment is presented in Figure 14.4. Age determinations of sediment layers are generally carried out by measuring the activity of ^{210}Po rather than that of the parent ^{210}Pb because the measurement of polonium is more accurate. It is assumed that the isotopes ^{210}Pb and ^{210}Po are in radiochemical equilibrium – fractionation has not occurred and that the activities of the two isotopes are the same. ^{210}Po, which comes directly from the atmosphere, decays in a few years because of its short half-life, and thus may occur in a higher concentration than that of ^{210}Pb only in the surface layer of the sediment. Moreover, the deposition of ^{210}Po is only 10–30% of the deposition of ^{210}Pb, so that even in the surface layer, ^{210}Po is not present in excess as compared with the equilibrium value. In deep layers, where ^{210}Po deposited from the atmosphere has decayed, it is assumed that fractionation has not occurred between the lead and polonium and that the two elements remain in place in each sediment layer.

The chemical separation of ^{210}Po starts by first adding a tracer (^{208}Po or ^{209}Po). The sample is then wet ashed in strong nitric acid or *aqua regia*. Hydrogen peroxide is added to the solution for the complete decomposition of the residual organic material. The insoluble mineral residue is removed by filtration; after dilution of the filtrate, the hydrolyzed metals, including polonium, are precipitated as hydroxides by adjusting the pH to 9 with ammonium hydroxide. The precipitate is filtered and dissolved in dilute acid. Hydrazine monohydrochoride ($NH_2OH{\cdot}HCl$) is added to reduce iron to the ferrous state to prevent it from precipitating on the surface of the

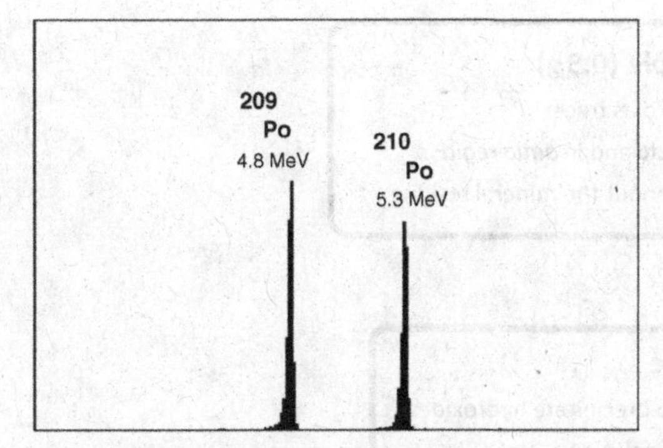

Figure 14.3 Alpha spectrum of polonium. On the left is the 4.88 MeV alpha peak of ^{209}Po, and on the right is the 5.31 MeV alpha peak of ^{210}Po.

silver disc. The reductant also assists the reduction of polonium to a metallic form on the surface of the silver disc. The solution is poured into a deposition vessel, the bottom of which is fitted with a polished and well-cleaned silver disc (Figure 5.4). Because polonium is a more noble metal than silver, it deposits spontaneously on the disc. By heating and stirring the solution during the deposition, the efficiency can be significantly improved, with recoveries as high as 99% being obtained. After the deposition, the silver disc is rinsed with alcohol, dried, and placed in the alpha spectrometer for measurement. Figure 14.3 shows the alpha spectrum of an analysis in which ^{209}Po was used as a tracer in the yield determination of ^{210}Po in water from a drilled well.

14.2 Lead – ^{210}Pb

14.2.1 Nuclear Characteristics and Measurement of ^{210}Pb

Lead has four stable isotopes, ^{204}Pb, ^{206}Pb, ^{207}Pb, and ^{208}Pb, the last three being the final nuclides of the uranium, actinium, and thorium decay series. At an average content of 13 ppm, lead is naturally widely occurring. In the determination of radioactive lead, it must be remembered that, unlike polonium, which has no stable isotopes, samples will contain abundant stable lead. While there are four radioactive isotopes of lead in the uranium, actinium, and thorium series, only ^{210}Pb is long lived ($t_{1/2}$ = 22 y), the half-lives of the other lead isotopes being only 0.5–11 h.

^{210}Pb decays by beta emission to ^{210}Bi, which, in turn, decays by beta emission with a five-day half-life to ^{210}Po (Figure 14.5). The maximum energy of the beta particles of ^{210}Pb is 63 keV (intensity 16%). Most of the beta decays (intensity 84%, maximum energy 17 keV) go through the 46.5 keV excited state of the daughter nuclide ^{210}Bi. This excited state is mostly relaxed by internal conversion, with the intensity of the

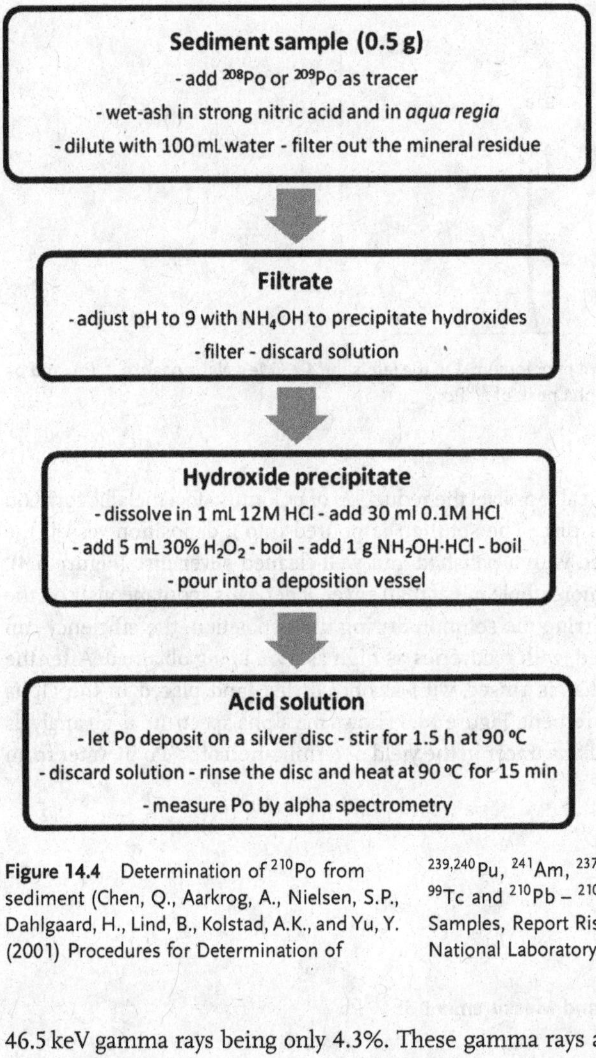

Sediment sample (0.5 g)

- add ^{208}Po or ^{209}Po as tracer
- wet-ash in strong nitric acid and in *aqua regia*
- dilute with 100 mL water - filter out the mineral residue

Filtrate

- adjust pH to 9 with NH_4OH to precipitate hydroxides
- filter - discard solution

Hydroxide precipitate

- dissolve in 1 mL 12M HCl - add 30 ml 0.1M HCl
- add 5 mL 30% H_2O_2 - boil - add 1 g $NH_2OH{\cdot}HCl$ - boil
- pour into a deposition vessel

Acid solution

- let Po deposit on a silver disc - stir for 1.5 h at 90 °C
- discard solution - rinse the disc and heat at 90 °C for 15 min
- measure Po by alpha spectrometry

Figure 14.4 Determination of ^{210}Po from sediment (Chen, Q., Aarkrog, A., Nielsen, S.P., Dahlgaard, H., Lind, B., Kolstad, A.K., and Yu, Y. (2001) Procedures for Determination of $^{239,240}Pu$, ^{241}Am, ^{237}Np, $^{234,238}U$, $^{228,230,232}Th$, ^{99}Tc and ^{210}Pb – ^{210}Po in Environmental Samples, Report Risoe R-1263(EN), Risoe National Laboratory, Denmark).

46.5 keV gamma rays being only 4.3%. These gamma rays are difficult to detect in solid samples, primarily because of their low energy and intensity, as well as the interference from many other radionuclides which cause strong Compton background at low energies. A major problem encountered with the gamma measurement of ^{210}Pb is the self-absorption into the sample and the low efficiency of ordinary semiconductor crystals for gamma-ray energies less than one hundred keV. If the ^{210}Pb content in the sample is high enough and a thin-window semiconducting detector capable of detecting gamma ray energies lower than 100 keV is available, ^{210}Pb can be measured directly with a gamma spectrometer from a solid sample. In determining the sediment's age, for example, much time is saved when ^{210}Pb determinations can be carried out directly with a gamma spectrometer. In addition, in direct measurements, any uncertainty about the fractionation of lead and polonium in the sediment is removed.

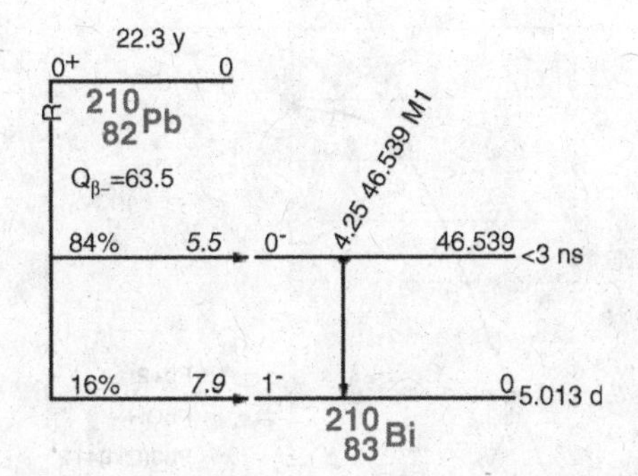

Figure 14.5 Decay chart for ^{210}Pb (Firestone, R.B., Shirley, V.S., Chu, S.Y.F., Baglin, C.M., and Zipkin, J., *Table of Isotopes*, Wiley-Interscience, 1996)..

14.2.2 Chemistry of Lead

Lead is located in the 14th group in the periodic table, and its chemistry for the most part matches the chemistry of tin. Like polonium, lead has full 4f and 5d shells, and there are $6s^2$ and $6p^2$ electrons in its outer shell. Its most common oxidation state is +II. In oxidizing conditions, oxidation state +IV is also possible. The Eh–pH diagram in Figure 14.6 shows that lead only achieves the +IV state in extremely oxidizing conditions.

Most lead compounds are only sparingly soluble, in particular lead sulfide (PbS) and lead sulfate ($PbSO_4$). The only lead compounds that dissolve in water are the acetate, nitrate, chlorate, and perchlorate. Sparingly soluble lead sulfate can be

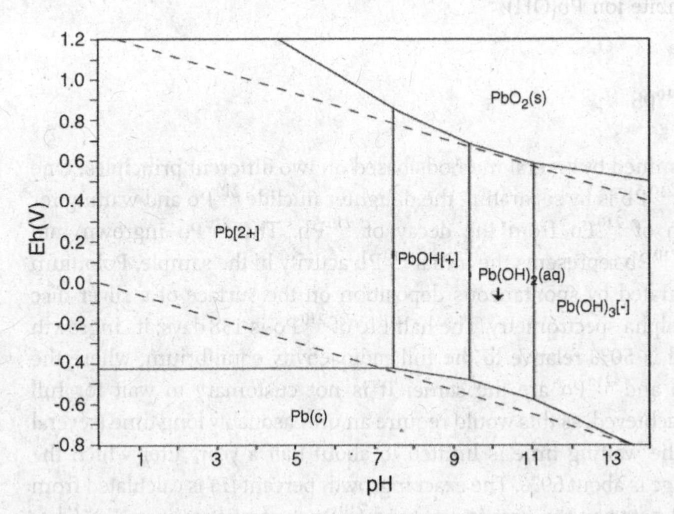

Figure 14.6 Eh–pH diagram for lead (Atlas of Eh – pH diagrams, Geological Survey of Japan Open File Report No. 419, 2005).

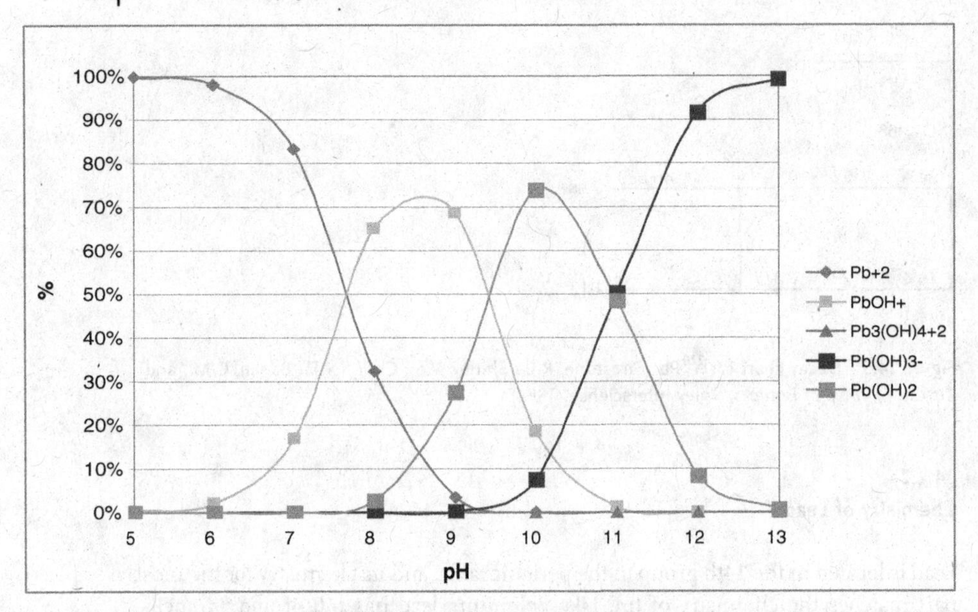

Figure 14.7 Hydrolysis of divalent lead in 10^{-6} M solution.

dissolved in excess acetate or hydroxide, giving rise to lead acetate or a lead hydroxide complex called a plumbite ion. Lead readily forms complexes, for example, with the above-mentioned acetate.

Lead in the oxidation state two does not hydrolyze very strongly (Figure 14.7): in 10^{-6} M lead solution, hydrolysis products begin to form only at pH 6. Pb^{2+}, precipitates as hydroxide $Pb(OH)_2$ at slightly alkaline pH and dissolves at high pH (>13) as plumbite ion $Pb(OH)_4{}^{2-}$.

14.2.3 Determination of ^{210}Pb

Radiolead is determined by several methods based on two different principles. One way to determine ^{210}Pb is by separating the daughter nuclide ^{210}Po and waiting for the new ingrowth of ^{210}Po from the decay of ^{210}Pb. The ^{210}Po ingrown into equilibrium with ^{210}Pb represents the initial ^{210}Pb activity in the sample. Polonium is then again separated by spontaneous deposition on the surface of a silver disc and measured by alpha spectrometry. The half-life of ^{210}Po is 138 days; its ingrowth during this period is 50% relative to the full radioactivity equilibrium, where the activities of ^{210}Pb and ^{210}Po are the same. It is not customary to wait for full equilibrium to be achieved, as this would require an unreasonably long time (several years). Typically, the waiting time is limited to about half a year, after which the ingrowth percentage is about 60%. The exact ingrowth percent (P) is calculated from the time delay (td) between the first and second ^{210}Po separations: $P = 1 - e^{-\lambda td} = 1 - e^{\frac{0.693}{t_{1/2}}td}$, where $t_{1/2}$ is the half-life of ^{210}Po.

Another way to determine ^{210}Pb is to separate lead from polonium and from other radionuclides, and determine its beta activity by either of two alternatives. In the first alternative, the beta particles of energy 64 keV emitted from ^{210}Pb, together with coemitting Auger electrons and X-rays, are measured with a liquid scintillation counter or a thin-window proportional counter immediately after separation. It is essential that the measurement is carried out immediately because the ingrowth of the beta-emitting daughter, ^{210}Bi, commences to grow in immediately and would cause excess pulses. In the second alternative, time is allowed for the ingrowth of ^{210}Bi, and its activity is measured with a liquid scintillation counter or a proportional counter. Instead of waiting the two months needed for full equilibrium to be achieved between ^{210}Pb and ^{210}Bi, it is also possible to separate ^{210}Bi from ^{210}Pb and measure the separated ^{210}Bi. The half-life of ^{210}Bi is 5 days, and the ingrowth sufficient for a successful measurement is obtained in just two weeks. The exact percentage of ^{210}Bi ingrowth (P) is calculated from the time delay (td) between the separation of ^{210}Bi and subsequent measurement as: $P = e^{-\lambda td} = e^{\frac{0.693}{t_{1/2}} td}$, where $t_{1/2}$ is the half-life of ^{210}Bi. The advantage of measurements made with the help of ^{210}Bi is that the measurement does not need to be made immediately after the separation of ^{210}Pb, and also that the beta energy of ^{210}Bi is much higher (1164 keV) than that of ^{210}Pb (64 keV). This greatly assists the determinations with a proportional counter, which more easily detects high-energy than low-energy beta particles, as the latter readily adsorb on the window of the tube.

14.2.3.1 Determination of ^{210}Pb from the Ingrowth of ^{210}Po

As mentioned above, the activity of ^{210}Pb in a sample can be determined by first carrying out a spontaneous deposition of ^{210}Po on the surface of a silver disc. The separation yield is determined by adding a tracer to the sample, either ^{208}Po or ^{209}Po. Typically, the chemical yield is 70–90%. ^{210}Pb remains in solution as well as some undeposited ^{210}Po, the proportion of which can be calculated from the yield. After about six months, a sufficient amount of new ^{210}Po will have been grown in the sample; a new spontaneous deposition of ^{210}Po on the silver disc can then be carried out. The exact ingrowth percent is calculated for the time delay between the two depositions. Before the deposition, a fresh ^{208}Po or ^{209}Po tracer is added to the sample to determine the chemical yield of the second precipitation. In handling the results, it should be noted that some ^{210}Po remains in the solution after the first precipitation, and this must be subtracted from the results obtained in the second precipitation. This excess ^{210}Po activity is calculated from the chemical yield of the first deposition. An alternative approach is to use a different tracer in the second deposition. Thus, if ^{209}Po was used in the first precipitation, ^{208}Po could be used in the second. In this case, after the second precipitation, there will be three peaks in the alpha spectrum: ^{208}Po (5.12 MeV), ^{209}Po (4.88 MeV), and ^{210}Po (5.34 MeV). The yield of the second deposition is measured with the first peak, and the proportion of ^{210}Po remaining in the solution after the first deposition with the second peak. This proportion can then be subtracted from the result of the second deposition. By carefully controlling the spontaneous deposition conditions, very high recoveries (>98%) can be achieved, and the problem of the incomplete removal of ^{210}Po in the first deposition can be avoided. In addition, the ^{210}Po left in the sample solution after spontaneous deposition can be completely removed by solvent extraction using triisooctylamine (TIOA)/xylene. As a result, no correction for the remaining ^{210}Po or the use of double tracers (^{209}Po/^{208}Po) is needed. (Chen *et al.* (2001) A rapid method for the separation of ^{210}Po from ^{210}Pb by TIOA extraction. *J. Radioanal. Nucl. Chem.*, **249**, 587).

14.2.3.2 Separation of ^{210}Pb by Precipitation

If, after the separation of polonium, it is not desired to wait for the ingrowth of polonium, lead can be separated by precipitation as sparingly soluble lead sulfide (PbS) or lead sulfate ($PbSO_4$). Before precipitation of the sulfide or sulfate, interfering bismuth is separated from the solution by ion exchange chromatography. As an example, Figure 14.8 presents a chemical procedure for the determination of ^{210}Pb and ^{210}Po in a soil sample.

In this method, ^{209}Po tracer and stable lead are added to the soil sample to determine the chemical yield of ^{210}Po and ^{210}Pb, respectively. As there is no suitable radioactive tracer for lead, stable lead is used in the yield determination. The soil sample is digested with acids, and ^{210}Po is separated from the leachate by spontaneous deposition on a silver disc. Thereafter, ^{210}Bi, the beta-emitting daughter of ^{210}Pb, is separated from the solution by ion exchange chromatography. The 1.5 M HCl solution remaining from the deposition of ^{210}Po is transferred to a strongly basic anion exchange column that has been pretreated with 1.5 M HCl. Bismuth is strongly fixed to the column, while lead is only loosely fixed (see Figure 4.6). Lead is eluted with water, and the time is noted, because the ingrowth of ^{210}Bi in the lead fraction begins immediately. The lead fraction is purified by sulfate and sulfide precipitations, after which the lead is finally precipitated with another sulfate precipitation. The precipitate is collected onto a filter paper, dried, and weighed to determine its yield. The ingrowth of ^{210}Bi is allowed for about one month, and the activity of ^{210}Bi is then determined with a proportional counter. The ingrowth percent is calculated from the time delay between the Bi separation from lead by an ion exchange step and the measurement. The activity of ^{210}Bi can also be measured with a liquid scintillation counter; however, for this the lead sulfate precipitate must be dissolved in ammonium acetate.

14.2.3.3 Separation of ^{210}Pb by Extraction Chromatography

The separation of ^{210}Pb by extraction chromatography relies on Eichrom/Triksem Pb Resin, which is almost identical to Sr Resin (see Chapters 4 and 7). As in the Sr Resin, 18-crown-6 crown ether is impregnated onto an inert organic polymer in the Pb Resin. In place of 1-octanol, the solvent in Pb Resin is isodecanol, which aids the stripping of Pb from the resin. Because these resins are very similar, Sr Resin is also widely used in the separation of lead. 18-crown-6 crown ether forms a very strong complex with lead, even stronger than that with strontium (Figure 14.9).

In the separation of ^{210}Pb from water by extraction chromatography, the pH of a 0.5–1 L aqueous sample is first adjusted to 2 with nitric acid, and a lead carrier is added. The lead is enriched either by coprecipitation with ferric hydroxide or by

Soil sample (3 g)
- add ^{209}Po tracer and 25 mg lead carrier
- wet ash with conc. HNO_3 and HF - filter - discard solid residue
- evaporate filtrate to near dryness, dissolve in 1.5M HCl
- deposit ^{210}Po onto a silver disc in 1.5M HCl

HCl solution
- pour into a strong anion exchange resin column conditioned with 1,5M HCl
- elute the lead with water - bismuth remains in the column

Effluent
- evaporate to 2 mL
- add 2 mL conc. H_2SO_4 to precipitate lead as $PbSO_4$
- centrifuge - discard the supernatant

$PbSO_4$ precipitate
- dissolve in 10 mL 6M ammonium acetate
- add 1 mL 0.5M Na_2S to precipitate lead as PbS
- centrifuge - discard the supernatant

PbS precipitate
- dissolve in 1 mL conc. HNO_3 - add 10 mL water
- add 2 mL conc. H_2SO_4 to precipitate lead as $PbSO_4$
- centrifuge - discard the supernatant

$PbSO_4$ precipitate
- filter onto a filter paper - dry at 110 °C
- weigh the precipitate - calculate the chemical yield
- allow ^{210}Bi to grow in for about one month
- measure the beta activity of ^{210}Bi with a proportional counter

Figure 14.8 Procedure for the determination of ^{210}Po and ^{210}Pb in a soil sample using spontaneous deposition for the separation of ^{210}Po and the precipitation of lead as sulfide and sulfate for the separation of ^{210}Pb (Guogang, Jia *et al.* (2001) Determination of ^{210}Pb and ^{210}Po in mineral and biological environmental samples. *J. Radioanal. Nucl. Chem.*, **247**, 491).

cation exchange in 0.1 M nitric acid. After enrichment, the lead is prepared in 1 M nitric acid (10 mL). This solution is poured into a 2-mL Pb Resin column, which is rinsed with additional 1 M nitric acid (10 mL) to remove Bi from Pb, the time for which should be noted. Lead is then eluted from the column with water (20 mL) and

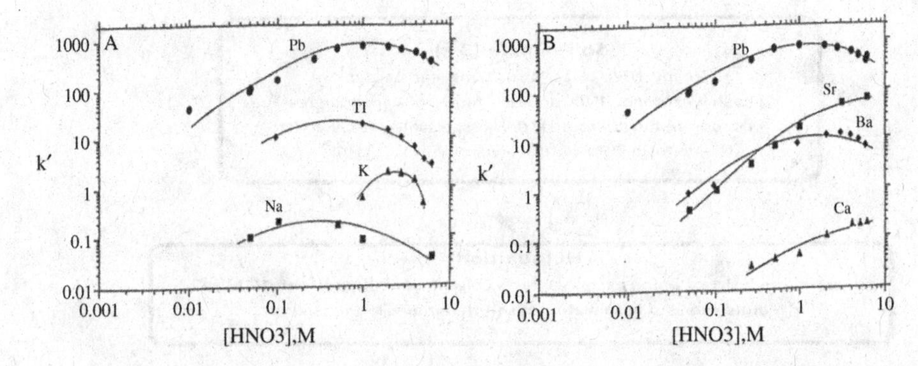

Figure 14.9 Sorption of lead and monovalent metals (A) and alkaline earth metals (B) to Pb Resin as a function of nitric acid concentration. *k′* values are column volumes in which the metal achieves its maximum value in the eluant when the column is eluted with acid of specified concentration (Horwitz, E.P. and Gale, N.H. *et al.* (1994) A lead-selective extraction chromatographic resin and its application to the isolation of lead from geological samples. *Analyt. Chim. Acta*, **292**, 263).

precipitated as $PbSO_4$ by adding concentrated sulfuric acid (4 mL). The precipitate is washed with water, collected onto a filter paper, and weighed after drying to determine the chemical yield of ^{210}Pb. After about one week's ingrowth of ^{210}Bi, the beta activity of ^{210}Bi is measured with a proportional counter. The ingrowth percent is calculated from the time delay between the separation of the bismuth from the lead and commencement of the measurement. The same method can be applied to determine ^{210}Pb from soil and sediment after wet ashing the solid samples with concentrated nitric acid.

14.3 Tin – ^{126}Sn

14.3.1 Nuclear Characteristics and Measurement of ^{126}Sn

Tin has 10 stable isotopes and 27 radioisotopes; however, only ^{126}Sn is important in nuclear waste disposal. This is not only because of its long half-life (2.3×10^5 years), but also because of its expected mobility in the environment. ^{126}Sn decays by beta emission with a maximum energy of 252 keV (Figure 14.10). It also emits low-energy gamma rays of 87.6 keV (37%), 86.9 keV (8.9%), 64.3 keV (9.6%), and 23.3 keV (6.4%). ^{126}Sn can thus be measured by beta counting with liquid scintillation counting or gas ionization detectors and by gamma spectrometry. In addition, ICP-MS can be used to measure ^{126}Sn. ^{126}Sn is a fission product of uranium and plutonium, with a fission yield of 0.06% of ^{235}U fission with thermal neutrons (and 1.76% by fast neutrons). ^{126}Sn mainly remains in spent nuclear fuel or in the high-level waste generated in the reprocessing of spent nuclear fuel. ^{126}Sn has been

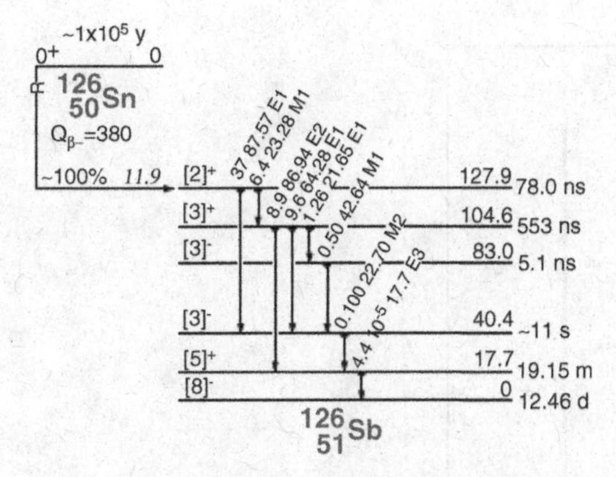

Figure 14.10 Decay chart of ^{126}Sn (Firestone, R.B., Shirley, V.S., Chu, S.Y.F., Baglin, C.M., and Zipkin, J. (1996) *Table of Isotopes*, Wiley-Interscience).

released into the environment from nuclear weapons testing and as discharges from nuclear facilities such as reprocessing plants and from nuclear accidents, but its activity concentrations in the environment are low.

14.3.2
Chemistry of Tin

Tin is located in the 14th group of the periodic table of elements just above lead. Thus their chemistries are somewhat similar. Tin is a relatively common element, its abundance in the Earth's crust being 2.1 ppm. Its typical oxidation states are +II (Sn^{2+}) and +IV (Sn^{4+}), the latter being more stable (Figure 14.11). Divalent tin (called stannous ions) oxidizes easily to tetravalent tin (called stannic ions). In acidic solutions, even air can cause this oxidation. The prevalence of the tetravalent state is biggest difference compared to lead, for which the divalent state is more stable. Tin is an amphoteric element, that is, metallic tin dissolves in both acid and in base. In alkaline solutions, tin forms soluble stannate ions SnO_3^{2-}, while in acidic solutions it forms Sn^{2+} or Sn^{4+}. Most tin salts, such as the chloride, bromide, sulfate, and nitrate, are soluble. However, Sn^{2+} and Sn^{4+} easily hydrolyze when acidic solutions are neutralized with a base and form insoluble $Sn(OH)_2$ or $Sn(OH)_4$. These dissolve in excess base and form the hydroxostannite complex, $Sn(OH)_3^-$, or stannate ions, SnO_3^{2-}. In HCl medium, Sn^{4+} forms a very stable anionic $SnCl_6^{2-}$ complex at HCl concentrations higher than 0.7 M. As with lead, sulfides of tin are insoluble. Stannic sulfide (SnS_2) can be precipitated quantitatively by the addition of H_2S to an acid solution of stannic salt. This cannot, however, be accomplished in strong HCl solutions because of the formation of $SnCl_6^{2-}$ complex. Therefore, the precipitation of SnS_2 has to be carried out in H_2SO_4, HNO_3, or dilute HCl (<1 M). Because of the high solubility of SnS_2 in strong HCl solutions, compared with sulfides of arsenic and antimony, tin can be separated from Sb and As by the subsequent precipitation of sulfides in the HCl medium. By forming neutral complexes with Cl^-, I^-, F^-, and SCN^-, Sn^{4+} can be extracted by many solvents

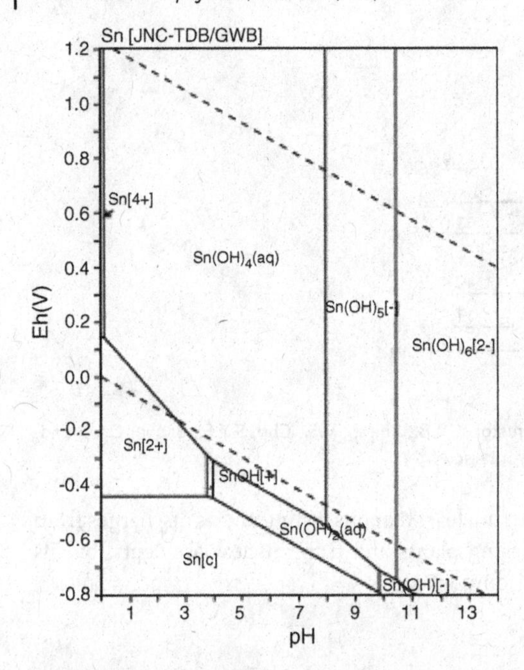

Figure 14.11 Eh – pH diagram of tin (Atlas of Eh – pH diagrams, Geological Survey of Japan Open File Report No. 419, 2005).

such as methyl isobutyl ketone, diethyl ether, and methyldioctylamine, which can be used to separate tin from other elements. On heating, $Sn(OH)_2$ or $Sn(OH)_4$ decompose to form oxides SnO or SnO_2. Chlorides and bromides of tin have low boiling points; $SnCl_4$ can be distillated at 300 °C and so separated from the matrix and other elements (Nervik, W.E. (1960) The radiochemistry of Tin, NAS-NS-3023, National Research Council Nuclear Science Series, National Academy of Sciences).

14.3.3
Determination of ^{126}Sn

To determine ^{126}Sn in nuclear waste, it has to be chemically separated from matrix components and from other radionuclides because of the very low concentrations of ^{126}Sn compared to other radionuclides and the low energies of gamma rays and beta emissions. Based on the chemical properties of tin, precipitation and extraction have been used to separate ^{126}Sn. Figure 14.12 shows a separation procedure for the determination of ^{126}Sn from reprocessing waste samples. In this method, a stable tin carrier and hold back carriers are first added to the sample. Thereafter, acids (HNO_3 and HCl) are added, and the sample is digested by heating to convert the tin to Sn^{4+}. The tin is then precipitated as $Sn(OH)_4$ by adding NH_4OH to pH 6–8. Transition metals are also precipitated in this step; however, alkali metals and alkaline earth metals, such as ^{137}Cs and ^{90}Sr, as well as ^{3}H, remain in the supernatant and are

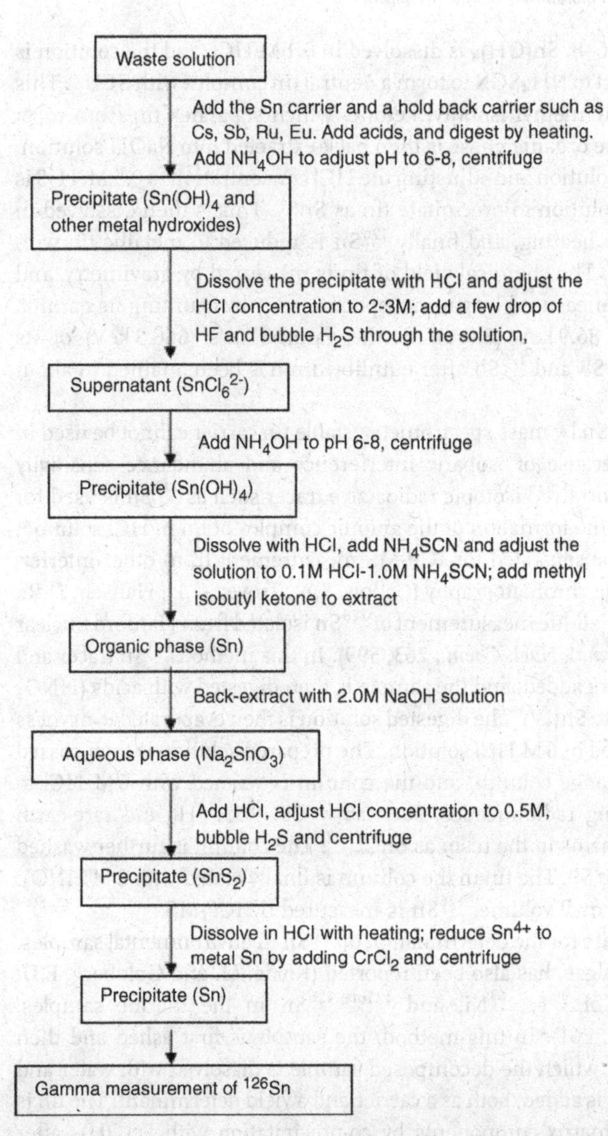

Figure 14.12 Procedure for the separation of ^{126}Sn from reprocessing waste (Zhang, S.D., Guo, J.R., Cui, A.Z., Li, D.M., and Liu, A.M. (1996) Measurement of the half-life of ^{126}Sn using a radiochemical method. *J. Radioanal. Nucl. Chem.*, **212**, 93).

separated from ^{126}Sn. The $Sn(OH)_4$ is then dissolved and prepared in 2–3 M HCl which contains some HF. The solution is bubbled with H_2S, resulting in the formation of insoluble antimony and ruthenium sulfides, removing their radionuclides such as ^{125}Sb and ^{106}Ru, while ^{126}Sn remains in the solution as $SnCl_6^{2-}$ and SnF_6^{2-} complexes. The tin in the solution is then reprecipitated as $Sn(OH)_4$ by adding NH_4OH to pH 6–8. $Sn(OH)_4$ is dissolved in 0.1 M HCl, and the solution is made 1.0 M with respect to NH_4SCN to form a neutral tin complex with SCN^-. This complex is extracted by methyl isobutyl ketone, which separates tin from most interferences. Tin in the organic phase is then back-extracted into NaOH solution. After neutralizing the solution and adjusting the HCl concentration to 0.5 M, H_2S is bubbled through the solution to precipitate tin as SnS_2. This is then dissolved in concentrated HCl with heating, and finally ^{126}Sn is reduced to metallic tin with $CrCl_2$ and precipitated. The chemical yield of tin is measured by gravimetry, and the activity of ^{126}Sn is measured by gamma spectrometry by counting its gamma rays (87.6 keV (37%), 86.9 keV (8.9%)) or the gamma rays (666.3 keV) of its daughter nuclides ^{126m}Sb and ^{126}Sb after equilibrium has been attained in about 140 days.

When measuring ^{126}Sn by mass spectrometry, stable tin carrier cannot be used in yield determination because of isobaric interference and abundance sensitivity (tailing) and instead short-lived isotopic radioactive tracer such as ^{113}Sn is used for this purpose. Based on the formation of the anionic complex of Sn in HCl solution, carrier-free ^{126}Sn can be separated for ICP-MS measurement from other interferences by anion exchange chromatography (Catlow, S.A., Troyer, G.L., Hansen, D.R., and Jones, R.A. (2005) Half-life measurement of ^{126}Sn isolated from Hanford nuclear defense waste. *J. Radioanal. Nucl. Chem.*, **263**, 599). In this method ^{113}Sn tracer and the hold back carriers are added, and the sample is then digested with acids (HNO_3 and HCl) to convert tin to Sn(IV). The digested solution is then evaporated to dryness and the residue dissolved in 6 M HCl solution. The prepared solution is then passed through an anion exchange column, and the column is washed with 6 M HCl to remove most interfering radionuclides, such as ^{137}Cs, ^{90}Sr, ^{3}H, and rare-earth elements, while tin remains in the resin as $SnCl_6{}^{-2}$. The column is further washed with 1 M HCl to remove Sb. The tin in the column is finally eluted with 2 M HNO_3. After evaporation to a small volume, ^{126}Sn is measured by ICP-MS.

An analytical procedure for the determination of ^{126}Sn in environmental samples, such as sediment and algae, has also been reported (Koide, M. and Goldberg, E.D. (1985) Determination of ^{99}Tc, ^{63}Ni, and $^{121m+126}Sn$ in the marine samples. *J. Environ. Radioact.*, **2**, 261). In this method, the sample is first ashed and then fused with Na_2O_2, after which the decomposed sample is dissolved with water and HCl solution. Stable tin is added, both as a carrier and a yield determinant. The tin is separated from some matrix components by co-precipitation with $Fe(OH)_3$ after the addition of NaOH. The precipitate is dissolved using HCl and re-precipitated using NaOH. After dissolution in HCl and adjusting the HCl concentration to 0.3 M, a Pb carrier is added and H_2S is bubbled through the solution to separate tin as a sulfide. The sulfide is dissolved in an HCl solution and the remaining Fe is removed by extracting it with isopropyl ether. Thereafter, the tin is extracted with methyl isobutyl ketone (MIBK), and the tin in the organic phase is back-extracted with H_2O. The extraction and back-extraction steps are repeated, and the final aqueous phase is mixed with scintillation cocktail to measure ^{126}Sn using a liquid scintillation counter.

14.4
Selenium – ^{79}Se

14.4.1
Nuclear Characteristics and Measurement of ^{79}Se

Selenium has five stable isotopes (^{74}Se (80.89%), ^{76}Se (9.36%), ^{77}Se (7.63%), ^{78}Se (23.78%), and ^{80}Se (49.61%)) and 26 radioisotopes. Of the radioisotopes, ^{75}Se and ^{79}Se are the two most important. ^{75}Se ($t_{1/2}$ = 120 days) is a neutron activation product of stable ^{74}Se and is widely used as a tracer for biomedical studies. ^{79}Se is a neutron activation product of stable ^{78}Se. It is produced in selenium-bearing materials which are exposed to neutron irradiation, such as nuclear fuel and reactor construction materials, and is thus widespread in nuclear waste. Because of its long half-life (3.7×10^5 years) and high mobility in environment, ^{79}Se is an important radionuclide for a waste depository. Its possible high mobility arises from its occurrence in nonsorbing anionic forms (SeO_3^{2-} or SeO_4^{2-}) in ground waters. As ^{79}Se decays by pure beta emission with a maximum energy of 151 keV (Figure 14.13), its activity concentration can thus be measured by beta counting with a liquid scintillation counter or with a proportional counter. Mass spectrometry, such as ICP-MS and AMS, has also been used to measure ^{79}Se. Because of the severe isobaric interference of ^{79}Br (50.69%), as well as abundance sensitivity (tailing of ^{78}Se and ^{80}Se), the detection limit and analytical accuracy of mass spectrometry for ^{79}Se is not normally satisfactory.

14.4.2
Chemistry of Selenium

In the periodic table of the elements, selenium is located in the 16th group – the same group as polonium. The electron configuration of selenium is $[Ar]3d^{10}4s^24p^4$. Chemically, selenium very closely resembles sulfur, which is located immediately

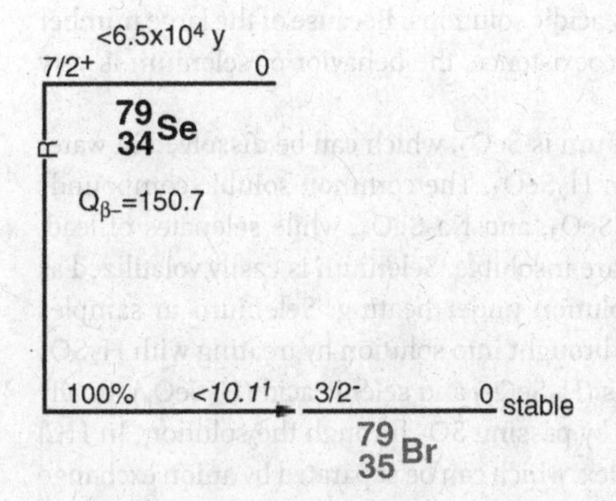

Figure 14.13 Decay chart of ^{79}Se (Firestone, R.B., Shirley, V.S., Chu, S.Y.F., Baglin, C.M., and Zipkin, J. (1996) *Table of Isotopes*, Wiley-Interscience).

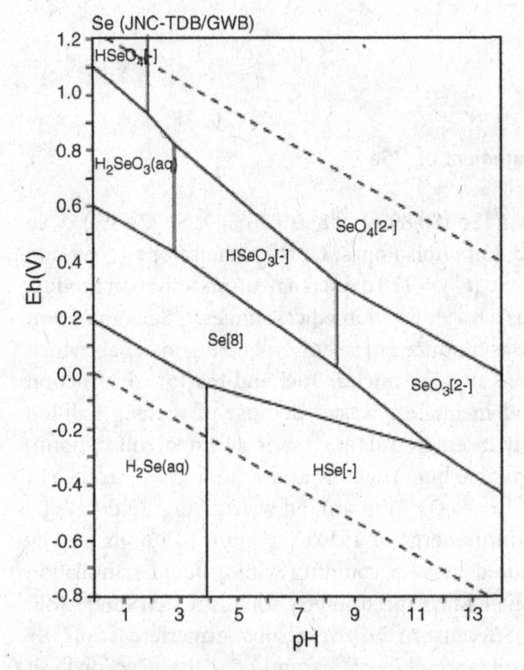

Figure 14.14 Eh – pH diagram of selenium (Atlas of Eh – pH diagrams, Geological Survey of Japan Open File Report No. 419, 2005).

above selenium in this group. Just like sulfur, selenium as a nonmetal forms basic anions selenite (SeO_3^{2-}) at oxidation state +IV and selenate (SeO_4^{2-}) at oxidation state +VI – the latter at higher redox potentials (Figure 14.14). In addition, similarly to sulfur, selenium forms selenide ions Se^{2-} at oxidation state –II at low redox potentials. At intermediate redox potentials, selenium also exists as elemental selenium. Selenite ions (SeO_3^{2-}) can be readily reduced to elemental selenium by SO_2, Sn^{2+}, Fe^{2+}, I^-, and Zn in slightly acidic solutions. Because of the large number of possible oxidation states and their coexistence, the behavior of selenium is very complex and difficult to explore.

The most stable compound of selenium is SeO_2, which can be dissolved in water and concentrated sulfuric acid to form H_2SeO_3. The common soluble compounds of selenium include SeO_2, SeO_3, Na_2SeO_3, and Na_2SeO_4, while selenates of lead, barium, and calcium, as well as SeS_2, are insoluble. Selenium is easily volatilized as $SeO_2{\cdot}2HCl$ in a concentrated HCl solution under heating. Selenium in samples such as ores, soil, and waste is usually brought into solution by treating with H_2SO_4 and/or HNO_3 to convert it to selenious (H_2SeO_3) and selenic acid (H_2SeO_4), finally precipitating it as elemental selenium by passing SO_2 through the solution. In HCl solution, Se^{4+} forms an anionic complex, which can be separated by anion exchange chromatography (Molinski, V.J. and Leddicotte, G.W. (1965) Radiochemistry of selenium, Nuclear Science Series, National Academy of Science – National Research Council, NAS-NS-3030, United States Atomic Energy Commission).

14.4.3
Determination of ^{79}Se

For the determination of ^{79}Se in nuclear waste and environmental samples, ^{79}Se has to be chemically separated from matrix elements and other radionuclides because of both the low concentration of ^{79}Se compared to other radionuclides and the pure beta emission. If selenium is to be measured by mass spectrometry, the effective removal of bromine is also essential to avoid the severe isobaric interference of ^{79}Br. Precipitation, solvent extraction, and ion exchange chromatography have been used to separate ^{79}Se from reprocessing waste and environmental samples. A scheme of the chemical separation of ^{79}Se from sample matrix and interfering radionuclides is given in Figure 14.15. Selenium carrier as $NaSeO_3$ and hold back carriers such as

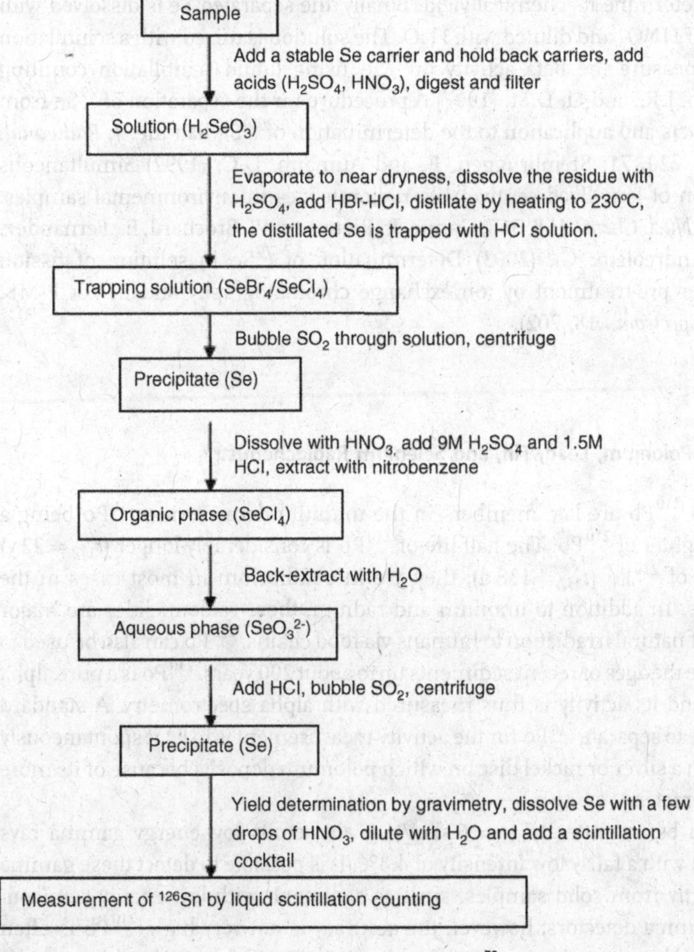

Figure 14.15 Chemical procedure for the separation of ^{79}Se from waste and environmental samples.

Cs, Sr, Eu, Ru, and Sb are first added. After adding concentrated HNO_3 and H_2SO_4, the sample is digested by heating. The solution is then filtered, the residue is discarded, and the solution is evaporated to near dryness. After adding H_2SO_4 and HBr-HCl, Se is distilled off as $SeBr_4$ at 230 °C; the distilled $SeBr_4$ is then trapped in HCl solution. SO_2 gas is passed through the solution to reduce selenium to elemental Se, which is precipitated and separated by centrifuge. The Se is dissolved in a few drops of HNO_3, after which H_2SO_4 and HCl are added to adjust the H_2SO_4 concentration to 9 M and the HCl concentration to 1.5 M. In this solution, Se^{4+} forms $SeCl_4$ complex, which is extracted with nitrobenzene. $SeCl_4$ in the organic phase is back-extracted using H_2O, in which the $SeCl_4$ complex is decomposed and selenium forms water-soluble SeO_3^{2-}. After the addition of HCl, the Se in the solution is further purified by treating with SO_2, which results in the reduction of Se to elemental selenium and precipitation. The separated elemental Se is dried and weighed to determine its chemical yield. Finally, the separated Se is dissolved with a few drops of HNO_3 and diluted with H_2O. The solution is mixed with a scintillation cocktail to measure the beta activity of ^{79}Se using liquid scintillation counting (Li, C.S., Guo, J.R., and Li, D.M. (1997) A procedure for the separation of ^{79}Se from fission products and application to the determination of ^{79}Se half-life. *J. Radioanal. Nucl. Chem.*, **220**, 71; Steinbruggen, E. and Aumann, D.C. (1992) Simultaneous determination of long-lived semivolatile radionuclides in environmental samples. *J. Radioanal. Nucl. Chem.*, **158**, 367; Comte, J., Bienvenu, P., Brochard, E., Fernandez, J.-M., and Andreoletti, G., (2003) Determination of ^{79}Se in solution of fission products after pre-treatment by ion exchange chromatography and ETV-ICP-MS. *J. Anal. At. Spectrom.*, **18**, 702).

14.5
Essentials of Polonium, Lead, Tin, and Selenium Radiochemistry

- ^{210}Po and ^{210}Pb are late members in the uranium decay series, ^{210}Po being a granddaughter of ^{210}Pb. The half-life of ^{210}Pb is considerably longer ($t_{1/2} = 22$ y) than that of ^{210}Po ($t_{1/2} = 138$ d); they are in equilibrium in most cases in the geosphere. In addition to uranium and radium, these radionuclides are major sources of natural irradiation to humans via food chains. ^{210}Pb can also be used to determine the ages of recent sediments up to about 200 years. ^{210}Po is a pure alpha emitter, and its activity is thus measured with alpha spectrometry. A standard procedure to separate ^{210}Po for the activity measurement is to let it spontaneously deposit on a silver or nickel disc on which polonium deposits because of its more noble character.
- ^{210}Pb is a beta emitter ($E_{max} = 63$ keV). It also emits low-energy gamma rays (46.5 keV) with a fairly low intensity of 4.3%. It is possible to detect these gamma rays directly from solid samples, such as sediment, with low-energy semiconductor gamma detectors; however, the accuracy is not very high. ^{210}Pb is often determined by measuring its grand daughter's (^{210}Po) alpha activity; this naturally requires that these nuclides are in equilibrium or the ingrowth level of ^{210}Po is

known. ^{210}Pb can also be measured by its beta emissions with LSC following its radiochemical separation; this can be accomplished by precipitations as lead sulfate and sulfide, and by extraction chromatography using 18-crown-6 crown ether-bearing resin such as Pb Resin from Eichrom/Triskem.

- ^{79}Se is a neutron activation product and ^{126}Sn a fission product of uranium and plutonium and both can be found in spent nuclear fuel. They are important in the evaluation of its final disposal strategies. ^{79}Se has a half-life of 3.7×10^5 years and ^{126}Sn 2.3×10^5 years. ^{79}Se is a pure beta emitter with a maximum beta energy of 151 keV. ^{126}Sn emits, in addition to 252 keV beta particles, gamma rays – the most intense (37%) being the 88 keV gamma rays. These gamma rays, however, cannot be used to measure ^{126}Sn directly from samples, since its activity is very low compared to those of other coexisting nuclides such as ^{137}Cs; further, a radiochemical separation of tin is needed before its measurement. Precipitation, solvent extraction, and ion exchange have been used for the separation of both ^{79}Se and ^{126}Sn.

15
Radiochemistry of the Actinides

The actinides are the group of 14 elements beginning with thorium ($Z = 90$) and ending with lawrencium ($Z = 103$). Actinium ($Z = 89$), the element preceding this series, is usually included in the group, and the actinides, the 'actinium-like' elements, are named after it. Actinides are characterized by the filling of the seven 5f orbitals, similarly to the lanthanides, in which the 4f orbitals are filled. The similarity between these two groups is indeed extensive, although considerable differences also exist. For the most part, the chemistry of the lanthanides is fairly straightforward, while the chemistry of the actinides is much more complex. Actinides are commonly denoted by the abbreviation An. They can be divided according to their origin into two groups. The light actinides actinium, thorium, protactinium, and uranium, which are members of the natural decay series, and the transuranium elements, artificial actinides neptunium, plutonium, americium, curium, and so on. Since this book's focus is on the chemistry of radionuclides appearing in nuclear waste and the environment, elements heavier than curium, rarely occurring in the environment and nuclear waste, are not discussed in detail.

15.1
Important Actinide Isotopes

Table 15.1 presents the most important actinide isotopes. The isotopes from actinium to curium appear most abundantly in the environment and nuclear waste and are most abundantly generated in nuclear reactors and explosions. For actinides from berkelium onwards, only the longest-lived isotope is listed.

15.2
Generation and Origin of the Actinides

Uranium and the lighter actinides protactinium, thorium, and actinium isotopes are naturally occurring radionuclides belonging to the uranium, actinium, and

Chemistry and Analysis of Radionuclides. Jukka Lehto and Xiaolin Hou

ISBN: 978-3-527-32658-7

Table 15.1 Most important long-lived actinide isotopes. All isotopes up to ^{237}Np are natural radionuclides except ^{236}U. All isotopes from ^{237}Np on and ^{236}U are artificial.

Element	Major radionuclides	Half-life	Decay mode
Actinium	^{227}Ac	22 y	β^-/α
Thorium	^{232}Th	1.4×10^{10} y	α
	^{230}Th	7.6×10^4 y	α
	^{228}Th	5.8 y	α
Protactinium	^{231}Pa	3.3×10^4 y	α
Uranium	^{238}U	4.5×10^9 y	α/SF
	^{235}U	7.0×10^8 y	α
	^{236}U	2.3×10^7	α
	^{234}U	2.5×10^5 y	α
Neptunium	^{237}Np	2.1×10^6 y	α
Plutonium	^{238}Pu	88 y	α
	^{239}Pu	2.4×10^4 y	α
	^{240}Pu	6500 y	α
	^{241}Pu	14 y	β^-
Americium	^{241}Am	433 y	α
Curium	^{244}Cm	18 y	α/SF
	^{242}Cm	0.45 y	α/SF
Berkelium	^{247}Bk	1380 y	α
Californium	^{251}Cf	898 y	α
Einsteinium	^{252}Es	1.3 y	α/β^+/EC
Fermium	^{257}Fm	0.27 y	α/SF
Mendelevium	^{258}Md	0.14 y	α/SF/β^+/EC
Nobelium	^{259}No	1 h	α/EC/SF
Lawrencium	^{262}Lr	3.6 h	SF

SF = Spontaneous fission.

thorium decay series. The uranium isotopes, ^{238}U and ^{235}U, and the thorium isotope, ^{232}Th, are initial elements in the uranium, actinium, and thorium decay series. The Figures 15.1–15.3 show those parts of the series with actinides. An exception in this category is ^{236}U, which is formed by neutron activation from ^{235}U in nuclear fuel.

Uranium is the initial member of two decay series and also appears as the important, long-lived isotope ^{234}U. Thorium appears in the series as six isotopes, of which ^{232}Th and ^{230}Th are extremely long-lived. Protactinium appears as two isotopes, of which ^{231}Pa is long-lived. Likewise, actinium appears as two isotopes, and, of these, ^{227}Ac is long-lived. Both uranium and thorium are widely dispersed. The concentration of thorium in the Earth's crust is 8.1 ppm and that of uranium 2.3 ppm. As early as in the nineteenth century both elements were separated in macro amounts for clarification of their chemical properties. Although all isotopes of these elements emit alpha radiation and are potentially radiotoxic, they do not pose a particular safety problem during handling since the half-lives of the dominant

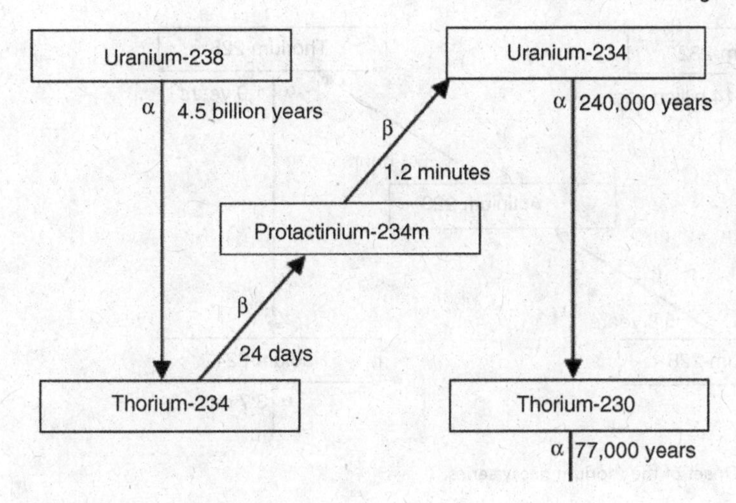

Figure 15.1 Onset of the uranium decay series.

isotopes of uranium and thorium are very long, and their specific activities are generally low.

Actinium and protactinium, in turn, are rare. One tonne of uranium ore contains only about 100 grams of protactinium, and the amount of actinium is about 1000 times less. Because the concentrations of Pa and Ac are so low, the separation of macro amounts from natural materials is laborious. Both elements, however, can be produced in a weighable amount in nuclear reactors by bombarding ^{226}Ra and ^{230}Th with neutrons. The activation products ^{227}Ra and ^{231}Th, produced in the irradiation, decay through beta emission to ^{227}Ac and ^{231}Pa, which can be separated from the

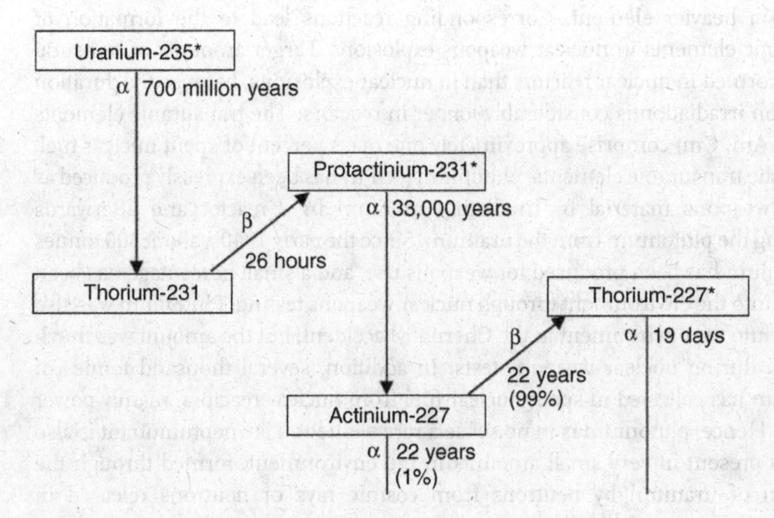

Figure 15.2 Onset of the actinium decay series.

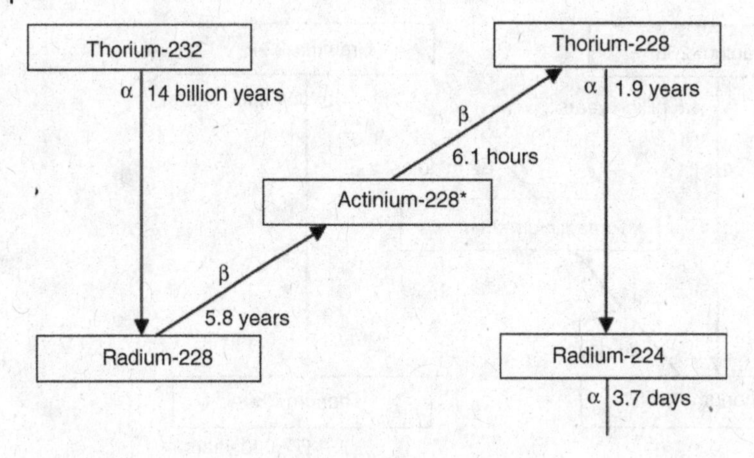

Figure 15.3 Onset of the thorium decay series.

mixture chemically after sufficient ingrowth.

$$^{226}\text{Ra}(\text{n},\gamma)^{227}\text{Ra} \underset{\beta^-}{\longrightarrow} {}^{227}\text{Ac}$$

$$^{230}\text{Th}(\text{n},\gamma)^{231}\text{Th} \underset{\beta^-}{\longrightarrow} {}^{231}\text{Pa}$$

The transuranic elements (those heavier than uranium), that is, neptunium, plutonium, americium, and curium are produced in nuclear reactors and nuclear explosions. In reactors they form through neutron activation of the uranium isotopes ^{235}U and ^{238}U, as shown in Figure 1.1, where neutron capture is followed by beta decay to a heavier element. Corresponding reactions lead to the formation of transuranic elements in nuclear weapons explosions. Larger amounts of transuranics are formed in nuclear reactors than in nuclear explosions, because the duration of neutron irradiation is considerably longer in reactors. The transuranic elements (Np, Pu, Am, Cm) comprise approximately one mass percent of spent nuclear fuel. Among the transuranic elements, plutonium (^{239}Pu) has been expressly produced as nuclear weapons material by irradiating uranium in a reactor and afterwards separating the plutonium from the uranium. Since the early 1940s, about 300 tonnes of plutonium has been produced for weapons use, and a small percentage has been released into the environment through nuclear weapons testing. Plutonium was also released into the environment in the Chernobyl accident, but the amount was much less than during nuclear weapons tests. In addition, several thousand tonnes of plutonium has collected in spent nuclear fuel from nuclear reactors, mainly power reactors. Hence, plutonium is in no sense a rare element. Like neptunium, it is also naturally present in very small amounts in the environment, formed through the activation of uranium by neutrons from cosmic rays or neutrons released in spontaneous fission of ^{238}U.

Table 15.2 Activities of transuranic elements in Finland in the environment and in spent nuclear fuel relative to ^{239,240}Pu activity. Environmental activities from nuclear weapons testing are estimated total activities at the time of deposition (Salminen, S. (2009) Development of analytical methods for the separation of plutonium, americium, curium, and neptunium from environmental samples. Doctoral Dissertation, University of Helsinki). The activities in spent fuel are mean values of four Finnish nuclear reactors calculated after 30 years of cooling (Smith, P., Nordman, Pastina, B., Snellman, M., Hjerpe, T., and Johnson, L. (2007) Safety Assessment for a KBS-3H Spent Nuclear Fuel Repository at Olkiluoto Radionuclide Transport Report, Report Posiva 2007-07).

Nuclide	Half-life (y)	Nuclear weapons tests fallout	Chernobyl fallout	Spent nuclear fuel
^{237}Np	2.1×10^6	0.004	0.001	0.0005
^{238}Pu	88	0.03	0.54	3
^{239,240}Pu	24 000 and 6500	1	1	1
^{241}Pu	14	16	95	36
^{241}Am	433	0.1	0.37	4
^{242}Cm	0.44	—	15	—
^{244}Cm	18	—	0.08	—
^{245}Cm	8500	—	—	0.0003
^{246}Cm	4700	—	—	0.00005

Neptunium, americium, and curium form in nuclear reactors in a combined amount about one-tenth that of plutonium in mass. The activity of americium in the environment is less than half that of plutonium, but the mass is only about one hundredth. Curium is present in the environment in exceedingly small amounts. A rough picture of the activities of the transuranic elements can be obtained from Table 15.2, which shows the activities in the environment and in spent nuclear fuel normalized to the activity of the long-lived plutonium isotopes ^{239,240}Pu. The highest activity in fresh nuclear fallout and spent fuel comes from the beta decaying ^{241}Pu isotope. Its half-life is, however, fairly short (14 y) and it more or less decays in about a hundred years. The decay of ^{241}Pu to ^{241}Am, in turn, considerably increases the activity of the latter during the seventy years following a fallout or the removal of nuclear fuel from a reactor. Regarding the alpha-emitting transuranium isotopes the highest activity is represented by ^{242}Cm, but it decays in a rather short time because of its short half-life. In the long-term, allowing for thousands of years, the activity of ^{237}Np also increases since it is the daughter of ^{241}Am.

Over several decades, these four elements have been prepared in weighable macro amounts for study of their chemical properties and for the preparation of a diversity of solid compounds. In particular, these elements have been obtained through chemical separation from the irradiated uranium dissolved in acids. Studies on the chemical properties of these actinides in macro amounts, however, are not easy. To begin with, they are alpha emitters and highly radiotoxic. Compared with uranium and thorium, their half-lives are notably shorter, and their specific activities are correspondingly greatly higher. The handling of macro amounts requires a safe environment, which usually means remote handling in hot cells. Another factor affecting the investigation of chemical properties is the effect of intense alpha radiation on the chemistry of the studied system, especially regarding ^{241}Am. One milligram of ^{241}Am releases about one hundred million alpha particles per second. An amount such as this causes significant changes in the crystal lattices of solid compounds of americium and causes strong radiolysis in the solution, in turn affecting the chemistry of americium in solution.

Of the *elements heavier than curium*, berkelium and californium were observed for the first time in the fallout from fusion bombs and afterwards have been synthesized by neutron activation in reactors and particle accelerators. Berkelium was first synthesized by the bombardment of americium with alpha particles, in the reaction ^{241}Am(α, 2n)^{243}Bk. Both berkelium and californium have been prepared and their chemical properties studied in microgram amounts. The heavier actinides, whose half-lives decrease with increasing atomic number, however, have only been synthesized in nanogram or picogram amounts (Es, Fm), and in rare cases in amounts of just a few atoms (Md, No, Lr). Neither the chemistry of these elements nor their separation is discussed in this book.

15.3
Electronic Structures of the Actinides

Besides the full shells $1s^22s^22p^63s^23p^64s^23d^{10}4p^65s^24d^{10}5p^66s^24f^{14}5d^{10}6p^6$, the actinides also have electrons in the 6d and 7s orbitals. The actinides are characterized by the filling of the seven 5f orbitals, whose energy level is a somewhat lower than that of the outer 6d and 7s orbitals. In ground state, the 7s orbital is full, containing two electrons, while the 6d orbital has one or two electrons or none at all. The electronic structures of the actinides are presented in Table 15.3. The outer 6d and 7s orbitals of the actinides shield the 5f orbitals, which are closer to the nucleus. This behavior is analogous to that of the lanthanides, by which the 4f orbitals are filled. The energy differences between the 6d, 7s, and 5f orbitals of the actinides are very small, however, being about the same as the energy of a chemical bond, and the 5f orbitals of the actinides are less shielded than the 4f orbitals of the lanthanides. Thus, whereas in the lanthanides only the 4f electrons of Ce participate in chemical bonding, in the lighter actinides, up to americium, the 5f electrons also take part. This makes the chemistry of the actinides considerably more complex than that of the lanthanides, and, unlike the lanthanides, the lighter actinides readily form complexes and form covalent bonds. With increase in the atomic number of the actinides, the charge on the nucleus grows and increasingly attracts the 5f electrons. Beginning with americium, the heavier actinides largely behave like the lanthanides, and the 5f electrons seldom participate in bond formation. This also applies to the lightest actinide actinium. The bonding of compounds of these heavier actinides, like that of the lanthanides, is mostly ionic. In contrast, the chemistry of the lighter actinides Th, Pa, U, Np, and Pu resembles that of the d-transition elements.

Table 15.3 Electronic structures of the actinides as gaseous atoms, gaseous M^+, M^{2+}, M^{3+}, and M^{4+} ions, and M^{3+} ions in aqueous solution. The electronic structures of gaseous M^{4+} ions, except for U, have been predicted but are not known experimentally.

	Atom (g)	M^+ (g)	M^{2+} (g)	M^{3+} (g)	M^{3+} (aq)	M^{4+} (g)
Actinium	$6d7s^2$	$7s^2$	$7s$			
Thorium	$6d^27s^2$	$6d7s^2$	$5f6d$	$5f$		
Protactinium	$5f^26d7s^2$	$5f^27s^2$	$5f^26d$	$5f^2$		$5f^1$
Uranium	$5f^36d7s^2$	$5f^37s^2$	$5f^36d$	$5f^3$	$5f^3$	$5f^2$
Neptunium	$5f^46d7s^2$	$5f^57s$	$5f^5$	$5f^4$	$5f^4$	$5f^3$
Plutonium	$5f^67s^2$	$5f^67s$	$5f^6$	$5f^5$	$5f^5$	$5f^4$
Americium	$5f^77s^2$	$5f^77s$	$5f^7$	$5f^6$	$5f^6$	$5f^5$
Curium	$5f^76d7s^2$	$5f^77s^2$	$5f^8$	$5f^7$	$5f^7$	$5f^6$
Berkelium	$5f^97s^2$	$5f^97s$	$5f^9$	$5f^8$	$5f^8$	$5f^7$
Californium	$5f^{10}7s^2$	$5f^{10}7s$	$5f^{10}$	$5f^9$	$5f^9$	$5f^8$
Einsteinium	$5f^{11}7s^2$	$5f^{11}7s$	$5f^{11}$	$5f^{10}$	$5f^{10}$	$5f^9$
Fermium	$5f^{12}7s^2$	$5f^{12}7s$	$5f^{12}$	$5f^{11}$	$5f^{11}$	$5f^{10}$
Mendelevium	$5f^{13}7s^2$	$5f^{13}7s$	$5f^{13}$	$5f^{12}$	$5f^{12}$	$5f^{11}$
Nobelium	$5f^{14}7s^2$	$5f^{14}7s$	$5f^{14}$	$5f^{13}$	$5f^{13}$	$5f^{12}$
Lawrencium	$5f^{14}6d7s^2$	$5f^{14}7s^2$	$5f^{14}7s$	$5f^{14}$	$5f^{14}$	$5f^{13}$

15.4 Oxidation States of the Actinides

The oxidation states of the actinides vary widely, between +II and +VII (Table 15.4). For actinium the most stable state is +III. The most stable oxidation state then

Table 15.4 Oxidation states of the actinides. The most stable oxidation states are indicated by X, others by x. Oxidation states indicated in parentheses occur only in solid-state compounds.

	+II	+III	+IV	+V	+VI	+VII
Actinium		X				
Thorium		x	X			
Protactinium		(x)	x	X		
Uranium		x	x	x	X	
Neptunium		x	x	X	x	x
Plutonium		x	X	x	x	x
Americium	(x)	X	x	x	x	
Curium		X	x			
Berkelium		X	x			
Californium	(x)	X	(x)			
Einsteinium	(x)	X				
Fermium	x	X				
Mendelevium	x	X				
Nobelium	X	x				
Lawrencium		X				

increases systematically, from +IV for thorium to +V for protactinium and +VI for uranium. Afterwards it decreases systematically to +V for neptunium, +IV for plutonium, and +III for americium. The most stable and typical oxidation state for the actinides heavier than plutonium is +III, these actinides otherwise closely resemble the lanthanides. The one exception is nobelium, whose most stable oxidation state is +II, where it obtains a full 5f shell (see Table 15.3).

15.5
Ionic Radii of the Actinides

The ionic radii of actinides in the same oxidation state systematically decrease with increase of atomic number. Figure 15.4 shows the ionic radii of actinides in the oxidation states +III, +IV, +V, and +VI. The reason the ionic radius decreases with the atomic number is the increasing nuclear charge attracting more effectively the 5f electron shell. This decrease, known as the actinide contraction, is analogous to the lanthanide contraction. Decrease in the ionic radius affects the hydrolysis and complex formation of the actinides. The smaller the radius, the more readily the ion hydrolyzes and forms complexes. Otherwise, actinide ions in the same oxidation state behave in a closely similar way. Those actinides that typically appear only in the trivalent state are separated from each other mainly on the basis of difference in ionic radius. As shown in Figure 10.2, in the chromatographic separation of lanthanides and actinides by cation exchange, the greater the atomic number of a lanthanide or

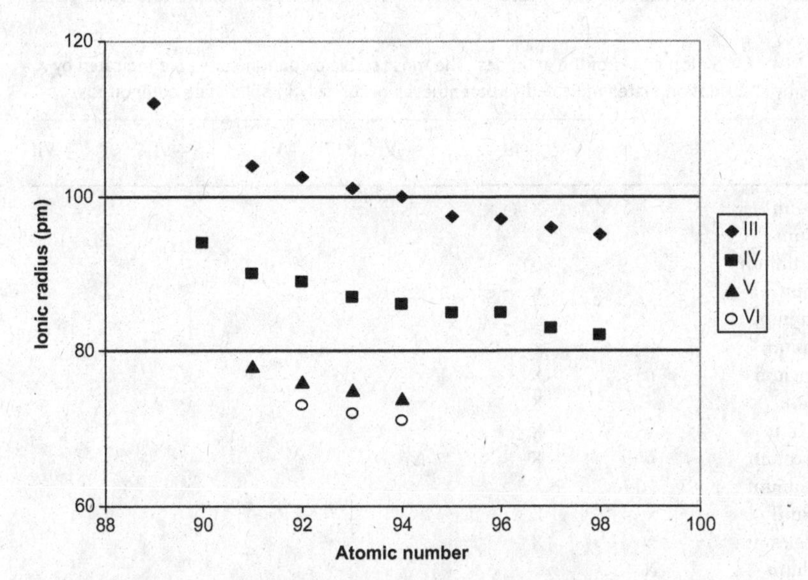

Figure 15.4 Ionic radii of actinides (Ac–Cf) in oxidation states three, four, five, and six at coordination six.

actinide, the smaller is its ionic radius, the stronger is the complex formed with ammonium α-hydroxyisobutyrate, and the more rapidly it elutes from the column.

15.6 Major Chemical Forms of the Actinides

Actinides appear in oxidation states +III to +VI as the following forms:

+III : An^{3+}
+IV : An^{4+}
+V : AnO_2^+
+VI : AnO_2^{2+}

The last two forms, known as actinyl ions, are formed by hydrolysis, in this case by forming covalent bonds with oxygen atoms in the hydration shell. This is a typical reaction for metal cations of high oxidation state (+V to +VII). Ionic radii of the actinides are large, however, and charge densities are not, therefore, as high as for smaller metal cations of corresponding oxidation state. As a result, they never bind more than two oxygen atoms, and in oxidation states +V and +VI do not form oxoanions such as AnO_3^- or AnO_4^{2-}, respectively. Hexavalent actinyl ions, AnO_2^{2+}, do not appear as distinct species in solution only, but also in solid compounds of U, Np, Pu, and Am, in AnO_2F_2, for example. In complexes, AnO_2^{2+} also appears as a distinct species. Actinyl ions, $(O{=}An{=}O)^+$ and $(O{=}An{=}O)^{2+}$, are symmetrical and linear. Because of the high charge of actinides in oxidation state six, the bonds with oxygen are stronger than the corresponding bonds in actinides in oxidation state five. Actinyl ions are only found for actinides Pa, U, Np, Pu, and Am. But, whereas actinyl ions are common for the first four, the actinyl ions of Am only appear in highly oxidizing conditions.

The gross charges of the above-mentioned actinide species are +3 (An^{3+}), +4 (An^{4+}), +1 (AnO_2^+), and +2 (AnO_2^{2+}). The positive charge is not, however, equally distributed in the actinyl ions but is localized on the metal, whose effective charge decreases in the order $An^{4+} > AnO_2^{2+} > An^{3+} > AnO_2^+$ and in the case of plutonium their charges are +4, +3.2, +3 and +2.2, respectively. Like the ionic radius, the charge on the metal affects the hydrolysis and complex formation occurring in solution. This will be discussed later.

Figure 15.5 shows the distribution of these species for U, Np, Pu, and Am as a function of redox potential (Eh) in 1 M $HClO_4$. More detailed information for each element is given later in their specific Eh – pH diagrams.

15.7 Disproportionation

Some of the aforementioned actinide species are not entirely stable in solution. Of the pentavalent actinyl ions, PaO_2^+ and NpO_2^+ are stable but PuO_2^+ and UO_2^+ are not, and change to other species by disproportionation. Pentavalent Pu easily

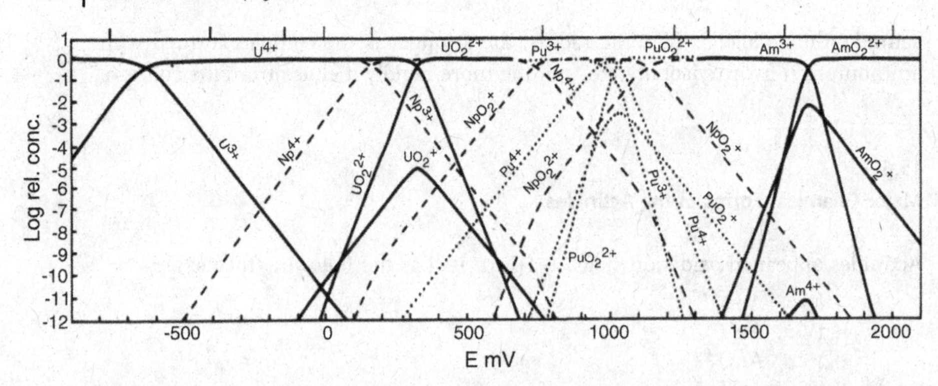

Figure 15.5 Distribution of U, Np, Pu, and Am oxidation states as a function of Eh in 1 M $HClO_4$ (Choppin, G., Liljenzin, J.-O., and Rydberg, J. (2002) *Radiochemistry and Nuclear Chemistry*, 3rd edn, Butterworth-Heineman).

disproportionates in acidic solutions and forms plutonium species in oxidation state four and five according to the following reactions:

$$2PuO_2^+ + 4H^+ \leftrightarrows Pu^{4+} + PuO_2^{2+} + 2H_2O$$

$$PuO_2^+ + Pu^{4+} \leftrightarrows Pu^{3+} + PuO_2^{2+}$$

UO_2^+ disproportionates to U^{4+} and UO_2^{2+} according to the reaction:

$$2UO_2^+ + 4H^+ \leftrightarrows U^{4+} + UO_2^{2+} + 2H_2O$$

Also, Pu^{4+}, despite being the most stable oxidation state of plutonium, is somewhat unstable in weakly acidic solutions and disproportionates to PuO_2^+ and Pu^{3+}. Figure 15.6 shows the disproportionation of the An^{4+} and AnO_2^+ ions of U, Np, Pu, and Am in a 1 M acid solution. In the oxidation state +IV only uranium and neptunium are stable while plutonium disproportionates partly and americium

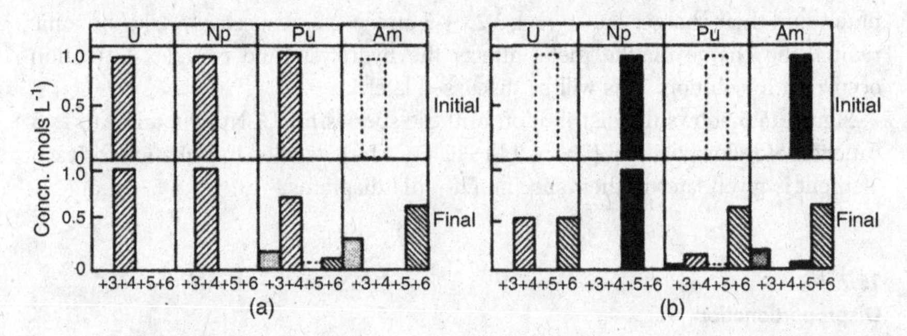

Figure 15.6 Disproportionation of An^{4+} and AnO_2^+ ions of U, Np, Pu, and Am in 1 M acid. The top part of the diagram shows the original concentration of species and the bottom part the concentrations at equilibrium. (Cotton, F.A. and Wilkinson, G. (1988) *Advanced Inorganic Chemistry*, John Wiley and Sons, New York).

completely. In the oxidation state +V only neptunium is stable, uranium more or less completely disproportionates, and plutonium and americium disproportionate to a great extent. In all respect plutonium is the most challenging, since it exists in several oxidation states at the same time.

15.8 Hydrolysis and Polymerization of the Actinides

Actinides in the same oxidation state hydrolyze according to their ionic radius: more strongly when the ionic radius is small and the metal more acidic. The strength of the coulomb interaction between the metal and oxygen atoms in water follows the charge density of the metal. Thus, hydrolysis and complex formation increase in the following order (only those actinides typically appearing in solution are included):

$$+\mathrm{III}: \quad Ac^{3+} < Pu^{3+} < Am^{3+} < Cm^{3+} < Bk^{3+} \text{ etc.}$$

$$+\mathrm{IV}: \quad Th^{4+} < U^{4+} < Np^{4+} \sim Pu^{4+}$$

$$+\mathrm{V}: \quad PaO_2^{+} > NpO_2^{+} < PuO_2^{+}$$

$$+\mathrm{VI}: \quad UO_2^{2+} < PuO_2^{2+}$$

Some exceptions to this order exist. Tetravalent Np hydrolyses just as strongly as Pu, which is heavier. In addition, pentavalent Pa is clearly more acidic than heavier actinides in the same oxidation state.

For the same actinide, hydrolysis and complex formation increase in the order

$$AnO_2^{+} < An^{3+} < AnO_2^{2+} < An^{4+}$$

because of the already noted growth of the effective charge of the metal. Again, it is a case of the coulombic interaction increasing with the charge density of the metal.

Trivalent actinides form the hydrolysis products $AnOH^{2+}$, $An(OH)_2{}^{+}$, and $An(OH)_3$, the last of which precipitates if the concentration of the actinide in solution is high enough. Tetravalent actinides form the following products: $AnOH^{3+}$, $An(OH)_2{}^{2+}$, $An(OH)_3{}^{+}$, and $An(OH)_4$. Once again, the last species will precipitate if the concentration of the actinide is sufficiently high. Since actinides are least soluble at oxidation state +IV the precipitation of hydroxides takes place at very low concentrations. Hydrolysis of the tetravalent actinides commences even at pH 2–3. Before precipitation occurs, the actinides form simple, monomeric $An(OH)_4$ complexes and, as the concentration increases, also dimeric and polymeric complexes, such as $Th_2(OH)_2{}^{6+}$ and $Th_6(OH)_{15}{}^{9+}$. Pentavalent actinyl ions, which only weakly hydrolyze and only at pH 8–9, form soluble AnO_2OH complexes. Hexavalent actinyl ions form the mono, di, and trimeric hydrolysis products AnO_2OH^{+}, $(AnO_2)_2(OH)_2{}^{2+}$, and $(AnO_2)_3(OH)_5{}^{+}$. In the polymeric hydrolysis products, bonds are formed between metal atoms, hydrogen bonds through the OH group (M–OH–M) or covalent bonds through oxygen (M–O–M). The hydrolysis of polyvalent actinides is complex. Monomeric hydrolysis products are formed in dilute solutions. As the concentration increases, so does the proportion of polymers, and when the concentration is high enough, precipitation as hydroxide occurs. Besides monomers and polymers, hydrolysis products appear in solution as colloids, small 1–100 nm particles of nonstoichiometric composition, which are kept in suspension by Brownian motion.

15.9
Complex Formation of the Actinides

The actinides form a large group of complexes. With respect to the oxidation states of actinides, the stabilities of complexes follow the same general rules as those that apply in hydrolysis. The strength of complexes formed with inorganic monovalent anionic ligands decreases in the following order:

$$F^- > NO_3^+ > Cl^- > ClO_4^-$$

With divalent anionic ligands the strength of complexes decreases in the order

$$CO_3^{2-} > C_2O_4^{2-} > SO_4^{2}$$

In natural waters, carbonate complexes forming alongside hydrolysis products of the actinides are major solution species in the neutral area. In weakly acidic solutions, ligands such as sulfate, phosphate, and fluoride also form complexes with the actinides. Important ligands forming complexes with the actinides in natural waters are humic and fulvic acids, which vary widely in composition and concentration. Humic and fulvic acids are large, complex, and heterogeneous organic molecules with complexing carboxyl groups and oxygen and nitrogen atoms with free electron pairs on the surface. The carboxyl groups and free electron pairs strongly bind metals in solution, especially polyvalent metals, such as the actinides, which readily form complexes.

The actinides form strong complexes with inorganic and organic phosphates. Several phosphate compounds, which strongly bind An^{4+} and $AnO_2{}^{2+}$ ions, are exploited in analytical separations by solvent extraction. The most important of the organic phosphates is tributyl phosphate (TBP), which is used in reprocessing facilities for spent nuclear fuel to separate uranium and plutonium from fission products and other actinides. Other solvent extraction reagents widely used in the separation of actinides have oxygen, nitrogen, or sulfur as complex-forming electron donors.

15.10
Oxides of the Actinides

The actinides form oxides with many different oxidation states (Table 15.5). In many cases, the oxidation state of the actinide in the most stable oxide is the same as it is in

Table 15.5 Oxides of the light actinides. The most stable oxides are indicated by X, others by x.

	Th	Pa	U	Np	Pu	Am	Cm
+6 (AnO_3)			x				
+4– +6			U_3O_8				
+5 (An_2O_5)		X	x	x	x		
+4 (AnO_2)		x	X	X	X	X	x
+3 (An_2O_3)						x	X

solution. This is true for Th, Pa, Pu, and Cm. Exceptionally, Np and Am mostly form dioxides, although their most stable oxidation states in solution are +V and +III, respectively. Thus the most typical stable oxide is AnO_2. The most stable oxides of uranium are UO_2 and U_3O_8, the mean oxidation state of U in the latter being +5.33. This is calculated to be the product of two hexavalent uranium atoms and one tetravalent atom per molecule. Many nonstoichiometric compounds are found among the actinide oxides, particularly for uranium. Several crystal phases are formed as the composition of the uranium oxides changes from UO_2 to UO_3 or the oxidation state varies from +IV to +VI. At first, the extra oxygen atoms joined to UO_2 are distributed randomly in the UO_2 crystal lattice, and the structure remains the same. When the concentration of oxygen increases to 2.4, however, a new crystal phase appears, and, proceeding further to UO_3, six new crystal phases are formed one by one. Protactinium also has several nonstoichiometric oxides between Pa_2O_5 and PaO_2, while plutonium has only one, $PuO_{1.61}$, between PuO_2 and Pu_2O_3. Only stoichiometric oxides have been reported for neptunium (NpO_2 and Np_2O_5) and thorium (ThO_2).

15.11 Actinium

15.11.1 Isotopes of Actinium

Actinium has just two isotopes in the natural decay series: in the thorium series ^{228}Ac (Figure 15.3) and in the actinium series ^{227}Ac (Figure 15.2). The first is a beta emitter and has a half-life of just 6.5 h, decaying through many excited states to ^{228}Th, with a maximum beta energy of 2.127 MeV. The energy of the most intense gamma ray of the tens of gamma rays emitted in relaxation of the excited states is 911 keV, with an intensity of 25.8%. Gamma emissions can be used in the determination of this nuclide. ^{228}Ac can also be used to measure the activity of its parent nuclide, the beta-emitting isotope of radium ^{228}Ra (see Chapter 7): the radium isotopes (^{226}Ra and ^{228}Ra) are separated from other nuclides, and after ingrowth into equilibrium with ^{228}Ra, ^{228}Ac is separated, for example, by Ln Resin, and its activity is measured by gamma spectrometry or by liquid scintillation counting.

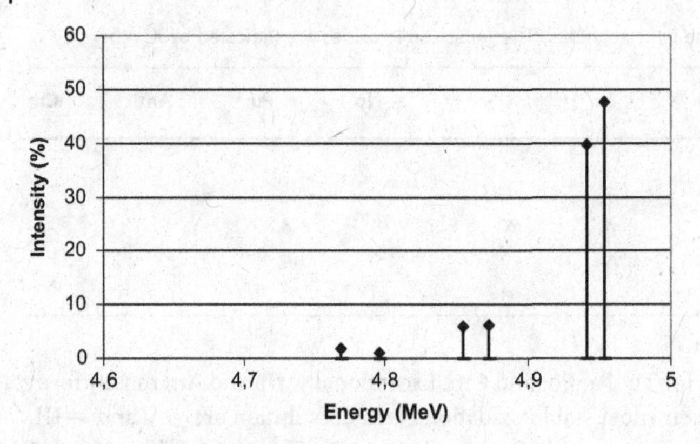

Figure 15.7 Simulated alpha spectrum of ^{227}Ac.

The other natural isotope of actinium, ^{227}Ac, is considerably longer lived, with a half-life of 21.7 years. It decays primarily (98.6%) through beta emission to ^{227}Th with a maximum energy of 45 keV, the rest (1.4%) of the decay occurring through alpha emission to ^{223}Fr (Figure 15.8). Figure 15.7 shows the simulated alpha spectrum. The alpha peaks of ^{227}Ac can be distinguished from those of the daughter nuclides.

After ^{227}Ac, the longest-lived isotope of actinium is ^{225}Ac, which has a half-life of 10 days. It is the daughter of beta-emitting ^{225}Ra and is used as a tracer in ^{227}Ac analyses. Neither ^{225}Ac nor its parent ^{225}Ra occur naturally, but the latter is milked from a generator of artificially produced ^{229}Th. For the utilization of ^{225}Ac as a tracer, its parent nuclide, ^{225}Ra, must be first separated from the solution. ^{225}Ac is an alpha emitter with several alpha and gamma transitions. The intensities of the gamma transitions are very low, the highest intensity being only 0.5%.

15.11.2
Chemistry of Actinium

Actinium is the first member of the actinide series. Its electron configuration is [Rn] $6d7s^2$ and its only known oxidation state is +III; in solution it appears as the Ac^{3+} ion. Although the general characteristic of the actinides is the filling of the 5f shell, in no form does actinium have electrons in its 5f shell. It closely resembles lanthanum in its chemistry, differing from it slightly in size: the ionic radius of the trivalent ion of actinium is 0.112 nm, whereas the radius of the trivalent lanthanum ion is 0.103 nm. Lanthanum is an effective carrier for actinium since its compounds quantitatively precipitate compounds of actinium. Of all trivalent ions (of the lanthanides, actinides, and groups 3 and 13), the Ac^{3+} ion is the largest. It is also, therefore, the least acidic of these, which means that its tendency to hydrolyze and form complexes is the lowest. Hydrolysis of Ac^{3+} begins at pH 6–7, and at pH 8, 74% of actinium is in the form of

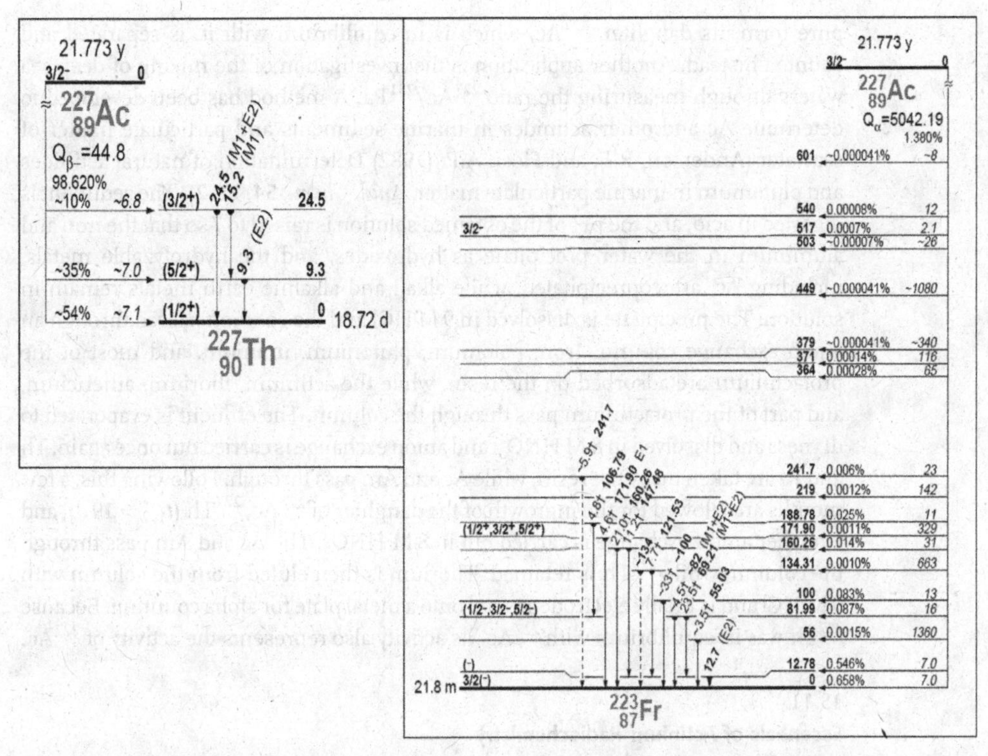

Figure 15.8 Decay schemes of ^{227}Ac. On the left is the beta decay (98.6%) and on the right the alpha decay (1.4%) (Firestone, R.B., Shirley, V.S., Chu, S.Y.F., Baglin, C.M., and Zipkin, J. (1996) *Table of Isotopes*, Wiley-Interscience).

$Ac(OH)^{2+}$ and 26% in the form of $Ac(OH)_2{}^+$. Actinium forms strong complexes with oxalate, citrate, and EDTA, and moderately strong complexes with fluoride and sulfate. The chemistry of actinium has been difficult to study because only microgram amounts are available. In weighable amounts, actinium has been studied only as its ^{227}Ac isotope. Although the half life of ^{227}Ac (21.7 y) is longer than that of other actinium isotopes, for example that of ^{228}Ac (6.5 h), it is still too short to make it possible to obtain it in larger amounts. Its specific activity is very high, and all operations must be carried out with special handling techniques and great care. The high specific activity also means that autoradiolysis is strong, which influences the chemistry of the investigated system.

15.11.3
Separation of Actinium

Separations of actinium are seldom required. ^{227}Ac is determined for the calculation of the $^{231}Pa/^{235}U$ ratio in geological dating studies. As ^{231}Pa is difficult to separate in pure form, its daughter, ^{227}Ac, which is in equilibrium with it, is separated and counted instead. Another application is the investigation of the mixing of deep sea waters through measuring the ratio $^{227}Ac/^{231}Pa$. A method has been developed to determine Ac and other actinides in marine sediments and particulate matter of seawater (Anderson, R.F. and Fleer, A.P. (1982) Determination of natural actinides and plutonium in marine particulate matter. *Anal. Chem.*, **54**, 1142). The sediment is digested in acid, and the pH of the obtained solution is raised to 7 so that the iron and aluminum in the water precipitate as hydroxides, and the hydrolyzable metals, including Ac, are coprecipitated, while alkali and alkaline earth metals remain in solution. The precipitate is dissolved in 9 M HCl and the solution passed through an anion exchange column. Iron, polonium, plutonium, uranium, and most of the protactinium are adsorbed on the resin, while the actinium, thorium, americium, and part of the protactinium pass through the column. The effluent is evaporated to dryness and dissolved in 8 M HNO_3, and anion exchange is carried out once again. Th and Pa are taken up by the resin, while Ac and Am pass through. Following this, a few months are allowed for the ingrowth of the daughter of ^{227}Ac, ^{227}Th ($t_{1/2} = 19$ d), and a further anion exchange is carried out in 8 M HNO_3. The Ac and Am pass through the column, while ^{227}Th is retained. Thorium is then eluted from the column with 9 M HCl and is finally electrodeposited onto a metal plate for alpha counting. Because ^{227}Th was in equilibrium with ^{227}Ac, its activity also represents the activity of ^{227}Ac.

15.11.4 Essentials of Actinium Radiochemistry

- Actinium has only two natural isotopes: ^{228}Ac in the thorium series and ^{227}Ac in the actinium series. The former is a pure beta emitter with a very short half-life of 6.5 h. The latter has a fairly long half-life of 21.7 years and decays primarily by beta emission (98.6%); the rest of the decays (1.4%) occur by alpha emission. Artificially produced ^{225}Ac, an alpha emitter with a half-life of 10 days, is used as a tracer in ^{227}Ac analyses.
- Actinium is the first member in the actinide series and very much resembles lanthanum in its chemistry. The only oxidation state of actinium is +III, the Ac^{3+} ion being formed. Actinium primarily forms ionic compounds and does not hydrolyze strongly.
- Actinium separations are rarely needed. While the determination of the shorter-lived isotope ^{228}Ac as such is of practically of no importance, it is commonly separated and measured to determine the activity of its parent, the beta-emitting radium isotope ^{228}Ra, presuming there is an equilibrium between these two nuclides. ^{228}Ac activity can be determined either by LSC, measuring the 2.127 MeV (maximum energy) beta particles or by gamma spectrometry, measuring the 911 keV gamma emission with 25.8 % intensity.
- The determination of the longer-lived isotope ^{227}Ac has some geological and environmental applications. The separation of actinium can be accomplished by ion exchange chromatography, for example. Nonhydrolyzable metals, including

radium, are first removed by hydroxide coprecipitation. Anion exchange in 9 M HCl removes iron, polonium, plutonium, uranium, and most of protactinium. Another ion exchange step in 8 M HNO_3 is carried out to remove thorium and the rest of the protactinium. After these steps, only americium follows actinium. To remove americium, ^{227}Th ($t_{1/2} = 19$ d), the daughter of ^{227}Ac, is allowed to grow in during a few months and one more anion exchange step is carried in 8 M HNO_3. Ac and Am pass through the column while ^{227}Th is retained. It is eluted from the column with 9 M HCl and electrodeposited onto a metal plate for alpha counting.

15.12 Thorium

15.12.1 Occurrence of Thorium

Thorium is a relatively common element – more common than tin and almost as common as lead. Its abundance in the Earth's crust is 8 ppm, about 3.5 times that of uranium. Thorium is homogeneously distributed in small concentrations in the overburden and bedrock. There are two pure minerals of thorium: thorianite (ThO_2, thorium dioxide), and thorium silicate ($ThSiO_4$). Both are rare and often occur as mixed minerals with uranium, $(Th,U)O_2$ and $(Th,U)SiO_4$. The main source of thorium is monazite, a lanthanide phosphate mineral, in which thorium may comprise as much as 12%. The solubility of thorium is very low and its concentration in natural water bodies is extremely low: in seawater, its concentration is 1.5 ng L^{-1} and in oxic groundwater less than 1 μg L^{-1}.

15.12.2 Thorium Isotopes and their Measurement

Thorium occurs naturally as six isotopes (Figures 15.1–15.3 and Table 15.6). The longest-lived isotope is the first member of the thorium series, ^{232}Th, which comprises 99.9995% of the total mass of thorium. Figure 15.9 shows the decay scheme of this isotope. Two of the isotopes of thorium, ^{231}Th and ^{234}Th, are beta emitters, and the

Table 15.6 Activities and masses of the six natural thorium isotopes relative to ^{232}Th.

Nuclide	Half-life (y)	Activity	Mass	Decay mode
^{227}Th	0.051	0.04	1.4×10^{-13}	α
^{228}Th	1.91	1	1.4×10^{-10}	α
^{230}Th	75 400	0.90	4.8×10^{-6}	α
^{231}Th	0.0029	0.04	8.2×10^{-15}	β
^{232}Th	1.4×10^{10}	1	1	α
^{234}Th	0.066	0.90	4.2×10^{-12}	β

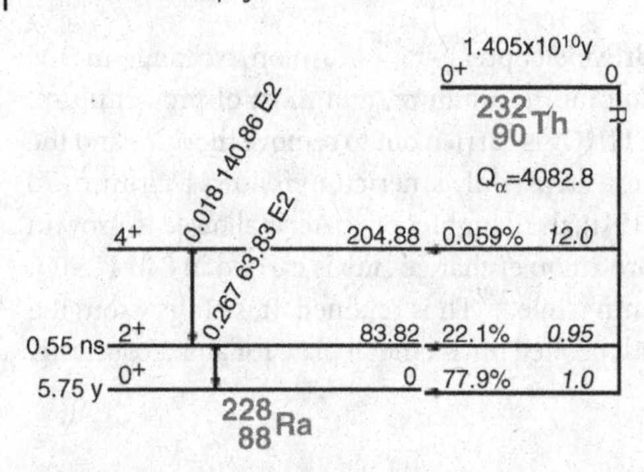

Figure 15.9 Decay scheme of ^{232}Th (Firestone, R.B., Shirley, V.S., Chu, S.Y.F., Baglin, C.M., and Zipkin, J. (1996) *Table of Isotopes*, Wiley-Interscience).

remaining four are alpha emitters. Yield determinations for thorium analyses make use of the artificial isotope ^{229}Th, an alpha emitter with a half-life of 7340 years. It is obtained as an alpha decay product of ^{233}U which, in turn, can be obtained by the neutron bombardment of ^{232}Th:

$$^{232}\text{Th}(n,\gamma)^{233}\text{Th} \rightarrow \beta^- + {}^{233}\text{Pa} \rightarrow \beta^- + {}^{233}\text{U} + \alpha \rightarrow + {}^{229}\text{Th}$$

Alpha spectrometry is the most utilized method to measure the alpha-emitting thorium isotopes. The alpha energies of the four natural thorium isotopes that decay by alpha emission vary between 3.95 and 6.04 MeV. The alpha energies of these isotopes, as well as the energy of the artificial isotope ^{229}Th used as marker, differ sufficiently for all five isotopes to be identified in the same spectrum from a semiconductor detector (Figure 15.10a). The activity of ^{227}Th, however, compared to other thorium isotopes, is very low, and only ^{228}Th, ^{230}Th, and ^{232}Th are detected in the alpha spectrum, as can be seen in Figure 15.10b.

The intensities of the gamma transitions of ^{232}Th are too low to allow the direct determination of ^{232}Th by gamma spectrometry. The 63.8 keV γ-ray has the highest intensity of 0.26%. Further, the energies of the gamma rays are so low that self-absorption occurs, causing a marked reduction in the counting efficiency and thus increased uncertainty in the results. ^{232}Th can, however, be measured through its granddaughter nuclide ^{228}Ac, which has a high intensity (48%) gamma ray at 74.7 keV energy. Again, this gamma ray is of such low energy that self-absorption is a problem, and the measurement requires a thin-window semiconductor detector. While the use of the granddaughter nuclide in the measurement of ^{232}Th obviously requires equilibrium with the parent nuclide, this can be assumed in rocks since the half-life of ^{228}Ac is just 6 h and the half-life of ^{228}Ra, which lies between ^{228}Ac and ^{232}Th, is 5.8 years (see Figure 15.3).

15.12.3
Chemistry of Thorium

The electron configuration of the thorium atom is [Rn]$6d^27s^2$. Practically its sole oxidation state is +IV, the Th^{4+} ion being formed. The chemistry of thorium

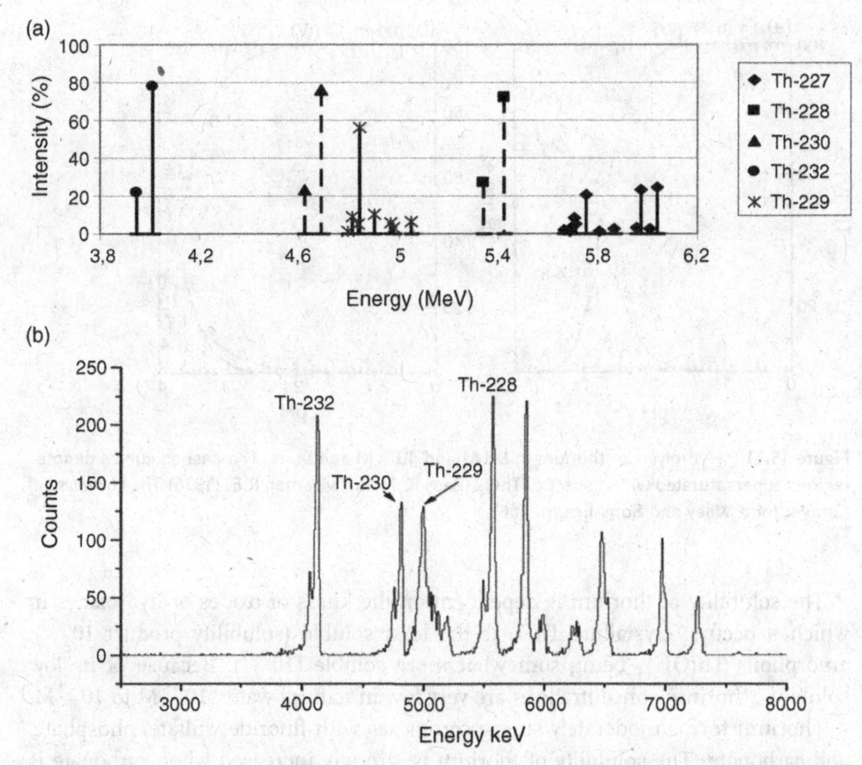

Figure 15.10 (a) Simulated alpha spectra of the natural isotopes of thorium and the artificial isotope ^{229}Th, and (b) a real alpha spectrum. The peaks on the left of ^{228}Th are due to the progeny of thorium isotopes.

strongly resembles that of the group 4 metals Ti, Zr, and Hf. Since thorium has only one oxidation state, it is used as a chemical analog for other tetravalent actinide ions (U^{4+}, Np^{4+}, Pu^{4+}) in research carried out in conditions where these other actinides may appear in several oxidation states. Although the chemistry of thorium is relatively straightforward compared to that of many other actinides, its research is still demanding. Many factors, such as low solubility and strong absorption on surfaces, create great uncertainty in solubility products, complex formation constants, and other values relevant to its chemistry.

As the largest of the tetravalent actinide ions (Figure 15.4), Th^{4+} forms the weakest hydrolysis products and complexes compared to the other actinides. Th(IV) begins to hydrolyze at pH 1, forming mononuclear mono-, di-, and trihydroxy species in weak solutions until, at pH 4, $Th(OH)_4$ predominates in the solution (Figure 15.11). At high concentrations of thorium, polynuclear species such as $[Th_2(OH)_2]^{6+}$, $[Th_4(OH)_8]^{8+}$, and $[Th_6(OH)_{15}]^{9+}$ appear, the latter being dominant in 0.1 M thorium solution at pH 4–6. Colloidal thorium also appears at high concentrations.

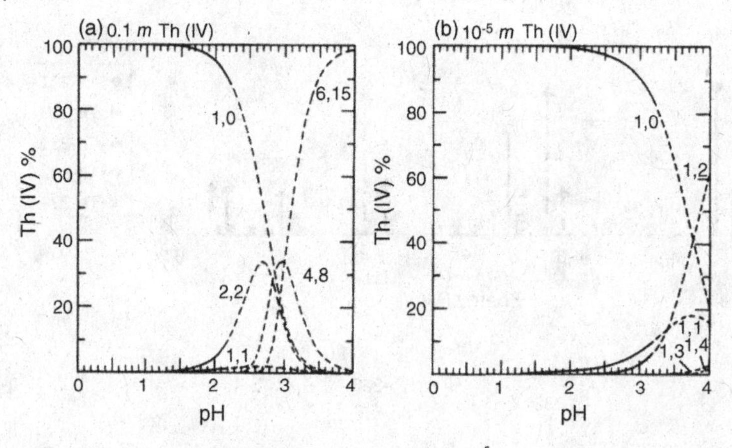

Figure 15.11 Hydrolysis of thorium in 0.1 M and 10^{-5} M solutions. The dashed curves denote regions supersaturated with respect to ThO_2 (Baes, C.F. and Mesmer, R.E. (1976) *The Hydrolysis of Cations*, John Wiley and Sons Inc., p. 166).

The solubility of thorium is dependent on the kinds of oxides or hydroxides in which it occurs: crystalline ThO_2 is the least soluble (solubility product 10^{-54}), amorphous $Th(OH)_4$ being somewhat more soluble (10^{-47}). Because of its low solubility, thorium concentrations are very low in natural water, 10^{-8} M to 10^{-9} M.

Thorium forms moderately strong complexes with fluoride, sulfate, phosphate, and carbonate. The solubility of thorium is strongly increased when carbonate is present.

15.12.4 Separation of Thorium

15.12.4.1 Separation of Thorium by Precipitation

Thorium forms sparingly soluble precipitates with hydroxide, fluoride, iodate, oxalate, and phosphate. Hydroxide precipitation separates thorium, along with other hydrolyzing metals, from alkali and alkaline earth metals and anions. Fluoride precipitation separates thorium from penta- and hexavalent actinides, which do not form fluorides; tri- and tetravalent actinides and lanthanides follow thorium in the fluoride precipitation. In fluoride precipitation, iron is also removed because it does not precipitate with fluoride in acidic solution. In addition, iron can be removed in an oxalate precipitation; iron does not precipitate as oxalate when in a trivalent state.

15.12.4.2 Separation of Thorium by Anion Exchange

Thorium does not form negatively charged complexes in hydrochloric acid systems and thus does not adsorb to anion exchange resins. Several natural alpha-active nuclides, uranium, protactinium, and polonium, are retained, however. In addition, iron, which interferes in the preparation of the counting source, and the transuranics plutonium and neptunium are retained in the column. When anion exchange is carried out in strong hydrochloric acid, only radium among the natural alpha emitters and americium and curium among the transuranics pass through the column along with thorium. These can be removed by carrying out a second ion exchange, this time in strong nitric acid. In 8 M nitric acid, thorium forms the strong nitrate complex $Th(NO_3)_6^{2-}$, which is effectively retained in the column, while radium, americium, and curium pass through. Thorium, separated from other alpha emitters, is eluted from the column with weak nitric acid, and a source for alpha measurement is prepared by coprecipitation with lanthanide fluoride or by electrodeposition onto a metal disk. Alternatively, the amount of thorium can be determined by ICP-MS.

15.12.4.3 Separation of Thorium by Solvent Extraction

Thorium can be extracted as a nitrate complex with a number of organic extraction reagents, including ethers, ketones, and phosphates. In nitric acid systems, thorium is effectively extracted into tributylphosphote (TBP), for example, together with uranium and other tetra- and hexavalent actinides. Thorium cannot be extracted from hydrochloric acid medium, however, because thorium does not form a chloride complex. Extraction in hydrochloric acid can therefore be used to remove uranium, since this forms a complex with Cl^-. The removal of iron is also accomplished in HCl extraction: in 6 M HCl solution, iron is extracted with diethylether as the $FeCl_3$ complex, while thorium remains in the aqueous phase.

15.12.4.4 Separation of Thorium by Extraction Chromatography

There are several extraction chromatography resins for the separation of thorium: Actinide, DGA, Diphonix, RE, TRU, UTEVA, and TEVA Resins. Various methods have been reported for the separation of thorium from water using TEVA Resin (Figure 4.13). The functional group in TEVA Resin is trialkyl methylammonium nitrate (or chloride). The actinides are first enriched from water by coprecipitation with calcium phosphate and the precipitate is then dissolved in 3M HNO_3. Plutonium is converted to Pu(IV) with iron sulfamate and sodium nitrite, and the solution is then poured into a TEVA column pretreated with 3 M nitric acid. Th^{4+}, along with Pu^{4+} and Np^{4+}, are taken up by the resin, while UO_2^{2+}, NpO_2^+, and Am^{3+} pass through. Thorium is eluted from the column with 9 M HCl, while plutonium and neptunium remain in the column. Finally, a counting source of the thorium is prepared for the alpha measurement.

15.12.5 Essentials of Thorium Radiochemistry

- Thorium is the most abundant natural radioactive element, constituting 8 ppm of the Earth's crust. There are two major thorium minerals, thorianite (ThO_2) and thorium silicate ($ThSiO_4$); however, the major source of thorium is monazite, a lanthanide phosphate mineral.
- The most long-lived isotope of thorium ^{232}Th ($t_{1/2} = 1.4 \times 10^{10}$ y) keeps up one of the three natural decay chain, called thorium series. Thorium has another long-lived isotope ^{230}Th ($t_{1/2} = 75\ 400$ y) in the uranium series. Three other natural thorium isotopes, ^{227}Th, ^{231}Th, and ^{234}Th, are short lived, the half-lives ranging

from 1.1 to 24 days, while ^{228}Th has a somewhat longer half-life of 1.91 years. ^{227}Th, ^{228}Th, ^{230}Th, and ^{232}Th isotopes are alpha emitters while ^{231}Th and ^{234}Th are beta emitters. As a yield determinant tracer in the analyses of the longer-lived alpha-emitting thorium isotopes (^{228}Th, ^{230}Th, ^{232}Th), ^{229}Th ($t_{1/2} = 7340$ y) is used. Alpha spectrometry is typically used to measure thorium isotopes, but mass spectrometry is also used to measure the two most long-lived isotopes.

- Thorium has only one oxidation state, +IV, and its chemistry resembles that of the group 4 metals Ti, Zr, and Hf. Thorium is readily hydrolyzable and highly insoluble. Therefore, its concentrations in natural waters are very low at 10^{-8} M to 10^{-9} M. High concentrations of carbonate in water considerably increase the solubility of thorium.
- Coprecipitation, ion exchange, solvent extraction, and extraction chromatography are used in thorium separations. Coprecipitations as hydroxide, carbonate, and phosphate are used in enrichments, oxalate precipitation in the removal of iron, and lanthanide fluoride precipitation to prepare a counting source for alpha counting.
- Separations of thorium by anion exchange, solvent extraction, and extraction chromatography are based on the fact that thorium forms a strong negative complex with nitrate anions in strong nitric acid but no chloride complex.
- In anion exchange chromatography, uranium, protactinium, polonium, plutonium, and neptunium are removed from thorium by carrying out the ion exchange in strong hydrochloric acid, where the above-mentioned metals are efficiently retained in the column while thorium, radium, and americium pass through. Radium and americium are removed in another anion exchange step in strong nitric acid where thorium is retained in the columns while the other pass through.
- Several extraction chromatography resins have been utilized in thorium separations, the TEVA Resin most commonly. This resin takes up tetravalent actinides (Th^{4+}, Pu^{4+}, Np^{4+}), while UO_2^{2+} and Am^{3+} pass through. Thorium is then eluted from the resin with strong hydrochloric acid, the other actinides remaining in the resin.
- Several extraction agents have been used in thorium separations, for example TBP in nitric acid to remove thorium and other tetra- and hexavalent actinides. Since thorium does not form a chloride complex it cannot be extracted in hydrochloric medium. However, an HCl medium can be used to remove interfering elements which form chloride complexes, such as uranium and iron.

15.13 Protactinium

15.13.1 Isotopes of Protactinium

Protactinium is represented in the natural decay series by just two isotopes: ^{234}Pa in the ^{238}U series and ^{231}Pa in the ^{235}U series (Figures 15.1 and 15.2). Of these, the former is a short-lived ($t_{1/2} = 6.75$ h) beta emitter, which decays to ^{234}U. ^{231}Pa, in turn,

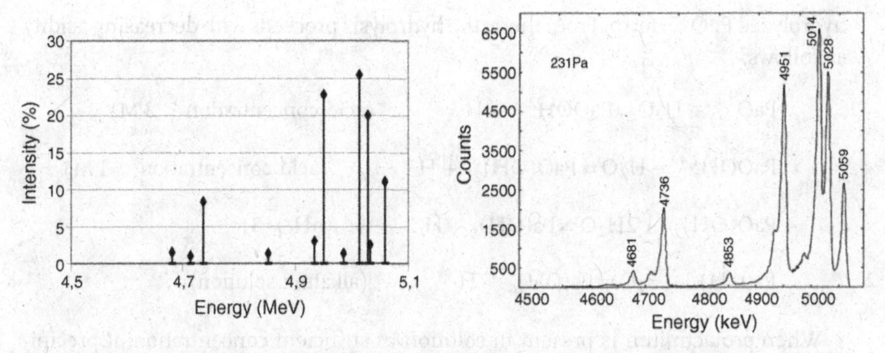

Figure 15.12 Simulated (a) and measured (b) alpha spectra of ^{231}Pa (*The Chemistry of the Actinide and Transactinide Elements*, vol. 1, 3rd edn, Springer, p. 167).

is a long-lived isotope with a half-life of 32 760 years. It decays by alpha emission to ^{227}Ac. Figure 15.12 presents the alpha spectrum of ^{231}Pa. In nature, the activity of the longer-lived ^{231}Pa is about 5% of that of ^{234}Pa, but the mass of ^{231}Pa is about two million times as great. In addition to these two natural isotopes, the artificial isotope ^{233}Pa is the third most important, and this can be used as a tracer in determining the chemical yield in ^{231}Pa analyses. ^{233}Pa is a relatively short-lived ($t_{1/2} = 27$ d) beta emitter with a maximum energy of 0.571 MeV and the most intense (38.6%) gamma emission is at 312 keV. To obtain ^{233}Pa, the best method is to separate it from the long-lived ^{237}Np ($t_{1/2} = 2.1 \times 10^6$ y), which decays by alpha emission to ^{233}Pa. The ingrowth of ^{233}Pa occurs during a few months' storage of ^{237}Np, after which ^{233}Pa can then be separated for use as a tracer. Milking ^{233}Pa from ^{237}Np can be repeated at intervals of a few months.

15.13.2
Chemistry of Protactinium

Protactinium is the third actinide and the first in the group to have electrons in the 5f shell. The electron configuration of its ground state is [Rn]5f^{2}6d^{1}7s^2, and its major oxidation state is +V. Although the oxidation state +IV is also possible, it is unstable and rapidly oxidizes to +V. Reduction to the oxidation state +IV requires strongly reducing conditions and a highly acidic solution.

Although protactinium is a rare element, it can be separated in amounts even greater than 100 grams in conjunction with the production of uranium; its chemistry, therefore, can be investigated in macro amounts. Nonetheless, studying the chemistry of protactinium is difficult, since it readily hydrolyzes and adsorbs on various surfaces. The chemistry of protactinium resembles that of the group 5 metals V, Nb, and Ta. The most common oxide of protactinium is Pa_2O_5; however, the oxide PaO_2 is also known.

The high charge on proctactinium causes it to hydrolyze readily. Thus, free Pa^{5+} ions do not occur even in highly acidic solutions. In acidic solutions in which hydrogen ion concentration is greater than 3 M, Pa probably appears as a partly hydrolyzed PaO^{3+} form. From there, the hydrolysis proceeds with decreasing acidity as follows:

$$PaO^{3+} + H_2O \leftrightarrows PaOOH^{2+} + H^+ \qquad \text{(acid concentration < 3 M)}$$

$$PaOOH^{2+} + H_2O \leftrightarrows PaO(OH)_2^+ + H^+ \qquad \text{(acid concentration < 1 M)}$$

$$PaO(OH)_2^+ + 2H_2O \leftrightarrows Pa(OH)_5 + H^+ \qquad (pH > 3)$$

$$Pa(OH)_5 + H_2O \leftrightarrows Pa(OH)_6^- + H^+ \qquad \text{(alkaline solution)}$$

When protactinium is present in solution in sufficient concentration, it precipitates as hydrous oxide at pH 5–6. Before the solubility product is exceeded, polymeric hydroxide species and colloids form in the solution. The formation of polymers is irreversible, and dilution of the solution does not break them back down into monomers.

In addition to hydroxyl complexes, Pa forms several inorganic complexes, their strengths decreasing in the following order:

$$F^- > OH^- > SO_4^{2-} > Cl^- > Br^- > I^- > NO_3^- > ClO_4^-$$

15.13.3
Separation of Protactinium

Protactinium is separated by precipitation, ion exchange, solvent extraction, and extraction chromatography. Precipitation is used for enrichment purposes and the removal of interfering radionuclides. Many protactinium compounds are sparingly soluble, and hydroxide as well as carbonate and phosphate are common coprecipitants. Pa adsorbs on anion exchangers from both nitric acid and hydrochloric acids, though only weakly from nitric acid (Figures 4.5 and 4.6). Thus, separations by anion exchange are mainly carried out in hydrochloric acid, where the best separation efficiency is obtained with an acid concentration of 9–10 M. Of the natural actinides, uranium follows Pa into the resin, while thorium and actinium do not. Further, iron, zirconium, tantalum, and niobium are adsorbed and, like uranium, are removed by chromatography by eluting the column with HCl–HF mixtures of different strengths. By way of example, Figure 15.13 describes a procedure for the separation of protactinium from rock. Coprecipitation with hydroxide is used to separate radium, which remains in solution while the protactinium is precipitated. To separate Pa from thorium and uranium, anion exchange in hydrochloric acid is used: in a strong HCl, thorium passes through the anion exchange column, while U and Pa are adsorbed. Pa is removed by eluting the column with a mixture of hydrochloric and hydrofluoric acids; fluoride forms a complex with Pa and leads it out of the column, while U remains in the resin. After an additional purification by anion exchange in the same conditions, the activities of ^{231}Pa and ^{233}Pa are counted. The alpha activity of ^{231}Pa is measured with a semiconductor detector and the beta activity of ^{233}Pa with either a liquid scintillation counter or a proportional counter. A

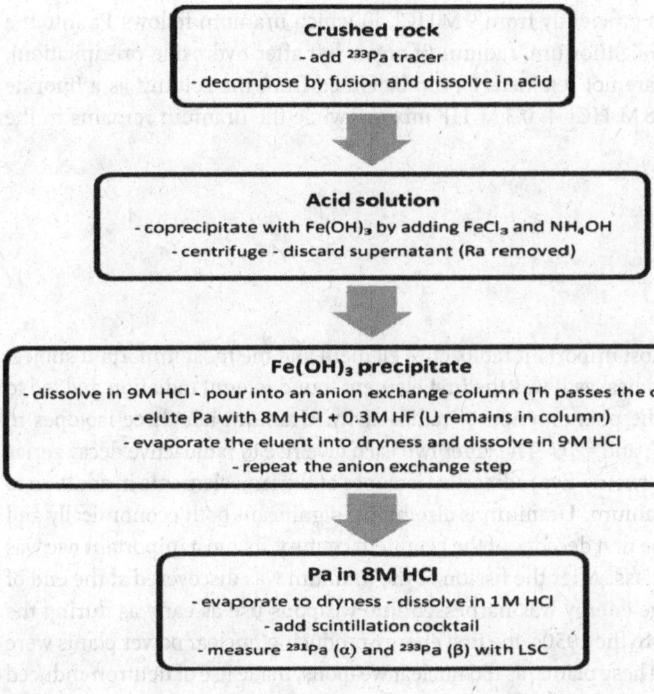

Crushed rock
- add ^{233}Pa tracer
- decompose by fusion and dissolve in acid

Acid solution
- coprecipitate with $Fe(OH)_3$ by adding $FeCl_3$ and NH_4OH
- centrifuge - discard supernatant (Ra removed)

$Fe(OH)_3$ precipitate
- dissolve in 9M HCl - pour into an anion exchange column (Th passes the column)
- elute Pa with 8M HCl + 0.3M HF (U remains in column)
- evaporate the eluent into dryness and dissolve in 9M HCl
- repeat the anion exchange step

Pa in 8M HCl
- evaporate to dryness - dissolve in 1M HCl
- add scintillation cocktail
- measure ^{231}Pa (α) and ^{233}Pa (β) with LSC

Figure 15.13 Determination of ^{231}Pa in rock by liquid scintillation counting after separation by anion exchange (Saarinen, L. and Suksi, J. (1992) Determination of Uranium Series Radionuclides Pa-231 and Ra-226 by Liquid Scintillation Counting, Report YJT-92-20, Nuclear Waste Commission of the Finnish Power Companies, Helsinki).

less laborious method is to determine both radioisotopes in the same sample with a liquid scintillation counter capable of alpha/beta discrimination.

15.13.4
Essentials of Protactinium Radiochemistry

- Protactinium has only two natural isotopes: ^{234}Pa in the uranium series and ^{231}Pa in the actinium series. The former is a short-lived ($t_{1/2} = 6.75$ h) beta emitter while the latter is an alpha emitter with a half-life of 32760 years. As a tracer in ^{231}Pa analyses, a beta-emitting ^{233}Pa ($t_{1/2} = 27$ d) is used.
- Most typically protactinium appears in oxidation state +V and is readily hydrolyzing even in strong acids, forming PaO^{3+} in acid concentrations lower than 3M and further $PaO(OH)_2^{+}$, $Pa(OH)_5$, and $Pa(OH)_6^{-}$ as the acid concentrations decreases (pH increases). The chemistry of protactinium closely resembles that of group 5 metals V, Nb, and Ta.
- Solvent extraction, as well as ion exchange and extraction chromatographies are used to separate ^{231}Pa. Coprecipitation as hydroxide is used for the enrichment and to remove nonhydrolyzable metals, such as radium. Pa absorbs in an anion

exchange resin efficiently from 9 M HCl, in which uranium follows Pa into the resin phase while thorium, radium (if any is left after hydroxide precipitation), and actinium are not retained. Pa can be eluted from the column as a fluoride complex with 8 M HCl + 0.3 M HF mixture while the uranium remains in the column.

15.14 Uranium

Uranium is the most important radioactive element and the most important subject in radioactivity studies, as it was the first element found to emit radiation and led to the discovery of the phenomenon of radioactivity. Uranium has three isotopes in nature, ^{234}U, ^{235}U, and ^{238}U. The latter two each give rise to radioactive decay series in which there are twenty-five radioactive isotopes of thirteen elements in addition to the isotopes of uranium. Uranium is also highly significant both economically and socially. During the first decades of the twentieth century, its most important use was as a pigment for glass. After the fission of the uranium was discovered at the end of the 1930s, its huge energy was harnessed into weapons use as early as during the following decade. In the 1950s, the first energy-producing nuclear power plants were brought into use. These plants, as did nuclear weapons, made use of neutron-induced fission of the ^{235}U isotope, which had been enriched from the natural uranium in the gas diffusion processes. In the use of both nuclear weapons and nuclear energy production, a large number of radioactive fission products and transuranics have been created – some released into the environment as fallout and other emissions. Uranium is also important because plutonium is made from it in reactors with the aid of neutron irradiation and the subsequent beta decays:

$$^{238}U(n,\gamma)^{239}U \xrightarrow{\beta^-} {}^{239}Np \xrightarrow{\beta^-} {}^{239}Pu.$$

15.14.1 The Most Important Uranium Isotopes

Uranium has three isotopes in nature: ^{234}U, ^{235}U, and ^{238}U, of which ^{235}U starts the actinium decay chain and ^{238}U the uranium decay chain (Tables 1.1 and 1.2 and Figures 15.1 and 15.2), while ^{234}U belongs to the uranium chain. Of natural uranium, 99.28% is ^{238}U and 0.72% ^{235}U, while the mass fraction of ^{234}U is very low. All these uranium isotopes decay through alpha emissions (Figure 15.14). ^{236}U and ^{232}U, which are also alpha emitters, are used as tracers for chemical yield determinations in uranium analyses. Of these two isotopes, ^{236}U is also generated in nuclear fuel from ^{235}U by neutron activation. The activity of ^{236}U in the spent fuel is at the same level as that of ^{238}U, but because of its much shorter half-life its mass is only less than a percent of that of ^{238}U. ^{233}U is a further important anthropogenic isotope of uranium, because it, as a fissile isotope, could be used as nuclear fuel in nuclear power reactors

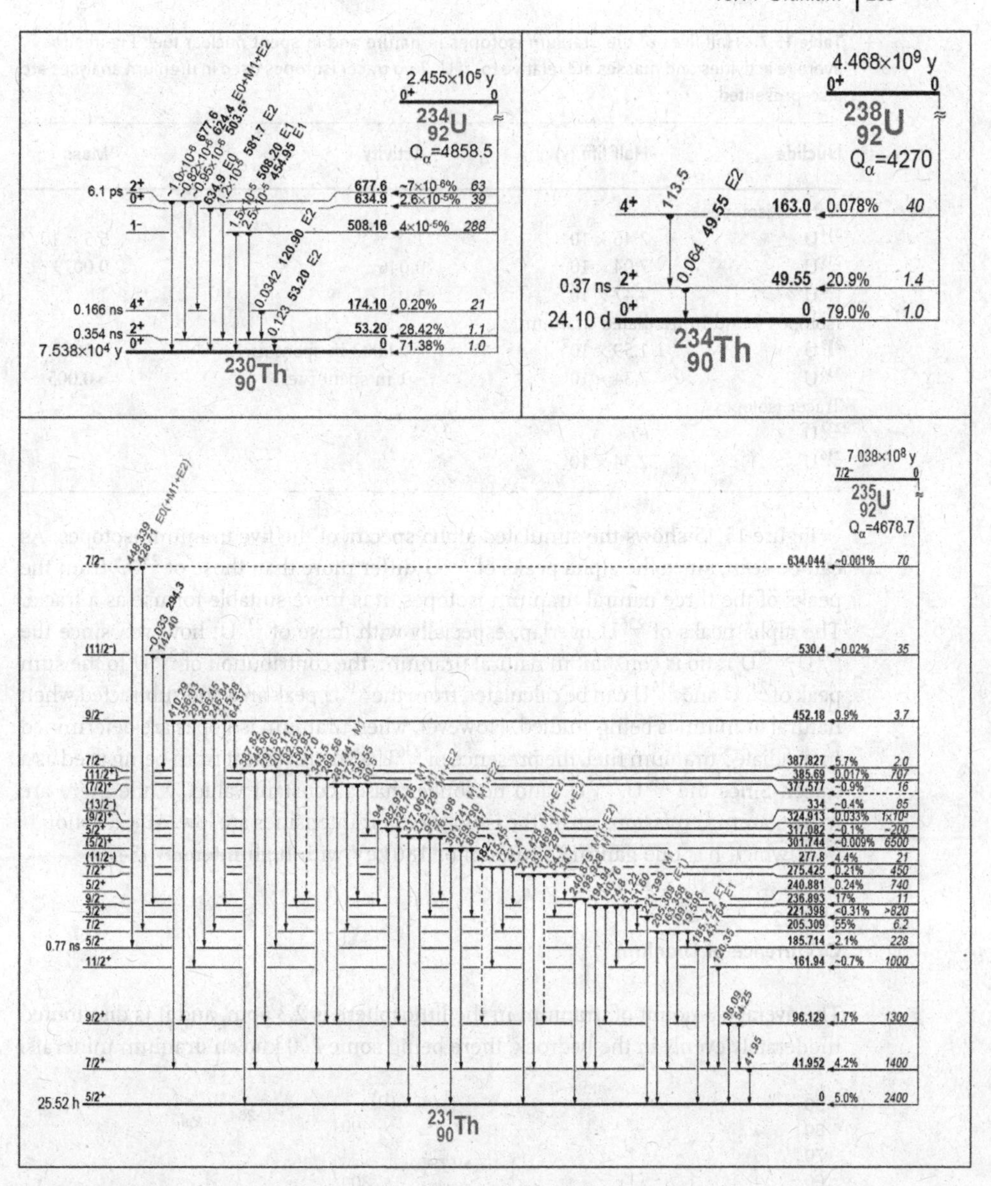

Figure 15.14 Decay schemes of ^{234}U, ^{235}U, and ^{238}U (Firestone, R.B., Shirley, V.S., Chu, S.Y.F., Baglin, C.M., and Zipkin, J. (1996) *Table of Isotopes*, Wiley-Interscience).

in addition to ^{235}U and ^{239}Pu. ^{233}U can be produced by neutron irradiation from ^{232}Th in a reactor through reaction:

$$^{232}\text{Th}(n,\gamma)^{233}\text{Th}(t_{1/2} = 22.3\,\text{min}) \xrightarrow{\beta^-} {}^{233}\text{Pa}(t_{1/2} = 27.0\,\text{d}) \xrightarrow{\beta^-} {}^{233}\text{U}(t_{1/2} = 1.59 \times 10^5\text{y})$$

^{233}U is also an alpha emitter with alpha energies of 5.789 MeV (13.2%) and 4.82 (84.4%) (Table 15.7).

Table 15.7 Half-lives of the uranium isotopes in nature and in spent nuclear fuel. Figures for average activities and masses are relative to ^{238}U. Two tracer isotopes used in uranium analyses are also presented.

Nuclide	Half life (y)	Activity	Mass
Natural isotopes			
^{234}U	2.46×10^5	1	5.5×10^{-5}
^{235}U	7.04×10^8	0.046	0.0073
^{238}U	4.47×10^9	1	1
Isotopes found in irradiated uranium			
^{233}U	1.59×10^5	~0.0002 in spent fuel	$\sim 10^{-8}$
^{236}U	2.34×10^7	~1 in spent fuel	~0.005
Tracer isotopes			
^{232}U	67		
^{236}U	2.34×10^7		

Figure 15.15 shows the simulated alpha spectra of the five uranium isotopes. As can be seen, since the alpha peaks of ^{232}U differ more than those of ^{236}U from the peaks of the three natural uranium isotopes, it is more suitable for use as a tracer. The alpha peaks of ^{236}U overlap, especially with those of ^{235}U; however, since the $^{235}U/^{238}U$ ratio is constant in natural uranium, the contribution of ^{235}U to the sum peak of ^{235}U and ^{236}U can be calculated from the ^{238}U peak area and subtracted when natural uranium is being studied. However, when uranium isotopes are determined in irradiated uranium fuel, the presence of ^{236}U also prevents it from being used as a tracer. Since the $^{235}U/^{238}U$ ratio no longer has a constant value. While there are many gamma emissions on all the isotopes, their intensities are low. An exception is ^{235}U which has one gamma emission of 186 keV with high intensity (54%).

15.14.2
Occurrence of Uranium

The average amount of uranium in the lithosphere is 2.3 ppm, and it is distributed moderately evenly in the bedrock, there being some 200 known uranium minerals.

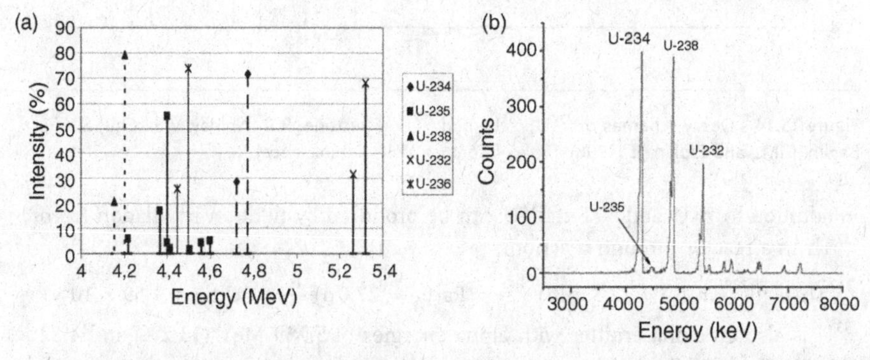

Figure 15.15 Simulated (a) and real (b) alpha spectra of uranium isotopes. In spectrum b no ^{236}U is present.

However, only a few uranium minerals are found as ores, of which uraninite, the ideal formula being UO_2, is the most important. In uraninite, uranium is ideally in the oxidation state +IV. In practice, part of the uranium in uraninite is partly oxidized; this mineral is called pitchblende (UO_{2+x}), formally a mixture of UO_2 and UO_3. Another important ore mineral is carnotite, $K_2(UO_2)_2(VO_4)_2 \cdot 3H_2O$, in which uranium has an oxidation state of +VI. To be economically usable, the uranium concentration should be at least 0.1%, that is, 100 times the average content of uranium, although poorer concentrations than this have been exploited. Uranium ores were formed when oxic ground waters oxidized the uranium to the hexavalent uranyl ion, which is highly soluble as carbonate complexes $UO_2(CO_3)_2^{-2}$ and $UO_2(CO_3)_3^{-4}$. When uranyl-containing waters meet a redox front, where oxic conditions have changed to reducing anoxic conditions, uranium is reduced back to its tetravalent state and reprecipitated as UO_2. The reducing conditions are mainly due to the presence of pyrite mineral (FeS) or organic material.

The concentrations of uranium in natural waters vary over a wide range. In seawater, the concentrations are moderately even at 2–4 $\mu g\,L^{-1}$. In other waters, however, concentrations vary greatly, from 0.1 $\mu g\,L^{-1}$ to 1 $mg\,L^{-1}$. The highest concentrations are found in groundwaters, where they can be even as high as milligrams per liter.

15.14.3 Chemistry of Uranium

The electron configuration of uranium is $[Rn]5f^36d7s^2$. Uranium can occur in solution in four oxidation states, +III to +VI (Figure 15.16), of which the oxidation states +IV and +VI are the most stable. Uranium in the higher oxidation states of +V and +VI does not occur as free U^{5+} and U^{6+} ions in solution; these hydrolyze because of their high charge, even in highly acidic solutions, and form the uranyl species UO_2^+ and UO_2^{2+}, which appear as linear O=U=O units. U^{3+} only occurs in very reducing circumstances and oxidizes rapidly in aqueous solutions, releasing hydrogen from water. Pentavalent uranium (UO_2^+), in turn, only appears in a very narrow redox potential range. It can be produced even in millimolar concentrations in solution by reducing UO_2^{2+} with hydrogen, for example. Pentavalent uranium is not thermodynamically stable but slowly oxidizes

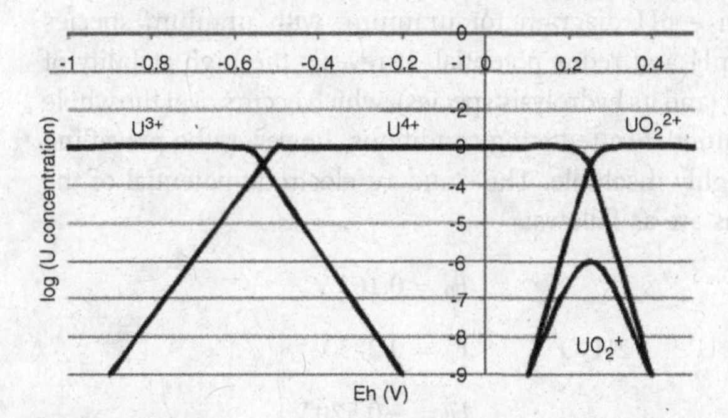

Figure 15.16 Distribution of the oxidation states of uranium in 1 M perchloric acid.

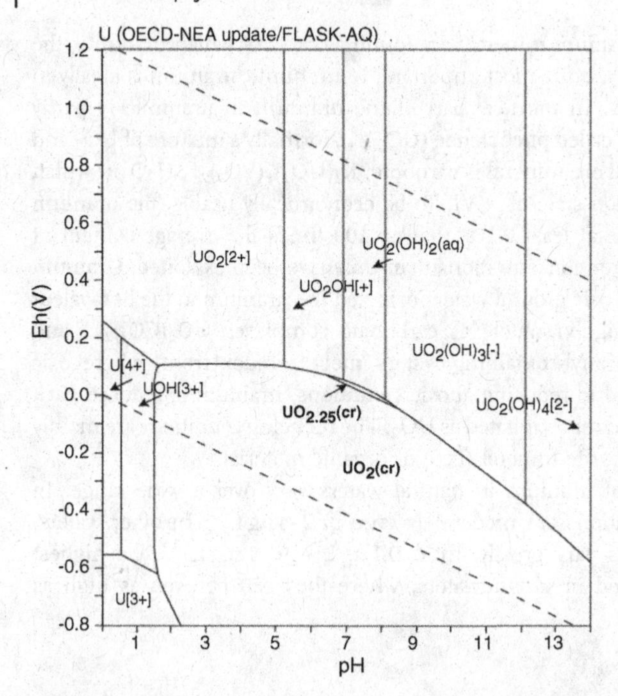

Figure 15.17 Eh – pH diagram of uranium in 10^{-10} M solution (Atlas of Eh-pH diagrams, Geological Survey of Japan Open File Report No. 419, 2005).

back to a hexavalent form. It also disproportionates into the tetravalent and hexavalent uranium as follows:

$$2UO_2^+ + 4H^+ \leftrightarrows U^{4+} + UO_2^{2+} + 2H_2O$$

Because trivalent uranium is not found in conventional chemical conditions and because the stability range of pentavalent uranium is narrow and its concentrations are very low, these two uranium species are not discussed further in this book. The following will focus on the chemistry and analysis of the two prevailing forms U^{4+} and $UO_2{}^{2+}$. Of these, the latter is more stable, and, in solution, U^{4+} oxidizes slowly to $UO_2{}^{2+}$. Although tetravalent uranium is more stable in acidic solutions, it oxidizes in them to the hexavalent form, especially when complex-forming agents are present. Figure 15.17 shows the Eh – pH diagram for uranium, with uranium species presented as a function of pH and redox potential. It reveals the high stability of hexavalent uranium, $UO_2{}^{2+}$ (and its hydrolysis species), which occurs over the whole pH range in oxidizing conditions. In reducing conditions, however, the prevailing species is UO_2, which is highly insoluble. The standard electrode potential of the uranium reduction reactions are as follows:

$$UO_2^{2+} + e^- \rightarrow UO_2^+ \qquad E_0 = 0.163\ \text{V}$$

$$UO_2^+ + 4H^+ + e^- \rightarrow U^{4+} + 2H_2O \qquad E_0 = 0.273\ \text{V}$$

$$U^{4+} + e^- \rightarrow U^{3+} \qquad E_0 = -0.520\ \text{V}$$

The first and the last reactions are fast since there is only an electron transfer involved, while the middle reaction is slow since it also requires breaking the covalent bond between uranium and oxygen.

15.14.4 Hydrolysis of Uranium

Tetravalent uranium U(IV) hydrolyzes strongly. In a strongly acidic solution, it occurs as U^{4+} ion but already starts to hydrolyze when the acid concentration decreases below 0.5 M. At even lower acid concentrations, the mono-, di- and trihydroxide species $U(OH)^{3+}$, $U(OH)_2{}^{2+}$, $U(OH)_3{}^{+}$ are formed; and at pH above 4, the prevailing species is the neutral soluble $U(OH)_4$(aq) complex. If the uranium concentration is high enough, precipitation of $U(OH)_4/UO_2$ takes place, the solubility minimum being at pH about 5. In the alkaline region, however, uranium does not form anionic hydroxide species such as $U(OH)_5{}^{-}$.

Hexavalent uranium also hydrolyzes readily, but not as strongly as tetravalent uranium. In 10 μM solution hydrolysis starts at pH 3, when cationic UO_2OH^{+} and its dimeric form $(UO_2)_2(OH)_2{}^{2+}$ are formed (Figure 15.18). When the pH increases to 5–6, the prevailing species will be $(UO_2)_3O(OH)_3{}^{+}$. It was previously assumed to be $(UO_2)_3(OH)_5{}^{+}$ and is presented as such in Figure 15.18.

15.14.5 Formation of Uranium Complexes

Uranium is a strong Lewis acid and thus readily accepts electrons from various ligands. It therefore forms strong complexes with fluoride and ligands that contain free electron pairs of oxygen and nitrogen. The stabilities of halide complexes

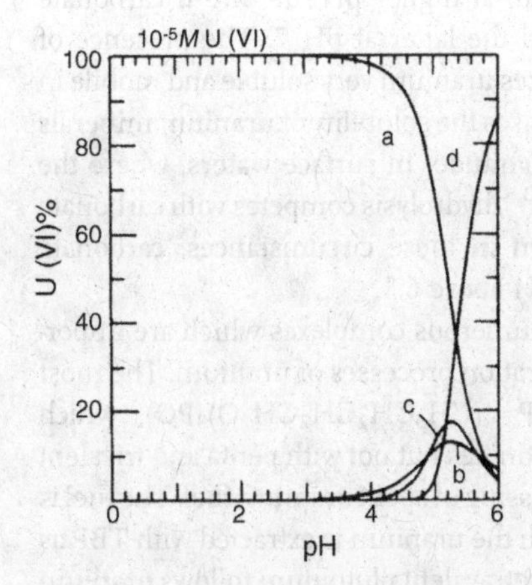

Figure 15.18 Hydrolysis of hexavalent uranium in 10^{-5} M solution. a: $UO_2{}^{+}$, b: UO_2OH^{+}, c: $(UO_2)_2(OH)_2{}^{2+}$, d: $(UO_2)_3O(OH)_3{}^{+}$. (Baes, C.F. and Messmer, R.E. (1976) *The Hydrolysis of Cations*, John Wiley & Sons).

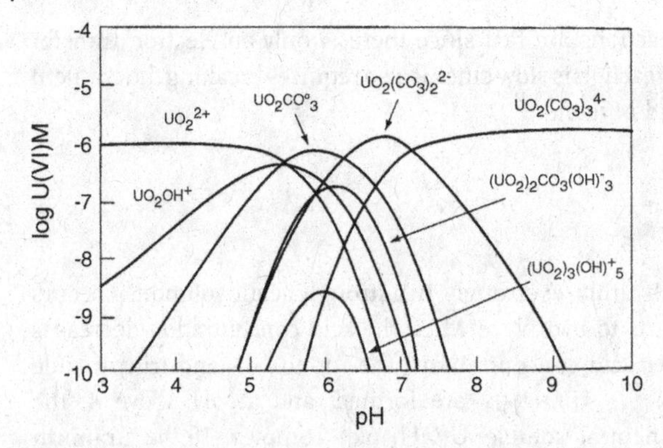

Figure 15.19 Carbonate species of UO_2^{2+} as a function of pH at the carbon dioxide partial pressure of 10^{-2} bar. Uranium concentration 10^{-6} M. (Langmuir, D. (1997) *Aqueous Environmental Geochemistry*, Prentice-Hall, Inc.).

decrease in the order of their increasing ionic size: $F^- >> Cl^- > Br^- > I^-$. While the fluoride complex is strong, complexes with other halides are rather weak. The complexes with oxoanions weaken in the order $PO_4^{3-} > CO_3^{2-} > SO_4^{2-} > NO_3^-$, which shows that the higher the ligand's valence and the stronger its basicity the stronger are the complexes.

The most important uranium complexes in natural waters are formed with carbonate. In acidic waters (pH < 6), uranium forms complexes, especially with sulfate, phosphate, and fluoride ions; however, in neutral and alkaline solutions, uranium occurs practically solely as carbonate complexes. Figure 15.19 shows the uranium species in solution as a function of pH when carbon dioxide is present at a partial pressure of 10^{-2} bar, a typical value for deep groundwaters. As can be seen, at pH 6 the prevailing species is UO_2CO_3, and, at higher pH, di- and tricarbonate complexes appear – the former at pH 6 and the latter at pH 7. The presence of uranium as anionic carbonate complexes makes uranium very soluble and mobile in groundwaters. Carbonate complexation increases the solubility of uranium minerals considerably – by two to three orders of magnitude. In surface waters, where the carbon dioxide partial pressure is lower at $10^{-3.5}$, hydrolysis competes with carbonate complexation to some extent; however, even in these circumstances, carbonate complexes are the predominant species at pH above 6.5.

With organic molecules, uranium forms numerous complexes which are important both in the analytical and industrial separation processes of uranium. The most notable example is tributylphosphate (TBP = $(CH_3CH_2CH_2CH_2O)_3PO$), which forms complexes with tetra and hexavalent actinides but not with penta and trivalent actinides. TBP is widely utilized in the reprocessing of spent uranium fuel. The fuel is first dissolved in strong nitric acid, after which the uranium is extracted with TBP as the neutral $UO_2(NO_3)_2(TBP)_2$ complex. Only tetravalent plutonium follows uranium into the organic phase, while most of the other actinides and fission products remain in the aqueous phase.

15.14.6 Uranium Oxides

The most common uranium oxide is UO_2, in which the oxidation state of uranium is +IV. As a mineral it is known as uraninite, but this is seldom found as pure UO_2, as described earlier. Although UO_2 is very stable over a wide pH range in reducing conditions, as the conditions become more oxidizing, it transforms into uranium oxides with higher oxygen content. The following stoichiometric uranium oxides, in addition to UO_2, are U_4O_9, U_3O_7, U_3O_8, and UO_3, which are listed in the order of increasing oxidation state. In the first three, uranium occurs formally in the two oxidation states +IV and +VI, which can also be presented as mean values of 4.5, 4.67, and 5.33. UO_3, in which the oxidation state of uranium is +VI, is the most oxidized uranium oxide. As a mineral it is known as schoepite. It is very soluble, forming uranyl ions UO_2^{2+} in aqueous solution, and thus it is a rare mineral, found only in oxic places with no water. The intermediate oxides between uraninite and schoepite (U_4O_9, U_3O_7, U_3O_8) are stable in the pH range typical for natural waters' 4–8; however, the Eh range where they are stable is rather narrow, at about 200 mV.

15.14.7 From Ore to Uranium Fuel

Uranium ore is first ground to a fine powder, and from it the lighter rock, called gangue, is removed by flotation, for example. The ground ore is then roasted at a high temperature to oxidize the uranium and thus enhance its solubility. The uranium is then leached from the ore by either sodium carbonate or sulfuric acid. To further oxidize the uranium, either manganese oxide (MnO_2) or sodium chlorate ($NaClO_3$) is added to the mixture. In sulfuric acid leaching, the $UO_2(SO_4)_3^{4-}$ species is mainly formed, while in carbonate leaching $UO_2(CO_3)_3^{4-}$ is formed. Carbonate extracts uranium much more selectively than sulfuric acid. Sulfuric acid leaching is also used as an *in-situ* method: sulfuric acid is pumped into holes bored in the bedrock and the uranium dissolved by the acid is extracted from adjacent holes.

The sulfate and the carbonate leachates are dried, and a crude uranium product, yellow cake, is formed, which contains mainly U_3O_8 as well as some UO_2 and UO_2SO_4. Yellow cake also contains many impurities. The crude product is dissolved in strong nitric acid, after which the uranium is purified by solvent extraction using TBP, with which uranium forms the complex $UO_2(NO_3)_2(TBP)_2$. The uranium is stripped back to the aqueous phase with water. The extraction is very selective, producing pure uranium – except for very small amounts of thorium, which follow it to the end product. The aqueous uranium solution is dried and calcined to UO_3 before being reduced to UO_2 with the help of hydrogen.

UO_2 obtained in the above process can be used as a fuel in reactors that use unenriched (natural) uranium. Most reactors, however, use enriched UO_2 as a fuel, in which the fraction of ^{235}U has been increased from its naturally occurring value of 0.72% to about 3%. The enrichment is performed by transforming uranium with the help of hydrogen fluoride into the gaseous UF_6, which is then conducted through several consecutive gas centrifuge units. The separation of ^{235}U from ^{238}U takes place because of their small mass difference. Finally, the uranium is converted to UO_2, and the end product of uranium dioxide enriched with respect to its ^{235}U content is obtained as fuel for nuclear reactors. In addition, as a side product, the process produces depleted uranium, in which the fraction of ^{235}U is lower than its natural value of 0.72%.

15.14.8
Measurement of Uranium

If the concentration of uranium is high enough, its concentration can be measured by conventional methods such as titration, gravimetry, X-ray fluorescence, and spectrophotometry. These methods are used when larger amounts of uranium are being treated, for example, in uranium fuel-producing industries. These methods, however, are not discussed further since the focus of this book is on radionuclides in the environment and nuclear waste, where the levels of uranium are usually so low that more sensitive methods are needed. For environmental and waste samples, radiometric and mass spectrometric methods are normally used.

Since all uranium isotopes are alpha emitters, alpha spectrometry is an obvious choice for measuring them. The alpha energies of the natural uranium isotopes differ from each other so much that they can all be detected from the same sample (Figure 15.15). The detection of ^{236}U, however, is impossible, since its peaks overlap with those of ^{235}U. For the alpha measurement, the uranium is either electrodeposited on a stainless steel plate or microcoprecipitated with lanthanide fluoride on a membrane. While the latter is somewhat more straightforward, the former results in a better peak resolution.

Gamma spectrometry can be used to a limited extent to measure ^{238}U and ^{235}U. The intensities of ^{238}U gamma rays are very low – one percent and below; however, the gamma rays of its short-lived progeny nuclides ^{234}Th (63.3 keV and 92.5 keV) and ^{234m}Pa (1001 keV) can be used in some cases to measure ^{238}U, provided there is an equilibrium between them. ^{235}U emits several gamma rays with fairly high intensities, the 186 keV gamma ray being the most intense at 54%. ^{226}Ra has a gamma emission at the same energy and therefore needs either to be separated or its contribution to the gamma intensity at this energy corrected by calculation. The latter method requires that there should be equilibrium between the uranium and radium. While liquid scintillation counting can also be used to measure uranium, because of its poor energy resolution it is more suitable for measuring the total uranium content from radiochemically separated samples.

Mass spectrometry is the most sensitive method for the measurement of uranium, especially when determining its isotopic ratios. The most accurate method to measure the isotopic ratio is thermal ionization mass spectrometry (TIMS); however, preparing the sample is most laborious and only one element can measured at a time. ICP-MS with a quadrupole mass analyzer is suitable for the measurement of the major isotopes ^{235}U and ^{238}U, while high-resolution ICP-MS with double-focusing mass analyzers is needed for the accurate determination of the minor uranium isotopes ^{234}U and ^{236}U. High-resolution ICP-MS is also the most sensitive method for uranium detection: a detection limit as low as 1 pg in a 1 mL sample can be obtained.

15.14.9 Reasons for Determining Uranium Isotopes

There are several reasons to determine uranium and its isotopic ratios. The choice of the analytical method very much depends on the type and amount of uranium in the sample. Typically, uranium analyses are carried out for the purposes described below.

Ratios of ^{234}U to ^{238}U, as well as of uranium to its progeny isotopes, are used to evaluate geological processes. In general, equilibrium prevails between ^{238}U and its progeny isotopes; however, during certain geological processes, such as the mineralization of uranium by the reduction of uranyl ions, this equilibrium may become disturbed. Determining the isotope ratios provides data relevant to the time when this disturbance occurred.

Studies of uranium behavior in uranium deposits are used as analogs for the final disposal of spent nuclear fuel from nuclear power plants. Other important research topics in this field include the dissolution, reprocessing, and final disposal of spent fuel.

Uranium, thorium, and their progeny are responsible for some 5% of the total radiation dose to humans because of the consumption of food and drinking water. The levels of these radionuclides are usually quite stable; in some cases, however, they are elevated and cause higher doses, for example, in wells drilled in the bedrock where the concentrations of uranium can be very high, even as high as the milligrams per liter range.

As an environmental pollutant, uranium is not typically a problem because of its low specific activity compared to that of most transuranium elements. Studies of environmental uranium from nuclear weapons tests fallout and from accidental or intentional releases are thus not as extensive as those of plutonium, for example. Studies on uranium particles, however, have been carried out from nuclear weapons tests and from Chernobyl and other accidents. The main environmental impacts of uranium, and especially its progeny, have been created by mining it by conventional methods and by *in-situ* leaching, which have mobilized uranium both in surface waters and groundwaters. In these studies, the main emphasis has been put on the levels of uranium, its solubility, and its chemical forms.

The latest concerns in uranium studies relate to nuclear forensics, nuclear trafficking, and undeclared nuclear activities violating the nonproliferation of nuclear weapons. The origin of illegal nuclear material caught by customs, for example, has been studied by mass spectrometry. A major tool for the detection of undeclared nuclear activities has been air sampling to collect uranium- and plutonium-bearing particles from the air. The origin and intended purpose of these materials can be determined by the uranium and plutonium isotope ratios in the particles.

15.14.10 Separation of Uranium

Radiochemical separation methods for the determination of uranium are based on three main methods: ion exchange, solvent extraction, and extraction chromatography, which are usually optional but also supplement each other. A major challenge in the separation of uranium for the radiometric measurement, mostly done by alpha spectrometry, is to remove other alpha-emitting radionuclides, ^{210}Po, ^{226}Ra, thorium isotopes, plutonium isotopes, and ^{241}Am. If studies are carried out with samples that contain no artificial radionuclides, such as rocks, there is no need to remove the transuranium isotopes. Various methods are used to separate uranium, the most important of which are presented below.

15.14.10.1 Separation of Uranium from Other Naturally Occurring Alpha-Emitting Radionuclides

As shown in Figure 4.7, uranium, in a solution obtained by leaching a rock sample with acid, can be separated from other natural alpha-emitting radionuclides by ion exchange using a strongly basic anion exchange resin. In this type of sample, there are no transuranium elements requiring separation. In this ion exchange method, uranium, as an anionic complex ($UO_2Cl_4^{2-}$) in 9 M HCl is retained in an anion exchange resin bed. Since thorium and radium do not form chloride complexes, they pass the column in the effluent. In addition to uranium, ^{210}Po is also retained in the column as the $PoCl_6^{2-}$ complex. When the uranium is eluted out of the column by 0.1 M HCl, which decomposes the anionic uranyl complex, polonium still remains in the column. The uranium can thus be removed from interfering radionuclides in a single step.

15.14.10.2 Determination of Chemical forms of Uranium in Groundwater

The same anion exchange method, used above for the rock sample, can be applied to determine uranium and its chemical forms in groundwaters that do not contain transuranium elements. The possible physical and chemical forms of uranium in groundwaters are hexavalent uranyl ion, usually present as anionic carbonate complexes $UO_2(CO_3)_x^{2-2x}$, tetravalent uranium as $UO_2(aq)$, and uranium attached to particle matter. In most cases, the first is the prevailing form. These three forms can be distinguished by using filtration, anion exchange in 9 M HCl, and precipitation with lanthanide fluoride. Filtration is used to remove particles – the pore size of the filter determining the minimum size of the particles. The filter is digested in concentrated acids and the amount of uranium in it is measured after the anion exchange separation (see above) to get the fraction of uranium associated with particles. To separate tetravalent and hexavalent uranium from each other in the filtrate, lanthanide fluoride precipitation is carried out – LnF_3 coprecipitates tetravalent but not hexavalent uranium. The precipitate is collected on a filter membrane and dissolved along with the membrane in aqua regia (a mixture of concentrated HCl and HNO_3), evaporated to dryness, and dissolved in 9 M HCl to separate the uranium by anion exchange to obtain the fraction of uranium in the tetravalent state. The supernatant from lanthanide fluoride precipitation is evaporated to dryness and dissolved in 9 M HCl, and the uranium is separated again by anion exchange to get the uranium fraction in the hexavalent state.

15.14.10.3 Separation of Uranium from Transuranium Elements by Anion Exchange or by Extraction Chromatography

Several environmental samples also contain transuranium elements, especially plutonium and americium, which need to be separated prior to the alpha measurement of uranium. These samples include soil, sediment, surface water, vegetation, animals, and aerosol particles collected by air filtration. In addition, transuranium elements coexist with uranium in many nuclear waste samples. Transuranium elements can be removed from uranium by both anion exchange and extraction chromatography using the same methodology, plutonium being reduced to the trivalent state. This (like americium) is not retained in the anion exchanger or in the used extraction chromatography column. Solid samples (soil, sediment, precipitate from enrichment of actinides from water, etc.) are first digested in concentrated acids and evaporated to dryness. The residue is dissolved in 9 M HCl for the anion exchange separation and in 3 M HNO_3 for the extraction by chromatographic separation using a UTEVA column; yield determinant tracers are then added for uranium and other radionuclides of interest. Plutonium is then reduced to the trivalent state by adding ferrous sulfamate and ascorbic acid. The 9 M HCl solution is poured into an anion exchange column preconditioned with 9 M HCl, and the 3 M HNO_3 is poured into a UTEVA column preconditioned with 3 M HNO_3. In both cases, the uranium remains in the column, as an anionic $UO_2Cl_4^{2-}$ complex in the anion exchange column and as a neutral $UO_2(NO_3)_2$ complex in the UTEVA column. The trivalent plutonium and americium, in turn, pass the columns without retention (see Figures 4.6 and 4.12). The whole procedure for the separation of U, Pu, and Am from air filters is represented in Figure 4.13.

15.14.10.4 Separation of Uranium by Solvent Extraction with Tributylphosphate (TBP)

As mentioned, tibutylphosphate (TBP) has been used in analytical and industrial processes for the separation of uranium. In the analytical separation, the sample is prepared in 8 M HNO_3 and plutonium is reduced to the trivalent state to prevent its extraction along with uranium. Only the thorium follows uranium into the organic phase when extracted with TBP. The thorium can be removed from the uranium by back extraction with 1.5 M HCl, while the uranium remains in the organic phase. Finally, the uranium is stripped into the aqueous phase by water.

15.14.11 Essentials of Uranium Radiochemistry

- Uranium is the most important radioactive element and has three isotopes in nature: ^{234}U, ^{235}U, and ^{238}U. The first isotope belongs to the decay chain beginning with ^{238}U, while the last two are parents in two natural decay chains.

In irradiated uranium, ^{236}U is also present, which is formed from ^{235}U by neutron activation. All these uranium isotopes decay by alpha emission.

- Uranium has been widely mined for both nuclear fuel and weapons material. The natural abundance of uranium is 2.3 ppm, and the most important uranium ore mineral is uraninite with an ideal formula of UO_2. In natural uranium, the abundance of the fissile isotope ^{235}U is 0.72%, while that of ^{238}U is 99.28%. To be used as nuclear fuel, ^{235}U is enriched to about 3% in gas centrifuge facilities; for weapons material, its enrichment percentage is above 90%. Uranium is also a source of ^{239}Pu, another fuel and weapons material formed by neutron irradiation from ^{238}U in nuclear reactors.
- Uranium has four oxidation states between +III and +VI, of which +IV and +VI are the most predominant. Trivalent uranium can only be found in very reducing conditions in acidic solutions. The pentavalent state only occurs in small proportions over a limited redox potential range and readily disproportionates to the tetra- and hexavalent states. Tetravalent uranium occurs as the U^{4+} ion in acidic solutions, while hexavalent uranium occurs as the uranyl ion, UO_2^{2+}.
- Tetravalent uranium readily hydrolyzes, forming $U(OH)_4$, even in slightly acidic solutions. Hexavalent uranium hydrolyzes less readily and forms UO_2OH^+ and $(UO_2)_2(OH)_2^{2+}$ at pH above 4 and $(UO_2)_3O(OH)_3^+$ at higher pH values.
- Uranium forms the following stoichiometric oxides: UO_2, U_4O_9, U_3O_7, U_3O_8, and UO_3, in which the oxidation state gradually increases from +IV with UO_2 to +VI with UO_3.
- The stabilities of complexes with halides decrease in the order: $F^- >> Cl^- > Br^- > I^-$; and with common oxoanions in the order: $PO_4^{3-} > CO_3^{2-} > SO_4^{2-} > NO_3^-$. In natural waters of pH 6 and above, uranium typically occurs as carbonate complexes $UO_2(CO_3)_2^{2-}$ and $UO_2(CO_3)_3^{4-}$, which makes uranium very soluble and mobile in natural waters.
- Uranium isotopes can be measured by both alpha spectrometry and mass spectrometry (TIMS and ICP-MS), while LSC is only used to measure total uranium concentrations. High-resolution ICP-MS and TIMS are the most sensitive methods with which to measure uranium isotopes.
- If alpha spectrometry is used to measure uranium, the key factor is to separate it from the other alpha-emitting radionuclides, ^{210}Po, ^{226}Ra, and ^{241}Am, as well as thorium and plutonium isotopes.
- If samples that have no transuranium elements present are studied, a single anion exchange step in 9 M HCl is needed to separate uranium as anionic $UO_2Cl_4^{2-}$ from ^{226}Ra and thorium isotopes, which do not form chloride complexes and are thus not retained in the column. Uranium is eluted from the column with dilute HCl solution but polonium still remains.
- To separate uranium from the transuranium elements Pu and Am, anion exchange and extraction chromatographies as well as solvent extraction can be utilized. For the separation, plutonium is reduced to Pu^{3+}, in which oxidation state americium is initially in the samples. Unlike uranium, trivalent Pu and Am are not retained in an anion exchange column in 9 M HCl, in a UTEVA Resin column in 3 M HNO_3, or in tributyl phosphate, and are thus removed from the uranium.

- The separation of oxidation states in solutions can be accomplished by coprecipitation with lanthanide fluorides (LnF_3), which coprecipitate tetravalent but not hexavalent uranium.

15.15 Neptunium

15.15.1 Sources of Neptunium

Neptunium is the first transuranium element. The most important long-lived isotope of neptunium is ^{237}Np, with a half-life of 2.14×10^6 years. Since the half-life is three orders of magnitude shorter than the Earth's age, all primordial neptunium decayed long ago and therefore no decay chain, such as shown in Tables 1.1–1.3, originating from this nuclide exists. Though ^{237}Np is naturally found in minute amounts in uranium ores it can be considered purely an artificial element. The main source of neptunium is the nuclear power industry, and it is formed by neutron activation and through beta decays from uranium in nuclear fuel in the following reactions:

$$^{235}\mathrm{U}(\mathrm{n},\gamma)^{236}\mathrm{U}(\mathrm{n},\gamma)^{237}\mathrm{U} \rightarrow \beta^- + {}^{237}\mathrm{Np}$$

$$^{238}\mathrm{U}(\mathrm{n},2\mathrm{n})^{237}\mathrm{U} \rightarrow \beta^- + {}^{237}\mathrm{Np}$$

Since the ^{237}U isotope is short-lived ($t_{1/2} = 6.8$ days), the formation of ^{237}Np is rapid. In spent fuel, the activity fraction of ^{237}Np is quite small. The activity of ^{237}Np is only one thousandth of that of the plutonium isotopes ^{239}Pu and ^{240}Pu, for example, mainly because of its considerably longer half-life. After a few hundred thousand years it will be the prevailing transuranic element in spent fuel. In spent nuclear fuel reprocessing, where plutonium and uranium are separated for further use, neptunium goes to the high-activity waste fraction for final disposal. Concern arises about the behavior of neptunium in final disposal not only because of its long half-life, but also because it is fairly mobile in oxic conditions as the neptunyl ion NpO_2^+.

^{237}Np is also formed in nuclear explosions, with a total of about 2500 kg having been released to the environment. This is approximately the same amount as that of plutonium isotopes, but because of the longer half-life of ^{237}Np its activity in the environment is a hundred to a thousand times lower. The total environmental release of ^{237}Np from the Chernobyl accident was about five thousand times lower than that from weapons tests fallout. Another source of environmental ^{237}Np is in the release from nuclear fuel reprocessing plants, especially La Hague in France and Sellafield in the UK, together resulting in increased ^{237}Np levels in the nearby seas by a factor of as much as one thousand compared to the levels caused by the nuclear weapons tests fallout (10^{-15} to 10^{-14} g L^{-1}). ^{237}Np is also formed by the alpha decay of ^{241}Am, and therefore its activity in nuclear fuel and the environment is increasing. In conclusion, the levels of ^{237}Np are high only in spent nuclear fuel. In the environment, in turn, the activity levels are very low compared to other actinides (Th, U, Pu, Am), and therefore the determination of ^{237}Np from environmental samples requires a high separation efficiency and a sensitive measurement technique.

15.15.2
Nuclear Characteristics and Measurement of ^{237}Np

^{237}Np decays by alpha emission to ^{233}Pa. The main alpha particles have energies of 4.788 MeV (intensity 51%) and 4.770 MeV (19%). Because of the proximity of the energies of these two emissions, alpha spectrometry cannot distinguish between them and only gives a sum peak. ^{237}Np also emits 29.4 keV gamma rays with an intensity of 15%, but these gamma rays cannot be used for direct measurement from environmental and waste samples because of their low energy and especially the low activity of ^{237}Np compared to other coexisting radionuclides. Therefore, ^{237}Np needs to be separated from samples for measurement by alpha spectrometry, for which detection limits around 0.1 mBq are achieved. Much lower detection limits than those achieved by alpha spectrometry can be obtained by neutron activation analysis, in which ^{237}Np is activated by neutrons to form short-lived ($t_{1/2} = 2.1$ days) gamma-emitting ^{238}Np. Neutron activation analysis requires separation of neptunium both before and after irradiation. Even lower detection limits, clearly below microbequerels, can be obtained by mass spectrometry, especially with high-resolution ICP-MS, which has become a favored method for measuring ^{237}Np. A critical factor in ICP-MS measurement is the removal of the neighbor isotope ^{238}U from the samples to avoid downmass tailings to the mass peak of ^{237}Np. A description of ^{237}Np measurement with ICP-MS is given in Chapter 17. The most sensitive method for the measurement of ^{237}Np is AMS, but its use is hindered by the limited accessibility and high costs.

15.15.3
Chemistry of Neptunium

The electron configuration of neptunium is $[Rn]5f^4 6d^1 7s^2$. It can occur in all oxidation states between +III and +VII. Of these the oxidation states +IV and +V are the most important and only these are discussed. Figure 15.20 gives the Eh-pH diagram for neptunium. In oxidizing conditions the prevailing form is the pentavalent dioxocation NpO_2^+, called the neptunyl ion. The neptunyl ion is fairly soluble, and it is the form in which neptunium typically occurs in oxic natural waters. As the redox potential decreases, neptunium is reduced at a rather high Eh value to tetravalent neptunium Np^{4+}. Both tetravalent and pentavalent neptunium hydrolyze, the former much more strongly than the latter. Hydrolysis of Np^{4+} starts at the low pH of about 1, while that of NpO_2^+ only at a pH above 7. At higher neptunium concentrations, both form hydrous oxides, $NpO_2 \cdot xH_2O$ and NpO_2OH. The former is highly insoluble and is the solubility-limiting solid phase occurring in final disposal conditions for spent nuclear fuel, that is, $NpO_2 \cdot xH_2O$ is formed in these conditions if

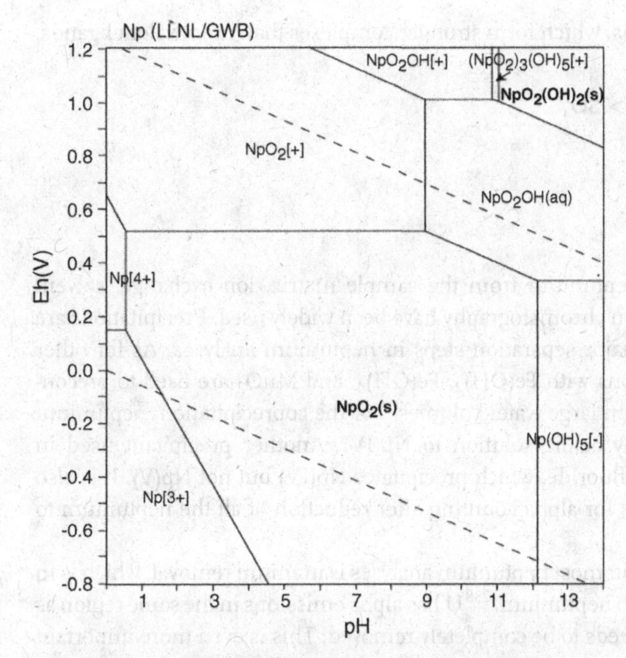

Figure 15.20 Eh-pH diagram of neptunium (Firestone, R.B., Shirley, V.S., Chu, S.Y.F., Baglin, C.M., and Zipkin, J. (1996) *Table of Isotopes*, Wiley-Interscience).

the neptunium concentration is high enough. Heating the hydrous tetravalent oxide results in the formation of anhydrous NpO_2, which is a very stable insoluble compound. Pentavalent neptunium also forms anhydrous Np_2O_5, but only under special conditions.

Pentavalent neptunium disproportionates in acidic solutions to tetravalent and hexavalent forms in the following way:

$$2NpO_2^+ + 4H^+ \leftrightarrows Np^{4+} + NpO_2^{2+} + H_2O$$

It is obvious from the equation that by increasing neptunium concentration and acidity the disproportionation reaction is promoted. The equilibrium constant of the reaction is fairly small, but the formation of complexes with Np^{4+} and ${NpO_2}^{2+}$ promotes the reaction to the right. For example, in acids, such as sulfuric acid, both tetravalent and hexavalent states form complexes with sulfate, driving the reaction to the right.

Because of its much higher ionic potential, tetravalent neptunium forms considerably stronger complexes than the pentavalent form, as is also the case in hydrolysis. The stabilities of the Np^{4+} complexes with monovalent ligands, decreases in the following order:

$$F^- > H_2PO_4^- > SCN^- > NO_3 > Cl^- > ClO_4^-$$

and with divalent ligands, which form stronger complexes than monovalent ligands, in the following order:

$$CO_3^{2-} > HPO_4^{2-} > SO_4^{2-}.$$

15.15.4
Separation of ^{237}Np

In the separation of neptunium from the sample matrix, ion exchange, solvent extraction, and extraction chromatography have been widely used. Precipitations are not typically used as major separation steps in neptunium analyses. As for other actinides, coprecipitations with $Fe(OH)_2$, $Fe(OH)_3$, and MnO_2 are used to preconcentrate neptunium from large water volumes. For the coprecipitation, neptunium is reduced in a slightly acidic solution to Np(IV). Another precipitant used in neptunium analyses is fluoride, which precipitates Np(IV) but not Np(V). It is also used to prepare sources for alpha counting after reduction of all the neptunium to Np(IV).

The major challenge in most neptunium analyses is uranium removal, which is in great excess compared to neptunium. ^{238}U has alpha emissions in the same region as ^{237}Np and, therefore, needs to be completely removed. This is even more important in ICP-MS measurement, since excessive ^{238}U causes downmass tailings to the ^{237}Np peak. Thorium, plutonium, and americium do not have major alpha peaks in the energy region of the alpha emissions of ^{237}Np, but they too are in great excess to neptunium and cause interference; hence removal is necessary. Removal of neptunium from these most important interferences makes use of the different behavior of neptunium, as well as the interfering actinides, in various oxidation states. Americium exists solely in oxidation +III and plutonium can also be converted to oxidation +III; consequently they are removed from neptunium in this oxidation state since it does not form oxidation state + III in the same conditions. Separation of uranium is typically carried out in redox conditions where neptunium is in a +IV state, while uranium is in a +VI state. This requires reduction of neptunium to Np (IV), leaving $UO_2{}^{2+}$ unreduced (see Figures 15.7 and 15.20). Thus, a major challenge in ^{237}Np analyses is a careful adjustment of the oxidation state of neptunium. Typically, reduction from a pentavalent state to a tetravalent state is done by addition of ferrous iron or iodide ions, hydrazine hydrochloride, sulfite, or sulfamate. Thorium is typically separated from neptunium based on differences in forming chloride and nitrate complexes in HCl and HNO_3 media. Both Np and Th form anionic nitrate complexes and adsorb on anion exchange resins, from which thorium is eluted with hydrochloric acid while neptunium remains.

15.15.4.1 Neptunium Tracers for Yield Determinations

In ^{237}Np separations, ^{239}Np is typically used as a tracer. It is a short-lived ($t_{1/2} = 2.4$ days) neptunium isotope that can be obtained by separation from its parent nuclide ^{243}Am in approximately one week intervals. ^{239}Np decays by emitting beta particles of 330 keV (40.5%) and 436 keV (45.3%) energies accompanied by gamma rays of

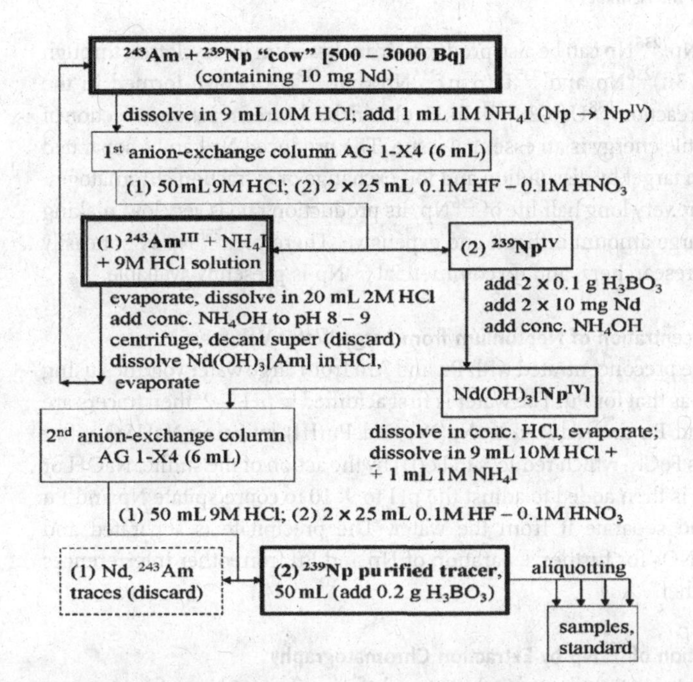

Figure 15.21 Preparation of an ^{239}Np tracer by anion exchange chromatography from ^{243}Am (La Rosa, J., Gastaud, J., Lagan, L., Lee, S.-H., Levy-Palomo, I., Povinec, P.P., and Wyse, E. (2005) Recent developments in the analysis of transuranics (Np, Pu, Am) in seawater. *J. Radioanal. Nucl. Chem.*, **263**, 427).

106.1 keV (27.2%) and 277.6 keV (14.4%). Measurement of ^{239}Np is easily carried out by gamma spectrometry. Figure 15.21 shows a separation procedure of ^{239}Np from its parent ^{241}Am using anion exchange chromatography. Neodymium is first added as a carrier, and neptunium is reduced to Np(IV) with iodide ions. Americium is then removed by two-step anion exchange chromatography in 9 M HCl; neptunium is retained as an $NpCl_6^{2-}$ complex, while americium passes through. Americium in the effluent from the first ion exchange step is recycled for further 'milking' of ^{239}Np after sufficient ingrowth time. The neptunium is eluted from the column with a dilute mixture of HNO_3 and HF and purified in a second ion exchange step. Addition of HF enhances the elution of Np by forming a complex with neptunium that is not retained in the anion exchange resin.

^{236}Np ($t_{1/2} = 1.54 \times 10^5$ y) is another potential tracer for ^{237}Np determinations. It decays through beta emission (12.5%), electron capture (87.3%), and alpha emission (0.16%). Because of its long half-life and very small alpha emission intensity, ^{236}Np is not an ideal isotopic tracer for ^{237}Np determination when using alpha spectrometry for the measurement of ^{237}Np. It is a good tracer for mass spectrometric measurement, however, since both ^{237}Np and ^{236}Np can be measured simultaneously. ^{236}Np can be produced by the neutron capture reaction of ^{237}Np in a nuclear reactor, through $^{237}Np(n, 2n)^{236,236m}Np$, but a problem is that the produced ^{236}Np cannot be separated from the target isotope ^{237}Np. In addition, the production rate of ^{236}Np is lower than that of its isomer ^{236m}Np. ^{236}Np can be also produced from uranium in a cyclotron through reactions $^{238}U(p, 3n)^{236}Np$ and $^{235}U(d, n)^{236}Np$, but ^{237}Np is also formed in the irradiation by the reaction $^{238}U(p, 2n)^{237}Np$. To obtain the desired reaction, selection of the proper projectile energy is an essential issue. The produced Np can be separated from the uranium target by dissolution and ion exchange or extraction chromatography. Because of the very long half-life of ^{236}Np, its production rate is very low, making production of a large amount difficult and expensive. Therefore, ^{236}Np is not easily available to most researchers, and no commercial ^{236}Np is presently available.

15.15.4.2 **Preconcentration of Neptunium from Large Water Volumes**

Neptunium can be preconcentrated with Pu and Am from large water volumes using the same method as that for Pu. The water is first acidified to pH 1–2, then tracers are added and Np and Pu are reduced to Np(IV) and Pu(III) by using $NaHSO_3$ after addition of iron as $FeCl_3$, which reduces to Fe(II) by the action of the sulfite. NaOH or NH_4OH solution is then added to adjust the pH to 9–10 to coprecipitate Np and Pu with $Fe(OH)_2$ and separate it from the water. The precipitate is separated and dissolved with HNO_3 for further separation of Np and Pu from other interferences and from each other.

15.15.4.3 **Separation of ^{237}Np by Extraction Chromatography**

Extraction chromatography allows a rather straightforward separation of neptunium from other actinides (Figure 15.22). Eichrom/Triskem TEVA Resin efficiently takes up tetravalent actinides, but not hexavalent or trivalent actinides (Figure 4.13). Thus, if

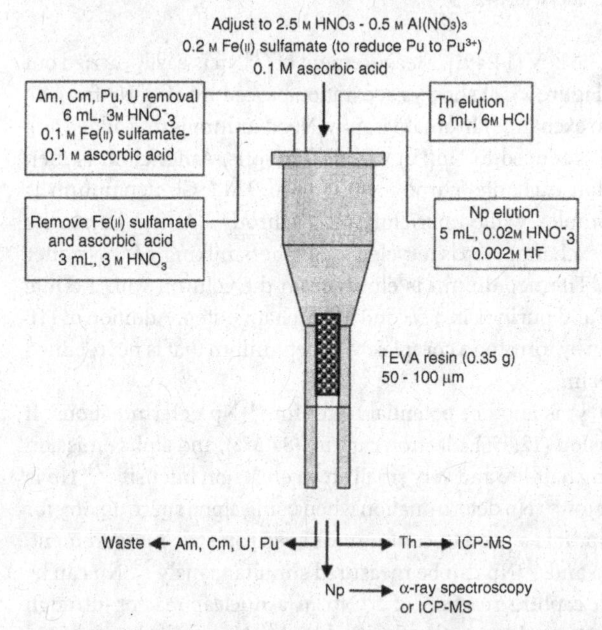

Figure 15.22 Separation of ^{237}Np from other actinides by extraction chromatography using Eichrom/Triskem TEVA Resin (Maxwell, S.L. III (1997) Rapid actinide-separation methods. *Radioact. Radiochem.* **8**, 36).

neptunium is adjusted to a tetravalent state by ferrous sulfamate, for example, it is very efficiently retained as an $Np(NO_3)_6^{2-}$ complex in a TEVA column in 2.5 M HNO_3. Sulfamate reduces plutonium to a trivalent state, which passes through the column with americium. Uranium, initially as uranyl ions, is not reduced by sulfamate and therefore not retained in the column. Thus, only neptunium and thorium remain in the column. Thorium can be removed by changing the eluant to 6 M HCl, and, since thorium does not form a chloride complex, it elutes out. Tetravalent neptunium forms a $NpCl_6^{2-}$ complex and remains in the column. Finally, neptunium is eluted from the column with dilute HNO_3 for measurement by alpha spectrometry or by ICP-MS.

15.15.4.4 Separation of ^{237}Np by Anion Exchange Chromatography

Anion exchange can be used to separate ^{237}Np in both HNO_3 and HCl media, where tetravalent neptunium forms anionic complexes $Np(NO_3)_6^{2-}$ and $NpCl_6^{2-}$. For both media, neptunium is first reduced to Np(IV) with ferrous sulfamate, for example. This reduces all plutonium species to Pu(III). In HNO_3, the solution is loaded into an anion exchange column at an acid concentration of 7–8 M (the nitrite ions present in concentrated nitric acid oxidize Pu(III) to Pu(IV)). Np(IV), Th(IV), and Pu(IV) are retained in the column, while Am(III) and U(VI) pass through and are completely removed by rinsing the column with 8 M HNO_3. Next, thorium is eluted with 8 M HCl, Th(IV) does not form a strong anionic complex with chloride. The Pu in the column is then reduced with hydrazine hydrochloride in 8 M HCl to Pu(III) and then eluted out. Finally, neptunium is eluted out with dilute nitric acid.

To carry out ion exchange in HCl media, neptunium is first reduced to Np(IV) and plutonium to Pu(III), and the solution is introduced into the column at an acid concentration of 9 M, where Th(IV), Pu(III), and Am(III) are not retained in the column. U(VI) is then rinsed out with 8 M HNO_3, and finally Np(IV) is eluted with 0.5 M HCl.

15.15.4.5 Separation of ^{237}Np by Solvent Extraction

Many solvent extraction agents have been used in the separations of neptunium. Of these, the most widely used agent is thenoyltriacetone (TTA). For extraction, neptunium is reduced to Np(IV) and the coexisting plutonium to Pu(III). Then neptunium is extracted into TTA in toluene in 1 M HNO_3. In these conditions, Pu (III) and U(VI) remain in the aqueous phase. Neptunium is back-extracted in either 8 M HNO_3 or 1 M HNO_3 after oxidation to Np(V).

If solid samples, such as soil and sediment, containing large amounts of iron, are analyzed for neptunium, the interfering iron can be removed by solvent extraction with isopropyl ether or with 4-methyl-2-pentanone (MIBK).

15.15.5 Essentials of Neptunium Radiochemistry

- Neptunium, the first transuranium element, has only one important isotope, ^{237}Np, an alpha emitter with a half-life of 2.14×10^6 years. ^{239}Np ($t_{1/2} = 2.4$ d) is used as a tracer in neptunium analyses. It can be obtained through separation from its parent, ^{243}Am.

- ^{237}Np can be measured by alpha spectrometry, but because of very low activity in environmental samples the measuring times are long. Activation analysis and especially high-resolution ICP-MS both offer shorter measurement times and lower detection limits.
- The most prevalent oxidation state of neptunium is +V, existing as neptunyl ion, NpO_2^+, in aqueous solutions. It is fairly easily reduced to the tetravalent Np^{4+} ion. The neptunyl ion is rather soluble and does not readily hydrolyze, while the tetravalent form hydrolyzes easily and forms the very insoluble hydrous oxide $NpO_2{\cdot}xH_2O$. NpO_2^+ also disproportionates in acidic solutions to form Np^{4+} and NpO_2^{2+}.
- For the separation of ^{237}Np, neptunium is typically first reduced to the tetravalent state. Preconcentration of neptunium is achieved by coprecipitation with $Fe(OH)_2$, $Fe(OH)_3$, or MnO_2.
- The major task in ^{237}Np analyses is uranium removal, since ^{238}U has alpha emissions at the same energies as ^{237}Np. Removal of uranium is also important in ICP-MS measurements, since excessive ^{238}U causes downmass tailings to the ^{237}Np peak. UO_2^{2+} can be separated from Np^{4+} by anion exchange and extraction chromatography as well as by solvent extraction. In 8 M HNO_3, Np^{4+} is retained in an anion exchange column while UO_2^{2+} passes through. The same separation occurs in 2.5 M HNO_3 in a TEVA column. In addition, neptunium can be extracted with TTA in 1 M HNO_3, while UO_2^{2+} remains in the aqueous phase.
- Thorium can be removed by absorbing Np^{4+} and Th^{4+} from nitric acid in an anion exchange or a TEVA column and then eluting Th^{4+} with hydrochloric acid, leaving neptunium in the column.
- Plutonium can be removed from Np^{4+} by reducing it to Pu(III), which absorbs in neither TEVA nor anion exchange columns nor is extracted with TTA. Americium, as a trivalent ion initially, follows plutonium in these treatments.

15.16 Plutonium

Plutonium is a fundamental element in the nuclear industry, employed in both nuclear weapons and nuclear power production. As a consequence, it is of essential concern in the investigation of environmental radioactivity and in waste treatment.

15.16.1 Isotopes of Plutonium

Plutonium is an anthropogenic element, though minute amounts occur naturally on the Earth, produced through neutron-induced reactions of uranium. Numerous

Table 15.8 Nuclear properties of the major plutonium isotopes in the environment and in nuclear waste.

Isotope	Half-life (y)	Specific activity ($Bq\,g^{-1}$)	Principal decay mode	Alpha/beta energy (MeV)	Example of production method
^{236}Pu	2.56	2.17×10^{13}	α	5.768 (69%), 5.721 (31%)	$^{235}U(\alpha, 3n)$
^{238}Pu	87.7	6.338×10^{11}	α	5.499 (70.9%), 5.456 (29.0%)	^{238}Np daughter
^{239}Pu	2.411×10^4	2.296×10^9	α	5.157 (70.77%), 5.144 (15.1%), 5.105 (11.5%)	^{239}Np daughter
^{240}Pu	6.561×10^3	8.401×10^9	α	5.168 (72.8%), 5.124 (27.1%)	Multiple n capture
^{241}Pu	14.35	3.825×10^{12}	β^->99.99%	20.8 keV	Multiple n capture
			α 2.45×10^{-3}%	4.896 (83.2%), 4.853 (12.2%)	
^{242}Pu	3.75×10^5	1.458×10^8	α	4.902 (76.49%), 4.856 (22.4%)	Multiple n capture
^{244}Pu	8.08×10^7	6.710×10^5	α	4.589 (81%), 4.546 (19.4%)	Multiple n capture

isotopes of Pu have been synthesized, with atomic numbers from 228 to 247, and all of them are radioactive, with half lives ranging from 1 second to 8×10^7 y. The most prevalent isotopes in the environment and nuclear waste are ^{238}Pu, ^{239}Pu, ^{240}Pu, and ^{241}Pu. Table 15.8 lists the physical properties of these four isotopes, and, in addition, those of ^{236}Pu, ^{242}Pu, and ^{244}Pu, which can be used as yield tracers in plutonium analysis. Isotope ^{242}Pu is also formed in nuclear reactors but in very small amounts compared with the four first-mentioned isotopes.

The most important isotope of plutonium is ^{239}Pu, which undergoes fission with thermal neutrons, providing a highly effective fuel for nuclear reactors. The critical mass of ^{239}Pu, that is, the minimum amount required to maintain a fission chain reaction, is 10 kg, or about one-fifth the amount of ^{235}U that is required. Because of its low critical mass, ^{239}Pu has also been used as weapons material in nuclear bombs. Most ^{239}Pu is produced by bombardment of ^{238}U with neutrons in a nuclear reactor. The ^{239}U that forms decays to ^{239}Np, and ultimately to ^{239}Pu:

$$^{238}U(n,\gamma)^{239}U(t_{1/2} = 23.5\,min) \xrightarrow{\beta^-} {}^{239}Np(t_{1/2} = 2.36\,d) \xrightarrow{\beta^-} {}^{239}Pu(t_{1/2} = 24110\,y)$$

Isotope ^{238}Pu, with a high power density of $6.8\,W\,cm^{-3}$ (or $0.57\,W\,g^{-1}$) because of its alpha decay, has been applied in power systems, where its alpha energy is transformed into electricity and used to power space satellites and remote instrument packages. It is produced in considerable amounts in nuclear reactors and nuclear explosions.

^{238}Pu is formed from ^{238}U in the following reactions:

$$^{238}U(n,2n)^{237}U(t_{1/2}=6.75\,\text{min})\xrightarrow{\beta^-}{}^{237}Np(n,\gamma)^{238}Np(t_{1/2}=2.12\,\text{d})\xrightarrow{\beta^-}{}^{238}Pu$$

and

$$^{235}U(n,\gamma)^{236}U(n,\gamma)^{237}U(t_{1/2}=6.75\,\text{min})\xrightarrow{\beta^-}$$

$$^{237}Np(n,\gamma)^{238}Np(t_{1/2}=2.12\,\text{d})\xrightarrow{\beta^-}{}^{238}Pu$$

The higher-mass Pu isotopes are formed as the result of successive neutron capture reactions of ^{239}Pu: $^{239}Pu(n,\gamma)^{240}Pu(n,\gamma)^{241}Pu(n,\gamma)^{242}Pu$, and so on.

The relative amounts, or isotopic ratios, of the isotopes produced in nuclear reactors and nuclear weapons explosions vary with the neutron flux and duration of the irradiation, and therefore the isotopic ratios of Pu is used for source identification.

15.16.2 Sources of Plutonium

World production of Pu (mainly ^{239}Pu) in 2005, existing in the form of spent nuclear fuel, nuclear weapons material, and nuclear waste, was approximately 2000 tonnes. In the reprocessing of spent nuclear fuel, uranium and plutonium are separated from fission products and transuranium elements (Np, Am, Cm) and can be used as mixed U–Pu fuel in nuclear power reactors. Only a small fraction of spent nuclear fuel has been reprocessed, however, and most of the plutonium remains in the spent fuel. The relative activities of plutonium isotopes in spent uranium fuel after 30 years' cooling time (normalized to the amount of the longest-lived isotope, ^{239}Pu), are ^{238}Pu: 8, ^{239}Pu: 1, ^{240}Pu: 1.6, ^{241}Pu: 100, ^{242}Pu: 0.007. Only ^{239}Pu and ^{240}Pu are relevant in the final disposal of spent nuclear fuel, since ^{238}Pu and ^{241}Pu are short-lived as compared with the expected life-time of the technical barriers preventing the release of radionuclides from spent fuel in geological conditions, and the fraction of ^{242}Pu is minor.

Weapons plutonium is produced by irradiating uranium in nuclear reactors and separating the plutonium by more or less the same process as that applied in reprocessing of spent nuclear fuel. Because the irradiation time in weapons production reactors is much shorter than that in power reactors, the isotopic composition is essentially different. Weapons plutonium contains practically no ^{238}Pu or ^{241}Pu, and the fraction of fissile ^{239}Pu is much higher. Furthermore, the mass ratio $^{240}Pu/^{239}Pu$ is only one-fifth of that in spent fuel, or about 0.05.

Plutonium is present in the environment as the result of nuclear weapons testing in the 1950s and 1960s, nuclear accidents (e.g., burn-up of a SNAP satellite in 1964, accidents of aircraft carrying nuclear weapons at Palomares in 1966 and at Thule in 1968, and the Chernobyl accident in 1986), and discharges from nuclear fuel reprocessing facilities and nuclear power plants. Table 15.9 lists the sources of Pu isotopes in the environment and shows that by far the largest source is nuclear weapons testing. The fallouts of 330 TBq of ^{238}Pu, 7.4 PBq of ^{239}Pu, 5.2 PBq of ^{240}Pu, 170 PBq of ^{241}Pu, and 16 TBq of ^{242}Pu from nuclear weapons tests are estimated to comprise about 85% of the total environmental plutonium activity. Releases from nuclear power plants (not included in Table 15.9) are minor compared with the amounts listed in the table.

15.16.3
Measurement of Plutonium Isotopes

Of the seven Pu isotopes listed in Table 15.8, all those except ^{241}Pu decay by alpha emission. ^{241}Pu decays by beta emission. The most common radiometric methods used for the measurement of plutonium isotopes are alpha spectrometry and liquid scintillation counting. Alpha spectrometry is used to determine ^{238}Pu, ^{239}Pu, and ^{240}Pu, and liquid scintillation counting for ^{241}Pu. Alpha spectrometry cannot separate the close-lying alpha peaks of ^{239}Pu (5.16 MeV) and ^{240}Pu (5.17 MeV), but measures the sum activity of ^{239}Pu and ^{240}Pu. Thus, the plutonium activities are presented in the form $^{239,240}Pu$ when measured with alpha spectrometry. Furthermore, the decay of ^{241}Am, ^{210}Po, ^{224}Ra, ^{229}Th, ^{231}Pa, ^{232}U, and ^{243}Am commonly interferes with the alpha-spectrometric determination of ^{238}Pu and $^{239,240}Pu$, and effective chemical separation of Pu from these interfering radionuclides is required before measurement. The detection limit of alpha spectrometry is about 0.02–0.1 mBq for ^{238}Pu and $^{239,240}Pu$, depending on the measurement time.

^{241}Pu is a low-energy beta emitter with maximum energy of 20.8 keV and can be measured by liquid scintillation counting (LSC). This technique requires a thorough chemical separation of Pu from the matrix and from all other radionuclides, especially interfering beta emitters. A detection limit of about 10 mBq is achievable for ^{241}Pu measured by LSC. A better detection limit can be achieved by measuring the ^{241}Pu activity indirectly, that is by measuring the alpha-emitting daughter ^{241}Am by alpha spectrometry. This requires fairly long ingrowth times, however. Detection limits down to 0.3 mBq have been reported with use of alpha-spectrometry and 13 years ingrowth time.

Another common technique used to measure plutonium isotopes is mass spectrometry (see Chapter 17). Methods include inductively coupled plasma mass spectrometry (ICP-MS), accelerator mass spectrometry (AMS), thermal ionization mass spectrometry (TIMS), and resonance ionization mass spectrometry (RIMS). Of these, ICP-MS is currently the most popular method. Attractive features are the relatively short measurement time and the provision of isotopic information on ^{239}Pu and ^{240}Pu separately. The most serious problem in employing ICP-MS to measure low-level samples for Pu isotopes is the interference from uranium, because the mass concentration of uranium is typically 10^6–10^9 times that of Pu. The tailing (abundance sensitivity) of ^{238}U to the mass range of ^{239}Pu and formation of the polyatomic ion $^{238}U^1H^+$ are major disturbances at mass 239. Thus, uranium must be removed before measurement. The detection limit of ICP-MS, especially of high-resolution

Table 15.9 Sources of plutonium isotopes in the environment (Bq).

Radionuclide	^{238}Pu	^{239}Pu	^{240}Pu	^{241}Pu	^{242}Pu	$^{239+240}$Pu	Total
Nuclear weapons testing	3.3×10^{14}	6.5×10^{15}	4.4×10^{15}	1.4×10^{17}	1.6×10^{13}	12.6×10^{15}	1.7×10^{17}
Burn-up of SNAP-9A satellite, 1964	6.3×10^{14}	—	—	—	—	—	6.3×10^{14}
Aircraft accident in Palomares, Spain 1966	—	—	—	—	—	5.5×10^{10}	5.5×10^{10}
Aircraft accident in Thule, Greenland, 1968	—	—	—	—	—	1×10^{13}	1×10^{13}
Nuclear power plant accident at Chernobyl, 1986	3.5×10^{13}	3×10^{13}	4.2×10^{13}	6×10^{15}	7.0×10^{10}	7.2×10^{13}	6×10^{15}
Reprocessing plant at Sellafield	1.2×10^{14}	—	—	2.2×10^{16}	—	6.1×10^{14}	2.2×10^{16}
Reprocessing plant at La Hague	2.7×10^{12}			1.2×10^{14}	1.7×10^{9}	3.4×10^{12}	1.4×10^{14}

ICP-MS, is comparable with that of alpha spectrometry for ^{239}Pu and ^{240}Pu (0.02–0.05 mBq), and with that of LSC for ^{241}Pu (10 mBq).

Plutonium isotopes can also be determined by AMS (see Chapter 17). AMS offers abundance sensitivity well suited for ultra trace detection of ^{239}Pu, ^{240}Pu, ^{241}Pu, ^{242}Pu, and ^{244}Pu, with a detection limit of 0.45–2 μBq for ^{239}Pu and 1.5 mBq for ^{241}Pu. The main advantage of AMS over other mass spectrometric techniques is that the interference of $^{238}U^{1}H$ with ^{239}Pu can be completely removed. The determination of ^{238}Pu remains difficult because of serious interference from ^{238}U.

TIMS measurements are made on a small volume (down to 1 μL) of aqueous solution containing the target nuclide. Careful separation of Pu from matrix elements and interfering radionuclides and concentration to a few microliters is required before measurement. Problems due to uranium are less severe in TIMS than in ICP-MS since uranium and plutonium have different ionization potentials and are emitted from the filament at different temperatures (plutonium leaves the filament first). Furthermore, the dry sample introduction considerably reduces the interference due to UH^{+}, and the abundance sensitivity is generally orders of magnitude better in TIMS than in ICP-MS. A detection limit of about 1 fg (2 μBq) ^{239}Pu has been reported with TIMS, which is a much lower level than for the radiometric method and comparable with that offered by AMS.

Of the various methods for measuring plutonium, alpha spectrometry is more straightforward than the mass spectrometric methods, and the risk of interfering signals is lower than with ICP-MS. There is also no interference from microgram amounts of stable elements in the activity measurement source after chemical separation and electrodeposition. The major disadvantage of alpha spectrometry is the lengthy measurement time, up to one week for low level samples. In ICP-MS, several polyatomic species, as well as tailing of ^{238}U, may appear in the mass range 230–245 if microgram amounts of uranium, lead, mercury, thallium, and rare earth elements are remaining in the sample after separation. Mass spectrometric methods such as AMS and TIMS are highly sensitive and allow measurement of ^{239}Pu, ^{240}Pu, and ^{241}Pu simultaneously. In contrast, use of mass spectrometric methods to measure ^{238}Pu in environmental samples is very difficult owing to the low mass concentration and serious isobaric interference from more abundant ^{238}U. For these reasons, the mass spectrometric methods should be seen as a complement to rather than a replacement for conventional alpha spectrometry for the measurement of Pu isotopes. Mass spectrometric methods are also inadequate for accurate measurement of short-lived ^{241}Pu ($k_{1/2}$ = 14.4 y) in environmental samples.

15.16.4 The Chemistry of Plutonium

15.16.4.1 Oxidation States and Plutonium

The electron configuration of plutonium is $[Rn]5f^{6}7s^{2}$, and plutonium can occur in all oxidation states between +III and +VII. Oxidation state +VII is rare, while +III, +IV, +V, and +VI are all common, and can even exist in solution simultaneously.

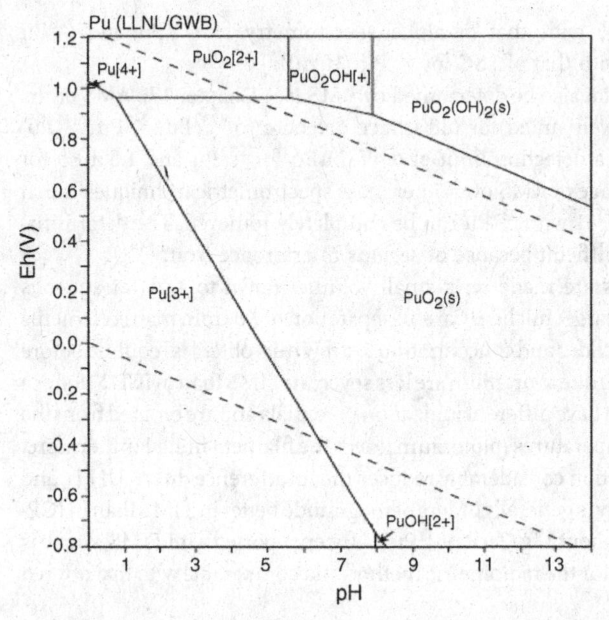

Figure 15.23 Eh-pH diagram of plutonium, at total Pu concentration of 10^{-10} M (Atlas of Eh-pH diagrams, Geological Survey of Japan File report No. 49, 2005).

Figure 15.23 shows the Eh-pH diagram for plutonium. The chemical properties of Pu greatly depend on its oxidation state. Pu ions in the lower oxidation states (III and IV) are more stable under acidic conditions, whereas Pu(VI) is more stable in alkaline media and Pu(V) in neutral media. Pu(IV) is the most stable and most thoroughly studied oxidation state. Ever since the discovery of plutonium in the early 1940s, studies on its solution chemistry have been motivated by the need to separate it from uranium and fission products for use in weapons production and as a fuel for power production. Under noncomplexing, strongly acidic conditions such as in perchloric or trifluoromethanesulfonic acid (triflic acid) solutions, both Pu(III) and Pu(IV) exist as the simple hydrated (or aquo) ions, Pu^{3+} (aq) and Pu^{4+} (aq), respectively, retaining their overall formal charge. Pu(V) and Pu(VI) cations have such large positive charges that they immediately hydrolyze in aqueous solution to form dioxo cations, $PuO_2{}^+$ and $PuO_2{}^{2+}$, which are commonly referred to as plutonyl ions. The effective charges of the plutonium forms decrease in the order

$$Pu^{4+}(+4) > PuO_2^{2+}(+3.3) > Pu^{3+}(+3) > PuO_2^{+}(+2.2)$$

15.16.4.2 **Disproportionation**

In acidic solutions in the absence of complexing ligands, Pu(IV) reacts by disproportionation, as follows:

$$3Pu^{4+} + 2\,H_2O \leftrightarrows 2Pu^{3+} + PuO_2^{2+} + 4H^+$$

The reaction probably takes place in two separate steps with a transient Pu(V) intermediate, as follows:

$$2Pu^{4+} + 2H_2O \leftrightarrows Pu^{3+} + PuO_2^{+} + 4H^{+} \text{ (slow)}$$

$$PuO_2^{+} + Pu^{4+} \leftrightarrows Pu^{3+} + PuO_2^{2+} \text{ (fast)}$$

The separate steps explain why the Pu^{3+}, $PuO_2{}^{+}$, and $PuO_2{}^{2+}$ species appear in the Pu^{4+} solution only after some period of time (Figure 15.6). In disproportionation, the first reaction is slow because formation of Pu=O bonds is involved in the formation of Pu(V). The Pu(V) produced in the first step then reacts further with Pu (IV) in a fast equilibrium. Since the first reaction is very slow, the entire disproportionation reaction of Pu(IV) is slow, and Pu(IV) is relatively stable in acidic solution.

In moderately acidic solution, Pu(V) is unstable with respect to two types of disproportionation reactions:

$$3PuO_2^{+} + 4H^{+} \leftrightarrows Pu^{3+} + 2PuO_2^{2+} + 2H_2O$$

$$2PuO_2^{+} + 4H^{+} \leftrightarrows Pu^{4+} + PuO_2^{2+} + 2H_2O$$

The disproportionation of Pu(V) depends on the conditions in the initial solution, and Pu(VI) disproportionation may follow both reactions. It is noteworthy that both reactions depend on the H^{+} ion concentration, with increasing H^{+} concentration promoting the disproportionation. This means that Pu(V) is stable only in solutions at nearly neutral pH.

15.16.4.3 **Hydrolysis**

Plutonium forms hydroxide complexes in all oxidation states, and its tendency to hydrolyze follows the effective charge, decreasing in the order

$$Pu^{4+} > PuO_2^{2+} \approx Pu^{3+} > PuO_2^{+}$$

The tendency to undergo hydrolysis is thus most pronounced for Pu^{4+} and least pronounced for $PuO_2{}^{+}$. Pu^{4+} starts to hydrolyze even at pH 0 (Figure 3.1) and can form complex polymeric and colloidal hydroxide species. Plutonium in other oxidation states forms more soluble and stoichiometric hydroxide species. The plutonyl species are 'double-hydrolyzed,': first they form the oxocations $PuO_2{}^{+}$ and $PuO_2{}^{2+}$, and these then form the hydroxide complexes PuO_2OH and PuO_2OH^{+}. In the reducing groundwater conditions that will prevail in geological repositories for spent nuclear fuel, the solubility limiting phase of plutonium will be amorphous Pu $(OH)_4$.

15.16.4.4 **Redox Behavior**

One of the most complex and fascinating aspects of the aqueous chemistry of Pu concerns the oxidation–reduction relationships of Pu species. Figure 15.24 shows the standard electrode potentials of the redox couples of Pu in different acid solutions.

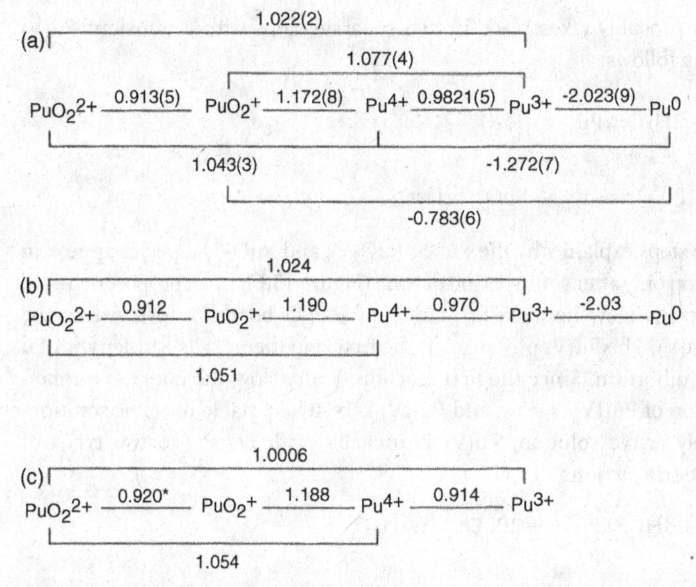

Figure 15.24 Redox potentials for selected plutonium couples at 25 °C in (a) 1 M $HClO_4$, (b) 1 M HCl, and (c) 1 M HNO_3 (Clark, D.L., Hecker, S.S., Jarvinen, G.D., and Neu, M.P. (2006) Plutonium, In: *The Chemistry of the Actinide and Transactinide Elements*, 3rd edn, Springer).

The redox couples of Pu(V)/Pu(III), Pu(VI)/Pu(III), Pu(V)/Pu(IV), and Pu(VI)/Pu (IV) are quasi-reversible or irreversible because they involve the breaking or forming of Pu=O bonds. In contrast, the redox coupling between species without forming or breaking of Pu=O bonds, such as Pu(IV)/Pu(III), Pu(VI)/Pu(V), and Pu(VII)/Pu(VI), are reversible.

Because of the instability of all plutonium oxidation states and the disproportionation of Pu(IV) and Pu(V), solutions of Pu are normally a mixture of several oxidation states (Figure 15.6). In studies of Pu chemistry and radiochemical analysis of Pu, however, there is generally a need to obtain a single oxidation state of Pu in solution. Study of the chemical behavior of plutonium in a certain oxidation state will usually require electrolytic oxidation or reduction at a certain potential. The oxidation state required for the separation of plutonium is Pu^{3+} or Pu^{4+}.

Chemical reductions needed to convert all Pu species to Pu^{3+} for purposes of chemical separation are most often carried out with hydroxylamine, hydrazine, sulfite, or sulfamate. This introduces extra salts to the solution, however, which may be unfavorable in further separation steps. In solutions of noncomplexing acid such as $HClO_4$, Pu^{3+} is stable with respect to reaction with oxygen in air. At pH 4 and above, however, it is rapidly oxidized by oxygen in air.

Pu^{4+} solutions are prepared by a combination of reduction and oxidation reactions. Pu is first reduced to Pu^{3+} using the one of the reductants mentioned above, and the formed Pu^{3+} is then oxidized to Pu^{4+} with nitrite. In HCl solution, Pu^{4+} is significantly more stable than Pu^{3+} owing to the formation of a chloride complex of Pu^{4+}, and in 3 M HCl solution Pu^{4+} is stable for several months.

A large number of reagents are suitable for the oxidation and reduction of Pu. Nitrite ions, in particular, play an important role in the aqueous redox processing of Pu. Nitrite ion is capable of oxidizing Pu(III) to Pu(IV), and at the same time it reduces Pu(VI) to Pu(IV). Since most aqueous processes (e.g., ion exchange chromatography and extraction chromatography) rely on Pu(IV), $NaNO_2$ is frequently employed to convert all Pu to the tetravalent state. Because the reduction of Pu(VI) to Pu(IV) by nitrite is a slow reaction, another reducing agent, such as ferrous iron, is frequently added to speed up the reaction.

15.16.4.5 Complex Formation

The coordination chemistry of Pu ions is generally characteristic of a 'hard' Lewis acid, which prefers complexation with hard ligands. Plutonium forms highly stable complexes with strong Lewis bases, such as carbonate, fluoride, and orthophosphate. Since plutonium ions have relatively large ionic radii, they give rise to complexes with high coordination numbers. For any given ligand, Pu^{4+} will normally form stronger complexes than the other plutonium ions because of its high ionic potential. The strength of complexes decreases with the effective charge:

$$Pu^{4+} > Pu^{3+} \approx PuO_2^{2+} > PuO_2^{+}$$

For a given plutonium oxidation state, the complexation strength with monovalent anionic ligands decreases in the order

$$F^- > H_2PO_4^- > SCN^- > NO_3^- > Cl^- > IO_3^- > ClO_4^-$$

and with multivalent anionic ligands in the order

$$PO_4^{3-} > CO_3^{2-} > SO_4^{2-}$$

15.16.5 Separation of Plutonium

Solvent extraction, ion exchange, and extraction chromatography, have been widely used to separate plutonium from interfering radionuclides and elements. Coprecipitation is the preferred method to prepare an alpha counting source but also to separate Pu from matrix components in the sample. As for other actinides, coprecipitations with $Fe(OH)_2$, $Fe(OH)_3$, and MnO_2 are used to preconcentrate plutonium from large water volumes. To assist the coprecipitation, plutonium is often reduced to Pu(III) in slightly acidic solution. Coprecipitation can also be used for the speciation analysis of Pu in water. Since Pu(III) and Pu(IV) are coprecipitated with lanthanide fluorides (LaF_3, CeF_3, NdF_3). They can be separated from water, while Pu(V) and Pu(VI) remain in solution. After removal of the LnF_3 precipitate, Pu(V) and Pu(VI) are reduced to Pu(III), and it is again coprecipitated with LnF_3. Plutonium is then determined in both LnF_3 fractions to obtain the separate proportions of Pu(III) + Pu(IV) and Pu(V) + Pu(VI).

Before determination of Pu isotopes by alpha spectrometry, all other alpha emitters must be removed, especially ^{241}Am, ^{210}Po, ^{224}Ra, ^{229}Th, ^{231}Pa, ^{232}U, and ^{243}Am, to prevent spectral interferences. In addition, most metals need to be removed in order to prepare a thin alpha source by electro-deposition. A thin source is important to improve the resolution of the alpha spectrum and the counting efficiency. In the determination of Pu isotopes by mass spectrometry, especially by ICP-MS, the major challenge is to remove uranium, which in most samples will be in great excess relative to the plutonium. Elements such as lead and mercury also need to be removed because they may form polyatomic ions with chlorine or argon in the mass range 239–241 and so interfere with the mass spectrometric measurement of Pu isotopes. Removal of salts and HCl improves the analytical sensitivity of ICP-MS.

15.16.6
Tracers Used in the Determination of Pu Isotopes

Chemical yield must be measured to correct for the losses of plutonium during the separation procedure. Measurements of chemical yield are made with ^{236}Pu or ^{242}Pu as tracer. ^{244}Pu would be suitable as well, but it is not readily available. Both ^{236}Pu and ^{242}Pu are alpha emitters and appropriate tracers for determinations of Pu isotopes by alpha spectrometry. Of the two, ^{236}Pu has the advantage that ^{242}Pu could be present in samples, in nuclear waste samples even in significant amounts. ^{242}Pu is seldom present in detectable amounts in environmental samples. ^{236}Pu ($t_{1/2} = 2.58$ y) is nevertheless relatively short-lived, and, owing to the difficult process involved in production, it is more expensive and less available than ^{242}Pu. The low mass (short half-life) of ^{236}Pu also makes it an unsuitable yield tracer when Pu isotopes are measured by mass spectrometry. Accordingly, the most widely used tracer in yield determinations today is ^{242}Pu.

^{236}Pu is produced by irradiation of ^{235}U in a cyclotron through the reaction

$$^{235}\mathrm{U}(\mathrm{d,n})^{236m}\mathrm{Np}(\beta^-, 22\,\mathrm{h})^{236}\mathrm{Pu},$$

and by irradiation of ^{237}Np in an accelerator through the reaction

$$^{237}\mathrm{Np}(\gamma,\mathrm{n})^{236m}\mathrm{Np}(\beta^-, 22\,\mathrm{h})^{236}\mathrm{Pu}.$$

An important task in the production of ^{236}Pu is to get rid of ^{238}Pu, which may be produced as a by-product in the irradiation of ^{235}U and ^{237}Np targets in the following ways:

$$^{237}\mathrm{Np}(\mathrm{n},\gamma)^{238}\mathrm{Np}(\beta^-, 2.12\,\mathrm{d})^{238}\mathrm{Pu}$$

and

$$^{235}\mathrm{U}(\mathrm{n},\gamma)^{236}\mathrm{U}(\mathrm{n},\gamma)^{237}\mathrm{U}(\beta^-, 6.75\,\mathrm{min})^{237}\mathrm{Np}(\mathrm{n},\gamma)^{238}\mathrm{Np}(\beta^-, 2.12\,\mathrm{d})\,^{238}\mathrm{Pu}.$$

Other alpha emitters, such as ^{237}Np, ^{235}U, ^{236}U, and ^{236}Np, will also need to be removed in order to obtain a pure ^{236}Pu tracer, because these impurities will interfere with the determination of U and Np isotopes when plutonium, uranium, and neptunium are determined in the same sample. ^{242}Pu is formed in nuclear fuel, and can be separated from the spent fuel. The production of pure ^{242}Pu is not easily accomplished, however, owing to the coexistence of other plutonium isotopes. Most of the ^{242}Pu presently used as tracer was produced in the United States and Russia in the 1950s and 1960s, but supplies are now almost exhausted, and it is uncertain whether any pure ^{242}Pu tracer will be commercially available in the future.

15.16.7
Separation by Solvent Extraction

Solvent extraction is frequently exploited to separate Pu from other actinides and fission products in the reprocessing of spent fuel and treatment of radioactive waste. HNO_3 is the preferred acid for solvent extraction separation of Pu because nitrate forms stable neutral complexes with Pu(IV), Pu(V), and Pu(VI). HCl, and acids such as H_2SO_4, H_3PO_4, and HF with strongly complexing anions, are seldom used in the solvent extraction of Pu because they form strong complexes with it. Tributyl phosphate (TBP) is well known for its use in the separation of Pu in the reprocessing of spent fuel in the PUREX process. In this process, plutonium in solution, after dissolution of the spent fuel in 1–3 M HNO_3, is first adjusted to Pu^{4+} with $Fe(NH_2SO_3)_2$ and NO_2 or with NO_2 alone. Pu(IV) is then extracted with TBP/kerosene as $Pu(NO_3)_4(TBP)_2$ complex. Uranium is also extracted, as $UO_2(NO_3)_2(TBP)_2$, while other actinides, such as Am, Cm, and Np, and fission products are not since they do not form neutral complexes with TBP. The extracted plutonium is then separated from uranium by back-extraction with $Fe(NH_2SO_3)_2$: plutonium is reduced by Fe^{2+} to Pu^{3+}, which does not form a neutral complex with TBP and does not remain in the organic phase. Some laboratories also use this plutonium separation procedure to separate Pu from nuclear waste and environmental samples.

A number of other extraction reagents, such as TIOA (triisooctylamine)/xylene, TTA (thenoyltrifluoroacetone)/benzene, HDEHP (di-2(ethylhexyl)-phosphoric acid) and TOPO (tri-*n*-octylphosphine oxide), have been used to separate and preconcentrate Pu. For instance, Pu has been preconcentrated from large volumes of seawater by solvent extraction with use of TTA/benzene, and a Pu recovery of 96% has been reported. TIOA/xylene has been used to separate Pu in pine needle, litter, and sediment samples: Pu in 8 M HNO_3 medium is twice extracted with TIOA/xylene and then back-extracted with a 0.1 M NH_4I/8.5 M HCl solution (NH_4I reduces Pu^{4+} to Pu^{3+}). In earlier decades, solvent extraction played an important role in the separation of Pu from environmental samples, but it has gradually been replaced by chromatographic methods. One clear advantage of solvent extraction is the availability of numerous extractants, which allow just the right selectivity to be achieved. However, the method is deemed too laborious for routine analysis because several consecutive extractions need to be performed to completely separate the analyte from the bulk solution. Difficulties in phase separation and the mutual solubility of the two phases could result in significant loss of the analyte. With the development of extraction chromatography and ion exchange chromatography, solvent extraction is now seldom used as a routine method for chemical separation of Pu in environmental samples.

15.16.8
Separation of Pu by Anion Exchange Chromatography

Separation of Pu by ion exchange chromatography is based on the formation of stable anionic complexes of Pu(IV) with NO_3^- or Cl^- in high concentrations of HNO_3 or HCl. Figure 15.25 shows a chemical separation procedure for Pu involving anion exchange chromatography. In this procedure, Pu is first adjusted to Pu(IV) by reducing all species of Pu to Pu(III) with a reductant such as sulfite; then Pu(III) is oxidized to Pu(IV) with nitrite. The Pu(IV) in 8 M HNO_3 is loaded into an anion exchange column. Since Am(III) and U(VI) do not form stable anionic complexes with NO_3^- in HNO_3, they pass through the column in the effluent. Minor amounts of Am(III) and U(VI) remaining in the column are removed by washing with 8 M HNO_3. Since almost all transition metals and many other elements, including mercury and lead, do not form anionic complexes with NO_3^-, they also pass through the column in the effluent and 8 M HNO_3 wash. Th(VI) forms anionic complexes with NO_3^- and so is also adsorbed in the column in the HNO_3 medium. It is eluted with 12 M HCl, which converts $Th(NO_3)_6^{2-}$ in the column to Th^{4+}. Finally, Pu remaining in the column is eluted by reducing it to Pu(III) with 2 M HCl/$NH_2OH{\cdot}HCl$ solution. Trivalent plutonium does not form anionic complex with nitrate.

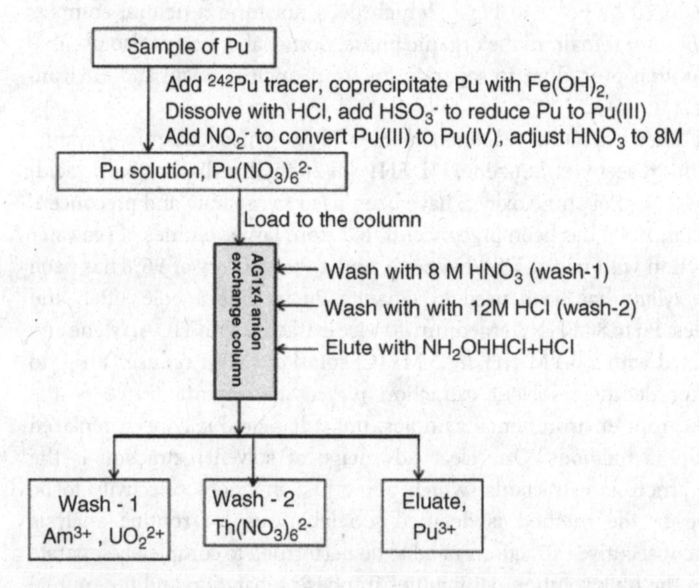

Figure 15.25 Separation of Pu by anion exchange chromatography (Chen, Q.J., Aarkrog, A., Nielsen, S.P., Dahlgaard, H., Lind, B., Kolstad, A.K., and Yu, Y.X. (2001) Procedures for determination of isotopes of Pu, Am Np, U, Th and ^{99}Tc, $^{210}Pb-^{210}Po$ in environmental materials. Risø report, Risø-R-1263).

15.16.9
Separation of Pu by Extraction Chromatography

Plutonium can be separated on Eichrom/Triskem TEVA, TRU, and UTEVA extraction chromatography resins, employed individually or with UTEVA and TRU columns connected in tandem.

Single-column extraction chromatography using TEVA Resin is widely employed for the separation of Pu. This resin separates tetravalent actinides in 1–4 M HNO_3 and HCl media. Since TEVA contains the same functionality as strongly basic anion exchangers (i.e., quaternary ammonium), the procedure is closely similar to anion exchange chromatographic separation. Plutonium in the sample solution is converted to Pu(IV) by reduction and oxidization steps, as described earlier, and the prepared Pu(IV) solution in 1–3 M HNO_3 is loaded onto a TEVA column. U(VI), Am(III), and most transition metals and other elements pass through the column, while tetravalent Th and Pu are retained. After rinsing with 3 M HNO_3, the Th(IV) retained in the column is selectively eluted with 6 M HCl. Finally, plutonium is stripped from the resin with 0.1 M HNO_3–0.1 M HF or 0.1–0.5 M HCl or 0.5 M $NH_2OH{\cdot}HCl$–2 M HCl.

TRU Resin, containing octyl(phenyl)-*N*,*N*-diisobutylcarbamoylmethylphosphine oxide dissolved in tri-*n*-butyl phosphate, is also used to separate actinides in HNO_3 and HCl media. Actinides in oxidation states +III, +IV, and +VI are strongly retained in this resin in HNO_3 solution stronger than 0.5 M, while most of the matrix constituents and interfering radionuclides are not. Am(III) and other radionuclides in the trivalent state can be eluted from the column with 4–6 M HCl since they do not form chloride complexes capable of adsorbing on the resin. Tetravalent and hexavalent actinides are strongly retained in the column, also in HCl medium. Pu adsorbed in the column is removed by an on-column reduction of Pu(IV) to Pu(III) and elution with 4–6 M HCl.

UTEVA, containing diamylamylphosphonate as the extractant, takes up actinides in tetra- and hexavalent state, such as Pu(IV), Np(IV), U(VI), and Th(IV) in HNO_3 and at high concentration of HCl. Since trivalent actinides such as Am(III) and Pu(III) are not retained, UTEVA can be used to separate Pu(III) (and Am(III)) from Np(IV), U (VI), and Th(IV) after adjustment of plutonium to the trivalent state.

A combination of UTEVA and TRU columns in tandem has been used to improve the separation efficiency for Pu from interfering radionuclides. In this procedure, plutonium is first reduced to Pu(III) with $Fe(NH_2SO_3)_2$ and ascorbic acid, and the sample, prepared in 3 M HNO_3, is passed through a UTEVA column. Pu(III) and Am (III) do not adsorb on UTEVA and pass through the column, while U(VI) and Th(IV) are efficiently taken up by UTEVA and separated from Pu. The effluent containing Pu (III) and Am(III) is then directly loaded to a TRU column connected to the UTEVA column. The adsorption of Pu on the TRU Resin is improved by converting Pu(III) in the column to Pu(IV), with nitrite. The column is then washed with 3 M HNO_3 and 9 M HCl to remove Am and other interfering elements and radionuclides. The Pu retained in the column is finally eluted with 0.1 M $NH_4HC_2O_4$ solution. Figure 15.26

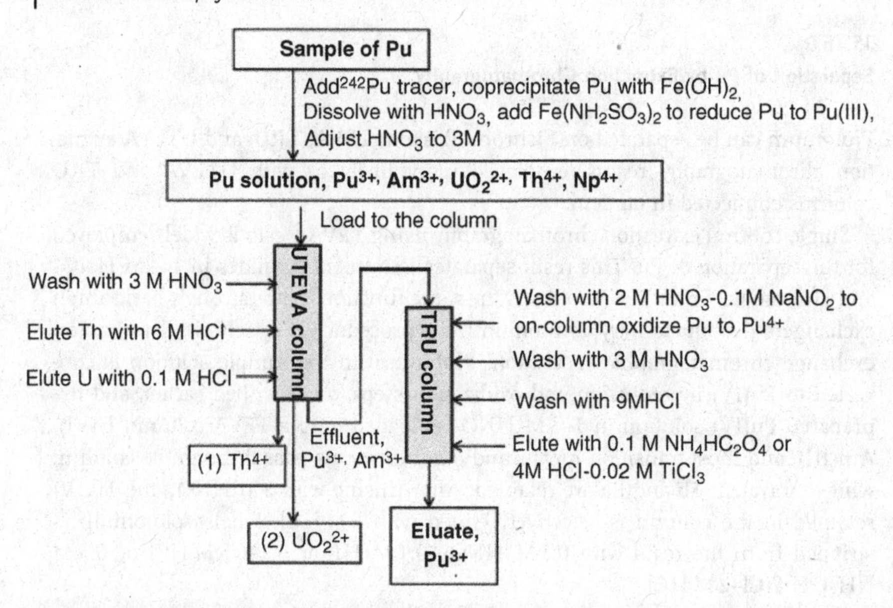

Figure 15.26 Extraction chromatography procedure using UTEVA and TRU for the separation of Pu, U, and Th (Toribio, M., Garcia, J.F., Rauret, G., Pilvio, R., and Bickel, M. (2001) Plutonium determination in mineral soils and sediments by a procedure involving microwave digestion and extraction chromatography, *Anal. Chim. Acta*, **447**, 179).

shows the separation procedure by which Pu, as well as U, Th, and Am, can be sequentially separated.

The extraction chromatographic column is normally much smaller (2 mL) than an anion exchange column (10 mL), which means that the volume of the final eluate is small (5–10 mL). This shortens the separation time significantly. In addition, the decontamination factor of extraction chromatography for uranium (10^{-4}–10^{-5}) is much higher than that of anion exchange chromatography (10^{-3}), which is an important consideration when Pu isotopes are measured by ICP-MS.

15.16.10 Separation of Pu from Large Volumes of Water

The very low concentrations of Pu in environmental samples make it necessary to preconcentrate the Pu from large volumes of seawater, as much as 100–1000 L. Preconcentration is normally carried out by $Fe(OH)_3$, $Fe(OH)_2$, or MnO_2 coprecipitation or by absorption onto MnO_2-impregnated fiber filters. Figure 15.27 shows a chemical procedure to separate Pu from a large volume of water. In this procedure, the water is first filtered, HCl is added to adjust the pH to 2, and ^{242}Pu tracer is added for chemical yield monitoring. $FeSO_4$ is added as a coprecipitation carrier, and then $NaHSO_3$ is added to reduce Pu to Pu^{3+} and Np to Np^{4+}, while Am, Th, and U remain as Am^{3+}, Th^{4+}, and UO_2^{2+}. Ammonium hydroxide is

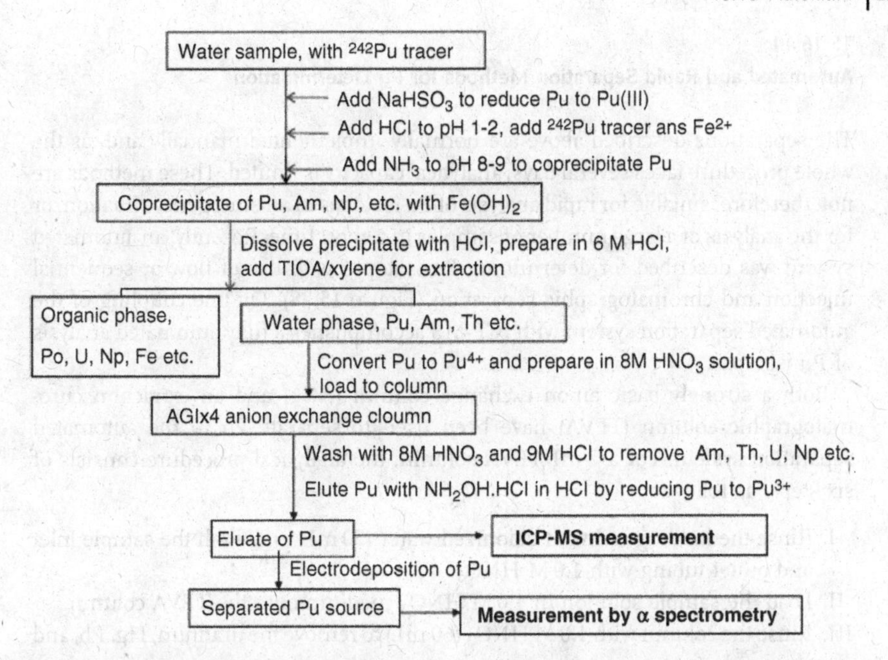

Figure 15.27 A procedure to separate Pu from large volume of seawater. (Chen, Q.J., Aarkrog, A., Nielsen, S.P., Dahlgaard, H., Lind, B., Kolstad, A.K., and Yu, Y.X. (2001) Procedures for determination of isotopes of Pu, Am, Np, U, Th, and ^{99}Tc. ^{210}Pb–^{210}Po in environmental materials. Risø report, Risø-R-1263).

added to adjust the pH to 8–9, which results in the coprecipitation of Pu, Np, Am, Th, and U with $Fe(OH)_2$. After the precipitate has settled on the bottom of the container, the supernatant is sucked out, and the precipitate slurry is transferred to a smaller container (5 L). Once again the precipitate is allowed to settle, the supernatant is removed, and the precipitate is finally completely separated from the supernatant by centrifugation. The precipitate is dissolved in HCl and prepared in 6 M HCl, after which triisooctylamine (TIOA)/xylene is added to extract U, Np, Po, and Fe, while Pu^{3+}, Am^{3+}, and Th^{4+}, which do not form stable complexes in HCl, remain in the aqueous phase. A small amount of $FeCl_3$ is added to the aqueous phase, and NaOH is added to adjust the pH to 8–9 to separate Pu by coprecipitation with $Fe(OH)_2$ (since $NaHSO_3$ remains in the solution, Fe^{3+} is reduced to Fe^{2+}). The precipitate is dissolved in a small amount of HCl, and concentrated HNO_3 is added to adjust the HNO_3 concentration to 8 M. The HNO_2 present in the concentrated HNO_3 oxidizes Pu^{3+} to Pu^{4+}. The prepared solution is then loaded into an anion exchange column, which is washed with 8 M HNO_3 to remove any remaining uranium, and eluted with 9 M HCl to remove Th^{4+}. In a final step, the Pu^{4+} in the column is eluted by reducing it to Pu^{3+} with 0.05 M $NH_2OH{\cdot}HCl$/2 M HCl solution.

15.16.11 Automated and Rapid Separation Methods for Pu Determination

The separations described above are normally implemented manually and, as the whole procedure takes several days, analytical capacity is limited. These methods are not, therefore, suitable for rapid analysis of Pu isotopes in an emergency situation, or for the analysis of a large number of samples in a short time. Recently, an automated system was described for determining Pu isotopes with use of flow or sequential injection and chromatographic separation (Figure 15.28). On-line coupling of the automated separation system with ICP-MS accomplishes a fully automated analysis of Pu isotopes.

Both a strongly basic anion exchange column (AG1) and an extraction chromatographic column (TEVA) have been used to separate Pu in the automated separation system. For a 2-mL TEVA column, the analytical procedure consists of six steps, as follows:

I. Rinse the holding coil with deionized water (50 mL) and wash the sample inlet and outlet tubing with 1.0 M HNO_3.
II. Load the sample solution in 1.0 M HNO_3 medium onto the TEVA column.
III. Rinse the column with 1.0 M HNO_3 (60 mL) to remove the uranium, Hg, Pb, and matrix elements.
IV. Elute thorium with 9.0 M HCl (60 mL).
V. Elute the plutonium with 0.1 M $NH_2OH{\cdot}HCl$ (10 mL) in 2 M HCl medium.
VI. Clean the column with 0.1 M $NH_2OH{\cdot}HCl$/2 M HCl (20 mL) and then 0.1 M HCl (60 mL) before loading the next sample.

After the addition of concentrated nitric acid and heating to decompose the hydroxylamine and eliminate the hydrochloric acid, the Pu eluate (10 mL) is

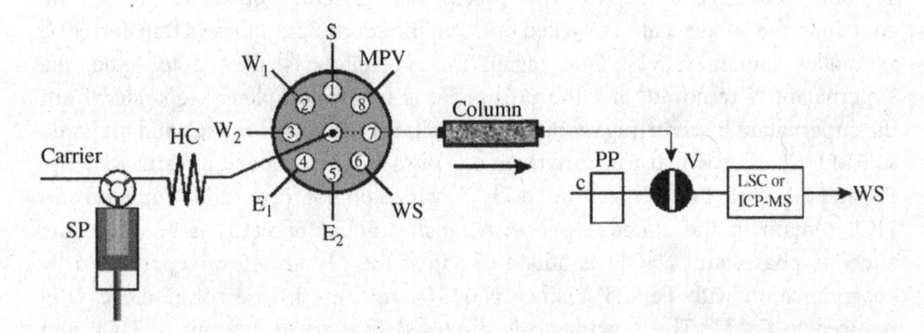

Figure 15.28 Diagram of an automated analysis system consisting of sequential injection and chromatographic separation column and connected to the analytical detection system. SP – syringe pump; HC – holding coil 1; S – sample; E_1, E_2 – eluent 1, eluent 2; W1,W2 –washing solution 1, washing solution 2; WS – waste; MPV – multiport valve; PP – peristaltic pump; C – carrier; V – valve (Qiao, J.X., Hou, X.L., Roos, P., and Miro, M. (2009) Rapid determination of Pu isotopes in environmental samples using sequential injection extraction chromatography and detection by ICP-MS. *Anal. Chem.*, **81**, 8185).

evaporated to dryness on a hot plate. The residue is dissolved in 0.5 M HNO_3 (5 mL) and the resulting solution is analyzed by ICP-MS to measure Pu isotopes and interfering nuclides. With use of this system, the separation time can be reduced from a few days to less than 1.5 h. Since the system can be operated automatically overnight, and the TEVA resin can be reused at least twelve times, the analytical capacity is significantly improved. Note, however, that the preconcentration and sample pretreatment steps cannot be included in the automated system; before automated separation can be carried out, plutonium has to be manually released from solid samples or preconcentrated from large water volumes, and the valence of Pu has to be adjusted to Pu^{4+}.

15.16.12
Essentials of Plutonium Radiochemistry

- Its use as a nuclear weapons material and a fuel in nuclear power reactors makes plutonium a key radioactive element in the nuclear industry as well as in environmental radioactivity and waste treatment and disposal. The most important plutonium isotopes are highly radiotoxic and have long half-lives.
- Plutonium has a number of isotopes, with atomic numbers from 228 to 245. The most important isotopes are ^{238}Pu, ^{239}Pu, ^{240}Pu, ^{241}Pu, and ^{242}Pu. Exceptionally, ^{241}Pu is a beta emitter, but all the others are alpha emitters. Measurements are therefore made by alpha spectrometry and liquid scintillation counting. Mass spectrometric methods, including ICP-MS, AMS, and TIMS, are also used to measure plutonium isotopes, with ICP-MS the most widely applied of the mass spectrometric methods. Alpha spectrometry can be used to measure ^{238}Pu and the total activity of ^{239}Pu and ^{240}Pu, but it cannot separate the activities of ^{239}Pu and ^{240}Pu. Mass spectrometry, in turn, can be used to measure ^{239}Pu and ^{240}Pu individually but cannot be used to measure ^{238}Pu. ^{242}Pu is the most widely used yield tracer in Pu determinations.
- Plutonium can occur in all oxidation states between +III and +VII, although oxidation state +VII is rare. Plutonium in its common oxidation states from +III to +VI occurs in the forms Pu^{3+}, Pu^{4+}, $PuO_2{}^{+}$, and $PuO_2{}^{2+}$. The redox chemistry of plutonium is highly complicated, and the four forms can exist in solution simultaneously. Pu^{3+} and Pu^{4+} are more stable in acid conditions, while $PuO_2{}^{+}$ is stable in neutral media and $PuO_2{}^{2+}$ in alkaline solutions. Pu^{4+} is the most stable and the most studied of the five oxidation states. Both Pu^{4+} and $PuO_2{}^{+}$ disproportionate in acidic solution and form all oxidation states of Pu (Pu^{3+}, Pu^{4+}, $PuO_2{}^{+}$, and $PuO_2{}^{2+}$).
- All oxidation states of Pu undergo hydrolysis. The tendency for hydrolysis follows the general order $Pu^{4+} > PuO_2{}^{2+} \approx Pu^{3+} > PuO_2{}^{+}$, which is the order in which the effective charges of the plutonium species decrease. Pu ions readily form complexes with ligands, the strength of the complexes decreasing in the same order as the tendency to hydrolyze. For a given oxidation state the strength of the plutonium complex with monovalent anionic ligands decreases in the order $F^- > H_2PO_4{}^- > SCN^- > NO_3{}^- > Cl^- > IO_3{}^- > ClO_4{}^-$.

- Coprecipitations with $Fe(OH)_3$, $Fe(OH)_2$, and MnO_2 are used to preconcentrate Pu from large water samples, while coprecipitation of Pu with LnF_3 is used to prepare alpha counting sources and to separate plutonium in different oxidation states.
- Pu is separated from matrix elements and interfering radionuclides by solvent extraction, ion exchange, and extraction chromatography. Of the three techniques, solvent extraction with TBP is successfully used to separate Pu in the reprocessing of spent nuclear fuel, while many other extractants, such as TIOA, are used to separate Pu in radiochemical analyses.
- Anion exchange chromatography is widely applied in the separation of plutonium. Pu is adjusted to Pu^{4+} and loaded into a column, after which 8 M HNO_3 and 12 M HCl are used to remove the matrix elements and interfering radionuclides. Finally, Pu is eluted by reducing it to Pu^{3+}.
- Extraction chromatography has become a popular technique to separate Pu, for example, by the use of a single TEVA column. With UTEVA and TRU resins combined in tandem mode, Pu and other actinides, including U, Th, Np, and Am, can be separated simultaneously.
- An automated separation system employing flow injection/sequential injection and chromatographic separation considerably improves the analytical capacity for Pu determinations.

15.17 Americium and Curium

15.17.1 Sources of Americium and Curium

Since americium and curium behave chemically in a very similar manner they are dealt with together in this part of the actinide chapter. Am and Cm are transuranium elements and are formed in the same ways as Np and Pu, that is, by successive neutron captures and beta decays from uranium and plutonium in nuclear fuel and in nuclear explosions. Also, their sources are more or less the same as those of Np and Pu. For studying their chemical properties, Am and Cm have also been produced using by nuclear reactions using reactors and accelerators.

Americium and curium both have some twenty isotopes, all radioactive. Only a few of them, however, are long-lived and produced in larger amounts (Table 15.2). Americium has only one important isotope, ^{241}Am, which is primarily formed by the beta decay of ^{241}Pu. This plutonium isotope is fairly short-lived, having a half-life of only 14.4 years. Because of the decay of ^{241}Pu, the amount of ^{241}Am in the environment is increasing and will continue to do so for the next several decades. Presently, the ratio of ^{241}Am to the most prevailing transuranium isotopes $^{239,240}Pu$ in environmental samples dominated by global fallout is around 0.4–0.5. In the nuclear fuel the activity of ^{241}Am will be even higher than that of $^{239,240}Pu$, but it will decay in a few thousand years. Therefore, it is not considered to be as critical a radiation hazard as plutonium in the final disposal of spent nuclear fuel, since the technical barriers and the bedrock will prevent the release of radionuclides for at least thousands of years.

In fresh nuclear fallout, the prevailing curium isotope (which is also the most prevailing transuranium isotope) is ^{242}Cm. Its half-life is only 163 days and it will decay in a few years following deposition. Thus there is no more ^{242}Cm from the nuclear weapons testing and the Chernobyl fallouts left in the environment. A somewhat longer-lived curium isotope in the fallout is ^{244}Cm ($t_{1/2} = 18$ y), but its amounts are very low. These two curium isotopes are not, because of their relatively short half-lives, relevant in the final disposal of spent nuclear fuel, in which the longer-lived curium isotopes ^{245}Cm and ^{246}Cm, having half-lives of several thousands years, are more important. Their activities are, however, only a very small fraction of those of plutonium isotopes ^{239}Pu and ^{240}Pu.

15.17.2
Nuclear Characteristics and Measurement of ^{241}Am, ^{242}Cm, ^{243}Cm, and ^{244}Cm

All the important americium and curium isotopes decay by alpha mode (Table 15.10). Their alpha energies differ enough from each other to be measured simultaneously by alpha spectrometry. Curium isotopes have a number of gamma emissions, but their intensities are too low to be feasibly measured for the activities of these nuclides. ^{241}Am, however, has one intense gamma emission at 59.5 keV, with 36% intensity. This gamma emission can be used to measure ^{241}Am directly from some samples, such as highly contaminated soil, where the shorter-lived radionuclides have decayed and thus do not create too high a Compton background, thus enabling detection of gamma rays in the low energy region. Most typically, americium and curium isotopes are measured by alpha spectrometry. An example of an alpha spectrum is given in Figure 15.29, where both ^{241}Am and ^{244}Cm peaks are seen together with the ^{243}Am isotope which was used as tracer for yield determination. ^{243}Am has a half-life of 7370 years and emits alpha particles with energies 5.275 MeV (87%) and 5.233 MeV (11%). ^{243}Am is normally produced by neutron activation of ^{242}Pu followed by a beta decay:

Table 15.10 Nuclear characteristics of ^{241}Am, ^{242}Cm, ^{243}Cm, and ^{244}Cm. All decay by alpha mode.

Nuclide	Half-life (y)	Alpha energies (MeV)	Intensities (%)	Decay product
^{241}Am	433	5.486	84	^{237}Np
		5.443	13	
^{242}Cm	0.44	6.113	74	^{238}Pu
		6.070	26	
^{243}Cm	29	5.785	73	^{239}Pu
		5.742	12	
^{244}Cm	18	5.805	77	^{240}Pu
		5.764	23	

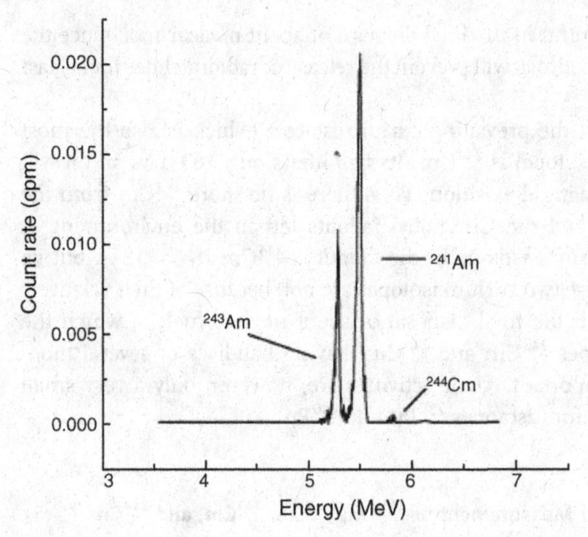

Figure 15.29 Alpha spectrum of ^{241}Am and ^{244}Am. ^{243}Am is a tracer.

$^{242}Pu(n,\gamma)^{243}Pu(t_{1/2} = 5\,h) \rightarrow \beta^- + {}^{243}Am$. ^{241}Am can also be measured by ICP-MS, but great care should be taken to remove interfering elements and isotopes, especially the isobaric ^{241}Pu. The detection limit of ICP-MS for ^{241}Am under the most favorable conditions is comparable with that of alpha spectrometry. The advantage of ICP-MS measurement is a shorter measurement time, but alpha spectrometry still gives more reliable results. ICP-MS measurement of ^{244}Cm results in a higher detection than alpha spectrometry and is thus not feasible.

15.17.3
Chemistry of Americium and Curium

Americium and curium behave chemically very similarly to each other. Their electron configurations are $[Rn]7s^2 5f^7$ and $[Rn]7s^2 5f^7 6d^1$, respectively. Their most typical oxidation state is +III, as it is for other actinides heavier than plutonium except for nobelium. Thus, americium and curium behave much like lanthanides. The closest lanthanide analog for americium is europium, and that for curium is gadolinium. Americium can occur in several oxidation states from +II to +VII, but only oxidation state +III is relevant in ordinary redox conditions prevailing in natural systems (Figure 15.30). Oxidation state +IV in solution is possible only in highly oxidizing conditions in alkaline media. Therefore, only oxidation state +III is considered here. For curium, the +III oxidation state is even less stable than it is for americium, and oxidation to Cm(IV) takes place only in the most oxidizing conditions. Evidence for curium existing at oxidation states higher than +IV is very rare.

Figure 15.30 also shows the hydrolysis behavior of Am and Cm. Like other trivalent actinides and lanthanides, Am(III) and Cm(III) are not very strongly hydrolyzed; the hydrolysis starts only at pH 5–6, depending on conditions. The monohydroxo complex is stable only in the narrow pH range between 7 and 8, while the dihydroxo

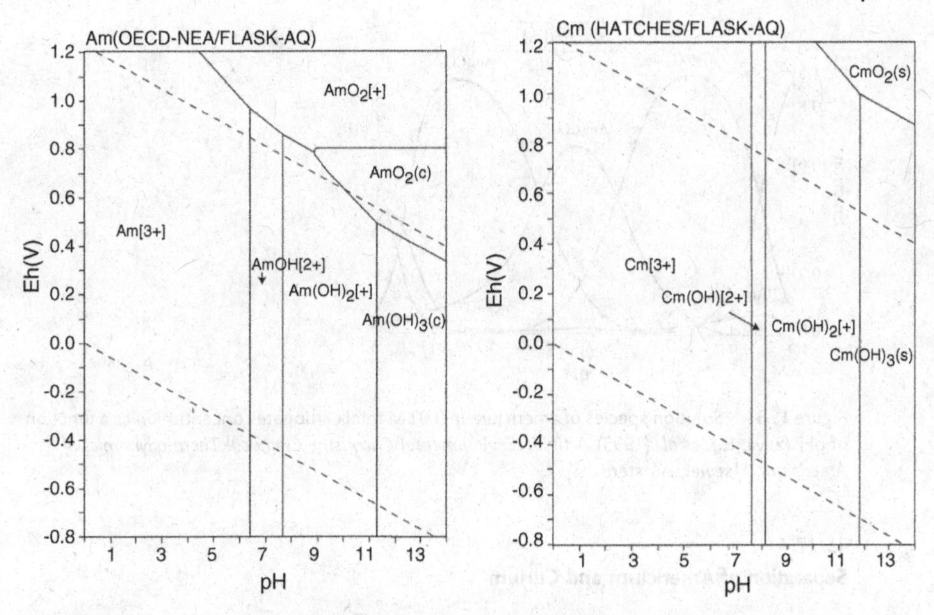

Figure 15.30 Eh – pH diagram of americium and curium in 10^{-10} M solutions (Atlas of Eh – pH diagrams, Geological Survey of Japan Open File Report No. 419, 2005).

complex is stable up to a pH of about 11. Precipitation of $Am(OH)_3$ and $Cm(OH)_3$ takes place at pH values between 11 and 12, provided that the concentrations are high enough to exceed the solubility products ($\log K_{sp}(Am) = -25$ to -26). Amorphous hydroxides formed by precipitation crystallize with aging, which results in decreased solubility. Dissolution of americium hydroxide as $Am(OH)_4^-$ at higher alkalinities and the formation of polynuclear hydroxocomplexes are highly questionable.

Both americium and curium form anhydrous sesquioxides, Am_2O_3 and Cm_2O_3, but the americium sesquioxide is not stable, oxidizing readily in air to form AmO_2. Americium dioxide also forms on heating many Am(III) compounds such as hydroxide, carbonate, and oxalate in air. Other water-insoluble compounds, in addition to hydroxides, are formed with fluoride, oxalate, iodate, and phosphate, while nitrates, halides (other than fluoride), sulfates, and chlorates are readily soluble. In conditions prevailing in sites used for the deep geological disposal of spent nuclear fuel, americium and curium carbonates $(Am/Cm)_2(CO_3)_3$, and especially their ternary compounds $(Am/Cm)Na(CO_3)_2 \cdot xH_2O$ and $(Am/Cm)OHCO_3$, are important solubility-limiting solid phases, that is, the compounds that are formed if the concentrations of americium and curium are high enough.

Stabilities of americium complexes with inorganic ligands decrease in the order:

$$OH^- > CO_3^{2-} > F^- > H_2PO_4^- > SCN^- > NO_3^- > Cl^- > ClO_4^-$$

In ground waters which are rich in carbonate, carbonate complexes of americium are the prevailing species in the pH range from 6 to 11 (Figure 15.31). Americium is also strongly associated with organic matter such as humic acids.

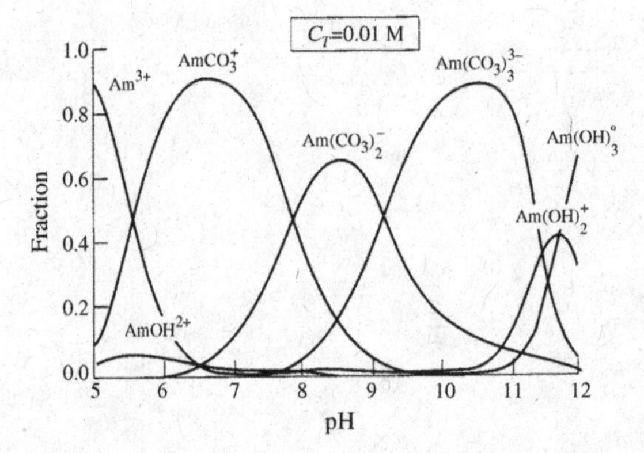

Figure 15.31 Solution species of americium in 0.01 M total carbonate concentration as a function of pH (Silva, R.J. *et al.* (1995) *Actinide Environmental Chemistry, Chemical Thermodynamics of Americium*, Elsevier, Amsterdam).

15.17.4
Separation of Americium and Curium

Since alpha spectrometry is the standard method to measure both Am and Cm, an essential task in their separation is to remove other alpha-emitting radionuclides, that is, other actinides, radium, and polonium. Ion exchange and extraction chromatography as well as solvent extraction are all used for this purpose. Precipitations are not used in primary separations but in preconcentrations and in the preparation of counting sources. Trivalent americium and curium are efficiently separated by coprecipitation with ferric hydroxide and manganese dioxide flocs in preconcentration processes for large volumes of natural waters. Lanthanide fluorides (LaF_3, CeF_3, NdF_3) are used to prepare counting sources for alpha spectrometry.

Separations of americium and curium from other actinides using ion exchange, solvent extraction, and extraction chromatography are carried out by adjusting oxidation states and making use of actinides' different behavior in different oxidation states. Am and Cm occur only in the oxidation state +III, while the other important actinides, Th, U and Np, do not have this oxidation state in conditions where the separations are typically done. Pu has a stable oxidation state +III, but it can easily be oxidized to higher oxidation states to enable the separation from americium and curium. The general idea in Am and Cm separations is that the other actinides are removed from the sample at oxidation state +IV (Th, Np and Pu) and +VI (U), while americium and curium remain in the initial solution. If plutonium is also to be determined it is usually separated with Am and Cm as Pu(III) and at a later stage oxidized to Pu(IV) for the separation from Am and Cm. Separation of radium is based on the fact that radium does not form complexes with nitrate and chloride that can be extracted or separated by ion exchange. Polonium, in turn, is effectively removed with tetra- and hexavalent actinides. An essential task in Am and Cm separations, especially when determining them from soil and sediment, is to remove interfering iron, which is typically accomplished by calcium oxalate coprecipitation in the first step of the chemical separation: trivalent iron does not form an oxalate and remains in the solution, while actinides, including Am and Cm, are coprecipitated as oxalate with calcium.

^{243}Am is used as a yield tracer for americium. Since Am and Cm behave in separations in a more or less identical manner and since their alpha peaks do not overlap, they are separated and measured together, and the yield for curium is typically calculated from the americium tracer. It is also possible to use ^{244}Cm as a tracer in curium analyses when only ^{242}Cm concentrations are being looked for and the amount of ^{244}Cm added is considerably higher than that of the ^{244}Cm initially present in the sample.

15.17.4.1 Separation of Am and Cm by Ion Exchange

Consecutive anion exchange separations in 8 M HNO_3 and 9 M HCl are an effective procedure to separate interfering alpha emitters from Am and Cm. Pu(IV), Th(IV), and polonium are removed from the 8 M HNO_3 solution, while Ra^{2+} and U(VI) follow Am and Cm in the effluent. In the second anion exchange step in 9 M HCl, U (VI) is removed. One more anion exchange step in 1 M HNO_3-CH_3OH is needed to remove lanthanides (described later). In this step Am and Cm are retained on the column when loaded into it, while radium is not absorbed and goes to the effluent.

15.17.4.2 Separation of Am and Cm by Extraction Chromatography

Removal of other actinides from Am and Cm by extraction chromatography is typically accomplished in 3 M HNO_3 by UTEVA Resin, which removes tetravalent actinides (Th, Pu, Np) and hexavalent uranium, while Am and Cm are not retained. Since Am and Cm analysis is most often combined with plutonium analysis, Pu is first reduced to Pu(III) with ferrous sulfamate, for example, which makes Pu follow Am and Cm in the effluent. Separation of plutonium is then carried out with a TRU Resin column. The solution is introduced into the TRU column in 3 M HNO_3 followed by oxidation of plutonium to Pu(IV) in the column by rinsing with $NaNO_2$-bearing nitric acid. This results in a much stronger absorption of plutonium in the column. Am and Cm are now eluted with 4 M HCl, while plutonium remains in the column. Finally plutonium is eluted with 4 M HCl containing $TiCl_3$, which reduces plutonium to Pu(III). An example of such a separation is given in Figure 4.14.

15.17.4.3 Separation of Am and Cm by Solvent Extraction

Solvent extraction is widely used to separate Am and Cm in the reprocessing of spent nuclear fuel, especially to separate Am, Cm, and other actinides from short-lived fission products. In radiochemical analyses, solvent extraction is not often used as the only method to separate Am and Cm from interferences, but is used in combination with chromatographic methods. Am and Cm can be separated with many extraction agents, such as tri-*n*-octylphosphine oxide (TOPO), mixed trialkylphosphine oxides (TRPO), bis(2-ethylhexyl)phosphoric acid (HDEHP), dihexyl-*N,N*-diethylcarbamoylmethyl phosphonate (DHDECMP), and carbamoylmethylenephosphine oxide (CMPO), which have been used to separate Am and Cm, particularly in waste treatment processes, but also in radiochemical separations. An example of the analytical use of solvent extraction in the determination Pu, Am, and Cm from soil is as follows. Am, Cm, and Pu are leached from the soil with 8 M HNO_3, and Pu(IV) is extracted from this solution with 0.2 M TOPO/cyclohexane. Am and Cm are then extracted with the same agent but at the lower acid concentration of 0.1 M. Am and Cm are back-extracted with 2 M HNO_3 and further purified by extraction chromatography using TEVA and TRU columns (Pimpl, M. and Higgy, R.H. (2001) Improvement of Am and Cm determination in soil samples. *J. Radioanal. Nucl. Chem.*, **248**, 537).

15.17.4.4 Separation of Lanthanides from Am and Cm

In the preparation of a counting source for alpha measurement, whether by electrodeposition or by lanthanide fluoride coprecipitation, it is essential to obtain a very thin source to avoid self-absorption, which weakens the counting efficiency and resolution of the spectrum. The separation procedures described above remove most other elements but not efficiently lanthanides, which as trivalent ions behave in a very similar way to Am and Cm both in separation processes and in the preparation of counting sources. Especially when analyzing Am and Cm from soils and sediments, an excess of lanthanides is always present. For the analyses of Am and Cm, several grams of soil or sediment should be taken to have enough Am and Cm to be measured. One gram of soil or sediment contains about 0.2–0.3 mg of lanthanides, which would result in too thick a counting source even if no lanthanide is added to prepare the counting source with fluoride precipitation. Typically 0.05 mg of lanthanide (La, Ce, Nd) is added to the sample to obtain an adequate fluoride precipitate for the counting source. Separation of lanthanides can be accomplished both by anion exchange and by extraction chromatography. Both methods make use of the greater stability of Am and Cm complexes with thiocyanate SCN^- compared to those with lanthanides.

In an anion exchange method, Am, Cm, and lanthanides are bound to the anion exchanger as nitrate complexes from a solution containing 1 M HNO_3 and 93% methanol. The lanthanides are eluted from the column with a solution of 0.1 M HCl/80% MeOH/0.5 M NH_4SCN. Under these conditions, the lanthanides do not form thiocyanate complexes and elute from the column, leaving americium and curium bound to the exchanger as anionic thiocyanate complexes $Am(SCN)_4^-$ and $Cm(SCN)_4^-$. Am and Cm are then eluted from the column with 1.5 M HCl/86% MeOH solution. Am and Cm do not form negatively charged chloride complexes under these conditions and elute from the column as cations.

Separation of lanthanides from Am and Cm by extraction chromatography is carried out by using TEVA resin (see Chapter 4), which is preconditioned with a mixture of 2 M NH_4SCN and 0.1 M formic acid. The sample solution, from which other actinides and polonium have been removed in earlier steps, is loaded into the column in the same solution as that used to precondition the column. Am and Cm are retained as anionic $Am(SCN)_4^-$ and $Cm(SCN)_4^-$ complexes, while lanthanides pass through the column and are found in the effluent and in the rinsing solution of 1 M NH_4SCN and 0.1 M formic acid together with radium that was not removed in earlier steps. Am and Cm are finally eluted from the column with 2 M HCl.

15.17.5
Essentials of Americium and Curium Radiochemistry

- Americium and curium are the transuranium elements next heaviest to plutonium. They are formed by neutron activation and subsequent beta decay processes in nuclear fuel and explosions. Americium has only one important isotope, ^{241}Am ($t_{1/2}$ = 433 y), which is primarily formed as a beta decay product of relatively short-lived plutonium isotope ^{241}Pu ($t_{1/2}$ = 14.4 y). Curium has two important isotopes: short-lived ^{242}Cm ($t_{1/2}$ = 0.44 y), which is the prevailing transuranium isotope in fresh nuclear fallout but decays in a few years following deposition, and somewhat longer-lived ^{244}Cm ($t_{1/2}$ = 18 y), the amounts of which are, however, very small compared to ^{242}Cm and ^{241}Am.
- ^{241}Am as well as ^{242}Cm, ^{243}Cm, and ^{244}Cm decay by alpha mode and thus can be measured by alpha spectrometry, which is the standard method used to measure them. ^{241}Am also emits 59.5 keV gamma rays with 36% intensity, and these can be used to measure ^{241}Am in some special cases. Since Am and Cm behave very similarly to each other, and since their peaks are separated from each other in the alpha spectra, they are typically measured from the same sample, and ^{243}Am is used as a yield determinant tracer for both elements.
- In conditions typically found in the environment and in nuclear waste, the oxidation state of americium and curium is +III, and they exist as Am^{3+} and Cm^{3+} ions, which closely resemble lanthanides in their behavior. Am and Cm do not hydrolyze very strongly, and their compounds are mainly ionic. In natural waters they occur mainly as carbonate complexes and associated with organic matter.
- Anion and extraction chromatography and solvent extraction are used in americium and curium separations. The key task in these separations is to remove other alpha-emitting radionuclides: plutonium, uranium, thorium, radium, and polonium isotopes. In anion exchange chromatography this is accomplished by first retaining Pu(IV), Th(IV), and Po in an anion exchange column in 8 M HNO_3 and then U(VI) in another column in 9 M HCl. Radium is removed in the step where lanthanides are removed. In extraction chromatography, the most interfering radionuclides of Th, Np, Pu, U and Po are removed by a UTEVA column in 3 M HNO_3. A number of solvent extraction agents can also be used in americium and curium separations.
- A special task in americium and curium separations is to remove lanthanides, which behave in a very similar manner in the separation procedures. An excess of lanthanides, especially present in soil and sediment samples, can yield too thick alpha counting sources that deteriorate resolution and counting efficiency. Removal of lanthanides is accomplished by both anion and extraction chromatographies. Am and Cm, together with lanthanides, are retained in an anion exchange column in 1 M HNO_3 and in a TEVA column in 3 M HNO_3. Elution with NH_4SCN removes lanthanides from the column since they do not form complexes with thiocyanate, while americium and curium remain in the column as anionic $Am(SCN)_4^-$ and $Cm(SCN)_4^-$ complexes.

16
Speciation Analysis

The speciation of elements refers to their distribution in different physical and chemical forms in natural or other systems of interest. The speciation of radionuclides in solutions was discussed in Chapter 3, while this chapter focuses on the speciation analysis of radionuclides in the environment: in surface water and groundwater, in the atmosphere, and in soil, sediments, and bedrock. Speciation analysis covers a wide range of analytical techniques by which the physical and chemical form of an element (or radionuclide) is determined. A list of papers which discuss this subject in more detail is given at the end of the chapter. Here, only the most important principles and methods are described.

16.1
Considerations Relevant to Speciation

A number of considerations are relevant to the speciation of nuclides:

- **What is the physical form?**: Radionuclides can be dissolved or be in a solid phase. They can also be suspended in solution as particles or colloids or adsorbed on the surface of particles suspended in solution. The possible physical forms in the atmosphere are aerosol particles and gases.
- **What is the chemical form?**: In solids, radionuclides can be present as a defined solid-phase compound, they can be coprecipitated with a compound of another element, or they can be adsorbed on the surface of some solid substance. In the solution phase, radionuclides can occur in ionic or molecular form or as part of a complex. The oxidation state of a radionuclide is important in determining its chemical state, especially in the case of the actinides.
- **What is the isotopic composition?**: While considerations of physical and chemical form are also relevant to the speciation of elements other than radionuclides, an additional factor for radionuclides is their isotopic composition, which may be informative of the source or intended use of the radionuclide, for example.

In some cases, the physical and chemical forms of a radionuclide in the environment are the same as those of stable isotopes of the element, which means that

Chemistry and Analysis of Radionuclides. Jukka Lehto and Xiaolin Hou

ISBN: 978-3-527-32658-7

information about the speciation of the stable isotopes can be utilized in determining the forms of radionuclides. It needs to be noted, however, that these forms are not typically the same. First, at the moment of generation in a decay event, radionuclides possess recoil energy and are kinetically and electronically highly excited. A radionuclide generated in a gas or in solution phase can reach equilibrium with the environment and finally form the same species as the corresponding stable isotope. Decay that occurs in the solid phase, however, causes chemical bonds to break and lattice errors to arise, and the radionuclides that are generated are not necessarily in the same chemical form as the stable isotopes. Furthermore, many radionuclides in the environment do not even have stable isotopes that could be used as a guide to their various forms. Radionuclides without stable isotopes include all elements heavier than bismuth and the lighter elements technetium and promethium. Further, the source of a radionuclide and the manner in which it was generated may cause it to behave differently from the corresponding stable isotope, or even from the same radionuclide derived from a different source. Thus, radionuclides in fallout particles may be in a very different form from isotopes of the same element already present in the soil. For example, radioactive cesium in nuclear fallout is attached to moderately soluble micrometer-size particles, whereas stable cesium, originally present in the soil, mostly appears in a virtually insoluble form trapped in the crystal lattices of minerals.

16.2 Significance of Speciation

The physical and chemical forms of a radionuclide largely determine how it will behave in the environment and in biological systems. The form of a radionuclide markedly affects its *transport in the environment*. The behavior of uranium in groundwater, for example, depends on its oxidation state: uranium in the oxidation state of four is only very slightly soluble, while uranium in the hexavalent form as the uranyl ion UO_2^{2+} may be highly soluble and mobile. The mobility of uranium in groundwater also depends on whether the uranyl ion is present as a free ion or in a complex form, most commonly as uranyl carbonate $UO_2(CO_3)_3^{4-}$. The free ion easily attaches to colloids in groundwater or to minerals on fracture surfaces, whereas the complex is highly mobile.

The speciation of radionuclides is also of considerable significance for their *transfer in food chains*. For example, the form of a radionuclide has a decisive impact on its bioavailability: radionuclides move into plants through their roots and leaves as ions or molecules. In estimating the bioavailability, it is important to know whether a radionuclide is mostly soluble or what proportion of it is soluble, or whether it is present in the soil entirely as an insoluble compound. Absorption through the roots is not necessarily the predominant mechanism by which radionuclides accumulate in plants – radionuclides may also be present as particles on plant surfaces. Soil particles containing radionuclides that have been resuspended by the wind from the ground and aerosol particles deposited during a fallout event may even be responsible for the greatest part of the radionuclides in plants, and, when digested, they thus contribute substantially to the body burden of humans and animals. Although radioactive particles deposited on plant surfaces add to the radionuclide uptake, they may be less bioavailable in the gastrointestinal tract than radionuclides taken up by plants through their roots. Further, plants can also absorb some radionuclides in the atmosphere as gas, such as gaseous elemental ^{131}I and methyl iodide (^{131}I). In this case, the adsorbed iodine is highly bioavailable, since it forms small molecular iodine compounds in the plants.

The form of radionuclides is also relevant to their *chemical toxicity and radiotoxicity*. The degree of radiotoxicity is affected by the type of radiation released by the radionuclide, its radiation energy, and the absorption of the radionuclide in the body. Insoluble forms are less easily absorbed from the gastrointestinal tract and cause a lower radiation dose. For radionuclides entering through the respiratory tract, the situation may be just the reverse. Radioactive particles that become lodged in the lungs may cause a long-term radiation body burden, whereas soluble forms may exit the body much faster. Within the body, natural uranium is more dangerous for its chemical toxicity than its radiotoxicity, damaging the kidneys. The chemical toxicity of uranium is strongly determined by the form of the uranium and the way in which it enters the body (through food and drink or through the respiratory tract). For example, permanent kidney damage was not firmly established for uranium mine workers who inhaled large amounts of uranium-bearing dust. Drinking water containing elevated concentrations of uranium, on the other hand, may cause damage to the kidneys at a much lower total intake. In particular, groundwater from drilled wells tends to contain high concentrations of uranium. However, since the concentrations of radionuclides in food, drinking water, and air are typically very low, little is known about their chemical forms either in the intake or in the body itself.

16.3
Categorization of Speciation Analyzes

Speciation analysis techniques can be categorized in several ways. First, they can be divided into methods based on the types of samples; speciation analysis can thus be divided into atmospheric, water, and geosphere speciation analysis techniques. In this book, however, the methods are divided into categories based on the type of data obtained by using them. Speciation analysis is later discussed under the following headings:

- fractionation techniques to identify and separate radionuclide-bearing phases;
- spectroscopic methods giving direct data on the chemical speciation of a radionuclide or on the matrix where the radionuclide is;
- methods that do not give direct data about the chemical forms, that is, indirect methods, from which conclusions can be drawn about the chemical form or assumptions can be made;
- wet chemical methods: solvent extraction, ion exchange, and coprecipitation techniques to separate various chemical forms;

- computational methods;
- radioactive particle (hot particle) characterization methods.

Selection of the speciation method depends on many factors: first, the sample type largely determines what method should be used; second, the type of data sought; and third (the most decisive factors), the type of radionuclide and its concentration level. While it is preferable to use methods that give direct information on the chemical form, this is not normally possible since the levels of radionuclides are mostly so low that the applicable methods are not sensitive enough to detect the forms of a radionuclide. For this reason, it is often necessary to use indirect methods even though the information obtained is not as valuable as that obtained by direct methods. Direct methods can also not be used for radionuclides that have no response to the used method, which is often the case in laser-induced fluorescence methods, as in the case of plutonium, for example. The most challenging factor in radionuclide speciation is their low concentrations in the systems studied, as in most cases this greatly limits the choice of speciation analysis technique.

16.4 Fractionation Techniques for Environmental Samples

In chemistry, fractionation refers to separation procedures in which a sample is divided into subsamples that have different compositions. For example, to study the occurrence of a radionuclide in particles of various sizes in water, the water is sequentially filtered through membranes of varying pore sizes. In phase separation, the phases are completely removed from each other; for example, all solid matter is removed from the water by ultrafiltration. Radionuclide fractionation serves several purposes depending on what information is being sought. In size fractionation, as mentioned above, various sizes of radionuclide-bearing particles are separated from water or from air, which are typical tasks in radionuclide speciation in natural waters and the atmosphere. If only information on the occurrence of a radionuclide in various sized particle is needed, its concentration is determined in various fractions. If further information is sought, the fractions are studied with proper analytical methods to obtain more detailed data on the occurrence forms of the radionuclide. Another form of fractionation technique aims at separating radionuclide species that have different charges, cationic, anionic, or neutral; this can be accomplished by ion exchange techniques, for example. All cations can be removed with a strongly acidic cation resin in hydrogen form, and likewise all anions with a strongly basic anion exchange resin in hydroxide form. Further speciation analysis of the fractions can be carried out if needed and if possible.

16.4.1 Particle Fractionation in Water

In natural waters, radionuclides may occur as simple cations or anions, molecules, polymerized hydrolysis species, and as attached to colloids or suspended particles.

Suspended particles are larger particles that sediment out in a reasonable time by gravity. Typically, 0.45 μm is used as the lower size limit for suspended particles. This limit is sometimes wrongly referred to in the literature as being the size which differentiates particles from soluble substances. Particles smaller than 0.45 μm are usually considered to be colloids; these remain in solution, that is, do not sediment out in a reasonable time because of Brownian motion. Colloids are mostly clay and oxide mineral particles and humic substances. Making the distinction between colloids and truly soluble species is not easy. In theory, soluble species, unlike colloids, have a definite composition, and their chemical activity can be defined. In practice, however, the complete separation of colloids is extremely difficult if not impossible in most cases. Typically, the lower experimental size limit of colloids is around one or a few nanometers – the smallest pore size achievable in ultrafiltration or dialysis membranes.

The simplest approach to radionuclide fractionation in particles is to separate all the particles by ultrafiltration and calculate the fraction of a radionuclide in particle form as the difference between its concentration in the initial solution and that in the filtrate. This procedure, however, is not strictly a fractionation technique, but rather phase separation. The water volumes which can be treated with ordinary ultrafiltration systems (pore sizes down to a few nanometers) are fairly small because of filter clogging. Since only small volumes can be filtered, this approach may not give the desired information since the concentrations of radionuclides in natural waters are usually very low and cannot be determined in the filtrate. To increase the filtered water volumes, a cross-flow (also called tangential flow) ultrafiltration system must be used. In cross-flow filtration, the water is pumped through a membrane tube with the water flow parallel to the membrane surface, not against it, which prevents the membrane from clogging. Species smaller than the pore of the membrane penetrate the filter and can be collected in the solution passing the membrane, called permeate. This kind of filter system can also be used in the field, which is advantageous since possible changes in the water composition, such as adsorption on transportation by container surfaces, can be mostly avoided. This technique has been successfully used to separate suspended particles and colloids (at least most of them) from natural water to study the radionuclides associated with them.

Another rather simple approach is to remove suspended particles with a 0.45 μm filter to find out the fraction of radionuclides attached to them. This is especially relevant after fallout situations to discover what fraction of radionuclides is expected to sediment out on lake bottoms, for example, in a reasonable time frame. Filtering through a 0.45 μm filter is much easier than ultrafiltration because of the much larger pore size, which prevents the filter from clogging so that much larger water volumes can be treated. During the filtration, however, the particles accumulating on the filter surface will reduce the effective pore size of the filtration system, which will still result in the filter clogging; further, the data on particle sizes is distorted, since particles smaller than 0.45 μm are increasingly retained on the filter during filtration.

To obtain data on the occurrence of radionuclides in various sized particles, successive filtrations need to be carried out; however, there are two major problems: first, as mentioned, the throughput of the membranes with the smallest pore size is

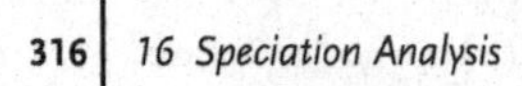

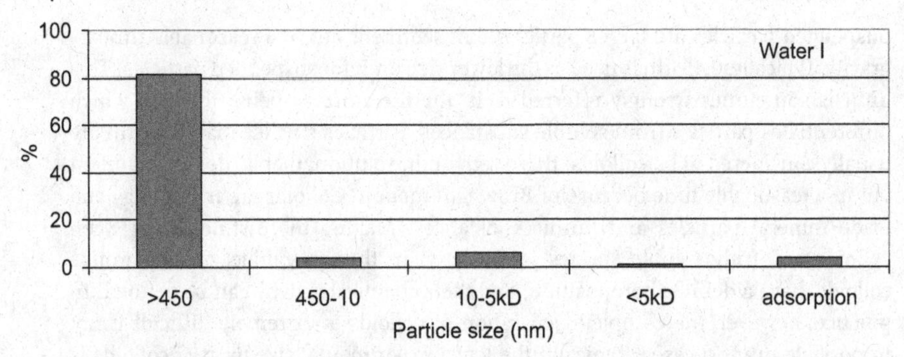

Figure 16.1 The occurrence of ^{210}Pb in particles of various sizes in a groundwater. The fraction below 5 kD is assumed to be soluble. Adsorption refers to the fractions adsorbed on the surfaces of the filtration system (Lehto, J. *et al.* (1999) Soluble and Particle-Bound ^{210}Po and ^{210}Pb in Groundwaters. *Radiochim. Acta.*, **85**, 149).

low and limits the activities that can be measured; second, some radionuclides may essentially be adsorbed on the surfaces of the filtration equipment. Figure 16.1 gives, as an example, distribution of ^{210}Pb in various sized particles in a sample of groundwater. In this water, ^{210}Pb mostly occurred in suspended particles with a diameter larger than 0.45 μm.

Instead of filtration, centrifugation can be used to remove particles from water; however, this is not applicable in the case of highly accurate fractionation, since the particle sizes removed from the solution cannot be controlled with the same accuracy as in filtration. Ultracentrifugation can also be used to separate colloids from water, but only small water volumes can be treated.

The determination of the particle size distribution in the solution is not part of fractionation, but is an essential part in radioactive particle characterization. This can be done, for example, by using the Dynamic Light Scattering (DLS) technique, in which a suspension of particles is placed in a glass cuvette and exposed to laser light. The scattered light fluctuates at a rate that is dependent on the size of particles; this enables the particle size distribution to be measured during one trial. Data on the particle distribution obtained in this way can be combined with the fractionation of particles by filtrations. Prior to the filtrations, it would be advantageous to get data on the particle sizes by using the DLS technique to plan the filtrations adequately.

16.4.2
Fractionation of Aerosol Particles

Radionuclides in the atmosphere occur either as gases or as aerosol particles. Only in special cases, such as fallouts from explosions or accidents, do radionuclides form particles with high activity contents, often called hot particles. Typically, the radionuclides in the atmosphere are attached to aerosol particles existing in the atmosphere in large concentrations. Smaller aerosol particles in the nanometer size range aggregate rapidly and form larger particles in the size range of 0.1–10 μm, which are not readily deposited onto the ground since their diffusion rate due to Brownian motion is higher than their deposition rate due to gravitation. As discussed in more detail in Chapter 18, aerosol particles can be removed by air filtration. In the simplest mode, all aerosol particles with a diameter larger than 0.1–0.2 μm are removed by an air filtration system. Radionuclides present in the particles are then identified and measured either by gamma spectrometry or by other methods after radiochemical separation; their initial concentrations in the air are then calculated from the filtered air volume. This process does not, however, deal with fractionation, that is, it does not give information on the presence of the radionuclide in various sized aerosol particles. Fractionation is carried out with cascade impactors, which can typically separate ten particle sizes in the approximate range of 0.2–20 μm. A cascade impactor consists of a chamber with successive plates. When larger particles hit the first plate they are removed from the air flow because of the impaction. The velocity of the flow increases from plate to plate, which results in the removal of smaller particles: the higher the air velocity before it meets the plate the smaller are the particles removed from the flow. The particles on the plates are then collected to identify and measure the radionuclides.

16.4.3 Fractionation of Soil and Sediments

Fractionation of soil and sediments can be done, as in the case of colloids and aerosol particles, to obtain data on the association of radionuclides with various sized grains. It is important to determine whether radionuclides are adsorbed on the grain surfaces only or if they are also absorbed into their inner structures. This is carried out by a correlation analysis between the radionuclide concentration and the grain size (or more specifically the surface area). The soil or sediment must be separated into various grain size fractions, which can be accomplished by sieving. A complicating factor in sieving is the attachment of the smallest grains, that is, clay particles of only a few to few tens of micrometers, onto the surfaces of the larger grains. To overcome this, sieving must be carried out in the presence of water, which removes the smaller grains from the surfaces of larger grains. Sieving is feasible for mineral soil but not for organic soil layers, which do not consist of grains. An example of the association of radionuclides in various sized grains in a sediment is given in Figure 16.2, which shows the correlation between the grain size and their plutonium concentration. Since a linear correlation was found between them, it was concluded that plutonium is adsorbed on the grain surfaces only because the surface area of grains per unit weight increases linearly with the grain size.

16.5 Analysis of Radionuclide and Isotope Compositions

The identification of radionuclides and the determination of isotopic composition is an essential part of radiochemical speciation analyses, although this cannot be

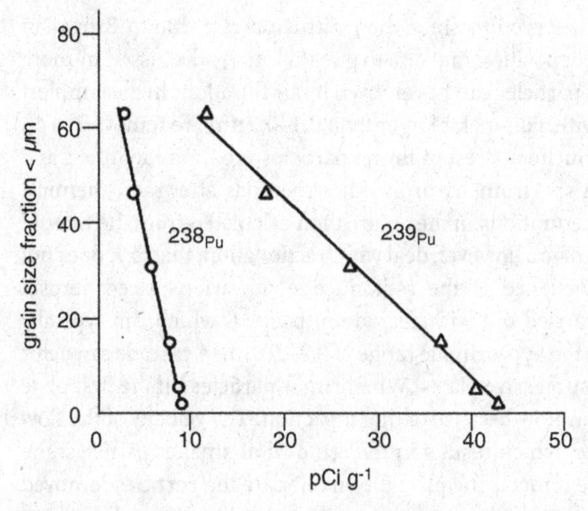

Figure 16.2 Correlation of grain size and the plutonium concentration in a sediment (Aston, S.R. *et al.* (1981) Plutonium occurrence and phase distribution in sediments on the Wyre estuary, Northwest England, *Mar. Pollut. Bull.*, **12**, 308).

considered as being part of chemical speciation. Based on radionuclidic and isotopic compositions, conclusions about the origin, source, or intended use can be drawn. Radionuclide identification and isotopic composition measurements are primarily done by gamma, beta, and alpha spectrometry and by various mass spectrometric techniques, especially ICP-MS. In normal situations, there is already much background information available on the radionuclides present in the sample; for this reason, a comprehensive analysis of the radionuclidic and isotopic composition is not required but is focused on a specific problem. In emergency situations, such as fresh fallout or a nuclear accident, however, complete characterization is needed. Typically, such a complete identification begins with gamma spectrometric and ICP-MS measurements of the sample, and, based on the results, further analyses are carried out. Since all these methods are described in other chapters, they are not discussed further here.

16.6 Spectroscopic Speciation Methods

There are a large number of spectroscopic methods that can be used in the chemical speciation analysis of the elements, such as UV/VIS, NMR, and IR spectroscopies. Many of these can be used in the speciation of radionuclides in macro amounts, which can only be found in the nuclear industry. Further, only a small number of specialized research laboratories can handle such high amounts of radionuclides that conventional spectroscopies can be applied for radionuclide speciation analysis. In most cases, especially in environmental radioactivity, the levels of radionuclides are so low that the detection limits of conventional spectroscopies are too high to even detect the forms of radionuclides which excludes most spectroscopic techniques when environmental samples are studied.

Major developments in two spectroscopic methods of sensitive radionuclide speciation have taken place since the 1990s. The first method is X-ray absorption spectroscopy (XAS), utilizing intense X-ray beams obtained from synchrotrons. This method, which includes XANES and EXAFS is discussed later in the context of radioactive particle characterization. The other method is laser-induced spectroscopy, which is based on the various effects of a short laser pulse on the studied material. Laser-induced (LI) spectroscopies include Thermal Lensing Spectroscopy (LITS); Photoacoustic Spectroscopy (LIPAS); Breakdown Spectroscopy (LIBS); and Fluorescence Spectroscopy (LIF), which is the most utilized and is thus the only method discussed here.

Laser-induced Fluorescence Spectroscopy (LIF) can be used to determine the chemical forms of radionuclides both in solution and on the surfaces of solid materials. In LIF, the sample is exposed to a nanosecond pulse of laser light, which results in the excitation of the sample atoms' electrons to higher energy levels. De-excitation may take place as the emission of light (fluorescence), and the light spectrum is detected with a CCD camera. The wavelength of the laser is tuned to correspond to the maximum cross section of excitation. The light emissions are observed at higher wavelengths, that is, at lower energies than the excitation energy, since part of the deexcitation takes place in a nonradiative way. As the electron levels of an atom are dependent on its chemical environment, LIF emission spectra thus yield information on the chemical species in which the atom is associated.

The advantage of LIF is primarily the very low detection limits compared to those of conventional UV/VIS spectroscopy. The detection limits achievable in speciation are down to about 10 nM, while the corresponding values for UV/VIS spectroscopy are several orders of magnitude higher. If just elemental concentrations are measured with LIF, the detection limits are orders of magnitude lower than 10 nM. A further advantage of LIF is that it can be used in a time-resolved mode (TRLIF), in which the emission spectra are recorded as a function of the delay time after the laser pulse. In the TR mode, the emission spectra, as detected in short nanosecond intervals, and the life-times of various species are determined from the attenuation of the emission peaks. A luminescence decay curve is obtained from the integrated emission spectra, and thus the luminescence lifetimes can be calculated for the various species present in the measured environment. Further, luminescence lifetimes can be correlated to the hydration state of the investigated metal ion. A time-resolved analysis mode is a good tool to resolve the spectra in more detail and thereby obtain more information on the speciation, especially when several species, the spectra of which are overlapping, occur in the sample

Figure 16.3. presents an example of a LIF study on uranyl phosphate complex speciation. Three species can be obtained from the spectrum, $UO_2H_2PO_4^+$, UO_2HPO_4, and $UO_2PO_4^-$, with lifetimes of 11 μs, 6 μs, and 24 μs, respectively.

As mentioned, LIF spectroscopy can also be used to study radionuclide association on solid surfaces. An example of such a study is presented in Figure 16.4, where the

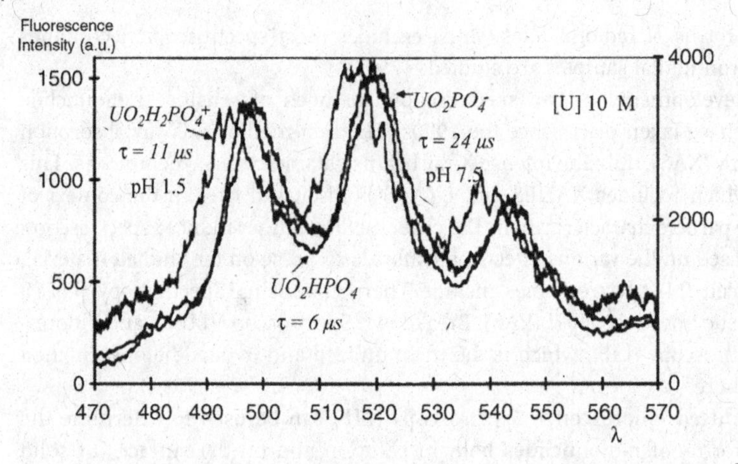

Figure 16.3 Laser-induced fluorescence spectra of uranyl phosphate species in 10^{-7} M uranium solution (Moulin, C. (2003) On the use of time-resolved laser-induced fluorescence (TRLIF) and electrospray mass spectrometry (ES-MS) for speciation studies. *Radiochim. Acta*, **91**, 651).

LIF spectra are shown for Cm in kaolinite mineral suspension at various pH values. At pH 5, only the spectrum of the curium aquo ion can be seen; however, as the pH is increased, the spectrum shifts increasingly to the higher wavelength values because of the sorption of curium on the kaolinite surface in various ways (Figure 16.4a). By deconvolution of the sum spectra, four different surface species of curium can be obtained in addition to the curium aquo ion (Figure 16.4b). These are inner-sphere bound curium species on the kaolinite surface and incorporation of curium in the kaolinite framework.

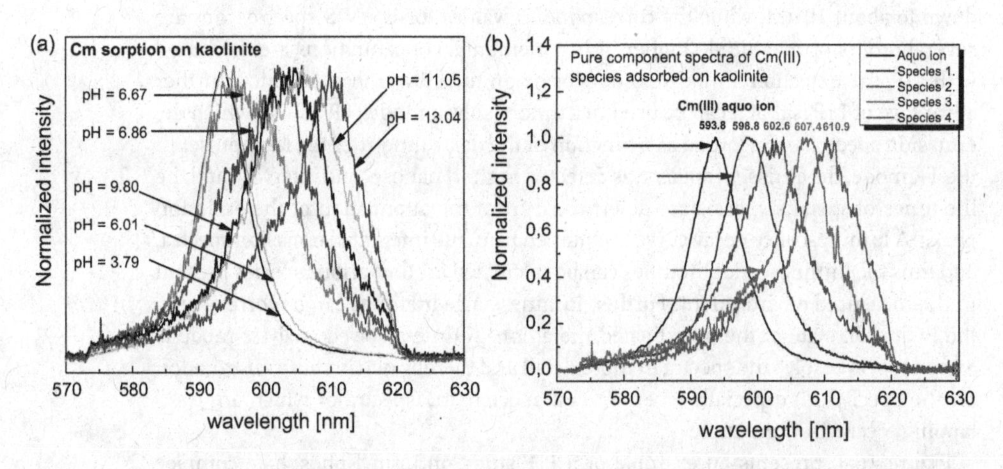

Figure 16.4 LIF spectra of curium in kaolinite mineral suspension at various pH values (left) and individual spectra of curium species in solution and on the kaolinite surface (right) (Huittinen, N., Rabung. Th., Andrieux, P., Lehto, J., and Geckeis, H. (in press) A comparative batch sorption and TRLFS study on the sorption of Eu(III) and Cm(III) on synthetic and natural kaolinite. *Radiochim. Acta*).

LIF spectroscopy is mainly used in actinide speciation studies. The shortcoming of LIF in this is that it can only be used for a few actinide species U(VI), Am(III), and Cm(III). The lowest concentration enabling U(VI) and Cm(III) speciation is around 10 nM, while for Am(III) it is a hundred times higher. The other actinide elements have no fluorescence response. Uranium is easily available in the amounts required for LIF studies, and, as its specific radioactivity is so low, there are no problems from a radiation protection point of view. This is not, however, the same in the case of americium and curium; these are not as easily available, and their specific activities are very much higher than that of uranium. Therefore, stable lanthanides (Eu and Gd) are often used as their analogs in these studies to overcome these problems.

16.7 Wet Chemical Methods

For the analysis of chemical forms or oxidation states of radionuclides in natural waters or aqueous solutions obtained by leaching solid substance, such as soil and sediment, wet chemical methods are normally used, including coprecipitation, solvent extraction, and ion exchange.

16.7.1 Coprecipitation

Since the concentrations of radionuclides in environmental samples are normally very low, coprecipitation is typically used to separate radionuclides, especially those that do not have stable isotopes. The separation of chemical forms of a radionuclide using coprecipitation is based on the difference in solubility of the radionuclide compounds in different chemical forms.

Coprecipitation has been successfully used for the determination of the oxidation states of actinides. In natural water samples or in solutions obtained by leaching solid samples, U, Np, and Pu can occur in the oxidation states III, IV, V, and VI. Uranium typically occurs as U^{4+} and $UO_2{}^{2+}$, neptunium as Np^{4+} and $NpO_2{}^{+}$, and plutonium as Pu^{3+}, Pu^{4+}, $PuO_2{}^{+}$, and $PuO_2{}^{2+}$ in environmental conditions. Lanthanide fluorides (LaF_3, CeF_3, NdF_3) coprecipitate the low (III and IV) but not the high oxidation states and thus can be used in their mutual separations.

For the coprecipitation of low oxidation states of Pu in seawater, the water is normally adjusted to 0.8 M in HNO_3, 0.25 M in H_2SO_4, and 0.7 mM in Nd^{3+} (or La or Ce); HF is then added to 0.25 M to coprecipitate Pu(III) and Pu(IV) with the forming NdF_3. One problem with this procedure is that HF is a reducing agent, which may reduce Pu(VI) and Pu(V) to low oxidation states. To prevent this, a holding oxidant $K_2Cr_2O_7$ is normally added in a concentration of 0.05 mM. Dichromate causes Pu(III) to oxidize to Pu(IV) and Pu(V) to Pu(VI), but not Pu(IV) to higher oxidation states since the latter process is very slow. Thus, the addition of $Cr_2O_7{}^{2-}$ does not affect the distribution of plutonium in lower and higher oxidation states. After the removal of the lower oxidation states by NdF_3, the high oxidation states Pu(V) and Pu(VI) in the supernatant are reduced to Pu(III) using $(NH_4)_2Fe(SO_4)_2$ or $NaHSO_3$, and coprecipitated with NdF_3 to determine the plutonium in higher oxidation states. By this method, the separation efficiency between the low and the high oxidation states of Pu is better than 97% (Nelson, D.M. and Lovett, M.B. (1978) Oxidation state of plutonium in the Irish Sea, *Nature*, **276**, 599).

In addition to actinides, coprecipitation can also be used to separate the forms of iodine radioisotopes (e.g., ^{129}I and ^{131}I). In seawater, iodine is present as iodide (I^-) and iodate (IO_3^-) as well as organic iodine as a minor component. The separation of iodide from iodate and organic iodine can be carried out by AgI + AgCl coprecipitation, which is based on the significant difference in solubility products between AgI (8.5×10^{-17}) and $AgIO_3$ (3.2×10^{-8}). HCl is first added to seawater to adjust the pH to 4–6, and then $AgNO_3$ is added to obtain an Ag/Cl ratio of about 0.1. This results in more than 85% coprecipitation of the iodide with the forming AgCl, while only less than 2% of the iodate goes into the precipitate. The total inorganic iodine, that is, the sum of iodide and iodate, is then determined from another seawater sample by first reducing the iodate to iodide with $NaHSO_3$ at pH 1–2; $AgNO_3$ is then added to coprecipitate the formed iodide with AgCl. The iodate in the seawater is obtained from the difference of these two AgCl coprecipitations, that is, the difference between the total inorganic iodine and the iodide.

16.7.2
Solvent Extraction

Solvent extraction is based on the transfer of neutral metal complexes from the aqueous into the organic phase. Because the stabilities of radionuclide complexes at various oxidation states vary, solvent extraction can be used to separate different chemical forms of radionuclides.

Distinguishing Pu(III) from Pu(IV) by lanthanide fluoride coprecipitation is not possible as described above. This can be done by solvent extraction based on the fact that at pH 0.4 only Pu(IV) is extracted by TTA while at a higher pH of 4.3 both Pu(III) and Pu(IV) are extractable. The pH of the plutonium-bearing solution is first adjusted to 0.4 and Pu(IV) is extracted into TTA-benzene (organic phase); the pH of the aqueous phase is then adjusted to 4.3 and Pu(III) is extracted into the organic phase while Pu(V) + Pu(VI) still remain in the aqueous phase. By this method, Pu(III), Pu(IV) and Pu(V + VI) can be identified.

Solvent extraction can also be used to determine ^{129}I and ^{131}I in water in four different iodine forms: organic iodine, molecular iodine (I_2), iodide (I^-), and iodate (IO_3^-). Organic iodine and molecular iodine are first removed by extraction with benzene or toluene; the molecular iodine in the organic phase is then back-extracted with $NaHSO_3$ solution, which reduces the molecular iodine to the water-soluble iodide. To further separate iodide and iodate remaining in the aqueous phase in the extraction step, the pH of this solution is adjusted to 5–7, and a small amount of NaClO is added to oxidize the iodide to I_2, which is then extracted to the $CHCl_3$ phase. The iodate still remaining in the aqueous phase is reduced to iodide using $NaHSO_3$ at pH 2 and further oxidized to I_2 with $NaNO_2$ for its extraction into $CHCl_3$.

Although solvent extraction has been successfully used to separate different chemical forms of radionuclides, it is not easy to use for environmental samples since the concentrations of radionuclides in natural water are normally very low. To overcome the low detection limits, the analyses would require large sample volumes; however, solvent extraction cannot handle water samples of more than one liter, and this is thus the main shortcoming of this method for the speciation analysis of radionuclides in natural waters.

16.7.3
Ion Exchange Chromatography

The separation of various forms of a radionuclide by ion exchange chromatography is based on their different affinities to the ion exchange resin. For example, iodide has a high affinity and is strongly adsorbed on strongly basic anion exchange resin. The affinity of iodate to anion exchange resin is much lower, and it is not adsorbed on the anion exchange column, especially when the resin is in nitrate form. By this method, ^{129}I in iodide and iodate forms can be separated. Figure 11.8 shows a separation procedure for the chemical speciation analysis of ^{129}I in environmental samples.

A mixed-bed ion exchange chromatography employing high performance liquid chromatography (HPLC) can also be used to separate chemical forms of actinides. In this method, the water sample, without any pretreatment, is loaded into the column and eluted with different solvents. To separate different chemical forms of plutonium, water, oxalic acid, diglycolic acid, and 1 M HNO_3 are used as eluents. Figure 16.5 shows the ion chromatogram of the chemical forms of Pu. The eluted radionuclides can be measured by ICP-MS connected to the HPLC. Since the sample is directly introduced into the column and the eluate is immediately measured, the analytical accuracy can be improved because of the lower risk of change of chemical forms during separation. However, this method is difficult to use for environmental samples since only a small sample (<1 mL), containing only very small amounts of radionuclides, can be loaded into the column.

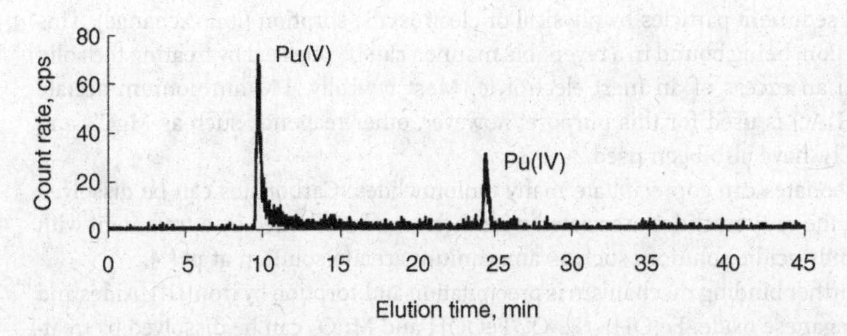

Figure 16.5 Chromatogram of Pu(IV) and Pu(V) in groundwater simulant using HPLC-ICP-MS by employing a mixed-bed ion exchange column (Dionex CS5) (Coates, J.T., Fjeld, R.A., Paulenova, A., and Devol, T. (2001) Evaluation of a rapid technique for measuring actinide oxidation states in a groundwater simulant. *J. Radioanal. Nucl. Chem.*, **248**, 506).

16.8
Sequential Extractions

Sequential extraction, often called selective extraction, is a widely used method to study the speciation of radionuclides in solid matter, especially in soil and sediments. This method is especially used to estimate the mobility and bioavailabity of radionuclides in the environment. The object of sequential extraction is to determine which geochemical phases in the soil or sediment the radionuclides are associated with. Thus, it does not give direct information on radionuclide speciation. In sequential extraction, the solid sample is sequentially equilibrated with various extracting solutions that have increasing power to release the radionuclides from the solid, partly simulating the various environmental conditions to which the soil may be subjected to. Single-step extractions with electrolytes, such as ammonium acetate, are also used, especially to study the loosely bound bioavailable fractions of the radionuclides in soil. After each extraction step, the solid and solution are separated from each other and the radionuclides in the solutions phase are determined to find out what fraction of the radionuclide was released in each step. To obtain a better picture of the solid phases from which the radionuclides were released, it would be desirable to measure also the stable elements from the extraction solution, with ICP-MS being the best option. While sequential extractions are mainly developed to study the occurrence forms of pollutant radionuclides in soil and sediment, they can also give information on the natural radionuclides.

The mechanisms by which the radionuclides are bound to soil and sediments vary greatly among each other.

- The most easily soluble fraction is the dissolved fraction initially present in intergranular water. Prior to sequential extractions, the soil and sediment samples are typically first dried, resulting in the retention of the solutes as salts on the grain surfaces. This fraction can be released with deionised water; however, it is most often very minor and its determination is not usually included in the procedures.
- Another easily removable fraction consists of radionuclides that are bound to soil and sediment particles by physical or electrostatic sorption (ion exchange). This fraction, being bound in a reversible manner, can be released by treating the solid with an excess of an inert electrolyte. Most typically, 1 M ammonium acetate (NH_4Ac) is used for this purpose; however, other reagents, such as $MgCl_2$ and $BaCl_2$, have also been used.
- Carbonates can coprecipitate many radionuclides. Carbonates can be dissolved and the radionuclides associated with them can be released by a treatment with slightly acidic solution, such as ammonium acetate solution at pH 4.
- A further binding mechanism is precipitation and sorption by iron(III) oxides and manganese oxide. $Fe(OH)_3/Fe_2O_3/FeOOH$ and MnO_2 can be dissolved by treating solid samples with a reducing agent, typically $NH_2OH{\cdot}HCl$, which reduces iron and manganese in acidic medium to soluble Fe(II) and Mn(II) ions, respectively; the radionuclides sorbed on them are thus released to the solution.

- Organic matter in soil and sediments bind trace elements and radionuclides in various ways: by ion exchange in carboxyl groups, by complexation, and by reducing some redox-sensitive elements to lower oxidation states, which can be less soluble than the initial oxidation state, as in case of technetium, for example. The ion-exchanged fraction in organic matter is most probably released already by the treatment with an electrolyte; some radionuclides, however, are still efficiently retained. These are released by treatment with an oxidizing agent, typically hydrogen peroxide in acidic medium.
- Another occurrence form in soil and sediments consists of the sparingly soluble oxides, such as uranium and plutonium oxides originating from nuclear fallouts. These can be dissolved with strong mineral acids. Many naturally occurring radioelements, such as uranium and thorium, are present in the lattices of soil and sediment minerals. These can only be dissolved by fusion or by a treatment with concentrated hydrofluoric acid. Some radionuclides, such as ^{137}Cs, often present in mineral lattices, can also only be released in this way.

There are many extraction procedures, depending on the extractants and the number of extraction steps. An example is given in Table 16.1, which shows an optimized sequential extraction procedure for radionuclides in sediments.

Figure 16.6 shows the distribution of plutonium and uranium in a sediment as determined by the method given above. In this sediment, the largest fraction of uranium was in fraction II, referring to the easily acid soluble fraction, probably carbonates. Plutonium was mainly in fraction IV which was intended to dissolve the organic matter.

There are many problems and uncertainties associated with sequential extractions. While the extraction steps are assumed to release radionuclides from certain geochemical phases, the overlapping of displacement and dissolution processes probably takes place in various extraction steps. Another problem is the incomplete attainment of equilibrium in releasing the electrostatically bound fraction (I) which would require several consecutive extractions with the same agent ($MgCl_2$ in the described case). Furthermore, extraction efficiencies vary according to the length of treatment and soil/sediment-to-extractant ratio. A major problem is that as there are

Table 16.1 Optimized sequential extraction procedure for the speciation of radionuclides in sediments (Outola, I. *et al.* (2009) Optimizing standard sequential extraction protocol with lake and ocean sediments, *J. Radioanal. Nucl. Chem.*, **282**, 321).

Fraction	Extractive reagent	Temperature (°C)	Treatment time (h)
I	0.1 M $MgCl_2$	25	1
II	1 M NH_4Ac in 25% HAc	50	2
III	0.1 M $NH_2OH \cdot HCl$ in 25% HAc	70	6
IV	H_2O_2 in 0.05 M HNO_3	70	3
V	4 M HNO_3	90	4

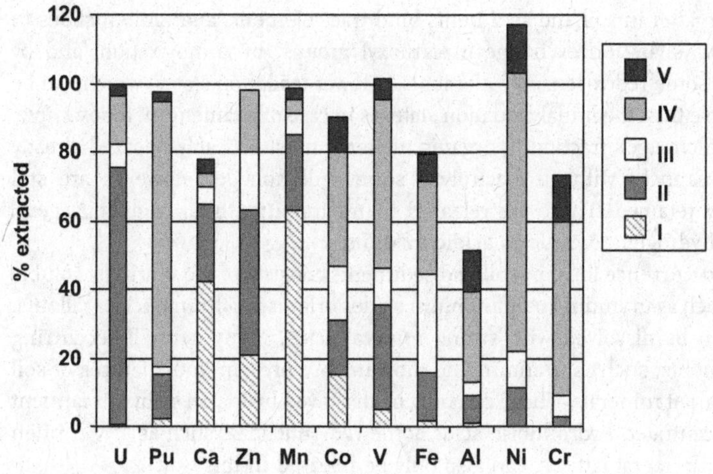

Figure 16.6 Release of uranium, plutonium, and some stable elements from NIST Lake Sediment SRM 4354 as determined with a sequential extraction procedure given in Table 16.1 (Outola, I. *et al.* (2009) Optimizing standard sequential extraction protocol with lake and ocean sediments, *J. Radioanal. Nucl. Chem.*, **282**, 321).

no internationally accepted standard procedures, so that the results of various sequential extractions are difficult to compare. Despite these shortcomings, sequential extraction will remain an important tool in radionuclide speciation, since the low concentrations of radionuclides do not allow direct methods to be used.

16.9 Computational Speciation Methods

Because of the fast development of computers, computational speciation methods have become an essential part of radionuclide speciation studies. However, for reasons stated later, they cannot be the primary speciation method, but are rather complementary techniques used in both planning experiments and evaluating experimental data. Computational methods are based on thermodynamic equilibrium constants or corresponding parameters of the reactions between the components in the studied system. Computational models are well developed for aqueous systems. A number of figures in this book have been produced utilizing such models. Typically, the aqueous models are used to calculate the metal species in the presence of one or several ligands forming complexes or ion associates with the metal. Although the models often calculate the distribution of metal complexes as a function of pH, other parameters can also be used. In addition to the effect of pH on metal complexation, the models also take into account the effects of temperature and redox potential. Figure 16.7 shows an example of such calculations for uranium speciation in a system containing carbonate as a complexing ligand in addition to hydroxide.

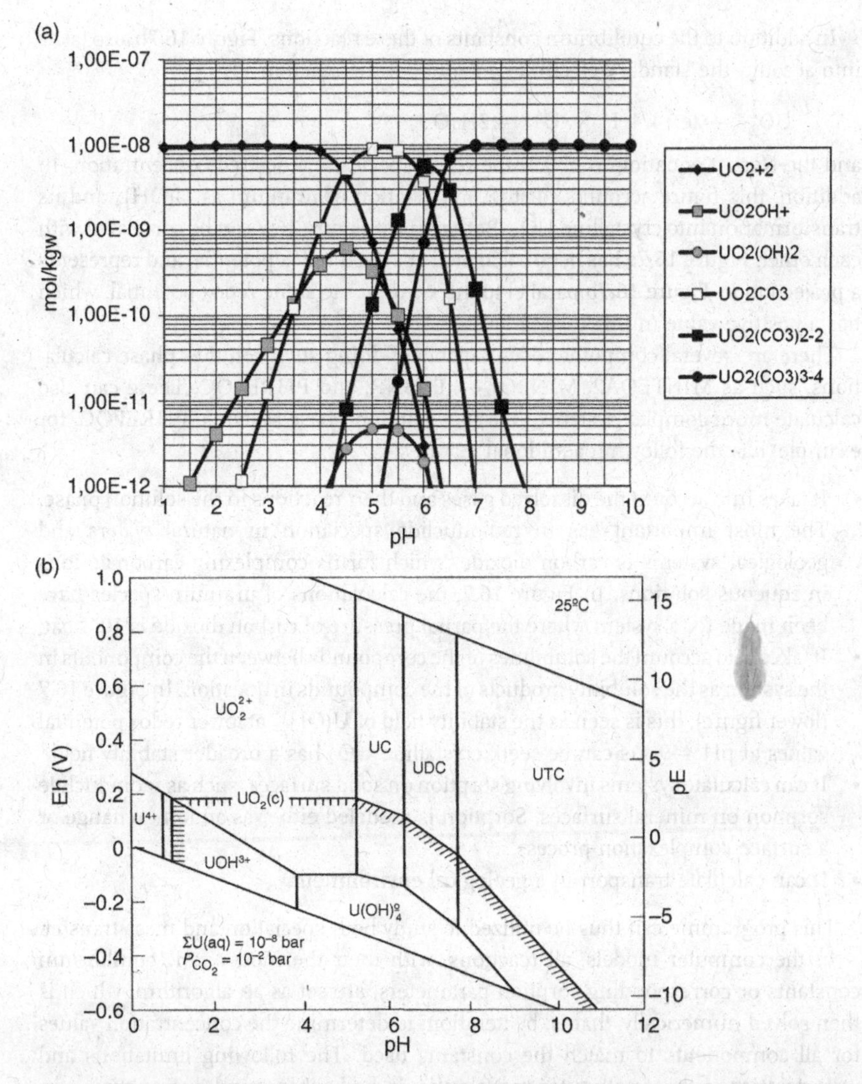

Figure 16.7 (a) Distribution of uranium species in 10^{-8} M uranium solution as a function of pH at a carbon dioxide partial pressure of 10^{-2} bar and (b) the Eh–pH diagram in identical conditions. UC = UO_2CO_3, UDC = $UO_2(CO_3)_2{}^{2-}$, UTC = $UO_2(CO_3)_3{}^{4-}$. (Lower figure from Langmuir, D. (1997) *Aqueous Environmental Geochemistry*, Prentice Hall).

Figure 16.7a shows the uranyl species as a function of pH and is computed based on equilibrium constants of two reactions

$$UO_2^{2+} + xH_2O \leftrightarrows UO_2(OH)_x^{2-x} + xH^+ \quad \text{(hydrolysis)}$$

$$UO_2^{2+} + xCO_3^{2-} \leftrightarrows UO_2(CO_3)_x^{2-2x} \quad \text{(carbonate complexation)}$$

In addition to the equilibrium constants of these reactions, Figure 16.7b also takes into account the standard electrode potential of the reaction

$$UO_2^{2+} + 2e^- + 4H^+ \rightarrow U^{4+} + 2H_2O$$

and the Nernst equation to obtain the redox species at a desired concentration. In addition, this figure accounts for the precipitation of uranium as $U(OH)_4$ and its transformation into crystalline UO_2. Both parts of Figure 16.7 can be correlated with each other. Figure 16.7a has been calculated at a fixed redox potential and represents a projection in Figure 16.7b parallel to the x-axis at the same redox potential, which has a positive value in this case.

There are several computer codes capable of doing such solution phase calculations, such as MINTEQA2, MINEQL+, EQ3NR, and PHREEQC. These can also calculate more complex systems than just solution phase systems. PHREEQC, for example, has the following additional features:

- It takes into account the dissolved gases and their reactions in the solution phase. The most important gas in radionuclide speciation in natural waters and geological systems is carbon dioxide, which forms complexing carbonate ions in aqueous solutions. In Figure 16.7, the calculations of uranium species have been made for a system where the partial pressure of carbon dioxide is 10^{-2} bar.
- It takes into account the solubilities of the compounds between the components in the system as the solubility products of the compounds in question. In Figure 16.7 (lower figure), this is seen as the stability field of $U(OH)_4$ at lower redox potential values at pH 4–9. As can be seen, crystalline UO_2 has a broader stability field.
- It can calculate systems involving sorption on solid surfaces, such as radionuclide sorption on mineral surfaces. Sorption is modeled either as an ion exchange or a surface complexation process.
- It can calculate transport in a geological environment.

This programme can thus be utilized to study both speciation and mass transfer.

In the computer models, all reactions, with their thermodynamic equilibrium constants or corresponding sorption parameters, are set as an algorithm, which is then solved numerically, that is, by iteration, to determine the concentration values for all components to match the constants used. The following limitations and uncertainties of these computer models should be kept in mind during their use. First, all the calculations are based on thermodynamic constants and usually do not take into account the kinetic parameters. Some reactions, although seemingly possible from a thermodynamic point of view, may be kinetically so slow that they do not essentially take place in the time frame considered. A wrong picture of the system could thus be obtained if the kinetic parameters are not taken into account. Kinetic parameters can be included in PHREEQC, for example, with special arrangements. The lack of equilibrium constants for many radionuclide complexes as well as the uncertainty of their values also places limitations on the models. Studies on equilibrium parameters on many radionuclides are still limited, and because of the complexity of the chemistry of rare radionuclides with very low solubility, such as plutonium, their solubility and complex formation constants are not very well defined. Most computer codes have been primarily designed for geochemical speciation and modeling for stable elements, and not for radionuclides. While these programmes include databases of equilibrium constants for most common geochemical reactions, additional information for radionuclides is needed from special databases, such as that of the Nuclear Energy Agency (NEA).

As discussed earlier, the computational methods cannot be the primary ones in radionuclide speciation since the results obtained from them are only indicative. They should, however, be used in both planning experiments and evaluating experimental data. Models are valuable tools in designing experiments to obtain optimum experimental conditions, since they give a good preliminary picture of stability fields of the components under study. Further, the post-experiment comparison of obtained data with the computed data can prevent drawing wrong conclusions. Computer programmes are also usually obtained in user-friendly interactive versions requiring only selection of components and their initial concentrations. This, however, may lead to serious errors if there is a lack of background knowledge of the reactions involved.

16.10 Characterization of Radioactive Particles

Some special events, such as nuclear explosions and accidents, bring larger particles with higher activity concentrations into the environment. These are often called hot particles. Radionuclides in hot particles behave very differently compared with those dispersed in a more homogeneous manner. In the Chernobyl accident, and especially in nuclear explosions, most of the radionuclides were volatilized, and, as the explosion cloud cooled, they condensed and formed particles, either by themselves or especially by association on pre-existing aerosol particles in the air. These particles, with diameters of around one micrometer, were deposited on the ground in a more or less homogeneous way. In nuclear bombs, a small part of the weapons material did not explode but was dispersed as fragments into the surroundings. In the Chernobyl accident, the larger particles, consisting of uranium fuel fragments, were released because of the explosion. The largest hot particles, up to several hundred micrometers in diameter, were deposited in the near surroundings of the explosion site. Smaller hot particles, around 10 micrometers in diameter, were transported by winds – even thousands of kilometres. In this section, the characterization of particles in the size range of ten to a hundred micrometers is the main scope. The motivation to study hot particles is that determination of only bulk concentrations of radionuclides in soil and sediments may, for example, give a wrong, or at least insufficient, picture of the mobility and bioavailability if the particles are not taken into account. Other application fields, in addition to environmental studies, are nuclear forensics and nuclear safeguards, since the data obtained from particles give valuable information about the source or intended use of the nuclear material.

The characterization of hot particles is naturally dependent on what analytical techniques are available. The following presents an idealized procedure, including the most advanced analytical techniques, and comprises the following steps:

- identification and isolation of the particles;
- nondestructive analysis of the particles;
- dissolution and further analysis of the particles.

16.10.1 Identification and Isolation of the Particles

The identification of hot particles in solid samples, typically soil or sediment, is usually accomplished by imaging techniques, film, or digital autoradiography, and solid state nuclear track detection (SSNTD). In autoradiography, a film or a phosphor plate is placed over the sample, and the hot particles are seen as dense dark points on the film. With phosphor plates, the radiation induces excitation of the phosphor molecules. The excitations are relaxed with a laser beam, and, by scanning the plate with the beam the hot particles can be localized by means of the light emissions released from plate in relaxations. The solid state nuclear track technique, which includes alpha and fission track techniques, is similar to autoradiography. A film, typically made of cellulose nitrate, allyl diglycol carbonate, or mica, is placed over the sample. In the alpha track technique, the alpha particles develop damage (tracks), which can be seen with an optical microscope after etching the film with hot alkaline solution. In the fission track analysis, the sample and film are irradiated in a reactor prior to alkaline etching to produce the tracks from fissile isotopes such as ^{235}U.

After identification, the particles are isolated, typically by sample splitting. The soil or sediment sample is divided into two or more parts, and their gamma spectra are measured with a semiconductor detector capable of measuring low-energy gamma rays. If ^{241}Am, the daughter of ^{241}Pu, is detected by its 59.5 keV gamma emission, this reveals that a plutonium-bearing particle is present in the sample aliquot. Additional splitting and gamma measurements are carried out with the part containing the hot particle, and, after some twenty steps, one single particle can be obtained. ^{137}Cs can also be measured to identify the particles in some cases. After isolation, the particle is attached to a conductive adhesive tape or metal surface for further treatment and analysis.

16.10.2 Scanning Electron Microscopic Analysis of the Particles

The second step in particle characterization is to carry out a microscopic study with a scanning electron microscope (SEM), which gives a picture of the particle's size and morphology (Figure 16.8). In addition, a semiquantitative elemental analysis of the particle surface is obtained with energy- or wavelength-dispersive X-ray spectrometers (EDX/WDX) attached to the SEM. Beam electrons ionize the sample atoms, resulting in the emission of element-specific X-rays, which are then detected by

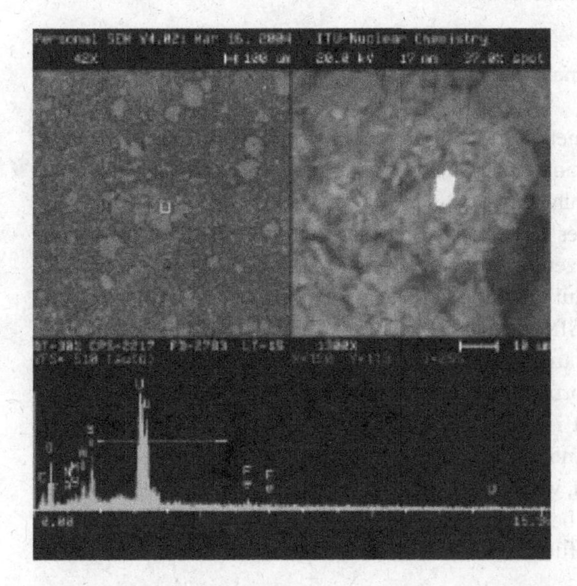

Figure 16.8 A scanning electron microscope image of a radioactive particle. The scale on the top left is about 1 cm and that on the top right about 50 μm. The lower part shows the EDX spectrum, which reveals that the particle mainly consists of uranium (Jernström, J. *et al.* (2004) Non-destructive characterization of low radioactive particles from Irish Sea sediment by micro X-ray synchrotron radiation techniques: micro X-rays fluorescence (μ-XRF) and micro X-ray absorption near edge spectroscopy (μ-XANES), *J. Anal. Atom. Spectrom.*, **19**, 1428).

silicon semiconductor detectors. By scanning the surface with a narrow electron beam, the distribution of elements and their concentration on the particle surface can be measured. Furthermore, heavy elements, such as uranium and plutonium, can be seen as bright areas in the SEM picture of the particle if the back-scattered electron imaging mode is utilized.

16.10.3 Gamma and X-ray Analysis of the Particles

Gamma and X-ray spectrometric analyses help in the preliminary identification of radionuclides occurring in the particle. The presence of gamma-emitting fission products, such as ^{137}Cs, ^{141}Ce, ^{144}Ce, and ^{154}Eu, may give information on the origin and source of the particle. If the plutonium concentration is high enough, the ^{239}Pu/^{240}Pu ratio can also be measured, which largely helps in identifying the source. The measurement of gamma and X-rays from a small solid particle is not very accurate, and thus only indicative information is obtained, but this is nonetheless important when planning further analyses. After dissolving the particle, the gamma and X-ray analysis can be repeated in well-defined standardized conditions to obtain more accurate data on radionuclide concentrations.

16.10.4
Secondary Ion Mass Spectrometry Analysis of Radioactive Particles

Secondary ion mass spectrometry (SIMS) offers a nondestructive method to study the isotopic composition of a radioactive particle. The particle is bombarded (sputtered) with high-energy ions, typically Cs^+ or O_2^+, which results in the release of various atomic and molecular species from the particle surface. Of these, the cations are transferred into a mass analyzer for isotope analysis. If a mass spectrometric analysis is carried out after the dissolution of the particle, a SIMS measurement may not be necessary. One advantage of SIMS is that the ion beam can be focused on a small spot, one micrometer in diameter, and thus several measurements of the same particle can be taken to get isotopic information from various parts of the particle. If a mass spectrometric measurement is taken from the dissolved sample, only the mean isotopic ratio is obtained. Another advantage is that repeated measurements can be taken from the same spot, which enables an isotopic composition analysis to be performed at various depths below the particle surface. This, however, could destroy the particle, preventing any further analysis from being carried out.

16.10.5
Synchrotron-Based X-ray Microanalyses

Synchrotron-based microanalysis is a powerful tool in the characterization of radioactive particles. Synchrotrons are large particle accelerators which produce monochromatic high-intensity X-ray beams. These X-rays have good penetrating power and can thus be used to study the interior parts of particles. The beam can be focused on a very small spot, even smaller than one micrometer, enabling X-ray studies on various parts of the particle to be carried out. There are several ways to use synchrotron-induced X-rays in particle analysis, and the most important of these are described below. There are only a few synchrotrons in the world and they are very expensive facilities, so that the main drawbacks of synchrotron techniques are their limited availability and high costs.

Micro X-ray fluorescence (μ-XRF) analysis gives information on the elemental composition of the whole particle instead of only a small part of the surface (as is the case in SEM-EDX or XRF analyzers using a low-intensity X-ray source). X-rays ionize the analyte atoms, which results in the formation of element-specific X-rays when the electron holes are filled with electrons from the upper electron shells. The emitted X-rays are recorded and analyzed with a silicon semiconductor detector. As in the case of SEM-EDX or conventional XRF, although the elemental analysis is only semiquantitative, the advantage is that information on the elemental composition of the whole particle can be obtained. An elemental composition map of the particle can be obtained by scanning the focused beam over the whole area of the particle.

Micro X-ray diffraction (μ-XRD) analysis gives information on the crystallographical phases in the particle. The advantage of synchrotron-based XRD compared to conventional XRD analysis using a low-intensity X-ray source is that information on the bulk of the material is obtained – not only the surface – and the particle can be scanned by focusing the beam to obtain the variation of phases within the particle.

Micro X-ray tomography gives information on the inner structure (pores, density) of the particle and is based on the attenuation of the X-ray beam in the particle. The particle is imaged from various angles, and a three-dimensional tomographic picture of the particle is obtained by complicated mathematical analysis of the transmitted X-ray spectra.

X-ray absorption near-edge spectroscopy(XANES) and extended X-ray absorption fine structure(EXAFS) provide information on the oxidation states as well as the atomic coordination and bond lengths in the particle. In XANES/EXAFS analysis, the particle is exposed to a highly monochromatized (narrow energy range) X-ray beam, and the intensity of the transmitted X-rays is recorded. The energy of the X-ray beam is gradually increased. When the X-ray energy is high enough to ionize the atoms in the sample, a steep increase and a peak in the absorption spectrum is observed (Figure 16.9). This increasing part of the spectrum is called the 'edge' and the position of the absorption maximum the 'white line.' The XANES analysis uses the information obtained from the position of the white line, while the EXAFS analysis uses the fine structure of the latter part of the spectrum.

The binding energies of the valence electrons are dependent on the oxidation state of an element, increasing with the oxidation state. A shift to higher X-ray energies of the white line is thus observed when the oxidation state increases. This feature is used to determine the oxidation states of radionuclides in radioactive particles, in particular those of uranium and plutonium. The X-ray absorption spectrum of an unknown sample is measured and compared to spectra of standards with a known oxidation state. Figure 16.10 shows the XANES spectra of three unknown uranium particles together with uranium oxide standards with oxidation states of +IV (UO_2), +VI (UO_3), and +IV/+VI (U_3O_8). The identification of the oxidation state in the

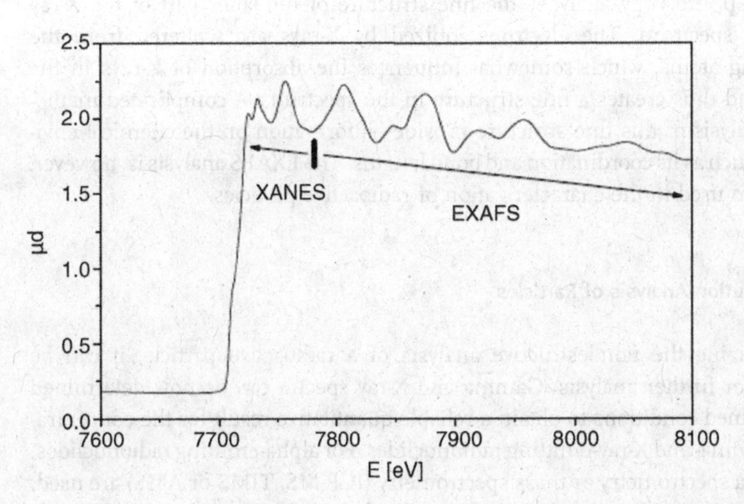

Figure 16.9 X-ray absorption spectrum and the ranges of XANES and EXAFS analyses.

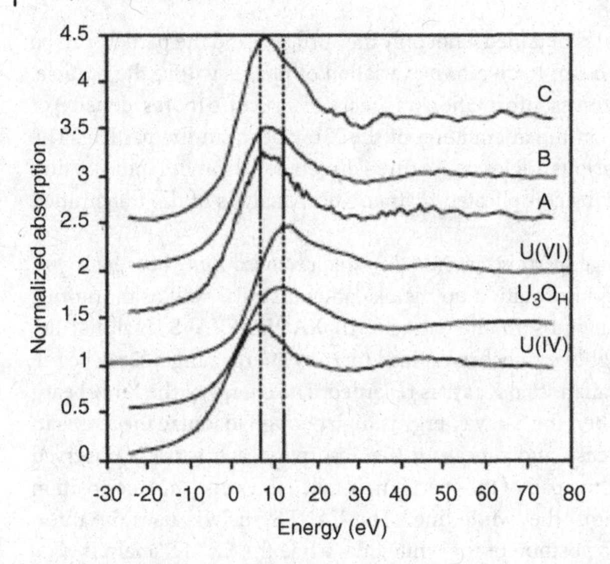

Figure 16.10 X-ray absorption spectra of three uranium particles from the Irish Sea together with the spectra of UO_2, UO_3, and U_3O_8 standards (Jernström, J. *et al.* (2004) Non-destructive characterization of low radioactive particles from Irish Sea sediment by micro X-ray synchrotron radiation techniques: micro X-rays fluorescence (μ-XRF) and micro X-ray absorption near edge spectroscopy (μ-XANES), *J. Anal. Atom. Spectrom.*, **19**, 1428).

unknown sample is based on the comparison of its spectrum to those of the standards. If several oxidation states are present in the unknown sample, a convolution process is needed, that is, combining standard spectra in proportions which yield the unknown spectrum. In this way, the proportions of various oxidation states in the unknown sample are obtained.

EXAFS spectroscopy analyzes the fine structure of the latter part of the X-ray absorption spectrum. The electrons ionized by X-rays are scattered from the neighboring atoms, which somewhat influences the absorption of X-rays in the material and thus creates a fine structure in the spectrum. A complicated mathematical analysis of this fine structure provides information on the chemical environment, such as its coordination and bond lengths. The EXAFS analysis is, however, very seldom used in the characterization of radioactive particles.

16.10.6
Post-Dissolution Analysis of Particles

After finalizing the nondestructive analysis of a radioactive particle, it can be dissolved for further analysis. Gamma and X-ray spectra can be now determined in well-defined conditions to obtain a reliable quantitative result for the concentrations of gamma- and X-ray-emitting radionuclides. For alpha-emitting radionuclides, either alpha spectrometry or mass spectrometry (ICP-MS, TIMS or AMS) are used, usually after chemical separations. To gather full data on all plutonium isotopes, both methods must be used, since mass spectrometry cannot measure ^{238}Pu and alpha spectrometry cannot separate ^{239}Pu and ^{240}Pu from each other. If ^{241}Pu is to be measured, LSC must also be applied.

Further Reading

Choppin, G.R. (2006) Actinide speciation in aquatic systems. *Mar. Chem.*, **99**, 83.

von Gunten, H.R. and Benes, P. (1995) Speciation of radionuclides in the environment. *Radiochim. Acta.*, **69**, 1–29.

Harvey, B.R. (1995) Speciation of radionuclides, in *Chemical Speciation in the Environment*, Blackie, Glasgow, pp. 276–306.

Madic, Ch. and Guillaumont, R.(eds) (2003) Special Issue, Speciation: databases and analytical methods. *Radiochim. Acta.*, **91**.

Moulin, V. and Moulin, C. (2001) Radionuclide speciation in the environment: a review. *Radiochim. Acta.*, **89**, 773–778.

Salbu, B. (2009) Fractionation of radionuclide species in the environment. *J. Environ. Radioact.*, **100**, 283.

Salbu, B., Lind, O.C., and Skipperud, L. (2004) Radionuclide speciation and its relevance in environmental impact assessment. *J. Environ. Radioact.*, **74**, 233.

17
Measurement of Radionuclides by Mass Spectrometry

17.1
Introduction

Radionuclides are most often measured by their characteristic radiation, that is, by radiometric methods, which were discussed in Chapter 1. In these methods, the decay rate of the radionuclide is measured by the number of the particles or photons emitted per unit of time. Depending on the decay mode, α decay, β decay, internal transition, spontaneous fission, and in particular on the types of radiation emitted in their decay processes, the radionuclides are measured by α-spectrometry, GM or proportional counting, liquid scintillation counting, X-ray spectrometry and γ-ray spectrometry.

Inorganic mass spectrometry is extensively used to determine element concentrations in the trace and ultra-trace concentration ranges by measuring the number of atoms. Since mass spectrometry differentiates isotopes of the elements based on their masses, it can be also used for the determination of radionuclides, that is, radioactive isotopes of elements. In mass spectrometric methods, the atoms of the radionuclides are directly measured and the activity (A) of a radionuclide is calculated by the equation representing radioactive decay rate: $A = N \times \lambda$, where A is the number of decays in a second, N the number of atoms of the radionuclide, and λ its decay constant. Instead of decay constant, the half-life ($t_{1/2}$) is usually used in calculations. Since $\lambda = \ln(2)/t_{1/2}$, $A = N \times \ln(2)/t_{1/2}$. For the same number of atoms (or mass) of the measured radionuclide, the longer the half-life of the radionuclide, the lower is its radioactivity and the more sensitive the mass spectrometric methods become compared to radiometric methods. Figure 17.1 plots the logarithm of the inverse of specific activity, that is, the mass corresponding one Bq for each radionuclide (ng/Bq) vs the half-lives of the radionuclides discussed in this book together with the detection limit for ICP mass spectrometry. Mass spectrometry is very suitable for measuring long-lived radionuclides with half-lives longer than 100 years. Various mass spectrometric methods have been used to determine radionuclides: inductively coupled plasma mass spectrometry (ICP-MS), accelerator mass spectrometry (AMS), thermal ionization mass spectrometry (TIMS), resonance ionization mass spectrometry (RIMS), secondary ion mass spectrometry (SIMS), and glow

Chemistry and Analysis of Radionuclides. Jukka Lehto and Xiaolin Hou

ISBN: 978-3-527-32658-7

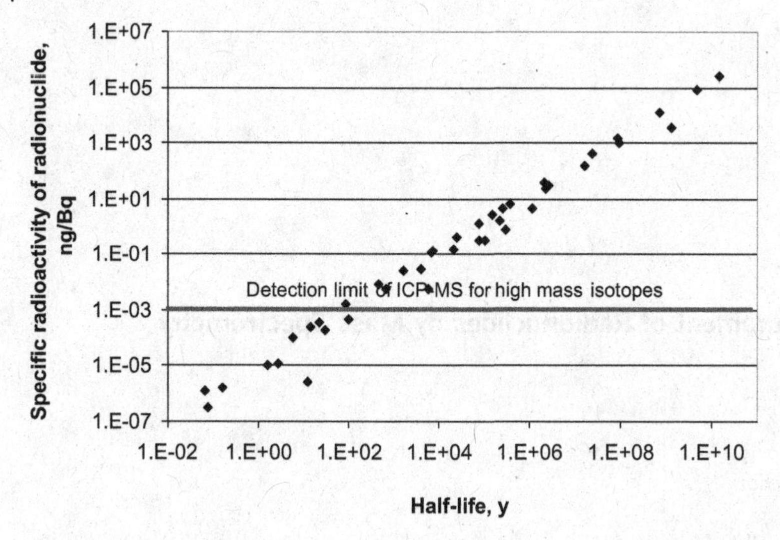

Figure 17.1 Relationship of the inverse of specific activity (i.e., the mass corresponding to one Bq for each radionuclide (ng Bq^{-1})) to the half-life of the corresponding radionuclide. Only radionuclides discussed in this book are included.

discharge mass spectrometry (GDMS). ICP-MS, because of improvements and easy accessibility, is an often used method to determine long-lived radionuclides. AMS, because of its extremely high sensitivity, has been used for the determination of many long-lived radionuclides at ultra-low levels. TIMS and RIMS are normally used to determine isotopic ratios, although these two techniques can also be used to determine radionuclide concentrations in environmental and waste samples. They are, however, not very often applied for this purpose because of their limited accessibility and high cost. SIMS and GDMS are the methods normally used for the surface analysis of solid samples; however, their detection limits are normally not good enough to analyze real environmental samples. Some review articles have addressed the application of mass spectrometric techniques for the determination of radionuclides (Becker, 2002, 2003; Hou and Roos, 2008). This chapter discusses the working principles and major applications of ICP-MS and AMS.

17.2
Inductively Coupled Plasma Mass Spectrometry (ICP-MS)

ICP-MS, developed in the 1980s, combines the easy sample introduction and quick analysis of ICP technology with the accuracy and low detection limits of a mass spectrometer. It has become the most frequently used mass spectrometric technique with which to determine the elements and isotope ratios in the trace and ultratrace concentration range. In recent years, it has also been increasingly used to determine radionuclides in environmental, biological, and waste samples. In contrast to conventional inorganic solid mass spectrometric techniques, ICP-MS allows a simple sample introduction in a normal pressure ion source and an easy quantification procedure using aqueous standard solutions. With a laser ablation introduction system, ICP-MS can also be used for the direct analysis of solid samples.

17.2.1 Components and Operation Principles of ICP-MS Systems

Figure 17.2 shows the scheme and components of the ICP-MS system. In ICP-MS, aqueous samples are normally introduced by way of a nebulizer, which aspirates the sample with argon at high velocity to form a fine mist (aerosol). The aerosol then passes into a spray chamber, where larger droplets are removed to a drain. Typically, only 2% of the original mist passes through the spray chamber with a cross-flow or concentric nebulizer. By using a high-efficiency nebulizer, such as an ultrasonic nebulizer or a desolvating micro-flow nebulizer, the sample introduction efficiency can be significantly increased. The nebulization process is necessary in order to produce droplets small enough to be vaporized in the plasma torch. The sample passes through the nebulizer, and the aerosol moves into the torch body, where the chemical compounds contained in the sample are decomposed into their atomic constituents in an inductively coupled argon plasma at a plasma temperature of approximately 6000–8000 K. The atoms are ionized in the torch to a high degree, >90% for most chemical elements with a low fraction of multiply charged ions (around 1%). The positively charged ions are extracted from the inductively coupled plasma at atmospheric pressure into the high-vacuum mass spectrometer via an interface (sample cone and skimmer cone). Figure 17.3 presents a diagram of a plasma torch and ICP-MS interface. Because atomization/ionization occurs at atmospheric pressure, the interface between the plasma and the mass analyzer compartment become crucial in creating a vacuum environment for the mass

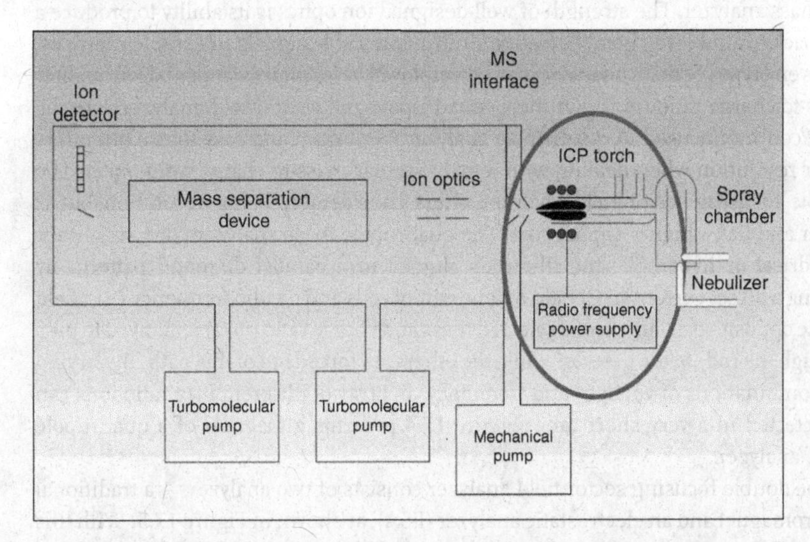

Figure 17.2 Diagram of ICP-MS system showing the components of the system.

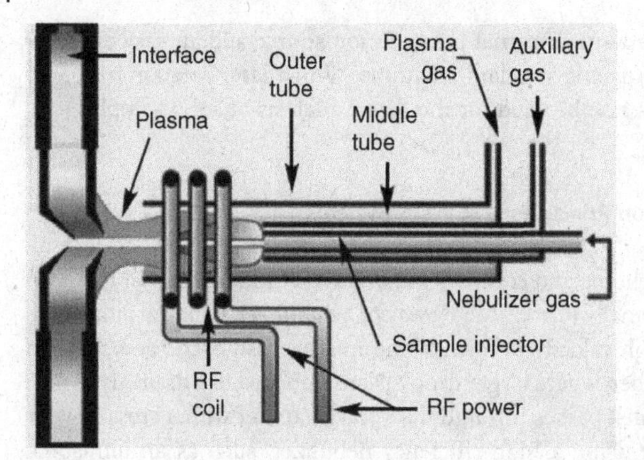

Figure 17.3 Diagram of ICP torch and interface to the mass detector.

spectrometry system. Ions flow through a small orifice, approximately 1 millimeter in diameter, into a pumped vacuum system. Here a supersonic jet forms and the sample ions pass into the mass analyzer system at high speed by expanding in the vacuum system. The entire mass analyzer must be kept in a vacuum so that the ions are free to move without colliding with the air molecules. Before entering the mass analyzer, the ions extracted from the plasma through the interface are focused by ion optics. The ion optics comprise several ion lens components, which electrostatically steer the analyte ions from the interface region into the mass analyzer by optimizing the voltage on every lens of the ion optics to achieve the desired ion specificity. The ion optics is made up of a series of metallic plates, barrels, or cylinders that have a voltage placed on them. The basic function of the ion optics is to stop particulates, neutral species, photons, and electrons, and only allow the analyte ions to be transmitted to the mass analyzer. The strength of well-designed ion optics is its ability to produce a low background level, good detection limits, and stable signals in sample matrices.

Several types of mass analyzers can be employed to separate isotopes based on their mass-to-charge ratio (m/z). Of these, quadrupole and sector field analyzers are the most commonly used. A quadrupole analyzer is compact and easy to use but offers lower resolution when dealing with ions of similar mass-to-charge ratio (m/z). The double focusing sector field analyzer offers considerably better resolution but is larger and has a higher capital cost. The quadrupole mass analyzer consists of four cylindrical or hyperbolic metallic rods aligned in a parallel diamond pattern. By placing a direct current (DC) field on one pair of rods and a radio frequency (RF) field on the opposite pair, an ion of a selected mass and charge ratio (m/z) is allowed to pass through the rods to the detector while the others are forced out of this path. By varying the combinations of voltages and frequency, an array of different m/z ratio ions can be detected in a very short time. Figure 17.4 presents a diagram of a quadrupole mass analyzer.

The double focusing sector field analyzer consists of two analyzers – a traditional electromagnet and an electrostatic analyzer (ESA), as shown in Figure 17.5. With this approach, ions extracted from the plasma through the interface are accelerated in the

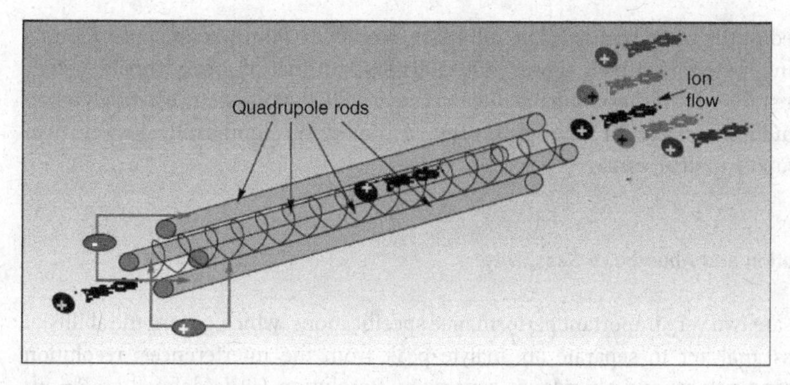

Figure 17.4 Schematic principle of quadrupole mass analyzer.

ion optics region; they are then focused in the magnetic field with diverging angles of motion from the entrance slit according to the energy and mass of the analyte ion. After entering the ESA, the ions are further focused with respect to ion energy onto the detector at the exit slit. If the energy dispersion of the magnet and the ESA are equal in magnitude, but opposite in direction, they will focus both ion angles (first focusing) and ion energy (double focusing). By changing the electric field in the opposite direction during the cycle time of the magnet, the mass is frozen (stopped) for detection when it passes through the exit slit. As soon as the magnetic field strength is passed, the electric field is set to its original value and the next mass is

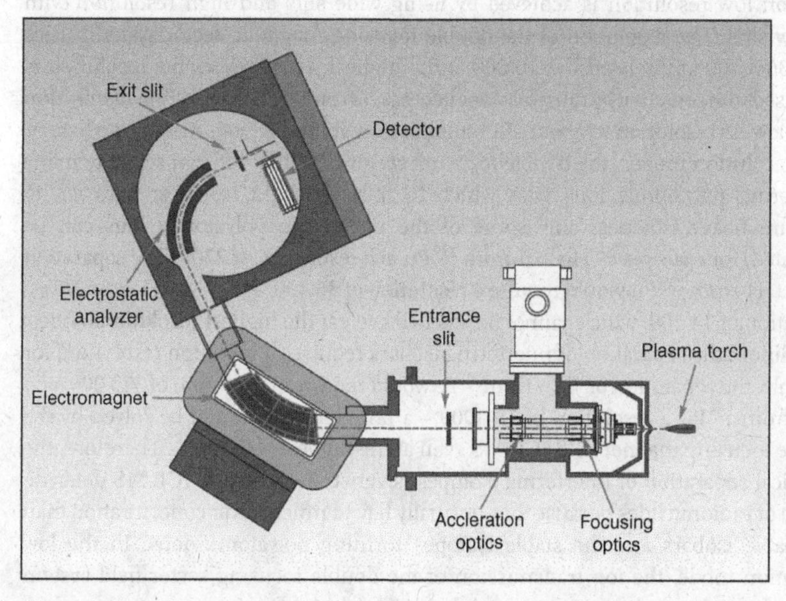

Figure 17.5 Schematic diagram of a reverse Nier-Johnson double focusing magnetic sector mass spectrometer.

stopped in the same manner. The full mass scan (0–250 amu) of a magnet (double focusing system) is much slower (400–500 ms) than that of a quadrupole system (100 ms). The double focusing magnetic sector field ICP-MS system, especially when the multiple resolution setting is used, is therefore significantly slower than quadrupole instruments.

17.2.2
Resolution and Abundance Sensitivity

There are two very important performance specifications, which govern the ability of a mass analyzer to separate an analyte peak from the interferences: resolution (resolving power) and abundance sensitivity. Resolution (R) is defined as $R = m/\Delta m$, where m is the nominal mass and Δm is the mass difference between two resolved peaks. In quadrupole mass spectrometry, the resolution is more commonly defined as the width of a peak at 10% of its height, and it can be adjusted between 0.3 and 3.0 amu. A higher resolution can remove more interferences but reduce the sensitivity. In practice, the resolution is normally set to 0.3–0.5 amu for the determination of radionuclides using quadrupole ICP-MS. Compared with the sector field ICP-MS, the resolution of a quadrupole mass analyzer is very low; therefore, it cannot be used to separate isobaric interference such as ^{93}Mo from ^{93}Nb and ^{93}Zr. The separation ability of quadrupole mass spectrometry for polyatomic ions from an analyte ion is also not sufficient, such as ^{238}U^{1}H from ^{239}Pu. In a double focusing magnetic sector instrument, mass resolution is achieved by using two mechanical slits, one at the entrance to the mass spectrometer and another at the exit to the detector: low resolution is achieved by using wide slits and high resolution with narrow slits. The resolution of the double focusing magnetic sector system varies from 300–400 at the lowest to 10 000 at the highest. However, as the resolution is increased, the sensitivity (transmission) decreases from a relative 100% transmission at the lowest resolution to about 2% transmission at the highest. At intermediate or high resolution modes, the double focusing sector field ICP-MS can separate many interfering polyatomic ions with which to measure light isotopes; however, to measure heavy isotopes, only some of the interfering polyatomic ions can be separated, for example ^{204}Hg^{35}Cl from ^{239}Pu at a resolution of 2200. The separation of ^{238}U^{1}H from ^{239}Pu would require a resolution of 36 400, and ^{98}Mo^{1}H from ^{99}Tc a resolution of 14 200, which cannot be reached even at the highest resolution modes. In addition, the separation of interfering isobars requires a very high resolution; for example, the separation of ^{99}Ru from ^{99}Tc would require a resolution of 313 000, and ^{238}U from ^{238}Pu a resolution of 194 000 – a problem that cannot be solved by the double focusing magnet field ICP-MS even at the highest resolution. Therefore, the chemical separation of interfering isotopes is very critical for the ICP-MS determination of radionuclides because they normally have a much lower concentration than the stable isobars and the stable isotopes forming polyatomic ions. In the low resolution mode, the ion transmission of the double focusing sector field system is much higher, while background levels resulting from extremely low dark current noise are typically very low (0.1–0.2 cps). Thus the detection limit, especially for high-mass isotopes, such as uranium, thorium, and plutonium isotopes, is much better than that in the quadrupole-based system.

The abundance sensitivity (also called tailing) is the signal contribution of the tail of an adjacent peak to the analyte peak, presented as the ratio of count rate at the mass of the analyte to that at the adjacent mass. The abundance sensitivity is always worse to the low mass side than to the high mass side, and is typically 1×10^{-6} at M–1 and 1×10^{-7} at M + 1. This is because the ions entering the quadrupole by the filtering process produce peaks with a pronounced tail or shoulder at the low mass end (M – 1) compared to the high mass end (M + 1). In the determination of radionuclides by ICP-MS, the abundance sensitivity is a very important factor to be considered because of the coexistence of stable isotopes at high concentrations close to masses of the analyte ions. For example, ^{238}U strongly interferes with the determination of ^{237}Np and ^{239}Pu; the contribution of the signal from ^{238}U to the signal at mass of ^{239}Pu is about 10^{-7} only because of the abundance sensitivity of the instrument.

17.2.3 Dynamic Collision/Reaction Cells

To solve the spectral interferences due to isobars and polyatomic ions generated by either argon, solvent, or sample-based ionic species, especially in the quadrupole system, a new approach called dynamic collision/reaction cell technology has recently been developed which virtually stops the formation of many of these harmful species before they enter the mass analyzer. In this technology, a collision cell, consisting of a multipole (quadrupole, hexapole, or octapole) and operating in the RF mode, is positioned between the interface and mass analyzer, normally after the ion optics. A collision/reaction gas, such as H_2, He_2, CO_2, NH_3, or CH_4, is injected into the cell to collide and react with different ion-molecules by focusing with the multipole device. The interferences are removed in two ways: either the polyatomic interfering ions are converted to harmless noninterfering species or the analyte is converted into another ion which is not subject to interference by the coexisting ions. For example, in the determination of $^{56}Fe^{+}$ by ICP-MS, the polyatomic ion $^{40}Ar^{16}O^{+}$ is a major interfering ion because of its m/z being very close to $^{56}Fe^{+}$. On injecting NH_3 into the reaction cell, NH_3 reacts with $^{40}Ar^{16}O^{+}$, and converts it into Ar and O and removed from the analyzer quadrupole, and thus interference of $^{40}Ar^{16}O^{+}$ is removed (Figure 17.6). In addition to removing the interfering polyatomic ions, the dynamic collision/reaction cell can also be used to remove the interfering isobars. This is based on different reaction rates of reactio gas with the analyte and the interfering isobar. For example, ^{129}Xe existing in the Ar carrier gas has a very serious interference to the ICP-MS measurement of ^{129}I. Since ^{129}I and ^{129}Xe have similar masses, a resolution of 618 000 would be required to separate them from each other. By using a mixture of oxygen and helium as collision/reaction gases in a hexapole collision cell, $^{129}Xe^{+}$ ions react with the gases, transfer their charge to the reaction gas, and become neutral ^{129}Xe atoms. These are removed before they enter the mass analyzer, while $^{129}I^{+}$ passes to the mass analyzer. At the same time, the polyatomic ion, $^{127}IH^{2+}$, is also removed in

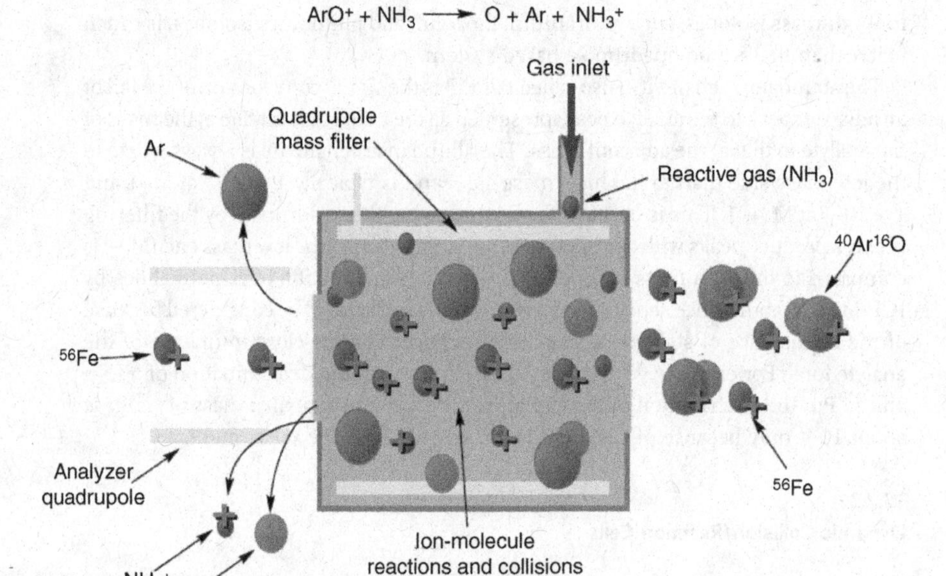

Figure 17.6 Working principle of dynamic collision/reaction cell technology to eliminate interfering polyatomic ions.

the collision/reaction cell, which improves the detection limit of ICP-MS for the measurement of ^{129}I.

17.2.4
Detectors

The number of ions emerging from the mass analyzer is finally counted by a counter. The detector converts the ions into electrical pulses, which are then counted by the integrated measurement circuitry. The magnitude of the electrical pulses corresponds to the number of analyte ions present in the sample. The quantitation of radionuclides of interest in an unknown sample is then carried out by comparing the ion signal with the known calibration or reference standards.

Different ion detectors have been designed and used; of these, electron multipliers for low ion count rates, and Faraday cup for high count rates are the most common. Most ICP-MS systems for ultra-trace analysis use detectors based on the active-film or discrete-dynode electron multiplier. Figure 17.7 illustrates how this detector works. When an ion emerges from the quadrupole and strikes the first dynode, secondary electrons are liberated; these secondary electrons go to the next dynode and generate more electrons. This process is repeated at each dynode, generating a pulse of electrons that is finally captured by the multiplier collector or anode.

In a Faraday cup detector, the ion beam from the mass analyzer is directed into a simple metal electrode or Faraday cup. As this approach can only be used for a high

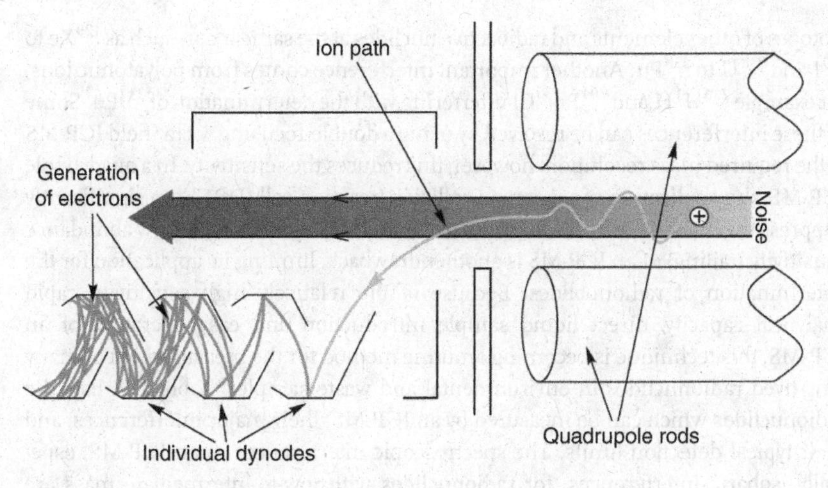

Figure 17.7 Schematic diagram of a discrete dynode electron multiplier.

ion current, the sensitivity of this detector is compromised. For this reason, Faraday cups are normally used in conjunction with a discrete dynode detector to extend the dynamic range of the instrument.

An ICP-MS that uses pulse-counting measurement is capable of measuring over about five orders of linear dynamic range, generally from ppt levels to as much as 100 ppb. The dynamic range of an ICP-MS can be extended to work from sub-part-per-trillion levels (0.1 ppt) to as high as 100 ppm with the aid of some approaches, of which the dual-stage discrete dynode detector is most often used in the present quadrupole ICP-MS. This technology uses measurement circuitry that allows both high and low concentrations to be determined, which is achieved by measuring the ion signal in both analog and pulse signals simultaneously in one scan.

17.2.5 Detection Limits

The detection limit of an ICP-MS varies from 10^{-15} g mL^{-1} (ppq) to 10^{-8} g mL^{-1} (10 ppb), depending on the background, interferences, the measured element, and the sensitivity of the instrument. In the low-resolution mode, the analytical sensitivity of a double-focusing sector field ICP-MS is higher than that in a conventional quadrupole ICP-MS. The precision for isotope ratio measurements in quadrupole ICP-MS varies between 0.1 and 0.5%, while a double focusing sector field ICP-MS with single ion detection allows isotope ratio measurements with a precision of 0.02%. A better precision of isotope ratio measurements, by one order of magnitude, can be achieved by introducing a multi-ions collector device in the sector field ICP-MS.

A major problem of an ICP-MS in determining radionuclides in nuclear waste and environmental samples is the appearance of isobaric interferences of stable isotopes of other elements and radioactive nuclides at the same mass, such as ^{129}Xe to ^{129}I and ^{238}U to ^{238}Pu. Another important interference comes from polyatomic ions, for example $^{238}U^{1}H$ and $^{204}Pb^{35}Cl$ interfering with the determination of ^{239}Pu. Some of these interferences can be resolved by using a double-focusing sector field ICP-MS at the required mass resolution; however, this reduces the sensitivity. In a quadrupole ICP-MS, the application of a dynamic collision/reaction cell (DRC) can significantly suppress the interference of isobaric ions of many radionuclides. The low abundance sensitivity (tailing) of an ICP-MS is another drawback, limiting its application for the determination of radionuclides. Because of the relatively high sensitivity, rapid analytical capacity, direct liquid sample introduction and easy operation of an ICP-MS, this technique is becoming aroutine method for the measurement of many long-lived radionuclides in environmental and waste samples. Table 17.1 lists the radionuclides which can be measured by an ICP-MS, their major interferences, and their typical detection limits. The spectroscopic interferences in an ICP-MS, especially isobaric interferences, for radionuclides with low-to-intermediate mass are much more serious than for the high-mass radionuclides. The background is also higher in the low-to-intermediate mass range. The detection limit of an ICP-MS is thus normally much better for high-mass radionuclides. The ICP-MS measurements for some important radionuclides are given below.

17.2.6 ^{90}Sr Measurement by ICP-MS

Combined with a chemical separation, a detection limit of $0.5\ Bq\ mL^{-1}$ can be obtained for a water sample. The main interference in the measurements of ^{90}Sr by an ICP-MS is the stable isobar ^{90}Zr, which has a natural abundance of 51%. An electrothermal vaporization (ETV) system can be used to discriminate against zirconium by taking advantage of the very high boiling point of Zr compared to Sr. By selecting an appropriate temperature, more than 95% of Zr can be separated from Sr. In addition, by using cold plasma, the Zr signal can also be significantly reduced, since the power required for the atomization of zirconium is significantly higher than that for strontium. For suppression of the isobaric interference from ^{90}Zr, a dynamic reaction cell technique can also be used, employing oxygen as a collision gas in a quadrupole ICP-MS, resulting in a significant removal of ^{90}Zr. In addition, this method also suppresses the interference from other molecular ions such as $^{50}Ti^{40}Ar^{+}$ and $^{50}Cr^{40}Ar^{+}$. In addition to isobaric interference, abundance sensitivity (tailing of stable ^{88}Sr) is another problem in the ICP-MS measurement of ^{90}Sr. The abundance sensitivity of ICP-MS for $^{90}Sr/^{88}Sr$ is normally about 10^{-6}, which means that the detection of $^{90}Sr/^{88}Sr$ ratios less than 10^{-7} is very difficult. The concentration of stable strontium is normally fairly high in environmental samples, for example $7–9\ mg\ L^{-1}$ in seawater. The signal at mass 90 contributed from ^{88}Sr because of the abundance sensitivity (10^{-6}) will be equivalent to $(4.8–6.1) \times 10^{13}$ atoms L^{-1} in seawater, corresponding to a ^{90}Sr concentration of $(3.7–4.7) \times 10^{4}$ $Bq\ L^{-1}$; this number is more than five orders of magnitude higher than the ^{90}Sr concentration in environmental seawater samples ($1–20\ mBq\ L^{-1}$). This makes the

Table 17.1 Radionuclides measured by an ICP-MS, interferences, and detection limits (quadrupole MS).

Radionuclide	Interferences		Detection limit (mBq mL^{-1})
	Isobaric	Polyatomic ions	
^{79}Se,	^{79}Br,	^{78}Se^{1}H, ^{40}Ar^{39}K, ^{16}O^{63}Cu, ^{14}N^{65}Cu, ^{12}C^{65}Zn	5
^{90}Sr,	^{90}Zr	^{89}Y^{1}H, ^{16}O^{74}Ge, ^{12}C^{78}Se, ^{14}N^{76}Se, ^{40}Ar^{50}Ti	400
^{93}Mo	^{93}Nb, ^{93}Zr	^{92}Mo^{1}H, ^{92}Zr^{1}H, ^{14}N^{79}Br, ^{12}C^{81}Br, ^{40}Ar^{53}Cr	300
^{93}Zr	^{93}Nb, ^{93}Mo	^{92}Mo^{1}H, ^{92}Zr^{1}H, ^{14}N^{79}Br, ^{12}C^{81}Br, ^{40}Ar^{53}Cr	1
^{99}Tc	^{99}Ru	^{98}Ru^{1}H, ^{98}Mo^{1}H, ^{18}O^{81}Br, ^{16}O^{83}Kr, ^{40}Ar^{59}Co, ^{36}Ar^{63}Cu, ^{64}Zn^{35}Cl, ^{62}Zn^{37}Cl,	0.1
^{126}Sn	^{126}Xe, ^{126}Te	^{125}Te^{1}H, ^{12}C^{114}Cd, ^{14}N^{112}Cd, ^{40}Ar^{86}Kr, ^{16}O^{110}Pd, ^{16}O^{110}Cd, ^{40}Ar^{86}Sr	0.5
^{129}I	^{129}Xe	^{128}Xe^{1}H, ^{127}I^{1}H$_2$, ^{128}Te^{1}H, ^{40}Ar^{89}Y, ^{14}N^{115}In, ^{16}O^{113}Cd	0.037
^{135}Cs	^{135}Ba	^{134}Ba^{1}H, ^{133}Cs^{1}H^{2}, ^{134}Xe^{1}H, ^{14}N^{121}Sb, ^{12}C^{123}Sb, ^{40}Ar^{95}Mo, ^{16}O^{119}Sn	0.01
^{210}Pb	^{210}Po	^{209}Bi^{1}H, ^{16}O^{194}Pt, ^{14}N^{196}Pt, ^{40}Ar^{170}Er, ^{12}C^{198}Hg, ^{12}C^{198}Pt, ^{40}Ar^{170}Yb, ^{1}H^{209}Bi	27
^{226}Ra		^{40}Ar^{186}W, ^{17}O^{1}H^{209}Bi, ^{40}Ar^{186}Os, ^{18}O^{208}Pb	0.3
^{231}Pa		^{40}Ar^{191}Ir, ^{36}Ar^{195}Pt	1
^{230}Th		^{229}Th^{1}H, ^{40}Ar^{190}Os, ^{36}Ar^{194}Pt	1
^{232}Th		^{231}Pa^{1}H, ^{40}Ar^{192}Os, ^{40}Ar^{192}Pt	0.00002
^{234}U		^{232}Th^{1}H$_2$, ^{40}Ar^{194}Pt	0.11
^{235}U		^{234}U^{1}H, ^{40}Ar^{195}Pt	0.00004
^{238}U		^{237}Np^{1}H, ^{12}C^{226}Ra, ^{40}Ar^{198}Hg, ^{40}Ar^{198}Pt, ^{36}Ar^{202}Hg	0.000006
^{237}Np		^{235}U^{1}H$_2$, ^{40}Ar^{197}Au	0.0001
^{239}Pu		^{238}U^{1}H, ^{40}Ar^{199}Hg, ^{37}Cl^{202}Hg, ^{35}Cl^{204}Hg	0.010
^{240}Pu		^{239}Pu^{1}H, ^{238}U^{1}H$_2$, ^{40}Ar^{200}Hg, ^{36}Ar^{204}Hg, ^{36}Ar^{204}Pb	0.040
^{241}Pu	^{241}Am	^{240}Pu^{1}H, ^{40}Ar^{201}Hg, ^{35}Cl^{206}Pb, ^{15}N^{226}Ra	20
^{241}Am	^{241}Pu	^{240}Pu^{1}H, ^{40}Ar^{201}Hg, ^{35}Cl^{206}Pb, ^{15}N^{226}Ra	0.6

determination of ^{90}Sr in environmental samples by an ICP-MS very difficult. For this reason, ^{90}Sr is mainly measured by a radiometric method.

17.2.7
^{99}Tc Measurement by ICP-MS

With a good chemical separation, the detection limit of an ICP-MS for ^{99}Tc can be as low as 0.1 mBq L^{-1}, which is better than that obtained with radiometric methods such as liquid scintillation counting. The main challenge for an ICP-MS determination of ^{99}Tc is the interferences from isobar and molecular (polyatomic) ions. Isobaric interferences from the stable isotope ^{99}Ru, the abundance sensitivity (tailing) of ^{98}Mo and ^{100}Mo, and the $^{98}MoH^{+}$ molecular ion are the major interferences in the analysis of environmental samples using an ICP-MS. Because of the high concentration of Ru (0.7 ng L^{-1} in seawater) compared with ^{99}Tc in environmental samples and a high isotopic abundance of ^{99}Ru (12.7%), a high decontamination factor (10^6–10^7) for Ru is required. Therefore, a very good chemical separation before it is measured is required. Because of the formation of $^{98}Mo^{1}H$, a high concentration of Mo in environmental samples (10 μg L^{-1}) and the low abundance sensitivity of the ICP-MS (10^{-6}), a high decontamination factor of Mo in the chemical separation ($>10^6$) is also required. Electrothermal vaporization (ETV) as the sample introduction system has been successfully used to remove ruthenium and Mo and to reduce the formation of MoH^{+} ions. However, chemical separation is still the key issue in order to concentrate ^{99}Tc from the sample and separate most of Ru and Mo.

17.2.8
Measurement of Uranium and Thorium Isotopes by ICP-MS

Because of the very long half-lives of ^{232}Th and ^{238}U, an ICP-MS is much more sensitive than radiometric methods for their measurement: detection limits of a few nBq mL^{-1} can easily be reached. These two radionuclides in environmental samples can be measured by using an ICP-MS without any chemical separation. The detection limits for ^{235}U, ^{236}U, ^{234}U, and ^{233}U are also very good: 0.00004 mBq mL^{-1}, 0.001 mBq mL^{-1}, 0.1 mBq mL^{-1} and 0.2 mBq mL^{-1}, respectively. However, because of the low concentrations of these isotopes (except ^{235}U) in environmental samples, a chemical separation and enrichment are normally needed. For ^{230}Th and ^{229}Th, detection limits down to 0.4–4 mBq mL^{-1} by an ICP-MS can be reached, which are comparable to the limits obtained by radiometric methods. The low detection limit of an ICP-MS for these two radionuclides compared to radiometric methods is attributed to the lack of isobaric interferences, very low background level, and high sensitivity for these two elements. In addition, interferences due to polyatomic ions are also very limited because of their high mass. The abundance sensitivity is a critical parameter affecting the analytical accuracy of ^{234}U, ^{236}U, and ^{230}Th results obtained by an ICP-MS because of the much higher abundances of the neighbor isotopes ^{235}U, ^{238}U, and ^{232}Th. The atom ratios of $^{236}U/^{235}U$ are usually at very low levels, 10^{-9}–10^{-6}, and therefore an ordinary ICP-MS is often not sensitive enough to detect

^{236}U because of the poor abundance sensitivity and the formation of interfering uranium hydride (^{235}UH). By using a sector field ICP-MS, an improved abundance sensitivity and minimized hydride formation can be obtained, which means that the ^{236}U/^{238}U atom ratios can be measured down to 10^{-7}.

17.2.9 ^{237}Np Measurement by ICP-MS

Because of its long half-life, that is, low specific radioactivity, an ICP-MS is a more sensitive method to measure ^{237}Np than radiometric methods. The main interference in the ICP-MS measurement of ^{237}Np is the tailing of ^{238}U. Because of the much higher concentration of uranium in almost any sample types compared with Np, a good chemical separation of Np from uranium is required. Several polyatomic ions such as ^{197}Ag^{40}Ar, ^{181}Ta^{40}Ar^{16}O, ^{183}W^{40}Ar^{14}N, combinations of thallium and mercury isotopes with sulfur and chlorine isotopes, and palladium isotopes with xenon, as well as ^{153}Eu^{84}Kr are possible interferences. With a thorough chemical separation, most of these interferences can be eliminated. The detection limit of an ICP-MS for ^{237}Np is much more dependent on procedure blank levels (uranium and polyatomic interferences) than on instrument sensitivity. Detection limits down to 0.1–1 μBq mL^{-1} (or total 0.01–0.1 μBq) of ^{237}Np are easily obtained by present ICP-MS systems with the chemical removal of uranium. As these levels are impossible to detect even with the best radiometric systems available, an ICP-MS is becoming a more common method with which to determine ^{237}Np in environmental samples.

17.2.10 Measurement of Plutonium Isotopes by ICP-MS

The ICP-MS is a sensitive method for the determination of ^{239}Pu and ^{240}Pu in environmental and waste samples. Although it can also be used to measure ^{241}Pu, a beta emitter, it is not sensitive enough for most environmental samples. The main difficulty when measuring low-level samples for Pu by ICP-MS is the interference from uranium; this is because the mass concentration of uranium is usually 10^6–10^9 times higher than that of Pu in environmental samples. The tailing of ^{238}U and the formation of ^{238}U^{1}H$^+$ represent a major disturbance at mass 239 unless the uranium is removed prior to analysis. In addition, polyatomic ions interference is another problem when measureing Pu isotopes. Of major concern are the combinations of Pb, Tl, and Hg isotopes with isotopes of Cl and Ar, since they all may contribute to the mass range 239–242. This problem may be serious when the Pu concentration is very low, and large-volume samples (e.g., 100–200 L of seawater or 100–200 g of soil and sediment) are analyzed. Chemical separations to remove interfering elements, especially lead, must thus be applied. In addition, chlorine also needs to be removed from the final solution before measurement to reduce the formation of polyatomic ions of chlorine and also to improve the counting efficiency. The removal of chlorine in the final solution after the separation of other elements can be carried out by heating the solution with the addition of concentrated HNO_3. In addition to good sensitivity, the ability of an ICP-MS to measure both ^{239}Pu and ^{240}Pu separately is another important favorable feature. Alpha spectrometry cannot do this since the peaks of ^{239}Pu and ^{240}Pu overlap. Because of the very strong isobaric interference of ^{238}U, an ICP-MS cannot be used to measure ^{238}Pu; for this reason, alpha spectrometry is still the best method. In order to get information on all environmentally relevant plutonium isotopes (^{238}Pu, ^{239}Pu, ^{240}Pu, ^{241}Pu), alpha and beta spectrometry are needed as well as an ICP-MS.

17.3 Accelerator Mass Spectrometry (AMS)

17.3.1 Components and Operation of AMS

AMS was developed in the late 1970s and has since been used as an ultra-sensitive mass spectrometric technique to measure isotopes. It is now mainly used for the determination of radionuclides, especially long-lived radionuclides, such as ^{3}H, ^{10}Be, ^{14}C, ^{26}Al, ^{32}Si, ^{36}Cl, ^{41}Ca, ^{53}Mn, ^{59}Ni, ^{63}Ni, ^{129}I, ^{182}Hf, ^{210}Pb, and actinide isotopes. Almost all AMS facilities consist of two mass spectrometers – first an 'injector' and then an 'analyzer' – linked with a tandem accelerator (Figure 17.8).

For AMS measurements, the radionuclide of interest is first chemically separated and then prepared as a solid target; this is then loaded into a sample holder made of copper or aluminum alloy. The sample is inserted through a vacuum lock into the ion source, where the atoms are sputtered from the sample by external ions such as Cs^+ which are produced on a hot spherical ionizer and focused to a small spot on the sample. The sputtered ions produced on the surface of the sample, having a negative charge, are extracted from the ion source and sent down the evacuated beam line toward the first magnetic mass analyzer, the injector. At this point, the beam is about 10 microamperes, corresponding to 10^{13} ions per second (mostly the stable isotopes). Since not all elements form negative ions, isobaric interferences can be effectively

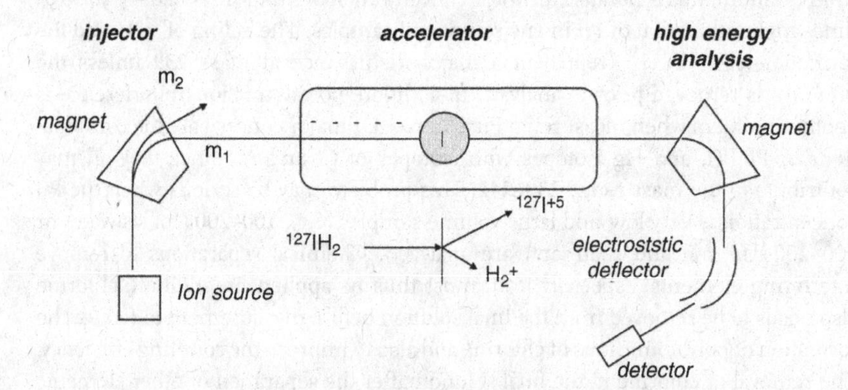

Figure 17.8 Schematic principle of accelerator mass spectrometry (AMS).

suppressed in some important cases: in the measurements of ^{14}C, ^{41}Ca, and ^{129}I, isobaric interferences because of ^{14}N, ^{41}K, and ^{129}Xe, respectively, are eliminated since these do not from stable negative ions.

The extracted negative ions are pre-accelerated and mass-analyzed in the injector magnet by bending them 90 degrees to select the mass of interest, that is, the radionuclide inserted in the sample holder, and to reject the much-more-intense neighboring stable isotopes. Polyatomic ions and isobars cannot, however, be completely removed in this step and thus require further separation. Several vacuum pumps remove all the air from the beam line to allow the beam particles a free path. The mass-analyzed negative ions from the injector magnet are directed to a tandem accelerator, which consists of two accelerating gaps with a large positive voltage in the middle. The center of the accelerator, the terminal, is charged to a voltage of up to 10^6 volts by two rotating chains. At the terminal, the accelerated ions pass through an electron stripper, either a gas or a very thin carbon foil, where several electrons are stripped off, converting the negative ions into multiply charged positive ions, for example, converting $^{129}I^-$ into $^{129}I^{5+}$. These are repelled from the positive terminal, accelerating again to the ground potential at the far end. The name 'tandem accelerator' comes from this dual acceleration concept. The final velocity of the ions is a small percentage of the speed of light or about 7×10^7 km h^{-1}. This stripping process has the advantage that it dissociates polyatomic (molecular) ions if enough electrons are stripped off, thus resulting in a further elimination of isobaric interferences.

The positively charged ions from the accelerator then pass through a magnetic analyzer, where the ions of interest with a well-defined combination of charge state and energy are selected. The stable and radioactive isotopes, for example ^{127}I and ^{129}I, ^{12}C and ^{14}C, ^{35}Cl and ^{36}Cl, are separated by this magnet; the stable isotope is then measured as a current by a Faraday cup. This again reduces the intensity of neighboring stable isotopes. In addition, it completely eliminates the molecules by selecting only the highly charged ions that are produced in the terminal stripper. Highly charged molecules are unstable since they lack the electrons that bind the atoms together. The higher energies of the ions after acceleration allow an additional separation of the wanted ions from possible background ions at the particle detector. The ions from the magnetic analyzer then pass through an electrostatic analyzer (ESA), which is a pair of metal plates at high voltage that deflect the beam by 20–30 degrees. This selects the particles based on their energy and thus removes the interfering ions. At the end of the AMS system, the ions are measured by an ion detector, and a range of detectors have been employed for this purpose. Of these, silicon detectors can only be used to measure the ion energy; the ionization chamber can measure both the total energy of an ion and its rate of energy loss in the detector; the time-of-flight detector is normally used to measure the energy of heavy ions such as ^{129}I. In addition, gas-filled magnets and X-ray detectors have been used as detectors in ASM measurements.

In general, AMS is a relative analytical method; a ratio of two isotopes of an element is normally measured, such as the ratios $^{10}Be/^{9}Be$, $^{14}C/^{12}C$, $^{36}Cl/^{35}Cl$, $^{41}Ca/^{40}Ca$, $^{129}I/^{127}I$, $^{233}U/^{238}U$, and $^{239}Pu/^{242}Pu$. Because of its unique features, AMS can be used to measure isotopic ratios well below 10^{-10}, and many interesting long-lived

Figure 17.9 A setup of a 3 MeV tandem accelerator mass spectrometer.

radionuclides occur in the environment in amounts that have this kind of ratio to the amounts of the corresponding stable isotopes. AMS is the most sensitive method with which to measure these radionuclides. Furthermore, with its capability of providing isotopic abundance ratios even as low as 10^{-15} for very small samples, AMS has been applied for the detection of extremely low concentrations of long-lived radionuclides in environmental, biological, geochronological, and archaeological studies. Figure 17.9 shows a setup of a 3 MeV AMS. In recent years, small AMS facilities with a maximum terminal voltage of 0.25–1.0 MeV have been commercialized. They can be operated as desk-top instruments and require less space and resources. The maintenance of these instruments is simplified and the operation is completely automated. The determination of some important radionuclides by the use of AMS is discussed below.

17.3.2 ^{14}C Measurement by AMS

AMS is the most sensitive method for the determination of ^{14}C, being able to detect activities of ^{14}C as low as 10^{-7} Bq or a ratio of $^{14}C/^{12}C$ of 10^{-15}. AMS has become a standard method in ^{14}C measurement for age dating, with samples up to 50 000 years old being dated. In addition, because of the very high sensitivity of AMS, only a very small sample (<1 μg carbon) is needed for the measurement; this makes it possible to analyze samples with very low carbon content, such as ice, water, ceramics, and metals. As well as having good sensitivity, the analytical precision of AMS (<0.2%) is also much better than that of radiometric methods – a very important factor in age dating. A compact AMS facility with smaller dimensions and drastically reduced investment and operating costs, specifically designed for radiocarbon dating, has recently been developed and introduced onto the market. In the AMS analysis of ^{14}C, carbon is first separated from the samples, mainly by combustion at high temperature. The released ^{14}C, as $^{14}CO_2$, is then normally converted into graphite using a catalyst such as iron and H_2, and this graphite is then used as a source for the AMS measurement. A new method for the direct introduction of CO_2 into the AMS for measurement is under development and will probably be in use in the near future. Three isotopes, ^{12}C, ^{13}C, and ^{14}C can be simultaneously determined, and the ratios $^{14}C/^{12}C$ and $^{13}C/^{14}C$ can be obtained with AMS.

17.3.3
^{36}Cl Measurement by AMS

AMS is the only technique useful for the determination of ^{36}Cl in low-level environmental samples. The reported detection limit of AMS is 10^{-9} Bq or a ratio of $^{36}Cl/^{35}Cl$ of 10^{-15}. In AMS, ^{36}Cl is normally prepared as an AgCl target, and the measurement is conducted using $^{36}Cl^{7+}$ ions. ^{36}S isobaric interference is the principal challenge in the AMS measurement of ^{36}Cl because sulfur forms negative ions as readily as chlorine, and the concentration of stable ^{36}S is normally several orders of magnitude higher in environmental samples than that of radioactive ^{36}Cl. By using 48 MeV $^{36}Cl^{7+}$ ion energy and a multi-anode ionization chamber as a detector, the ^{36}S signal can be highly suppressed (by a factor of 10^4). The chemical separation of sulfur from chlorine by the precipitation of sulfur as $BaSO_4$ is still the main method used to separate most ^{36}S from ^{36}Cl. The remaining sulfur, as sulfate, may co-precipitate together with AgCl as Ag_2SO_4. However, the solubility of Ag_2SO_4 is much higher than that of AgCl, and Ag_2SO_4 can therefore be removed by washing the precipitate with an acidic solution.

17.3.4
^{41}Ca Measurement by AMS

AMS is also the most sensitive method for ^{41}Ca. In the AMS measurement, calcium is injected as a molecule ion, typically as CaH_3^-, instead of as Ca^- ion because the formation of the latter in the sputter source is very low. Using CaH_3^-, the interfering ^{41}K isobar is effectively eliminated because the KH_3^- ion is not stable. CaF_3^- has also been used to inject Ca into the AMS system since the KF_3^- ion is unstable compared to CaF_3^-. Ideally, a $^{41}Ca/^{40}Ca$ ratio as low as 10^{-15} can be measured by AMS. The detection limit of ^{41}Ca is mainly affected by the blank and ^{41}K interference, and detection limits of about 10^{-12}–10^{-13} for the $^{41}Ca/^{40}Ca$ ratio or 0.1 mBq g^{-1} for the ^{41}Ca concentration are normally obtained for real samples. In addition, the existence of potassium impurity in the Cs ion sputter source makes an additional contribution to the background level of the AMS instrument, worsening the detection limit of AMS for ^{41}Ca.

17.3.5
^{63}Ni and ^{59}Ni Measurement by AMS

^{63}Ni and ^{59}Ni normally co-exist in the same sample, but the radioactivity concentration of ^{59}Ni is around 100 times lower than that of ^{63}Ni. It is thus difficult to determine ^{59}Ni by a radiometric method, such as liquid scintillation counting, except in samples with a high concentration of ^{59}Ni, for which measurement by X-ray spectrometry can be applied. AMS is the only method for the determination of ^{59}Ni in environmental and in most waste samples. In AMS, isobaric interferences from ^{63}Cu and ^{59}Co are the main challenge when measuring ^{63}Ni and ^{59}Ni, respectively, because Cu, Co, and Ni easily form negative ions in the ion sputtering source. To minimize these interferences, a combination of chemical separation methods and AMS instrument settings is used. A characteristic X-ray detector is used for ion detection and identification. After separation in the AMS, the Ni ions are detected via the X-rays emitted when they pass through a thin foil close to the detector. This allows the ions to be identified by atomic number, thereby separating the isobars. Using this method, a suppression of the ^{59}Co interference by a factor of 10^7 is obtained. Instrument background is another source of interference in the AMS when measuring ^{59}Ni and ^{63}Ni, since the sample holder and other parts of the ion source devices are normally made of copper and stainless steel. An improvement in the background level can therefore be achieved by using copper- and cobalt-free target holders and by thoroughly cleaning the ion source. However, the chemical separation of Ni from the matrix components and interferences (mainly from Co and Cu), for example, by using extraction chromatography with Ni Resin, is still the principal task when determining ^{59}Ni and ^{63}Ni by AMS, and. A chemical purification method that utilizes the volatility of the $Ni(CO)_4$ compound has proven to be very effective for the sample preparation when measuring ^{63}Ni and ^{59}Ni by AMS. In this method, the Ni in ammonia solution is mixed with $NaBH_4$ solution in a flask, and a mixture of CO and He is bubbled through the Ni solution to produce volatile $Ni(CO)_4$, which is then collected in a cold trap. The collected $Ni(CO)_4$ is heated and transferred to the AMS holder by He flow, where it is thermally decomposed to metallic Ni. By this process, a very low Cu/Ni ratio of $<2 \times 10^{-8}$ in the sample can be obtained, resulting in a detection limit of 2×10^{-11} for $^{59}Ni/Ni$ ratio or 0.05 mBq of ^{59}Ni and 45 mBq of ^{63}Ni.

17.3.6 ^{99}Tc Measurement by AMS

AMS potentially offers a very high sensitivity for the determination of ^{99}Tc. In this measurement, technetium is chemically separated as TcO_4^- ions and mixed with iron oxide or niobium oxide to prepare a target. The sputtered TcO^- anions are stripped to TcO^{14+} in a tandem accelerator, separated in a mass analyzer, and measured by a gas ionization chamber for ^{99}Tc. The stable ^{99}Ru isotope is the main isobaric interference, limiting the detection limit of AMS for ^{99}Tc. Discrimination between ^{99}Tc and ^{99}Ru requires an ionization detector that makes multiple measurements of the energy loss of the ions as they slow down, since it is only effective when the ion energy is more than 120 MeV. Therefore a large accelerator operating at more than 10 MeV is required to determine ^{99}Tc. The TcO^- ion is used in the ion injection instead of the Tc^- ion because Ru easily forms an interfering Ru^- anion. The effective suppression of Ru can be achieved by using the difference between the energy losses of ^{99}Tc and ^{99}Ru in the detector. With optimal settings, >90% of ^{99}Ru can be rejected; however, for environmental samples, ^{99}Ru concentration is normally many orders of magnitude higher than that of ^{99}Tc, and chemical separation is the key process before the AMS measurement. The sensitivity of the AMS method depends very much on the amounts of Ru in the target. A detection limit of 6–10 μBq for water and environmental samples is normally obtained. Because technetium has no stable isotopes, rhodium is usually added to the target for normalization purpose: the ratio of ^{99}Tc to ^{103}Rh is measured, and the ^{99}Tc concentration is calculated according to the amount of Rh added to the target.

17.3.7 ^{129}I Measurement by AMS

In AMS, iodine is first chemically separated from the sample and is then typically prepared as AgI precipitate, which is then mixed with metallic Ag or Nb powder and pressed into a copper sample holder. I^- ions are easily formed in the sputter source, stripped to I^{3+}, I^{5+}, or I^{7+} in the tandem accelerator, and then separated from the interferences such as $^{128}TeH^-$ and $^{127}IH_2^-$. The isobaric interference from ^{129}Xe in AMS is not a problem because the amount of $^{129}Xe^-$ ion formed is extremely low. The separated ^{129}I is then detected by a combination of time-of-flight and a silicon charged particle detector, or a gas ionization detector. The instrumental background of $^{129}I/^{127}I$ down to 10^{-14} can be obtained. The detection limit of ^{129}I very much depends on the level of procedure blank. By using a stable iodine (^{127}I) carrier with very low level of ^{129}I, a $^{129/127}I$ ratio of 1×10^{-13} has been obtained for procedure blank; this corresponds to 10^{-9} Bq ^{129}I for 1 mg ^{127}I carrier. Because of the very high sensitivity, most measurements of ^{129}I in environmental samples are now carried out by AMS, especially in low-level geological samples. AMS is the only method for the determination of ^{129}I in the pre-nuclear age samples ($^{129}I/^{127}I < 10^{-10}$).

17.3.8 Measurement of Plutonium Isotopes by AMS

In the AMS determination of plutonium isotopes, mainly ^{239}Pu and ^{240}Pu, the separated Pu is dispersed in an iron oxide and mixed with metallic Al or Nb in the target. The PuO^- molecular anion is selected from the ion sputtering source because Pu^- is very weakly produced. To efficiently discriminate against interferences, Pu^{3+} or Pu^{5+} ions are selected after the stripping process. The separated Pu ions are finally detected using a gas ionization detector. Because of the nonexistence of stable plutonium isotopes, ^{242}Pu is added before the chemical separation as a yield tracer, also acting as an isotopic spike. The ratios of $^{239}Pu/^{242}Pu$ and $^{240}Pu/^{242}Pu$ are measured and the concentrations of ^{239}Pu and ^{240}Pu are calculated according to the ^{242}Pu added before the chemical separation. Being similar to Pu, uranium can also easily form a UO^- anion, which follows Pu as U^{3+} or U^{5+} ions after the stripping and is directed to the ion detector. As the analyzer magnet cannot completely discriminate uranium, AMS cannot be used to measure ^{238}Pu. In addition, uranium is also a major interference for the determination of ^{239}Pu and ^{240}Pu, because $^{238}U^{17}O^-$ with $^{239}Pu^{16}O^-$, and $^{238}U^{18}O^-$ with $^{239}Pu^{17}O^-$ are injected into the accelerator. After stripping, the $^{238}U^{5+}$, which has the same momentum as $^{239}Pu^{5+}$, can pass through the analyzing magnet. $^{238}U^{5+}$ and $^{239}Pu^{5+}$ differ in energy by only 0.4%, which is much lower than the 3% resolution of the ionization detector. A high-resolution electrostatic analyzer (ESA) may be able to prevent most of the $^{238}U^{5+}$ from reaching the detector. As other elements such as Pt, Mo, Sm, and Ti can also interfere with the AMS determination of Pu isotopes, the chemical separation and purification procedure is a critical step that affects the analytical accuracy and detection limit. A detection limit of 0.1–1 μBq for ^{239}Pu can be obtained using AMS, depending on the levels of interferences.

17.4
Thermal Ionization Mass Spectrometry (TIMS)

For several decades, TIMS has been used for the determination of isotopic composition and the concentrations of different elements including some radioisotopes of elements, such as uranium, thorium, and plutonium. The ions produced by thermal ionization are focused into an ion beam and then passed through a magnetic field to separate them by their mass-to-charge ratios (m/z) (Figure 17.10). The relative abundances of different isotopes can then be measured, giving the isotope ratios.

In TIMS, a small volume (as little as 1 μL) of aqueous solution containing the target nuclide or element in the nanogram to microgram range is deposited on a cleaned filament surface (usually high-purity Re) and evaporated to dryness. The most

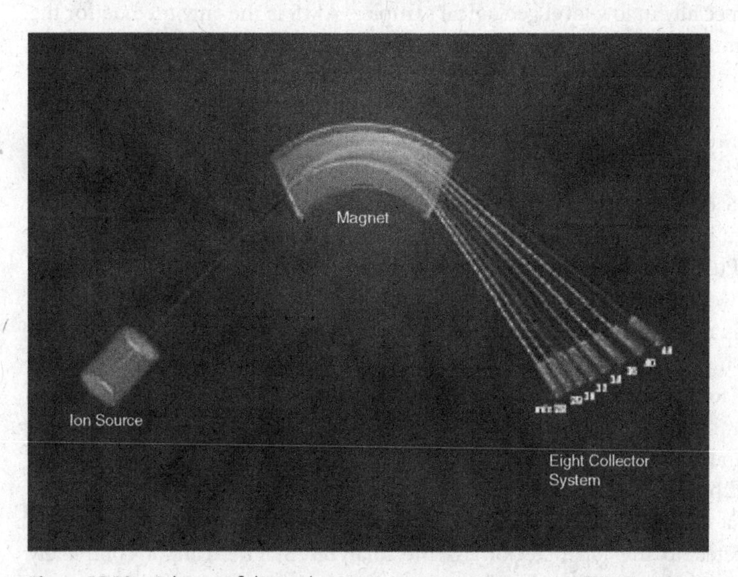

Figure 17.10 Scheme of thermal ionization mass spectrometry (TIMS).

frequently applied technique in TIMS works with two heated filaments arranged opposite to each another: one for evaporation of the sample and the other for ionization of evaporated atoms. Because of the low initial energies (0.1–0.2 eV) of the ions formed on the hot thermal surface, single magnetic sector field mass analyzers have mostly been used for ion separation. The limiting factors for the accuracy of measured isotopic ratios in TIMS are mass discrimination in the TIMS instrument (e.g., ion optical system or ion detector) and mass fractionation effects (caused during the evaporation of the sample, where the measured isotope ratio changes with time). These inherent effects, limiting the accuracy of isotope ratio measurements by TIMS, can be minimized by different internal calibration techniques or by using isotopic standard reference materials with well-known isotopic ratios for an element. In addition to U, Th, and Pu, TIMS has also been used for other radionuclides including ^{41}Ca, ^{90}Sr, ^{137}Cs, ^{99}Tc, ^{237}Np, ^{241}Am, ^{243}Cm, ^{126}Sn, ^{226}Ra, and ^{228}Ra.

The sensitivity and detection limit of TIMS for long-lived radionuclides (e.g., Pu, Np, and U) is normally better than ICP-MS and comparable with AMS; for example, a detection limit of 1 fg for plutonium isotopes can be obtained by loading separated solution directly to the Re cavity filament: 1 fg of plutonium corresponds to the activity detection limits of 2 μBq for ^{239}Pu and 8 μBq for ^{240}Pu. By loading the sample in resin bead form to the filament, the efficiency of the sample introduction and counting can be improved by 1–2 orders of magnitude and the detection limit for Pu isotopes can thus be improved to <0.1 fg (corresponding to <0.2 μBq for ^{239}Pu), which is the same as AMS. In this case, the separated Pu is absorbed on one or a small number of grains of resin, for example anion exchange resin, and more Pu can be loaded into the mass analyzer, resulting in a higher counting efficiency and a lower detection limit. In addition to isobaric interference, the abundance sensitivity of 10^6–10^8 is a major drawback in the determination of radionuclides by TIMS, thus making the determination of ^{238}Pu impossible. Alpha spectrometry as a complementary technique to mass spectrometry is therefore normally used to get the information on ^{238}Pu. TIMS has also been reported to be applicable for the determination of other relatively short-lived radionuclides, such as ^{90}Sr, ^{137}Cs, and ^{241}Am, with a quite low mass detection limit of 0.1 fg. The activity detection limits are still high: 0.5 mBq, 0.3 mBq and 0.01 mBq for ^{90}Sr, ^{137}Cs, and ^{241}Am, respectively, which are comparable with radiometric methods. When considering the serious isobaric interferences (e.g., ^{90}Zr, ^{137}Ba), and the abundance sensitivity interference of adjacent stable isotopes (^{88}Sr, ^{89}Y, ^{136}Ba, ^{138}Ba), the detection limit of TIMS for these radionuclides will be much poorer than that of radiometric methods. The low detection limit of TIMS for long-lived radionuclides makes it an often used technique for bioassay of Pu, U, and Np isotopes, since very low concentrations are needed to measure biological samples such as urine and feces to estimate internal radiation doses to humans. Compared to ICP-MS and AMS, another major drawback of TIMS is the sample preparation. Because of the very small sample load (normally 1 μL), the analytes separated from a large sample (0.5–1 L urine) by chemical procedures must be purified and concentrated to a very small volume of solution (<10 μL). To further improve the detection limits, the separated analytes need to be adsorbed on a single resin bead or on a small number of them. These post-separation processes are often time consuming – taking some 4–8 h per sample. This significantly reduces the analytical capacity of TIMS and makes it difficult to analyze a large number of samples in a short time.

17.5
Resonance Ionization Mass Spectrometry (RIMS)

RIMS, a highly selective and sensitive mass spectrometric technique for ultratrace element and isotope analysis, has been used for the determination of many radionuclides such as ^{41}Ca, ^{90}Sr, ^{99}Tc, ^{135}Cs, ^{210}Pb, ^{236}U, ^{238}Pu, ^{239}Pu, ^{240}Pu, ^{242}Pu, and ^{244}Pu in environmental and waste samples. In RIMS, the separated radionuclide is first vaporized and atomized by an atomic beam source, for example, in an atomic beam oven by thermal vaporization on a hot Re filament or by evaporating a sample using an electron beam. The evaporated atoms are then ionized by resonance optical excitation using one or more lasers. This process is normally completed in several steps by precisely tuning the required wavelength for excitation and ionization. The ion is then separated by a mass analyzer, for example, a magnetic sector field selector, time-of-flight mass analyzer, and quadrupole ion traps, and is measured by an ion detector such as single-channel secondary electron multipliers or discrete multi-stage secondary electron multipliers. Two to three steps of excitation/ionization processes with ultraviolet, visible, or infrared light are normally used for resonance laser ionization. These processes start from the ground state or a low-lying state of the element, and the atoms are step-wise excited to a high-lying state by absorption of one or two resonant photons. The highly excited atoms are finally ionized by the absorption of an additional photon. Since the energy between the ground state and the excited state (high-lying state) is a uniquely characteristic feature of an element, the ionization of an element can be completed by precisely tuning the wavelength of the laser light to match the energy required to excite the atom of interest. The isobaric interference can be completely removed, since the probability of a transition of an unwanted atomic species is extremely low, which make the selectivity of RIMS very high. Figure 17.11 shows an RIMS setup.

Unlike other mass spectrometric methods, RIMS can simultaneously measure ^{238}Pu, ^{239}Pu, and ^{240}Pu without serious interference from ^{238}U to the measurement of ^{238}Pu. For the RIMS determination of plutonium isotopes, the separated plutonium is electrodeposited on a tantalum filament as $Pu(OH)_4$ and covered with a 1 μm thick sputtered layer of titanium as a reducing reagent. During the thermal vaporization, $Pu(OH)_4$ is first converted to PuO_2, which diffuses through the titanium layer and reduces to Pu, and finally evaporates from the titanium surface. With this filament type, the evaporation is completed at a relatively low temperature, which prevents the ionization of Pu. Most of the evaporated Pu atoms remain in the ground state and are then resonance-ionized as they cross the laser beams, subsequently mass-separated in the TOF mass analyzer, and finally registered by multi-channel plate detectors.

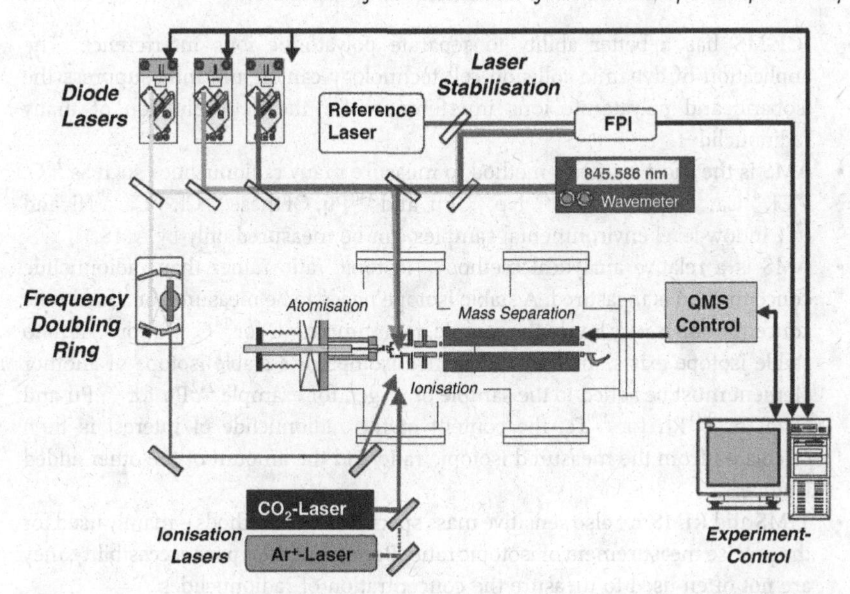

Figure 17.11 Scheme of the working principle of RIMS (Klaus *et al.* RIMS for sensitive and selective ultratrace analysis, http://www.physik.uni-mainz.de/forschungsbericht/exakt/fobe5.pdf).

RIMS offers a number of outstanding properties compared to other mass spectrometric techniques. First, RIMS provides nearly complete isobaric suppression, even better than 10^8, which is obtained by unique optical transitions, especially multi-step excitations. RIMS also has a high overall sensitivity, with detection limits in the 10^{-15}–10^{-18} g range, enabled by the high ionization efficiency, high transmission to mass spectrometers, and low background ion detection. Further, it has a high isotopic selectivity, with values of 10^{13} and higher, which is achieved by combining the isotope abundance sensitivity of the mass spectrometer and the optical isotope selectivity of the laser excitation process. Because of these advantages, RIMS has found broad application in various analytical fields, particularly the ultratrace determination of long-lived radioisotopes. However, since there is no commercial RIMS available, its accessibility limits the routine determination of radionuclides in environmental and waste samples.

17.6 Essentials of the Measurement of Radionuclides by Mass Spectrometry

- Mass spectrometric methods are suitable for the determination of long-lived radionuclides, especially radionuclides with a half-life longer than 100 years.
- ICP-MS is the most often used mass spectrometric method for the determination of radionuclides such as ^{99}Tc, ^{237}Np, ^{239}Pu, ^{240}Pu, ^{238}U, ^{235}U, ^{233}U, and ^{232}Th. The most important interferences in the ICP-MS measurement of radionuclides are isobars, polyatomic ions, and abundance sensitivity. The sector field type

ICP-MS has a better ability to separate polyatomic ions interference. The application of dynamic collision cell technology can significantly suppress the isobaric and polyatomic ions interferences for the determination of many radionuclides.

- AMS is the most sensitive method to measure many radionuclides such as ^{14}C, ^{36}Cl, ^{41}Ca, ^{59}Ni, ^{63}Ni, ^{129}I, ^{237}Np, ^{239}Pu, and ^{240}Pu. Of these, ^{36}Cl, ^{41}Ca, ^{59}Ni, and ^{129}I in low-level environmental samples can be measured only by AMS.
- AMS is a relative analytical method – isotopic ratio rather than radionuclide concentration is measured. A stable isotope needs to be measured or added if its concentration is very low in the sample, for example ^{12}C for ^{14}C, ^{127}I for ^{129}I. If no stable isotope exists, another radioactive isotope or a stable isotope of another element must be added to the sample or target, for example ^{242}Pu for ^{239}Pu and ^{240}Pu or ^{103}Rh for ^{99}Tc; the content of the radionuclide of interest is then calculated from the measured isotopic ratio and the amount of the other added isotope.
- TIMS and RIMS are also sensitive mass spectrometric methods – mainly used for the precise measurement of isotopic ratio. Because of their poor accessibility, they are not often used to measure the concentration of radionuclides.

Further Reading

Becker, J.S. (2002) ICP-MS: determination of long-lived radionuclides. *Spectrosc. Europe*, **14**, 8.

Becker, J.S. (2003) Mass spectrometry of long-lived radionuclides. Spectrochim. *Acta Part B*, **58**, 757.

Hou, X.L. and Roos, P. (2008) Critical comparison of radiometric and mass spectrometric methods for the determination of radionuclides in environmental, biological and nuclear waste samples. *Anal. Chim. Acta*, **608**, 105.

18
Sampling and Sample Pretreatment for the Determination of Radionuclides

18.1
Introduction

Sampling is the first and very important step in the investigation of radionuclides both in the environment and in nuclear waste. Uncertainty in the radiochemical analysis of a radionuclide can be controlled to a certain extent. Variability and uncertainty affecting the determination of the radionuclide concentration in a system may derive from unrepresentative sampling, contamination, loss of analytes during the sampling, and unsatisfactory storage and pretreatment of samples.

Before designing a sampling process, the objectives and purposes need to be specified to set up the sampling strategies, intensities, methods, and tools. Statistical sampling is pertinent and necessary in environmental radioactivity investigations because of the stochastic variation normally occurring in all environmental media. This sampling system also increases the degree to which the samples represent what is being investigated. In addition, information about whatever area is being investigated needs to be obtained or estimated, such as the expected behavior and environmental properties of the radionuclide of interest, the expected pattern and magnitude of variability in the observations, the suitability of the sampling tool and sampling techniques employed, and the spatial extent and temporal stability of the radionuclides in the sample.

In general, designing a sampling process includes five steps:

1) defining the objectives and purposes;
2) summarizing the environmental context for the quantities being sampled and measured;
3) identifying the target system including spatial and temporal extent;
4) selecting an appropriate sampling process; and
5) documenting the sampling design and its rationale.

There are normally three sampling approaches: simple random sampling, stratified sampling, and systematic sampling. Simple random sampling means that

Chemistry and Analysis of Radionuclides. Jukka Lehto and Xiaolin Hou

ISBN: 978-3-527-32658-7

a random selection of samples is collected for a measurement. While this operation is simple and cost effective, the results may be poorly representative. Systematic sampling, on the other hand, usually involves sampling at regular spatial and/or temporal intervals and is the commonest method used in environmental radioactivity investigations, the results providing information on spatial distribution and temporal variation. In waste analysis, spatial sampling is also often used to obtain information on spatial distribution as well as to estimate the total inventory of the radionuclide under study. Usually, the crucial decisions in planning a sampling process include the number of sampling sites, the sampling frequency, and the number of samples; these decisions can be made based on the degree of variability and level of the analyte of interest. If a speciation analysis of radionuclides is also required, special planning is needed to prevent any chemical change of the different species of radionuclide or the loss of any species during sampling and storage. In general, the radionuclide concentrations in the environment are very low, and therefore a large amount of sample is normally required to accurately determine the analyte. To reduce the costs of sample transportation and to obtain large number of samples, *in-situ* preconcentration/treatment is often used to enrich the radionuclide by employing coprecipitation, selective adsorption, ion exchange, air drying, and evaporation.

Sampling theory has been widely discussed in the literature, including the sampling concept, the design of sampling processing, the selection of sampling sites and sampling amounts, and the statistical aspects of sampling. Some typical methods and techniques of sampling and sample pretreatment in environmental radioactivity investigations are discussed in this chapter and illustrated by some examples.

18.2 Air Sampling and Pretreatment

Radionuclides may exist in the air in both gaseous and aerosol particle-associated forms. Most metals are mainly associated with the aerosol particles in the air, while the major form of radon in the air is as a gas, whereas iodine exists in aerosol-associated, inorganic gaseous, and organic gaseous forms. The sampling methods for both aerosol particles and gaseous components are discussed below.

18.2.1 Sampling Aerosol Particles

There are two major types of aerosol particle sampling techniques: integral and size selective. In integral sampling, which is a simple and widely used technique used in many laboratories, air is pumped through a filter to collect all the aerosol particles of a size bigger than the pore size of the filter. The size-selective sampler is used to separate and collect the aerosol particles in different size fractions, the most often used sampler of this type of being the cascade impactor.

18.2.1.1 Radioactive Aerosol Particles

An aerosol is a suspension of fine solid particles and/or liquid droplets in air, for example smoke, oceanic haze, smog, and suspended dust. The size of aerosol particles ranges from a nanometer to a few micrometers. Because of the short residence time of large size particles in the air and the difficulties in collecting particles of very small size ($<100\,nm$), the particles between 0.1 and $10\,\mu m$ in diameter are the most important with respect to the measurement of radionuclides in the air. The two major sources of aerosol particles are 'natural' and 'anthropogenic'. The natural sources of aerosol particles are various, and include wind-blown dust and salt particles formed from evaporating seawater droplets, forest fires generating carbon-containing particles, volcanic dust, pollen; and bacteria. Small particles can also be produced by the conversion of element species, for example, from I_2 to iodine oxides (e.g., IO). Anthropogenic processes, which also release large amounts of aerosol particles, include automobile emissions and combustion products released from power plants and other factories by burning coal, oil, and biomass. It is estimated that about 3.5×10^{15} g particles are released into the air annually. All aerosol particles, both radioactive and nonradioactive, are generated in two ways: anthropogenic releases and natural production. Naturally produced radionuclides from cosmic-ray-induced nuclear reactions with gaseous elements in air, for example, the cosmic ray reaction with gases in the air such as N_2 to form ^{7}Be, ^{10}Be, ^{14}C, ^{22}Na, ^{41}Ca, and so on, and the progeny radionuclides from the decay of radon isotopes in the air, including ^{218}Po, ^{214}Po, ^{210}Pb, and ^{210}Po, can attach to naturally occurring particles to form radioactive particles. Anthropogenic processes are a major source of radioactive aerosol particles. For example, the releases from the atmospheric nuclear weapons testing during the 1950s and 1960s are the the dominant source on the global scale, while nuclear accidents such as Chernobyl are another large source of radioactive aerosol particles. Nuclear facilities, including nuclear power plants, also release radioactive particles to the air to a small extent. The anthropogenic releases of radionuclides can also attach to nonradioactive particles after they have been released to the atmosphere, changing their physical and chemical forms. For example ^{131}I released from nuclear power plants and a nuclear accident is normally in a gaseous molecular iodine form when released; however, this form is not stable and can quickly change to other forms such as iodide and iodate, which easily attach to other particles. Larger radioactive particles with higher specific activities – or hot particles – have been released in nuclear explosions and in nuclear accidents. These particles are formed either by mechanical fragmentation of the nuclear fuel or the weapons material or by condensation processes in the cooling of explosion or release clouds. These larger particles are typically tens of micrometers in diameter and are deposited at relatively short distances away from the release point because of gravity.

18.2.1.2 Integral Aerosol Particle Sampling

Since the 1950s, following the nuclear weapons testing at that time, many countries have conducted surface air sampling programmes. The primary objective of these programmes is to monitor the radioactivity levels in the air for radiation protection, and to investigate the temporal and spatial distribution of anthropogenic and natural radionuclides in the air. A suitable air filter and air mover (pump) are the key components in aerosol particle sampling. The filter should have the following features: good collecting efficiency for submicron particles, high particle and mass loading capacity, low flow resistance, low cost, high mechanical strength, low background activity, good compressibility, low ash content, temperature stability, and suitable size. Two types of filters are normally used: a polypropylene or polyvinyl chloride (PVC) filter and a glass fiber filter. A glass fiber filter, such as the TFAGF 810 (Staplex), has good collecting efficiency (99%) for submicron particles as small as 0.3 μm as well as having high mechanical strength and good temperature stability. However, this type of filter is not ashable, and cannot be dissolved in normal acids. It therefore forms a large volume when compressed for the gamma measurement of radionuclides and consequently gives lower counting efficiency. The PVC filter, on the other hand, which can be ashed and thereby converted to a small size for gamma measurement, has good geometry and counting efficiency, which significantly improve the detection limit for radionuclides at low concentrations in the air. However, this does not affect the measurement of volatile radionuclides, such as ^{131}I and ^{99}Tc, as the ashing of the filter would result in a significant loss of these radionuclides. Glass fiber filters also cause problems in the dissolution of radionuclides from the filter, especially for the radiochemical analysis of pure beta- and alpha-emitting radionuclides, because the dissolution of the entire glass fiber filter requires large amounts of HF which is highly hazardous. In addition, the concentration of uranium impurity is normally high in glass fiber filters, which is a disadvantage in the mass spectrometric measurement of uranium and plutonium. A polyvinyl chloride filter (Petrianov FPP-15-1.5) has been confirmed to be a good filter for collecting integral aerosol samples for the determination of radionuclides. With two layers, one upon another, it can collect more than 90% of the 0.3 μm particles; further, the flow rate is not decreased significantly by the mass loading after a sampling time of up to 20 days (Valmari, T., Tarvainen, M., Lehtinen, J., Rosenberg, R., Honkamaa, T., Ossintsev, A., Lehtimäki, M., Taipale, A., Ylätalo, S., and Zilliacus, R. (2002) Aerosol Sampling Methods for Wide Area Environmental Sampling (WAES), STUK-YTO-TR 183).

The air mover (pump) used to sample large air volumes must have a high operational flow rate, durability for continuous operation of up to a month, temperature resistance in field operations, and low maintenance. The portability of the air sampler is another factor to be considered if it needs to be moved often. In addition, the sampling volume has to be measured accurately and monitored in the air sampler to calculate the concentration of radionuclides in the air. Normally, the air sampler used for measuring environmental radioactivity has a flow rate of 1–10 m^3 min^{-1} and is capable of continuous sampling for one day to one week with a filter size of 500–5000 cm^2. The air sampling facility operated at the Risø National Laboratory, Denmark, which has been collecting weekly samples since 1957, is shown in Figure 18.1. The aerosol particle sample collected on a polypropylene filter (3000 m^3) is compressed into a 50 mL box for gamma measurement (Figure 18.2). Thereafter, it is ashed at 500 °C for the radiochemical analysis for the beta- and alpha-emitting radionuclides. Figure 18.3 gives the measured variation of ^{137}Cs and ^{90}Sr in the aerosol samples collected in Denmark since 1957.

Figure 18.1 Aerosol particle sampling at the Risø National Laboratory, Denmark.

18.2.1.3 Size-Selective Aerosol Particle Sampling

Multistage cascade impactor aerosol samplers are normally used to study the size distribution of aerosol particles rather than for long-term monitoring of activity levels in the air. A high-volume instrument is typically needed to collect enough particles for

Figure 18.2 Preparation of an aerosol particle sample by compressing a filter into a 50-mL box for gamma measurement.

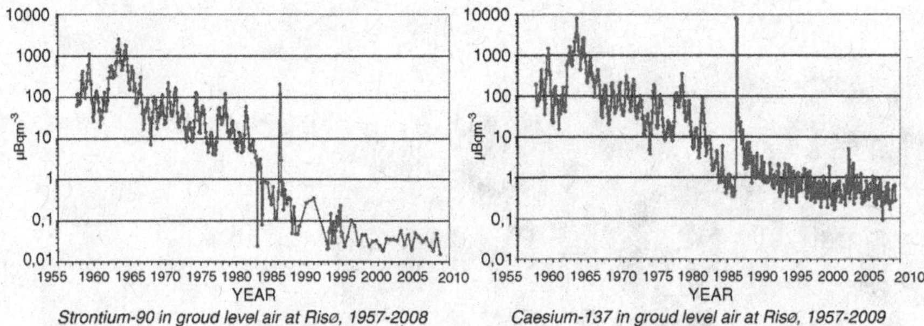

Figure 18.3 Temporal variation of ^{137}Cs and ^{90}Sr in ground level air at Risø, Denmark. The highest levels of both radionuclides, which were observed in the 1950s and 1960s, were due to the extensive atmospheric nuclear weapons testing during this period; the peak value observed in 1986 is attributed to the Chernobyl accident on 26 April 1986.

radionuclide measurements. The commonly used high-volume multistage cascade impactor (Andersen air sampler) can separate particles into different sizes, such as $>7\,\mu m$, $3.0–7.0\,\mu m$, $2.0–3.0\,\mu m$, $1.0–2.0\,\mu m$, and $<1.0\,\mu m$ (backup filter), and the sampling can take place continuously for more than one week with a flow rate of $0.5–2\,m^3\,min^{-1}$. Aerosol particles are collected on filters according to particle size, after which the filters are folded and compressed into a vial for gamma measurement. They are then ashed for the radiochemical analysis of radionuclides with beta and alpha emission. Figure 18.4 shows an Andersen cascade impactor which can collect eight particle size fractions.

18.2.1.4 **Passive Aerosol Particle Sampling**

In addition to integral and size-selective air samplers, a passive aerosol sampling technique is also used, which is especially useful to investigate the air particles resuspended from the ground. This can be carried out by hanging a sticky foil at the sampling site at a height of 1.5 m above the ground to trap the particles suspended in the air (Figure 18.5). The foil is then shipped to the laboratory for radiochemical analysis of the radionuclide concentrations and/or separation of hot particles from the foil.

18.3
Sampling Gaseous Components

In air, some radionuclides exist as gas – the most important being tritium, radioisotopes of radon, and krypton – and cannot be collected using a filter. The state-of-the-art equipment for the determination of radon and krypton in the air is normally an integrated system consisting of an air sampler for the collection and preconcentration of the analyte combined with on-line measurement instruments.

Figure 18.4 Andersen cascade impactor (Thermo Scientific) which can collect eight aerosol particle size fractions from the air.

The methods for sampling and determining radon and radioactive krypton are presented in Chapter 12 and are therefore not discussed here.

Tritium is an important radionuclide in the environment, originating from both natural processes and anthropogenic releases; however, the current main source of tritium in the environment is from releases at nuclear facilities, especially from nuclear reactors (particularly heavy water reactors) and nuclear fuel reprocessing plants because of the large amounts of tritium formed by neutron activation. Tritium monitoring is normally conducted as an on-line measurement of the tritium level in indoor air, but the analytical sensitivity of this kind of instrument is normally low and does not meet with the requirements needed for radiation protection and environmental investigations. However, the collection of airborne tritium and its measurement with sensitive liquid scintillation counting can measure its levels in the air with sufficient sensitivity and accuracy. Tritium in the air of nuclear facilities is generally in the form of water vapor, and the collection of tritium-bearing water in the air can thus be carried out by cooling the air. For this purpose, the air sampler consists of a U-type tube cooled with liquid nitrogen with an aerosol filter in the head of the system. The air is sucked through the system with a vacuum pump, the aerosol particles are

Figure 18.5 A passive aerosol particle sampling setup using sticky foil to investigate the radioactive particles suspended from contaminated land.

removed on the filter, and the tritium-bearing water is collected in the U-tube. After sampling, the U-tube is disconnected and the collected water is measured for its tritium concentration. For tritium in outdoor air, molecular sieve instead of cooled U-tube is often used to collect the tritiated vapor from large air volumes.

Iodine is a volatile element existing in air in gaseous forms such as elemental iodine (I_2) and volatile organic compounds, mainly alkyl iodides (e.g., CH_3I and CH_2I_2). During the operation of nuclear reactors and spent nuclear fuel reprocessing plants, large amounts of radioactive iodine, such as short-lived ^{131}I, ^{132}I, ^{133}I, and ^{134}I, and long-lived ^{129}I, are released into the atmosphere. In an accident, much larger amounts of these isotopes can be released into the atmosphere. Because of the high enrichment of iodine in the human thyroid gland, ^{131}I and other short-lived radioiodine isotopes form a major radiation risk to people exposed to the contaminated environment. While some of the iodine may associate onto aerosol particles after being released into the atmosphere, a large fraction still exists in gaseous form, and, of course, both aerosol particle-associated and gaseous iodine must be measured. Figure 18.6 shows an air sampling system which collects iodine in three forms: aerosol particles, inorganic gaseous iodine, and organic gaseous iodine. The air passes first through a filter to collect the aerosol particles, then a cellulose filter impregnated with NaOH to collect inorganic gaseous iodine (i.e., I_2, HI, and HIO), and finally a cartridge filled with activated carbon impregnated with an amine, for example, triethylene diamine (TEDA), to collect the organic gaseous iodine. The iodine in these different forms is then subjected to gamma spectrometry to measure ^{131}I and other short-lived gamma-emitting iodine radioisotopes, or used for the radiochemical analysis of long-lived ^{129}I (see Chapter 11).

Figure 18.6 Air sampler for collecting aerosol particle-associated iodine, inorganic gaseous iodine, and organic gaseous iodine.

18.4 Atmospheric Deposition Sampling

Radionuclides in the atmosphere are ultimately deposited onto the ground either by dry deposition due to gravity and interactions with the ground materials or by wet deposition (rainfall precipitation wash-out). Because of the low concentration of radionuclides in precipitation, an *in-situ* preconcentration is normally carried out to avoid the collection and transportation of large volumes of precipitation samples. Depending on the purpose, the collection of the deposition can be carried out by different methods.

18.4.1 Dry/Wet Deposition Sampling

For the collection of dry deposition, a high-walled pot collector is normally used. This is usually made of stainless steel, since this has a smooth surface which does not adsorb radionuclides, is durable, and has rounded corners which allow for easy cleaning (Figure 18.7). The pot collector is placed in an open area away from buildings or overhanging trees and shrubbery. If only dry deposition is to be collected, the pot is covered during periods of precipitation and opened again afterwards. At the end of the collection period (normally one month), the deposited material is transferred to a container by careful brushing. Distilled water is then added to the pot, the sides and bottom are scrubbed, and the slurry produced is combined with the rest of the deposited material in the container. The pot is further washed with HNO_3 and water, and this too is added to the container. The collected sample is then measured for

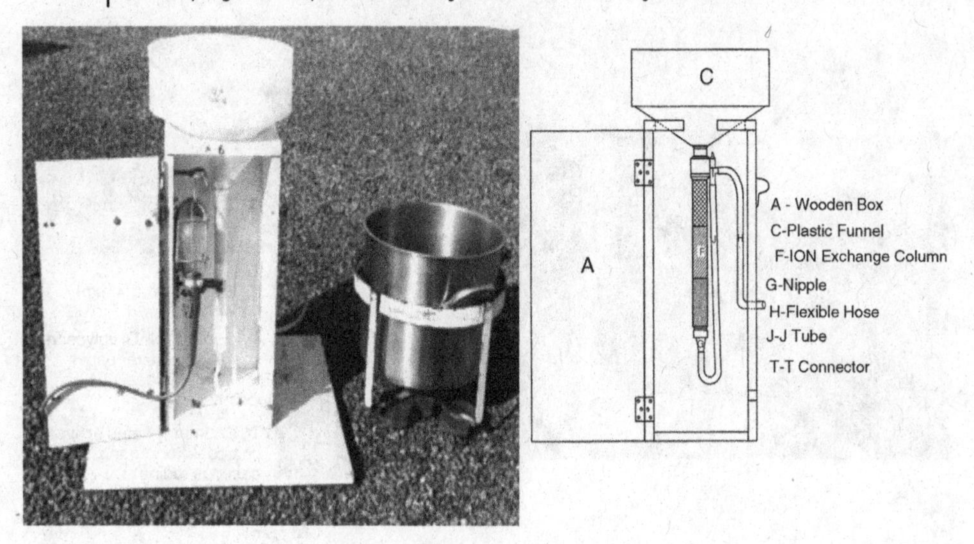

Figure 18.7 Ion-exchange fallout collection and fallout pot collector.

gamma-emitting radionuclides, after which the radiochemical analysis is carried out following dissolution of the sample.

Wet deposition can also be collected with the pot collector, but to avoid the loss of some of the radionuclides by evaporation a funnel type collector is normally used. If only wet deposition is to be collected, the mouth of the funnel is covered with a plastic film or sheet, which is removed during precipitation. In this case, single precipitation as well as integrated monthly precipitation can be separately collected. The surface of the funnel is typically made of smooth stainless steel and the container of plastic. The funnel is normally covered with a net to prevent debris such as tree leaves, dead insects, and grasses from entering. The collected precipitation is filtered through a filter paper/membrane and transported to the laboratory for storage and analysis.

18.4.2 Ion Exchange Collector

An ion exchange collector consists of a funnel to collect precipitation on top of an ion exchange column through which the water passes into a container to collect the effluent (Figure 18.7). If tritium is to be measured, the effluent is shipped to the laboratory for further treatment and analysis (see Chapter 13). The ion exchange column contains a mixed bed consisting of both anion and cation exchange resins, by means of which most of the important radionuclides can be absorbed from the water to the resin bed. After sample collection (e.g., one month), the column is shipped to the laboratory where the resin is taken for gamma measurement and radiochemical analysis. Using this type of collector, transportation of the precipitation is avoided and the analysis of a large amount of precipitation becomes easier. The experimental results (Table 18.1) show that more than 90% of the most important

Table 18.1 Collection efficiency of a mixed-bed ion exchange column (2.5 cm diameter × 40 cm height) for important fallout radionuclides (Risø National Laboratory for Sustainable Energy, Denmark).

Radionuclide	Efficiency, %	Radionuclide	Efficiency, %
^{7}Be	97	^{125}Sb	70
^{54}Mn	87	^{131}I	84
^{90}Sr	79	^{137}Cs	99
^{95}Zr	89	^{140}Ba	98
^{95}Nb	89	^{133}Ce	88
^{103}Ru	85	^{210}Pb	77

radionuclides in monthly precipitation collected in a 10 m^3 collector can be trapped using a column of diameter 2.5 cm and height 40 cm.

18.5 Water Sampling

Water samples collected from seas, lakes, rivers, and sometimes ice, are often analyzed in the investigation of environmental radioactivity. Water samples are relatively homogeneous in most circumstances. In most cases, the aims are to study the levels of radionuclides in water systems and determine their spatial distribution (vertical and horizontal). This is often required to investigate the transport of radionuclides and to apply them as tracers for the investigation of the circulation and transportation of water masses. Water sampling is mainly performed on surface waters and waters at different depths. In most cases, the sample is filtered *in situ* through a membrane filter (0.45–1.0 μm) to remove any suspended materials and is then stored in a polyethylene bottle. For the determination of most radionuclides, the water sample is acidified with HNO_3 to pH 1–2 to reduce their adsorption on the container walls. In the analysis of certain radionuclides, such as iodine isotopes, the samples should not be acidified, because iodide and iodate can form the volatile elemental iodine (I_2), which could get lost in the atmosphere and/or strongly adsorbed on the polyethylene material.

18.5.1 Surface Water Sampling

Surface water can be easily collected using various methods. For a small sample (<20 L), it can be taken directly into a polyethylene bottle or barrel. For a large sample (>20 L), a submersible pump is often used. To avoid any cross contamination of the samples, the first 10 L is normally discarded. Also, the sample container is first washed with the water sampled and the wash is discarded. Most expedition ships are equipped with a surface water sampling device, so that continuous surface water sampling can easily be carried out. In addition to a pump, these devices are also equipped with a number of water sensors to measure the physical and chemical parameters of the water, for example, pH, Eh, temperature, dissolved oxygen concentration, H_2S concentration, and salinity.

18.5.2
Water Core (Depth Profile)

When the depth distribution of radionuclides is studied, the water core needs to be sampled. For the collection of depth profile water, especially seawater to a depth of up to 4000 meters, a Nansen bottle – more precisely a metal or plastic cylinder – is often used (Figure 18.8). The bottle is lowered into the water by a cable until it reaches the required depth, when a brass weight – a 'messenger' – is dropped down the cable.

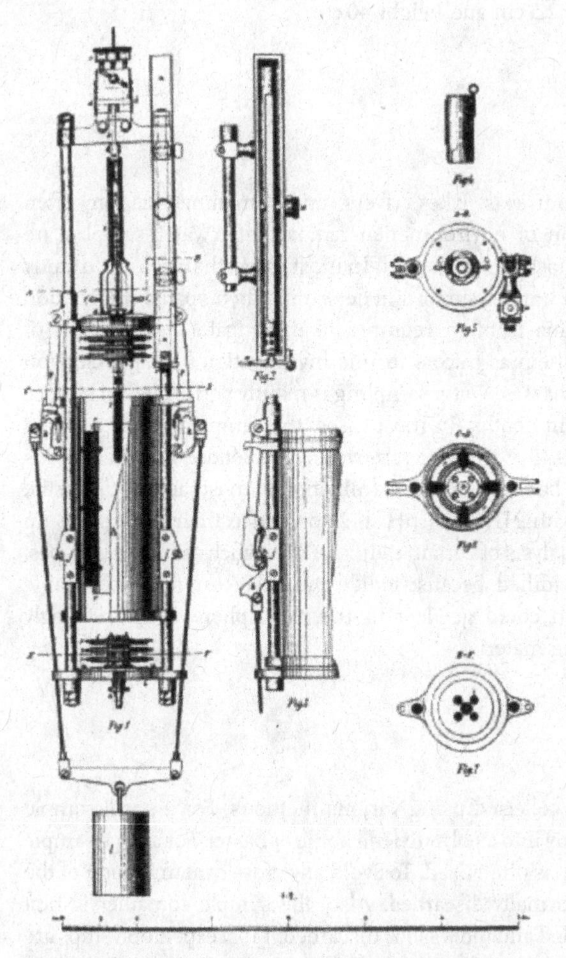

Figure 18.8 Nansen bottle.

When the weight reaches the bottle, the impact tips the bottle upside down and trips a spring-loaded valve at the end, trapping the water sample inside. The bottle with the sample is then retrieved by hauling in the cable. By fixing a sequence of bottles and messengers at intervals along the cable, a series of samples at different depths can be taken.

The Niskin bottle (Figure 18.9) was developed from the Nansen bottle and has largely superseded the latter, which is no longer in production. Instead of a metal bottle sealed at one end, the 'bottle' is a tube, usually plastic to minimize contamination of the sample, and open to the water at both ends. Each end is equipped with a cap which is either spring-loaded or tensioned by an elastic rope. The action of the messenger weight is to trigger both caps shut and seal the tube. A modern variation of the Niskin bottle uses actuated valves that may either be preset to trigger at a specific

Figure 18.9 Niskin bottle.

depth, detected by a pressure switch, or remotely controlled to do so via an electrical signal sent from the surface. This arrangement allows a large number of Niskin bottles to be mounted together in a circular frame termed a *rosette* (Figure 18.10). As many as 36 bottles may be mounted on a single rosette. A depth sensor/recorder, thermometer, pH-Eh meter, and other sensors are normally installed on the side of the bottle to measure various parameters of the water.

For sampling the depth profile of lake water, especially shallow lakes, a simple submersible pump can be used. In this case, the sampling depth is normally measured manually by the length of the tube from the top of the tube to the water surface. A heavy object is normally attached to the sampling tube/cable to keep it straight and ensure an accurate depth measurement.

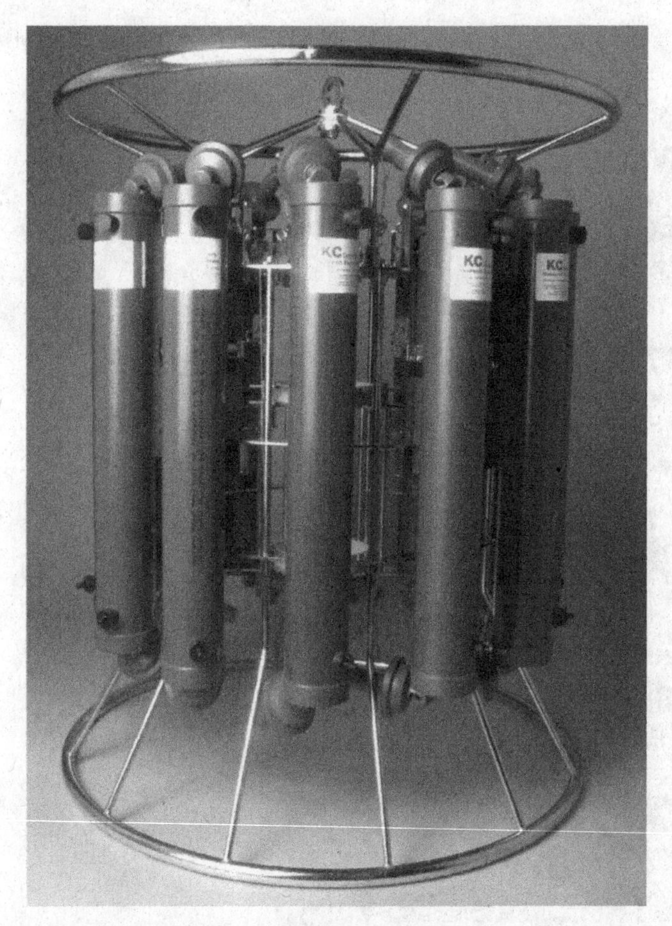

Figure 18.10 A rosette with 12 Niskin bottles mounted together in a circular frame. Each bottle can be triggered to close the caps at both ends at different depths to sample depth profile water samples.

18.5.3
Preconcentration of Radionuclides from Natural Waters

As the concentrations of radionuclides in natural waters, especially seawater, are normally very low, a large volume of water must be collected and analyzed to enable the radionuclide content to be measured accurately. For example, the volume of water needed to determine the activity concentrations of ^{137}Cs, ^{99}Tc, ^{237}Np, ^{241}Am, and plutonium isotopes in seawater is 200–1000 L. The shipment of such large volumes of water would be very expensive and would restrict the number of tests. To overcome this problem, radionuclide preconcentration methods have been developed and applied. *In-situ* preconcentration is very helpful in reducing the volume of the sample to be transported.

18.5.3.1 Preconcentration of Radiocesium (^{137}Cs and ^{134}Cs)

A widely used separation method for radiocesium in natural water is ammonium phosphomolybdate (AMP) precipitation. In this method, the water is first acidified to pH 1.5–2.0, AMP powder is then added, and the water is stirred for one hour by air bubbling. The cesium-bearing AMP precipitate is then collected by sedimentation on the container bottom. A chemical yield of 90–100% can be obtained for a 20–50-L sample of water. This method is, however, impractical and expensive for the treatment of larger volumes of water (>50 L). An *in-situ* sampler based on AMP-impregnated silica gel in a column has also been used for the separation of radiocesium. However, because of the microcrystalline structure of AMP and the high pressure build-up during filtration, this method cannot be used to treat volumes of water greater than 100 L.

Various transition metal hexacyanoferrates, such as copper hexacyanoferrate ($Cu_2Fe(CN)_6$), nickel hexacyanoferrate, iron(II) hexacyanoferrate, zirconium hexacyanoferrate, and potassium cobalt hexacyanoferrate ($K_2CoFe(CN)_6$) have also been used to extract radiocesium from water by precipitation or by adsorption on filters impregnated with a hexacyanoferrate. By using a cartridge filtration system (Figure 18.11), volumes of seawater or fresh water as large as 1000 L can be treated with two sequentially connected filters with a chemical yield higher than 80%. The main advantages of the $Cu_2Fe(CN)_6$ cartridge filtration system compared with the AMP method are the large volumes of water treated, the fact that it functions at natural water pH (6–8) (i.e. without any acidification), less adsorption of interfering ^{40}K, and low cost. The $Cu_2Fe(CN)_6$-impregnated cotton-wound cartridge filter can be home-prepared as follows: (i) the cartridge filter is first immersed in an adequate amount of 0.1 M potassium hexacyanoferrate ($K_4Fe(CN)_6$) solution for 1 h; (ii) the cartridge filter is dried and immersed in 0.2 M copper chloride ($CuCl_2$) solution for 1 h to form the copper hexacyanoferrate precipitate coating on the cartridge filter; and (iii) the impregnated cartridge filter is dried at 60–80 °C. A prefilter is normally added before the impregnated cartridge filter to remove any suspended materials in the water. This is especially necessary for fresh waters, as these usually contain large amounts of suspended matter. With this system, a flow rate of 4–6 L min^{-1} can be used and the treatment of 1000 L of water takes less than 3 h. It is therefore very

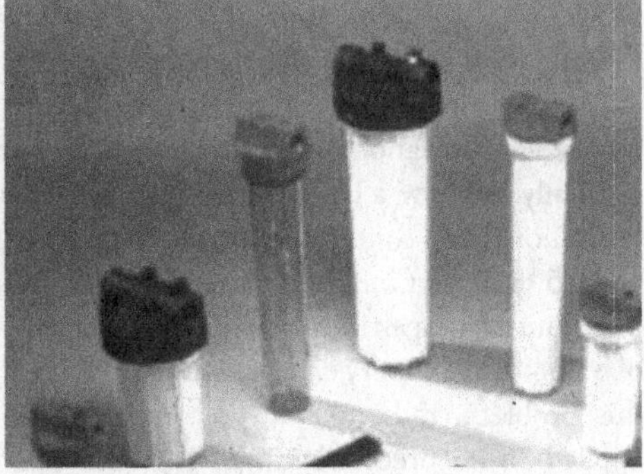

Figure 18.11 Cotton-wound cartridge filters (left) and filter housings (right) (Westdale Filters Ltd, Selby, UK) used for preparing $Cu_2Fe(CN)_6$-impregnated cartridge filters to separate radiocesium from large water volumes.

suitable for *in-situ* and on-board operations. After filtration, the cartridge filter is taken out of the filter housing and put into a plastic bag for storage and transportation. In the laboratory, the cartridge filter is dried and ashed at 420 °C, after which the ash is collected for the gamma measurement of radiocesium (Roos, P., Holm, E., and Persson, R.B.R. (1994) Comparison of AMP precipitation method and impregnated $Cu_2Fe(CN)_6$ filters for the determination of radiocesium concentration in natural waters. *Nucl. Instr. Method A*, **339**, 282).

18.5.3.2 Preconcentration of Pu, Am, Np, and ^{99}Tc

In natural waters, plutonium isotopes may exist at various oxidation states: Pu^{3+}, Pu^{4+}, $PuO_2{}^+$, $PuO_2{}^{2+}$, neptunium as Np^{4+} and $NpO_2{}^+$, americium as Am^{3+}, and technetium as $TcO_4{}^-$. For the preconcentration of these radionuclides from large water volumes (100–500 L), HCl is first added to adjust the pH to 2, and the tracers (^{242}Pu, ^{243}Am and ^{99m}Tc) are added for the chemical yield measurement. In addition, $FeCl_3$ is added as a coprecipitation agent. Thereafter, $Na_2S_2O_5$ is added to the water to reduce the plutonium to Pu^{3+}, Np to Np^{4+}, and ^{99}Tc to Tc^{4+}. Iron(III) is also reduced to iron(II). After mixing for 30 min with air bubbling, NaOH or NH_4OH solution is added to adjust the pH to 8–9, which results in the coprecipitation of Pu, Np, Am, and Tc with the forming $Fe(OH)_2$. After the precipitate settles, the supernatant is discarded and the precipitate is collected in a plastic container. HCl is finally added to the container to dissolve the precipitate, and the solution, now in a small volume (<5 L), is shipped to the laboratory for further radiochemical separation. It is important to dissolve the $Fe(OH)_2$ precipitation before storage and shipment, otherwise the Pu, Np, Am, and especially Tc will form insoluble oxides which are difficult to dissolve completely in the laboratory. As a consequence, a low chemical yield will be obtained for these radionuclides.

MnO_2 coprecipitation is also used for *in-situ* preconcentration of Pu, Am, and Np from large water volumes (100–500 L). In this method, the water sample is collected in a large container, HCl is added to adjust pH to 2, the yield tracers are added, and the water is stirred by bubbling. $KMnO_4$ is then added (for each 100 L of water, 2 g of

$KMnO_4$ is added), which results in the oxidation of Pu to PuO_2^+ and Np to NpO_2^+. After the addition of NaOH to adjust the pH to 8–9, $MnCl_2$ solution ($MnCl_2$: $KMnO_4 = 1$: 1 g/g) is added to form MnO_2 through the reaction $2MnO_4^- + 3Mn^{2+} + 2H_2O \rightleftarrows 5MnO_2 + 4H^+$. Because of the increase in hydrogen ion concentration, the pH may need to be re-adjusted to 8–9 for fresh waters. The MnO_2 suspension is stirred for 2–3 h by air bubbling. Pu, Am, Np, and many other radionuclides, such as Ra and Th, will coprecipitate with MnO_2. The formed precipitate is collected after settling down to the bottom of the container. The MnO_2 precipitate is finally dissolved using HCl with the assistance of H_2O_2 in a small volume (<5 L) and transferred to a plastic bottle, which is stored and shipped to the laboratory for radiochemical analysis. An MnO_2-impregnated filter or cartridge is also used for the preconcentration of radionuclides from large volumes of water. A similar method as that for $Cu_2Fe(CN)_6$-impregnated cartridge filters is used to prepare an MnO_2-impregnated cartridge filter. In this case, $KMnO_4$ and $MnCl_2$ are used to form MnO_2 on the cartridge filter. Two cartridges housed in a cartridge holder are connected sequentially to obtain a high trapping efficiency of radionuclides.

18.6 Sediment Sampling and Pretreatment

Sea and lake sediments are unique samples for the investigation of the occurrence, transport, accumulation, and inventory of anthropogenic as well as naturally occurring radionuclides since they act as a main sink for radionuclides entering water systems. To properly assess the geochronology of sediments, an undisturbed sediment core with a suitable depth has to be collected – the commonest procedure in sediment sampling. Sometimes, only top layer (surface) sediment is collected to investigate the contamination on the surface of the sea bed or lake bottom. In this case, accurate evaluation of the inventory of radionuclides is difficult. Surface sediment sampling, on the other hand, enables a larger number of samples to be collected to map the horizontal distribution of contamination on the bed of the investigated sea or lake.

18.6.1 Surface Sediment Sampling

Sediment at a depth of 0–15 cm is normally considered as surface sediment, which can be simply collected by using spades, shovels, trowels, and scoops from shallow water; however, this method is limited by the depth and movement of the water layer. For shallow waters (0–0.5 m), a tube auger (Figure 18.12a) or core tube (Figure 18.12b) is the most suitable device. By using additional extensions with the tube auger, surface sediment at water depths of 0.5–2.5 m can be collected. To sample surface sediment from deep waters, Ekman or Ponar dredges (Figure 18.13) should be used. The Ekman dredge is a lightweight sediment sampling device used for collecting moderately consolidated fine-textured sediment. The Ponar dredge, on the other hand, is heavy and suitable for consolidated fine- to coarse-textured sediment.

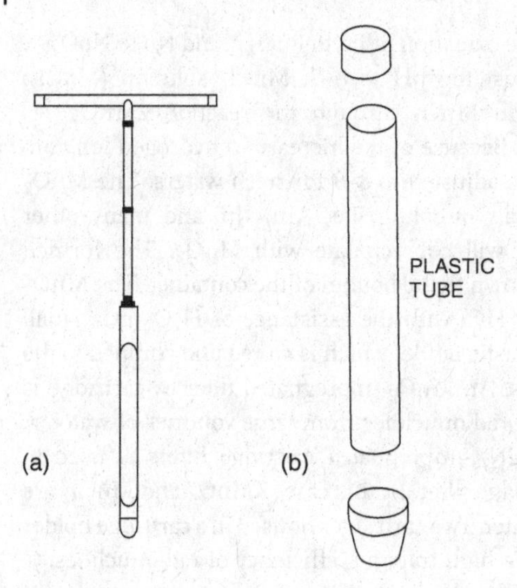

Figure 18.12 Sediment samplers: (a) tube auger, managed by hand or driver; (2) core tube, lowered into the sediment by weight.

To use the Ekman dredge, a sturdy nylon rope or stainless steel cable is first attached to the bracket, and springs are then attached to both sides of the jaws, which are fixed so that they are in the open position by placing trip cables over the release studs. The sampler is then lowered to 10–15 cm above the sediment surface and dropped. The jaw release mechanism is trigged by lowering a messenger down the line or by depressing the button on the upper end of the extension handle. The sampler is then raised, and any free liquid is slowly decanted through the top of the sampler; the dredge jaws are finally opened to transfer the sample into a container.

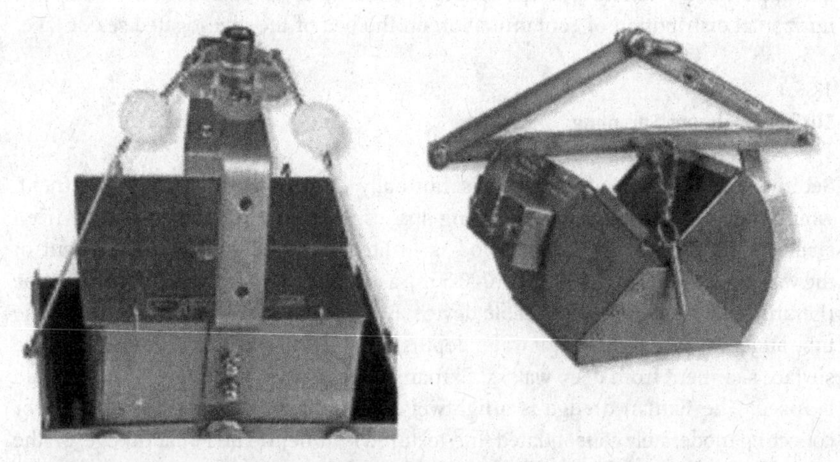

Figure 18.13 Surface sediment sampler (a) Ekman dredge; and (b) Ponar dredge.

18.6.2 Sediment Core Sampling

Sediment core sampling is required to obtain detailed information on the vertical distribution of the radionuclides in the sediment. Different types of sampling devices can be used depending on the requirements and sampling situation. The tube auger (Figure 18.12a), described above for surface sediment sampling, can also be used for short-depth sediment core sampling and is operated by hand or hammer. It is also possible to take samples from hard sediments with sand and clay, and extension handles can be used to take samples in slightly deeper waters. A simple core tube (Figure 18.12b) is commonly used to sample soft sediment from water depths of less than 100 m. The cap on the device must be tightened on the top of the tube before it is raised out of the water to prevent the sediment from dropping down.

The Limnos sampler is lightweight and suitable for fairly loose sediments in lakes with water depths not much deeper than 50 m (Figure 18.14). The diameter of the tube is 10 cm, and it collects up to 50-cm deep sediment cores. The tube consists of fifty 1-cm plastic rings by which the slicing is done. The sampler is lowered to the lake bottom on a rope by hand, penetrating the sediment by gravity aided by extra weights. After reaching the sediment, the lower end of the tube is closed with a lid and raised to the surface.

The box corer (Figure 18.15) is the simplest and most commonly used device for taking large sediment samples from deep water, having the dimensions 50 cm × 50 cm × 75 cm. A depth pinger or other depth indicator is generally used to determine when the box is completely filled with sediment. Once the core box is full, the sample is secured by moving the spade-closing lever arm to lower the cutting

Figure 18.14 Sampling a sediment core with a Limnos sampler at Lake Päijänne, Finland.

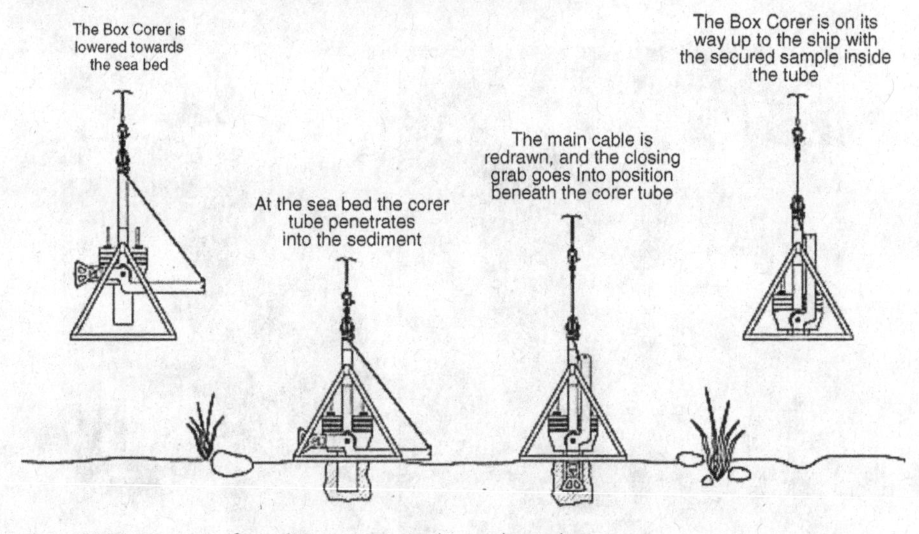

Figure 18.15 Box corer for sediment core sampling and sampling steps.

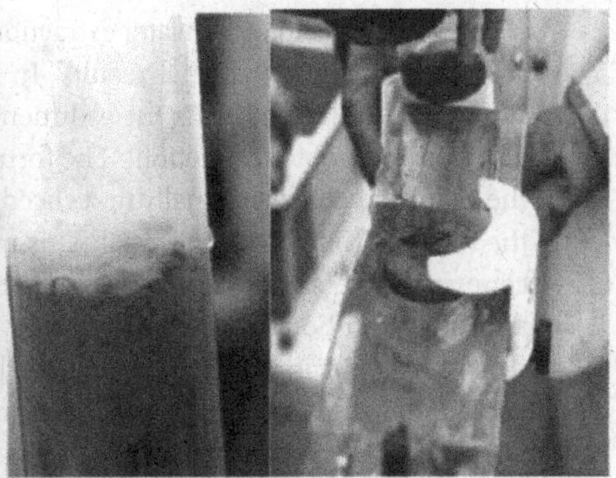

Figure 18.16 Gemini sediment sampler, sediment core, and core slicing.

edge of the spade into the sediment until the spade completely covers the bottom of the sediment box. Compared with other sediment corers, the depth of the sediment core taken by the box corer is shorter (<75 cm).

Figure 18.16 shows a Gemini sediment core sampler, which can be used to take different types of sediments from deep waters. The core barrel is lowered by a rope and driven into the sediment by the force of mounted weights. The core barrel stops its downward travel when it reaches a certain depth or when the resistance of the sediment is sufficient to stop the downward penetration. The top valve is then closed and the corer is pulled out of the sediment at the slowest winch speed.

Care must be taken to avoid the loss of the upper sediment layer, which can be caused either by washout at the core top because of a shorter length of the core tube compared with the penetration depth of the core barrel or by sweeping away due the shock wave of a rapidly falling core device. This problem often occurs with narrow diameter core tubes (<10 cm) and soft sediments. Methods used to minimize the sediment disturbance and the loss of the top layer include (i) choosing a larger diameter tube, (ii) keeping the sediment sampler in as vertical position as possible while it penetrates the sediment, and (iii) lowering the sampler very slowly.

Selecting a suitable point on the sea bed is a challenge in sediment core sampling, Sediment mapping by CHIRP (compressed high intensity radar pulse) and side-scan sonar are good methods which can provide useful information such as bathymetry, sediment profiles, and surface mosaic for sediment core sampling. On-line information from the CHIRP system assists in selecting suitable sampling points.

18.6.3
Sediment Pore Water Sampling

Information about radionuclides in sediment pore water is important for many reasons. Knowledge of the concentrations and forms of radionuclides present gives insight into the behavior and mobility of the radionuclides in water/sediment systems and helps in understanding the mobility of the radionuclides and in determining the value of distribution coefficients (K_d) in actual environmental conditions.

Many systems for pore-water extraction have been developed to obtain reliable and representative experimental results. In general, there are two approaches to separating pore water: (i) coring the sediment and separating the water in the laboratory, and (ii) *in-situ* water extraction. The former procedure is simple and easy to operate: the sediment core is normally first sliced and then centrifuged or squeezed to extract the water from the solid material. This method may, however, induce compositional changes in the sample due to pressure and temperature changes, redox variations, and solid-solution interactions during extraction, transport, and storage. These disadvantages can be partly overcome by taking the pore water directly from the core tube. In this case, small holes are drilled on the wall of the core tube at different depths and then sealed by a soft material before sediment sampling. The sediment core taken by the tube is stored and transported to the laboratory; and the pore water is finally extracted by using a syringe and further filtered through a membrane.

The best method is *in-situ* extraction of the pore water, which can minimize many of the sampling deficiencies mentioned; however, this operation is much more complex, and it is difficult to take pore water samples from deeper than a few meters. Figure 18.17 shows an *in-situ* pore water sampler for fine grained sediment in shallow waters. This device consists of a rectangular based nylon block in which V-shaped

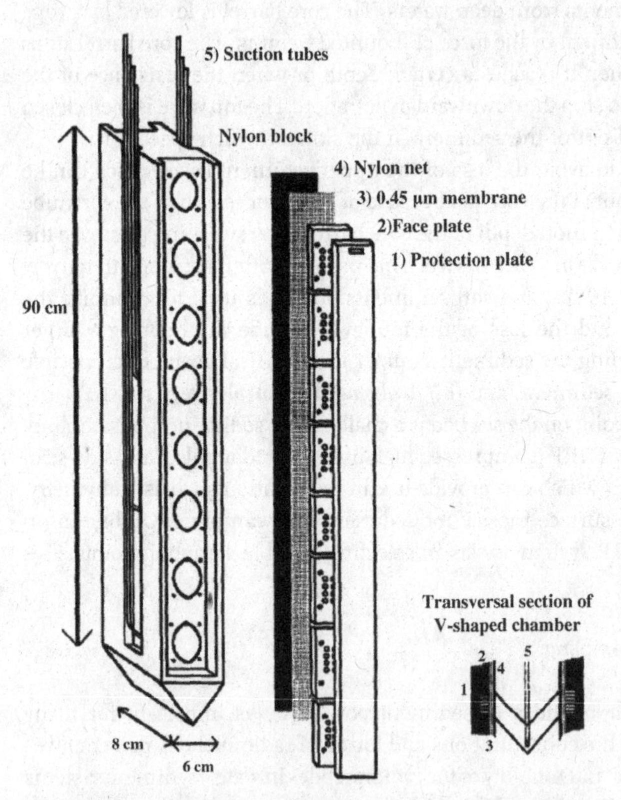

Figure 18.17 Schematic diagram of an *in-situ* pore water sampler (Bertolin, A., Rudello, D., Ugo, P. (1995) A new device for *in-situ* pore water sampling. *Marine Chemistry*, **49**, 233).

chambers for collecting pore-water are milled at 10 cm intervals. The sampling ports are entirely covered by a membrane, the inner side of which is protected by a nylon net (item 4 in Figure 18.17) to prevent membrane breakdown during sample suction. A series of silicone tubes (item 5 in Figure 18.17) connects each chamber with the surface to allow the collection of pore waters without having to remove the sampler. The outer side of the membrane is covered by a nylon face plate (item 2 in Figure 18.17) with holes drilled in correspondence with each sampling port so that the total free inlet area of each port is restricted. Each nylon face plate is fixed to the main frame by four nylon screws. The face plates obstruct the formation of preferential vertical paths, preventing waters belonging to different levels from percolating along the sampler walls. The protection plates (item 1 in Figure 18.17) are removed only after the final placing of the sampler into the sediment; the residual air present in the chambers is then removed by suction with a hand pump before the silicone tubes are sealed by closing the series of stopcocks. Sampling is carried out, after a suitable equilibration time, by opening one stopcock and gently extracting the material with a syringe for analysis. At the end of the extraction, there is no free volume left in the syringe, which is then sealed by closing a stopcock at the syringe inlet. No further manipulation or filtration of the pore-water is needed.

18.6.4
Pretreatment of Sediments – Storage, Drying, Homogenizing

If the chemical speciation of radionuclides is carried out for sediments, care should be taken not to change their chemical form by exposure to a different environment after sampling. The intrusion of oxygen into the samples, in particular, should be avoided so that the oxidation states of redox-sensitive radionuclides are not affected. To minimize the effects of the changing chemical environment, the collected sediment can be stored in nitrogen-filled containers, stored at a low temperature (for example, 4 °C or frozen at −20 °C), and handled in glove box with a nitrogen atmosphere. Such protection is not necessary, however, for the investigation of total concentrations of radionuclides in the sediment.

To investigate the depth distribution of radionuclides, the collected sediment core must be sliced into segments, which can be done either *in situ* or in a laboratory. *In-situ* slicing can be carried out by removing the sediment from one side of the sampler tube by inserting a valve of the correct size into the sampler tube, which is then precisely driven forward by a screw. In most cases, the concentrations of artificial radionuclides or other pollutants are higher in the upper layers of sediment than in the lower layers. Removing the sediment from the bottom of the tube thus prevents the contamination of the lower sediment layers. When removed from the sampler tube, a certain depth of sediment is collected into a container by shaving it off from along the top rim of the sampler tube. Since the sediment core has been squeezed, the density of the core sediment may have been changed, the pore water may have been moved across the different layer, and the depth of the collected sediment may not be precise. To overcome this problem, slicing the sediment in the laboratory after freezing is recommended: the sediment core is first deep-frozen and then sliced by sawing it along with the plastic sampler tube. The sliced sediment segments are then separated from the tube material after thawing and stored in plastic bags or bottles.

Before analysis, the sediment samples must be dried and homogenized. Freeze drying is normally used to dry the sediment samples, as the sediment is then obtained in the form of a soft substance which is easy to crush, grind, and homogenize.

18.7 Soil Sampling and Pretreatment

18.7.1 Planning the Sampling

Soil sampling is an essential method of investigating (i) the deposition levels of radionuclides on the ground from different sources, such as nuclear accidents, operation of nuclear facilities, weapons testing, and (ii) the migration of radionuclides in different soil types and environmental conditions. As in sediment sampling, the selection of area, sites, sampling points, and the number and volume of samples are issues to be considered in order to obtain representative and reliable results. Compared with sediment sampling, soil sampling is relatively easy to manage, and the devices are simpler to use since the operation is on land.

The purpose of a project, whether it deals with radionuclide deposition, resuspension, root uptake, or some other question, dictates the way of sampling. Sampling can be carried out to study the total inventory, surface concentration, or depth profile of radionuclides. Site characteristics, such as soil type, topography, and sources must also be considered when designing the sampling strategy. Radionuclide measurements in soil can be used to estimate their inventory and distribution over a given area. Figure 18.18 shows, as an example, the distribution of ^{137}Cs deposited from the Chernobyl accident on soil over Europe. In the inventory investigations, the sample should be taken to a sufficient depth so that all the deposited material is sampled. Without previous knowledge of the depth of radionuclide penetration, a great depth, certain to be adequate, must be selected. An optimum depth for sampling should contain more than 95% of the total radionuclide of interest.

When accumulated deposition over a given time period is to be estimated by soil sampling, the selected sampling area must have been undisturbed for this time period. To be representative of an area, the sampling site should be at the center of a large, flat, open area. An area at the foot of a slope, in a low spot, or in a flooded area would not be suitable. Also, the site should not be near to buildings or trees that shelter the area during rains and should be 100 m or more from a dusty road. The sample should preferably be collected from a large area to make it more representative. The selected area is better if it has vegetation and has moderate to good permeability to reduce the runoff of the deposited radionuclides. Soils with high earthworm activity should be avoided because of uneven mixing of the soil to considerable depths. Rodent activity also makes an area unsuitable for sampling.

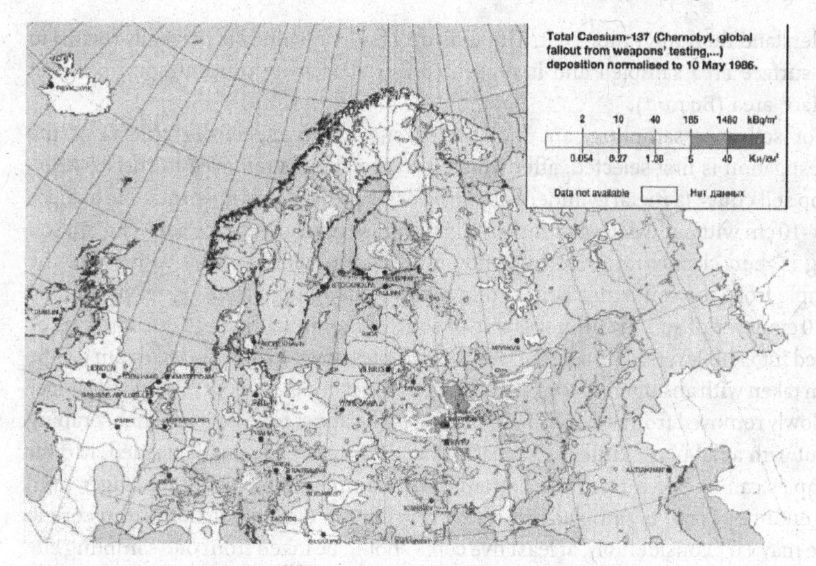

Figure 18.18 Distribution of ^{137}Cs in soil from the Chernobyl fallout over Europe.

To investigate contamination from an acute release or an accident at a specific facility, surface soil sampling soon after the event can be used to define the contamination contours or distribution pattern. Sampling the top soil to a depth of 5 cm is normally enough. In the event of accidental or operational releases, the amount deposited may vary according to the direction and distance from the release point. Airborne particle dispersion is affected by particle size and meteorological conditions. Gaussian plume models, which take into account meteorology, stack height, topography, and deposition velocity of the particles, can be used to map local dispersion patterns and design the sampling points. The horizontal dispersion of liquid effluent releases is influenced by the composition and quantity of the liquid, the topography, the soil type, and the properties of the released material.

When evaluating soil to estimate the uptake availability of a contaminant radionuclide, it is not necessary to measure the total deposition, only the amount in the root zone available to the plant or crop of interest. In most cases, this would be the depth of the plowing layer, that is, approximately 30 cm. In addition to the root zone concentration, the extent to which the nuclide is chemically available for uptake must be considered.

For practical soil sampling, three main soil sampling methods: core, template, and trench methods, are presented below. In all these methods, the sampling procedure has to be optimized to reduce any potential cross contamination.

18.7.2
Soil Core Sampling

The core method is suitable when investigating the total inventory of radionuclides from fallout to the ground, and the depth distribution of radionuclides in the soil to understand their migration in it. The analytical figure obtained is normally related to the surface area sampled and is presented as radioactivity concentration per unit surface area (Bq m^{-2}).

For soil core sampling, an undisturbed site which meets the criteria of the investigation is first selected, after which the vegetation is removed to surface level. A top soil cutter (a metal cylinder with handles) is pressed into the ground to a depth of 5–10 cm without twisting or disturbing the grass cover or surface soil. The top soil plug is then cleanly removed by gently twisting the handle of the cutter, and the sample from the soil cutter is placed in a plastic sampling bag to represent the top 5–10 cm layer of soil. If thinner layers of soil are required, the 5–10 cm core can be sliced into sub-layers. The sub-surface soil samples down to the desired depth can be then taken with an auger (Figure 18.19). When the cylinder is about 3/4 full, the auger is slowly removed from the soil. The soil core in the auger tube is sliced after scraping it out with a flat blade knife into a plastic bag. If deeper sampling is needed, further samples can be taken from the bottom of the sampling hole using the auger until the entire soil core is sampled. Since the distribution of radionuclides from core to core may vary considerably, at least five cores should be taken from one sampling site to allow for statistical variation in the results. The core method may be not suitable for sampling powdery, dry, loose, or single-grain soils. An alternative method for sampling loose soils is to leave the corer in place and scoop out the contents. However, only one composite depth can be taken, since once the corer is removed the integrity of the core is lost.

Another type of soil coring device is shown in Figure 18.20. This is designed to make holes on golf courses but is well suited to collect surface soil samples. The tube diameter is 10 cm and the length of the tube (consisting of two half-cylinders) is 22 cm. The advantage of this device is that the structure of the core can be seen and photographed. If deeper sampling is needed, further samples can be taken from the bottom of the sampling hole by an auger, for example.

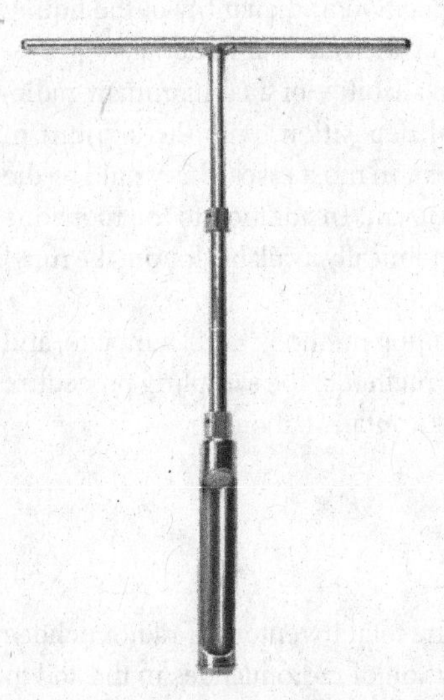

Figure 18.19 Soil core sampling device (auger).

Figure 18.20 Sampling podzolic forest soil in Finnish Lapland.

18.7.3
Template Method

The template method is normally used for sampling rocky soils. In this method, a circle or square template is used, and the soil and rocks are removed down to the desired depth with a chisel and scoop. Large rocks are discarded while the remaining smaller rocks are crushed as part of the sample. A depth profile may be drawn using this method by repeating the steps for each subsequent depth to be sampled.

18.7.4
Trench Method

The trench method of determining the depth profile of radionuclides can yield larger samples than the two methods described above. In this method, a sampling site is first selected and a trench is dug (of an appropriate size for ease of access) adjacent to the selected area. The face of the trench adjacent to the selected area is smoothed with a flat blade shovel or mortar trowel. A flat-bottomed, three-sided pan with sharpened edges on the open side is pressed into the face from ground surface to 5 cm. The first cut is removed and sealed in a small plastic bag, and the top soil is cut away on either side of the cut to make a shelf. Then the open-ended cutting pan is again pushed into the side and cuts out the next incremental sample. The procedure is continued until the desired depth is reached. The actual depth of each cut can be determined by placing a two by four on the surface and measuring it to each subsurface. Comparing with other methods, the trench method is more time consuming. However, if care is taken, there will be very little cross contamination and the data collected in terms of the depth profile will be more accurate. In addition, this method is useful for collecting a deep soil profile. In the investigation of the deposition of a natural radionuclide (such as ^{14}C, ^{10}Be and ^{129}I) over a very long time scale (>1000 years), a depth profile of several meters has been collected using this method.

18.7.5
Pretreatment of Soil Samples

Before radionuclide analysis, the collected samples must be processed to obtain them in a suitable form for analysis and archiving. The processing method depends on the radionuclide, the size of the sample, and the amount of processing already undertaken in the field. In general, the soil is dried, crushed, sieved, milled, and blended (in that order). Global fallout radionuclides from nuclear weapons testing are relatively homogeneous with respect to the particle size and the distribution in the sample. The radionuclides from accidental or operational releases, in contrast, are typically inhomogeneous, and thus special care must be taken to get representative samples. Multiple sub-sampling and multiple analyses is probably the only way to adequately define the content of radionuclides in heterogeneous samples. In addition, care must be taken in all stages of processing to avoid cross-contamination of samples and contamination arising from the circumstances of working. Starting the process with those samples that have the lowest radionuclide concentrations can effectively minimize cross-contamination.

In many cases, vegetation, root mat, large organic pieces, and large stones can be discarded from the soil sample if they are known not to contain the radionuclide of interest. Soil aggregates are broken down, and if the vegetation and root mat is included in the top soil sample, they should be cut up using scissors or clippers to ensure their homogenous distribution throughout the sample. Air-drying is normally used to dry the soil in the field and in the laboratory to remove most of moisture. The samples can then be dried at 95–100 °C overnight in laboratory and then weighed.

If the soil sample contains rocks or pebbles, the big aggregations of soil are first crushed and the rocks or pebbles are removed. The sample is then blended for 15–30 min and a sub-sample is taken to be ground or ball milled before being sieved through a 2 mm screen and stored in a container. Care must be taken to avoid cross-contamination in these steps – all machinery should be dismantled and decontaminated by washing it between samples.

18.8
Essentials in Sampling and Sample Pretreatment for Radionuclides

- Sampling is an important part in environmental radioactivity investigations. The representativity of the sample depends on the selection of the sampling location, area, site, sampling method and tools, and contamination, these being the main sources of uncertainty.

- Aerosol particle sampling is widely used for investigating and monitoring radioactivity in the atmosphere. Both glass fiber and polyvinyl chloride filters are used for collecting aerosol particles from large air volumes. Of these, the polyvinyl chloride filter is favored because it can be decomposed by ashing to reduce the sample size, and it is easier to digest by acids for radiochemical analysis. A cascade impactor is used for aerosol particle sampling to separate particles of different sizes; however, it is mainly used for research rather than for routine monitoring.
- For the radioisotopes of iodine and tritium, gaseous components in air are also sampled. Gaseous tritium, existing as water vapor in air, is normally collected by condensing it at lower temperatures. Gaseous iodine is trapped by an NaOH-impregnated cellulose filter, while organic gaseous iodine fractions are trapped by activated charcoal.
- Atmospheric deposition can be collected as dry and wet deposition. In some cases, both dry and wet deposition are obtained by collecting the integrated precipitation. The radionuclides in the precipitation are normally concentrated *in situ* by passing the water through a mixed-bed ion exchange column filled with both anion and cation exchange resins.
- Surface water is normally easy to collect, while water core samples are taken with a special device such as a Niskin bottle from seas or lakes. A simple submersible pump can be used to take water core samples from a depth of less than 100 m.
- *In-situ* preconcentration of radionuclides in water is normally carried out for radionuclides such as ^{137}Cs, Pu isotopes, ^{237}Np, and ^{99}Tc. Radiocesium in a large volume of water can be effectively concentrated using a cotton-wound cartridge impregnated with copper hexacyanoferrate; the Pu isotopes, ^{237}Np and ^{99}Tc can be preconcentrated by coprecipitation with $Fe(OH)_2$.
- Surface sediment is often sampled for mapping the distribution of radionuclides on the bed of a sea or lake. Spades, shovels, trowels, and scoops are used for shallow waters and a dredge for deep waters. A sediment core is normally sampled to investigate the depth distribution and total inventory of radionuclides. Different devices can be used for core sampling, for example an auger for shallow waters and a core tube (a Gemini device) for deep waters.
- Soil sampling is similar to sediment sampling but easier because it is carried out on land. Three methods are often used: core, template, and trench methods. An auger is an often-used device for soil core sampling. The trench method is more time consuming but suitable for large samples and leads to less contamination during sampling.
- Sediment and soil samples are normally pretreated *in situ* before transport, storage, and analysis. The sediment and soil can be sliced in the field, and the soil sample can also be air dried to remove most of the moisture. The sediment sample is normally dried by freeze drying in order to obtain a soft dried material. Cross-contamination during sampling and sample pretreatment must be minimized to reduce the total uncertainty of the results.

19
Chemical Changes Induced by Radioactive Decay

Radioactive decay leads to three kinds of chemical changes in the medium containing the radionuclide. First, the radiation emitted in the radioactive decay process causes chemical changes due to radiolysis; in this case, this is called autoradiolysis because the radiation originates from the medium itself and does not come from outside as is usual in radiolysis. Second, except for internal transition (IT), the radioactive decay always leads to the birth of a new element (a transmutation) and thus changes the system chemically. Third, the daughter nuclide receives the recoil energy, which leads to the breaking of the chemical bonds of the daughter atom itself and to the ionization of the surrounding medium. This is especially typical in alpha decay and highly energetic beta decay.

19.1
Autoradiolysis

The particles and rays created in radioactive decay have kinetic energy, which they lose when colliding with the atoms of the surrounding medium, causing their excitation and ionization. The chemical changes caused by radiation in a medium are the domain of radiation chemistry, and only those chemical changes which have an effect on the chemical behavior of the radioactive atoms or compounds containing them are discussed here. Autoradiolysis is mainly significant for alpha and beta radiation, where the radiation energy is absorbed within a short range from the point of decay – especially in the case of alpha radiation. In contrast, gamma radiation does not usually cause significant autoradiolysis because it is penetrating, and the interactions with the atoms in the medium take place at rather long distances from each other. The extent of autoradiolysis is also dependent on both the energy of the particle or ray and the activity level of radionuclides: the higher the activity, the more radiolysis takes place. Thus the autoradiolysis caused by the gamma radiation will also become significant with samples containing high activities. The significance of the effects of the autoradiolysis in the case of gamma radiation also depends on the size of the system to be examined: the majority of the gamma rays originating from

Chemistry and Analysis of Radionuclides. Jukka Lehto and Xiaolin Hou

ISBN: 978-3-527-32658-7

a sample in a 20 mL vial escape from the system; the proportion of the gamma rays absorbed in the system, however, is much higher in a 300 m^3 waste tank. In the case of bedrock, practically all the gamma rays will be absorbed into it.

19.1.1 Dissolved Gases

If oxygen is dissolved in water, the formation of hydrogen peroxide takes place because of radiation. Hydrogen peroxide is a strongly oxidizing agent and can cause the dissolution of radionuclides and their oxidation to higher oxidation states. Dissolved nitrogen is also oxidized by the action of the radiation; further, nitrogen oxides are formed in solution. These turn into HNO_3 and HNO_2 acids, the latter of which is an especially oxidizing acid which can cause the dissolution of radionuclides and their oxidation to higher oxidation states.

19.1.2 Water Solutions

In aqueous solutions the effect of the radiation on the forms of the radionuclides is indirect. The radiation causes the ionization and excitation of water molecules. The electrons created in the ionization of water form hydrated electrons e_{aq}^-. In turn, the excited water molecule forms H· and OH· radicals; in the subsequent reactions, hydrogen peroxide, among others, is created. OH· radical and hydrogen peroxide are strong oxidants, while the hydrated electron and H· are strong reducing agents. These species can have a strong effect on the oxidation states of redox-sensitive radionuclides (radionuclides with several oxidation states the energy difference of which are small), especially to the oxidation states of actinides. For example, Pu can be reduced by the effect of the hydrated electron from its higher oxidation states (VI,V) to lower ones (IV,III). The changes in the oxidation states can be possible even with moderately low activity concentrations. For example, in a ^{239}Pu solution with an activity concentration of 100 kBq L^{-1}, all the plutonium will change its oxidation state during one year's storage because of autoradiolysis.

19.1.3 Organic Compounds Labeled with Radionuclides

In organic compounds containing radionuclides an increasing number of bond breakings take place over time when the emitted beta and alpha particles collide in the sample medium. The energies of alpha particles in the radioactive decay are several MeVs, while those of beta particles are at least tens of keVs. These energies are extremely high compared to the energies of chemical bonds, which are in the range of 1–5 eV. Bond breaking results in the formation of new compounds from the fragments of the decomposed compounds. The amount of bond breaking is dependent on the specific activity of the compound, the exposure time, the type and energy of the radiation, and the radiation sensitivity of the compound.

19.1.4 Solid Compounds

In solid compounds the radiation causes crystal lattice damage and increases the amorphity of the material in time. For example, the change of crystalline minerals into amorphous ones is brought about by U or Th even at contents as low as 1% in geological time scales. In water-containing compounds, the radiation produces radiolysis of the water (see above) and changes in the compound itself through the radiolysis products generated.

19.2 Transmutation and Subsequent Chemical Changes

Excluding internal transition, the daughter nuclide created in radioactive decay is always of another element (transmutation), and thus the chemical state of the system changes somewhat. In alpha decay, the atomic number becomes smaller by two units, while in beta decay it either becomes smaller or larger by one unit. It is only in internal transition – where the metastable excited state of the nucleus relaxes by gamma emission or by internal conversion – that no transmutation takes place; however, the created nucleus is of the same element as the parent nucleus.

The creation of new elements that takes place in solutions and gases because of radioactive decay does not cause large changes in the chemical state of the system – assuming, of course, that the activity concentrations are not high. The created daughter nuclide rapidly forms a compound or ion typical of the new element and corresponding to the prevailing conditions (pH, Eh, ligands). In the solid state, however, the created daughter nuclide can end up in an atypical chemical state, especially in crystal lattices.

In alpha decay, the electron cloud of the daughter nuclide enlarges because two protons which attract electrons leave the nucleus. The daughter atom is thus not in an excited state but takes energy from outside. The oxidation state of the daughter atom is formally lower by two units compared to the oxidation state of the parent atom

$$M_1^{n+} \rightarrow M_2^{(n-2)+} + He^{2+}$$

However, such reductions of oxidation states are not usually perceived. For example, ^{218}Po, created in the alpha decay of ^{222}Rn, is not negative but neutral or even positive. Similarly, when U^{4+} decays into Th the valence of thorium should be +2; however, it is the same as that of the uranium. These examples indicate that the reorganization of electrons in the daughter atom takes place very rapidly; furthermore, the daughter atom is balanced with the electrons of its environment, quickly seeking the valence state which is characteristic of it. If the created atom has the same characteristic oxidation state as the parent atom and their sizes are of the same order, it is probable that the daughter atom will stay in the crystal lattice on the same site and in the same oxidation state as the start atom. If the created atom does not have the same allowed oxidation state as the parent atom, it can still stay in the crystal lattice but in another oxidation state.

In beta decay, the atomic number either decrease or increases by one unit. However, in all three beta decay modes, the electron balance remains. In the beta minus decay, where the atomic number increases by one unit, the daughter nuclide needs one extra electron. This is available as the electron, the β^- particle emitted by the decaying nucleus. In positron decay, the atomic number decreases by one unit. Here, the daughter nuclide requires one less electron, which is lost by annihilation of the positron particle with an electron in the surrounding medium. In electron capture, the atomic number also decreases by one unit. Since the nucleus takes one electron from the parent atom's inner electron shell, the electron balance is maintained in this case also. If the beta decaying atom is part of a compound, the chemical bond may break because of the decay. For example, when tritium in methane decays it causes a bond break since the products cannot form a compound $CH_3T \rightarrow CH_3^+ + {}^3He + \beta^-$. In a radiocarbon-labeled methane, however, no bond break is observed since the formed compound is stable ${}^{14}CH_4 \rightarrow {}^{14}NH_4^+ + \beta^-$.

In positron decay, the electron cloud enlarges because one proton leaves the nucleus. Thus the daughter atom is not in an excited state but takes energy from outside. In β^- decay, however, strong excitation states are created since the electron cloud shrinks. When the excitation relaxes, Auger electrons are emitted from the electron shells. The holes in the electrons shells that are created are filled by electrons from the upper shells. This causes the formation of X-rays which further induce the formation of Auger electrons. All these events cause an electron emission cascade and strong ionization of the daughter atom – even charges as high as $+20$ have been observed. Eventually, the daughter atom takes electrons from outside and the electron structure is balanced to a form typical for the daughter atom. A similar series of events will be created when a hole in the inner electron shell is filled with an electron from the upper shell.

19.3
Recoil – Hot Atom Chemistry

The daughter nuclide created in a radioactive decay receives recoil energy when emitting an alpha or beta particle or a gamma ray. In this process, the momentum is preserved, as shown in the equation

$$E_1 \times m_1 = E_2 \times m_2 \tag{19.1}$$

where E_1 and E_2 are the kinetic energies of the daughter atom and the emitted particles respectively, and m_1 and m_2 are their masses. As seen from the equation, the recoil energy is bigger the closer the mass of the daughter atom is to the emitted particle or ray. Thus the recoil is bigger in alpha decay compared to beta decay and especially to gamma decay. Similarly, the lower the mass of the daughter atom and the higher the kinetic energy of the emitting particle or ray the larger is the recoil energy. Table 19.1 presents the recoil energies for daughter atoms of various mass numbers

Table 19.1 Recoil energies of the daughter atoms of various mass numbers for alpha, beta and gamma emissions of different energies.

Energy of the emitted particle (MeV)	Mass number of daughter atom (A)	Recoil energy of daughter atom (keV)		
		α (keV)	β (eV)	γ (eV)
0.1	10	40	6	0.54
	50	8	1.2	0.11
	100	4	0.6	0.05
	200	8	0.3	0.03
0.3	10	120	21	4.83
	50	24	4.3	0.97
	100	12	2.1	0.48
	200	6	1.1	0.24
1.0	10	400	110	53.7
	50	80	22	10.7
	100	40	11	5.4
	200	20	5	2.7
3.0	10	1200	648	483
	50	240	130	97
	100	120	65	48
	200	60	32	24
10.0	10	4000	5920	5400
	50	800	1180	1100
	100	400	590	500
	200	200	300	300

for emitted particles or rays with different energies. In radioactive decay, the alpha particle energy is in the range 1.4–11.7 MeV, for which energies the recoil energy is at least 20 keV. Because the energy of the chemical bond is in the range of 0.4–5 eV, the recoil always leads to a bond break in alpha decay. For example, in the decay of ^{238}U to ^{234}Th, the kinetic energy of the emitting alpha particle is 4.202 MeV while the recoil energy of the daughter ^{234}Th is 0.072 MeV. The latter is more than 10 000 times higher than the energy of a chemical bond. It is obvious, therefore, that the bond breaks. In cases of beta and gamma radiation, while the recoil energy is smaller than the above, it is often sufficient to break the chemical bond. The daughter atoms which have high recoil energy are called hot atoms and they experience hot atom chemistry. The temperature cannot be defined for one single particle; however, hypothetically the above mentioned kinetic energy of 0.072 MeV corresponds to a temperature of nearly one billion Kelvin degrees. From this comes the definition 'hot atoms.'

When passing through a medium, hot atoms lose their electrons, ionize themselves, and ionize the medium. The latter phenomenon was dealt with in connection with autoradiolysis. In the solution and gas phases, hot atoms rapidly regain the lost electrons; in solid materials, even permanent changes can be caused both in the sites of atoms and in their oxidation states. In the decay of ^{238}U to ^{234}Th, for example, ^{234}Th further decays to ^{234}Pa. At radiochemical equilibrium, the activities of all these nuclides will be the same – in other words, the same as that of the parent nuclide ^{238}U. In minerals, this is nearly always the case. In ground waters, however, an excess of ^{234}U is often observed compared to ^{238}U. This could be explained by the fact that whereas uranium mainly exists as its sparingly soluble +IV oxidation state in minerals, it oxidizes in the decay process to the readily soluble uranyl ion with an oxidation state of +VI and the uranium is dissolved in mineral-groundwater interfaces. In crystalline materials, the recoil causes defects in the crystal framework because of recoiling atoms. In organic compounds, the recoil causes bond breaking and the formation of new compounds from the decomposition fragments.

Index

Chemistry and Analysis of Radionuclides. Jukka Lehto and Xiaolin Hou

ISBN: 978-3-527-32658-7